Economic and Social Issues in Agricultural Biotechnology

Dedicated to

Molleen Wright

Economic and Social Issues in Agricultural Biotechnology

Edited by

R.E. Evenson

Yale University
USA

V. Santaniello

University of Rome 'Tor Vergata'
Italy

and

D. Zilberman

University of California at Berkeley
USA

CABI *Publishing*

CABI *Publishing* is a division of CAB *International*

CABI Publishing
CAB International
Wallingford
Oxon OX10 8DE
UK

Tel: +44 (0)1491 832111
Fax: +44 (0)1491 833508
E-mail: cabi@cabi.org
Web site: www.cabi-publishing.org

CABI Publishing
10 E 40th Street
Suite 3203
New York, NY 10016
USA

Tel: +1 212 481 7018
Fax: +1 212 686 7993
E-mail: cabi-nao@cabi.org

A catalogue record for this book is available from the British Library, London, UK.

A catalogue record for this book is available from the Library of Congress, Washington, USA.

ISBN 0 85199 618 3

Typeset by Columns Design Ltd, Reading
Printed and bound in the UK by Biddles Ltd, Guildford and King's Lynn

Contents

Contributors

D. Alker, University of Hohenheim, Institut 490A, D-70593 Stuttgart, Germany.

K. Anderson, Centre for International Economic Studies, University of Adelaide, Australia.

V.M. Aquino, Institute of Plant Breeding, University of the Philippines at Los Baños, College, Laguna 4031, Philippines.

A.F.D. Avila, Embrapa, Parque Estação Biológica, Final Av. W3 Norte, 70770-901, Brasilia, DF, Brazil.

S. Babu, International Food Policy Research Institute, 2033 K Street NW, Washington, DC 20006, USA.

L. Beyers, Imperial College of Science, Technology and Medicine, RSM Building, Prince Consort Road, London SW7 2BP, UK.

M. Blakeney, Queen Mary Intellectual Property Research Institute, CCLS, University of London, Mile End Road, London E1 4NS, UK.

L.J. Butler, Department of Agricultural and Resource Economics, University of California, One Shields Avenue, Davis, CA 95616, USA.

F. Castillo, Department of Environmental Science, Policy and Management, University of California, Berkeley, California, USA.

J.I. Cohen, International Service for National Agricultural Research, PO Box 93375, 2509 AJ The Hague, The Netherlands.

E. Contini, Embrapa, Parque Estação Biológica, Final Av. W3 Norte, 70770-901, Brasilia, DF, Brazil.

K. Dreher, Economics Program, Centro International de Mejoramieonto de Maiz y Trigo (CIMMYT), Apdo. Postal 6-641, 06600 Mexico, DF, Mexico.

H. Egelyng, International Rice Research Institute (IRRI), PO Box 3127 (MCPO), 1271 Makati City, Philippines.

Z. Escobar, Universidad del Valle, Cali, Colombia.

R.E. Evenson, Economic Growth Center, Department of Economics, Yale University, New Haven, CT 06520, USA.

C. Falconi, InterAmerican Development Bank, 1300 New York Avenue NW, Washington, DC 20577, USA

E.L.R. Filho, Embrapa, Parque Estação Biológica, Final Av. W3 Norte, 70770-901, Brasilia, DF, Brazil.

S. Flasinski, Monsanto, BB51, 700 Chesterfield Parkway North, St Louis, MO 63198, USA.

T. Goeschl, Department of Land Economy, University of Cambridge, 19 Silver Street, Cambridge CB3 9EP, UK.

P. Goldsmith, Food and Agribusiness Management Group, Department of Agricultural and Consumer Economics, University of Illinois, 433 Mumford Hall, 1301 West Gregory Drive, Urbana, IL 61801-3605, USA.

V. Gottret, CIAT, Apartado aereo 6713, Cali, Colombia.

G. Graff, Department of Agricultural and Resource Economics, University of California at Berkeley, 201 Giannini Hall, Berkeley, CA 94720-3310, USA.

R.A. Hautea, ISAAA-SE Asia Center, c/o IRRI, MCPO Box 3127, 1271 Makati City, Philippines.

F. Heidhues, University of Hohenheim, Institut 490A, D-70593, Stuttgart, Germany.

A. Heiman, Department of Agricultural Economics and Management, Hebrew University, Rehovot, Israel,

R. Hu, Chinese Center for Agricultural Policy, CAS Institute of Geographical Sciences and Natural Resources Research, China.

J. Huang, Chinese Center for Agricultural Policy, CAS Institute of Geographical Sciences and Natural Resources Research, China.

Y. Ismaël, Department of Agricultural and Food Economics, University of Reading, Earley Gate, Whiteknights Road, Reading RG6 6AR, Berkshire, UK.

W.K. Kaniewski, Monsanto, BB51, 700 Chesterfield Parkway North, St Louis, MO 63198, USA.

M. Khairallah, Applied Biotechnology Center, Centro International de Mejoramieonto de Maiz y Trigo (CIMMYT), Apdo. Postal 6-641, 06600 Mexico, DF, Mexico.

J. Komen, International Service for National Agricultural Research (ISNAR), PO Box 93375, 2509 AJ The Hague, The Netherlands.

B. Koo, International Food Policy Research Institute, 2033 K Street NW, Washington, DC 20006, USA.

N.D. Lam, National Center of Natural Science and Technology, Institute of Biotechnology, Hoang Quoc Viet, Caugiay, Hanoi, Vietnam.

A.C. Marco, Department of Economics, Vassar College, Poughkeepsie, NY 12604-0708, USA.

M. Morris, Economics Program, Centro International de Mejoramieonto de Maiz y Trigo (CIMMYT), Apdo. Postal 6-641, 06600 Mexico, DF, Mexico.

C.P. Nielsen, Danish Institute of Agricultural and Fisheries Economics and University of Copenhagen, Denmark.

C.A. Ong, Malaysian Agricultural Research and Development Institute, Strategic, Environment and Natural Resources Research Center, MARDI, GPO Box 12301, 50774 Kuala Lumpur, Malaysia.

D. Pachico, CIAT, Apartado aereo 6713, Cali, Colombia.

S. Pandey, Maize Program, Centro International de Mejoramieonto de Maiz y Trigo (CIMMYT), Apdo. Postal 6-641, 06600 Mexico, DF, Mexico.

P.G. Pardey, International Food Policy Research Institute, 2033 K Street NW, Washington, DC 20006, USA

D. Parker, Department of Agricultural and Resource Economics, University of Maryland, College Park, Maryland, USA.

E.S. Perez, Universidad del Valle, Cali, Colombia.

J. Piesse, Birkbeck College, Malet Street, London WC1E 7HX, UK.

V. Pillai, Malaysian Agricultural Research and Development Institute, Strategic, Environment and Natural Resources Research Center, MARDI, GPO Box 12301, 50774 Kuala Lumpur, Malaysia.

C.E. Pray, Department of Agricultural, Food and Resource Economics, Rutgers University, 317 Dennison Street, Highland Park, NJ 08904, USA.

F. Qiao, Department of Agricultural Resources and Economics, University of California, Davis, CA 95616, USA.

T.R. Quirino, Embrapa, Parque Estação Biológica, Final Av. W3 Norte, 70770-901, Brasilia, DF, Brazil.

G. Ramos, University of Illinois, 203 S Third Street Apt 304, Champaign, IL 61820, USA.

G.C. Rausser, Department of Agricultural and Resource Economics, University of California at Berkeley, 201 Giannini Hall, Berkeley, CA 94720-3310. USA.

V. Rhoe, International Food Policy Research Institute, 2033 K Street NW, Washington, DC 20006, USA.

J.M. Ribaut, Applied Biotechnology Center, Centro International de Mejoramieonto de Maiz y Trigo (CIMMYT), Apdo. Postal 6-641, 06600 Mexico, DF, Mexico.

L. Rivas, CIAT, Apartado aereo 6713, Cali, Colombia.

F.S. Robinson, International Food Policy Research Institute, 2033 K Street NW, Washington, DC 20006, USA.

K. Romyanon, Plant Genetic Engineering Unit, Kasetsart University, Kamphaengsaen, Nakhon Pathom 73140, Thailand.

S. Rozelle, Department of Agricultural Resources and Economics, University of California, Davis, CA 95616, USA.

S. Salazar, PO Box 91-3100, Santo Domingo de Heredia, Costa Rica.

V. Santaniello, University of Rome 'Tor Vergata', 00133 Rome, Italy.

S. Shantharam, International Food Policy Research Institute, 2033 K Street NW, Washington, DC 20006, USA.

C.S. Srinivasan, Department of Agricultural and Food Economics, University of Reading, Earley Gate, Whiteknights Road, Reading RG6 6AR, Berkshire, UK.

G. Srinivasan, Maize Program, Centro International de Mejoramieonto de Maiz y Trigo (CIMMYT), Apdo. Postal 6-641, 06600 Mexico, DF, Mexico.

C. Steiger, Agribusiness Program, Universidad de Belgrano, Buenos Aires, Argentina.

T. Swanson, Department of Economics, Faculty of Laws and CSERGE, University College, Gower Street, London WC1E 6BT, UK.

S. Taba, Centro International de Mejoramieonto de Maiz y Trigo (CIMMYT), Apdo. Postal 6-641, 06600 Mexico, DF, Mexico.

C. Thirtle, Environmental Policy and Management Group, T.H. Huxley School of Environmental, Earth Sciences and Engineering, Imperial College of Science, Technology and Medicine, RSM Building, Prince Consort Road, London SW7 2BP, UK.

R. Tripp, Overseas Development Institute, 111 Westminster Bridge Road, London SE1 7JD, UK.

J. Valkoun, International Center for Agricultural Research in Dry Areas (ICARDA), Aleppo, Syria.

B.D. Wright, Department of Agricultural and Resource Economics, University of California, Berkeley, CA 94720, USA.

D. Zilberman, Department of Agricultural and Resource Economics, University of California at Berkeley, 201 Giannini Hall, Berkeley, CA 94720-3310, USA.

Acknowledgement

This book has arisen from a conference entitled 'Biotechnology, Science and Modern Agriculture: a New Industry at the Dawn of the Century', convened by the International Consortium on Agricultural Biotechnology Research (ICABR), and held in Ravello, Italy on 15–18 June 2001.

The editors would like to acknowledge the sponsorship of the following institutions:

- Ceis–University of Rome 'Tor Vergata',
- Center for Sustainable Resource Development, University of California at Berkeley,
- Economic Growth Center, Yale University.

The editors would also like to thank the Italian National Research Council (CNR) for their financial support of the conference.

Introduction and Overview

R.E. Evenson, V. Santaniello and D. Zilberman

The introduction and promotion of biotechnology in agriculture (ag biotech) are part of a vision that views this technology as a solution to major societal problems. Applications of new molecular cellular techniques in agriculture were predicted to create genetic materials that would help to alleviate the food scarcity problems in developing countries, alleviate the environmental side effects associated with commercial agriculture and improve the quality of foods in the developing world. To date, however, consumer resistance to genetically modified organisms (GMOs) and concerns about environmental side effects have created significant obstacles to development of this technology (see Santaniello *et al.*, 2002). In this volume we will address some of the major economic and institutional problems that affect the creation and international exchange of agricultural biotechnologies and of the capacity to create these technologies in developing countries.

Ag biotech products were introduced to farmers in the 1990s. During this era, agriculture in the developed economies was undergoing major structural transition and experiencing globalization. As agricultural sectors in the United States and Western Europe were becoming more industrialized, more emphasis was placed on vertical integration and contractual relationships with growers in attempts to develop a variety of products and plants in agriculture. The role of the state, even in agricultural research and development (R&D), was being diminished.

The private sector is now engaged in research activities that were formerly conducted by the public sector. The structure of agriculture is being affected, and it has influenced these processes. While the public sector has played a relatively minor role in developing mechanical and, to some extent, chemical technologies in agriculture, it has played a dominant role in developing genetic materials, particularly seeds. The existing production of genetic materials and seeds, especially in developing countries, was established around utilization of publicly collected and preserved germplasm. The International Agricultural Research Centers (IARCs) of the Consultative Group for International Agricultural Research (CGIAR) system, working in collaboration with National Agricultural Research Systems (NARS), were the main source of new seeds for the developing world. The ownership of rights to genetic knowledge and materials is not as acutely important when most R&D activities are carried out by the public sectors as it is in a system where R&D is conducted by private companies.

The first part of this volume will be devoted to issues of intellectual property rights (IPR). Those rights are important determinants of private sector investment in biotechnology research. They are also important for the marketing of ag biotech products. The second part of the volume addresses the changing nature of public sector research. Changing IPRs and changes in investment behaviour of private firms require responses by public sector research organizations in terms of their research portfolios and their relationship with the private sector. The third part of the volume is oriented to biotechnology tools. Much of the debate over biotechnology to date has revolved around biotechnology products yet, in the long run, biotechnology processes, including the capacity to use and modify tools and methods, will be important.

Part IV of the volume is directed toward developing countries' experience to date with agricultural biotechnology. This experience is limited both by the state of IPRs and related institutions supporting the purchase and sale of biotechnology products (or the associated rights to their use), and by the state of capacity in research units to effectively use the methods of biotechnology.

The final part of the volume addresses international exchange of biotechnology products and biotechnology capacity utilizing international market models.

The first chapter in the volume sets the stage for other chapters by drawing comparisons between the emerging Gene Revolution in developed countries and the Green Revolution that has been a prominent part of developing country experience. The Green Revolution in developing countries is measured by the production and diffusion of 'modern' crop varieties (MVs) using conventional breeding methods (crossing and selection) enhanced by wide-crossing methods. The population boom that took place as child mortality rates declined after World War II was perhaps the major factor in inducing investment in both the IARCs and NARS crop genetic improvement programmes. Chapter 1 summarizes the major features of production and diffusion of MVs in developing country agriculture.

Chapter 1 provides evidence that the extent of the diffusion of MVs to farmers has major welfare implications. If only a few farmers have access to cost-reducing technology but face declining prices, then significant inequalities in income (and in other welfare indexes) emerge. It is therefore important that access to MVs be extended to as many farmers as possible. The central message of Chapter 1 is that, in the Green Revolution, the incorporation of single gene traits (host plant resistance to insects and diseases and host plant tolerance to abiotic stresses) using conventional breeding techniques was a key determinant of MV diffusion to farmers *and* that further incorporation of these traits is possible using Gene Revolution or biotechnology methods. Thus there is a high degree of congruity between the Green and Gene Revolutions and, because of this access expansion dimension, the Gene Revolution has important welfare improvement prospects beyond those associated with producing more food at lower cost.

Chapter 1 also provides a review of the International Rice Biotechnology Programme (supported by the Rockefeller Foundation), the most important effort to support ag biotech capacity in developing countries to date, and utilizes survey data from scientists to develop an estimate of the increased productivity that biotechnology methods bring to rice research programmes. (The economic consequences of increased research productivity are evaluated in the final chapter in this volume.)

Part I of the volume includes four chapters dealing with intellectual property rights (IPRs). IPRs affect the purchase and sale of ag biotech products and of genetic resources (some of which are products of research programmes and some of which are farmer-selected landraces and related species, the basic materials held in genebanks). Chapters 2 (Butler) and 3 (Egelyng) address issues associated with World Trade Organization (WTO) requirements that a *sui generis* IPR system be implemented for plants in developing countries. These IPRs are generally expected to be forms of plant breeders rights (PBRs) and they affect exchanges of ag biotech products.

Two chapters in this part of the volume, Chapter 4 (Blakeney) and Chapter 5 (Alker and Heidhues), deal with a second and newer IPR affecting plants, the 'farmers rights' provision in the Convention on Biodiversity. These rights are designed to protect genetic resources *per se*, not ag biotech products.

Chapter 2 discusses the international obligations that developing countries face in implementing IPR systems for crops. The chapter discusses both plant variety protection (PVP) systems, as regulated under the UPOV convention, and farmers' rights (FRs) as regulated under the Convention on Biodiversity (CBD). The chapter argues that a modified PVP system with stronger farmers rights and breeders rights (relative to the 1991 act of UPOV) would be in the interest of developing countries.

Actually, a reasonable interpretation of the World Trade Organization Trade-related Intellectual Property (WTO-TRIPs) agreement calling for a *sui generis* system, would allow for some creativity on the part of lawmakers in developing countries. But this creativity has not been manifest to date in IPR system development. Perhaps this is because there is also tremendous pressure for 'harmonization' of IPR systems associated with trade law. The chapter also notes that, for most farmers in developing countries, any form of IPR is new. Most developing countries have staged battles to weaken and resist international IPR pressures rather than to be creative in developing IPRs suited to their needs.

Chapter 3 (Egelyng) addresses the implementation of PVP regimes in Asia by discussing PVP experiences in China, Indonesia, Korea, Bangladesh, Malaysia, Pakistan, Thailand and Vietnam. The author notes that the Bangladeshi law is the one most deserving of the description, *sui generis*. This law speaks of economic values and related 'social good' criteria. It is not clear, however, how this will be operationalized. In practice, many clauses of this type serve to render impotency to IPR laws (which is often the objective of countries passing laws under pressure).

Chapter 3 addresses questions of economics of scale in IPR administration and convergence between the TRIPs and CBD perspectives. A section on international exchange of germplasm raises concerns that exchange may be restricted in future years. In reality, because of the high degree of location specificity of crop varieties, relatively few varieties cross international borders (generally less than 10% of NARS-crossed varieties, even IARC-crossed varieties, are not widely diffused internationally (Evenson and Gollin, 2001). Thus, most concerns about incurring obligations to foreigners are not very relevant. Most PVPs will be awarded to domestic developers of varieties, whether public research systems or private firms.

However, neither Chapters 2 nor 3 address the role of conventional utility patent protection, also part of TRIPs, where patents are increasingly being applied to ag biotech products. The 'genes for rent' products, such as the Roundup Ready product from Monsanto or the *Bt* products, are protected by utility patents, not PVPs. Firms offering these products for sale are influenced by the IPR conditions of the potential buyer. Even when the country of the buyer has grudgingly implemented IPRs, the sellers may shun them because they perceive the countries to be lax about administering the laws, leaving them vulnerable to piracy.

Chapters 4 and 5 (and Chapter 2 as well) deal with an IPR that has very different implications for public research systems than utility patents or PVPs. These rights protect traditional agricultural knowledge (Chapter 4, Blakeney) and farmers' rights embodied in farmers' varieties (Chapter 5, Alker and Heidhues). These IPRs apply directly to the genetic resources that have formed the basis for conventional breeding programmes in the past and that will be important to ag biotech development as well. There is little doubt that the free exchange of these basic genetic resources (typically described as landraces) and wild relatives (i.e. from non-cultivated species in the same genus) has been important to plant breeding programmes (Evenson *et al.*, 1998). There is also little doubt that these resources have been valuable.

Chapter 4 reports a number of cases demonstrating this value, as well as efforts by private firms and public-sector research programmes to claim IPRs for traditional plant resources and associated traditional knowledge. This discussion highlights the evolutionary nature of IPRs and the role of 'case law'. In fact, most of the expansion of IPRs to plants (and animals) has taken place because of court decisions (case law) rather than because of changes in written laws. Such IPR scope expansions are characterized by opportunism on the part of claimants (not entirely confined to greedy private firms) to be adjudicated by court cases. It is not clear at this point whether the IPR claims on genetic resources described in Chapter 4 will stand up in court.

Chapters 4 and 5 both discuss the farmers' rights provision in the CBD. These rights are described in Chapter 4 as a counterbalance to conventional IPRs. Farmers' rights are supported by most developing countries and opposed by developed countries, in contrast to conventional IPRs. Failure to resolve farmers rights issues does represent a threat to the traditional system of free exchange of crop genetic resources.

Chapter 5 addresses the conflicts between farmers' rights and PVP rights, and provides a history of the development of each. The chapter then offers a discussion of the obstacles to a market solution enabling continued free exchange of plant genetic resources. Alternative institutional arrangements for reconciling farmers rights and genetic resource exchange are discussed. The most promising appears to be the current efforts by the International Undertaking on Plant Genetic Resources.

One of the features of IPR laws is that IPRs applied to inventions, including biological inventions, have a monopoly right that is limited in time, typically 20 years. The farmers' rights being debated are more like copyright and trademark IPRs, which generally are not time limited. This distinction is one of several that courts will be called upon to address as IPRs evolve.

The evolution of agricultural biotechnology depends largely on the division of responsibilities between the private and public sectors. Chapter 6, 'Universities, Technology Transfer and Industrial R&D,' by Greg Graff *et al.*, provides an overview of a large body of literature that suggests that there are significant differences in the incentives and outcomes of private vs. public research. Public research in most cases generates concepts that need further development for commercialization. A new law that gave universities the power to control their intellectual property has increased commercialization of university knowledge over the last 20 years. The chapter analyses the performance of Offices of Technology Transfer and transfer of technologies from the public to the private sector in the United States. This process has resulted in the establishment of new companies in biotechnology and has led universities to directly contribute to changes in industrial structure and economic development. The chapter addresses factors that explain the successes of technology transfer and some of the problems that these processes may lead to.

Chapter 7, 'Mergers and Intellectual Property in Agricultural Biotechnology,' by Alan Marco and Gordon C. Rausser, quantitatively analyses patterns of mergers among major agricultural business suppliers. These mergers are associated with the introduction of new agricultural biotechnology products and are perceived as part of the realignment and adjustment in response to these new technologies. The chapter argues that mergers and takeovers were designed to expand the intellectual property rights portfolio. It shows that the likelihood with which a firm will be involved in acquisitions or takeovers increases the more enforceable its patent portfolio is. Furthermore, firms with complementary intellectual property rights are more likely to match up with one another.

Part III of the volume includes four chapters featuring the methods of ag biotech invention. Two of the chapters (9 and 10) deal with genetic use restriction technologies (GURTs) which have important implications for incentives for private sector R&D because they provide a type of IPR similar to that afforded by 'hybrid' varieties. Farmers cannot save their own seed with hybrid varieties: GURTs can be used for a similar purpose. A third chapter (11) provides a cost analysis for alternative research techniques to achieve a common objective. The fourth chapter in this part reviews biotech capacity building in the IARCs and NARs programmes.

Srinivasan and Thirtle provide (Chapter 9) a general discussion of GURT technologies. (They use the more common term 'terminator technologies.') They also provide a set of economic calculations of innovators (= quasi rents) for alternative scenarios. These simulations are compared with data for hybrid seed crops (maize and sorghum) and non-hybrid seed crops (wheat, soybeans, cotton) in the United States. R&D investments in the hybrid crops were significantly higher (per unit product) than for non-hybrid crops and a higher proportion of seed sales was invested in R&D for hybrid crops. The authors discuss implications for developing countries where the proportions of seed saved by farmers are much higher than in the United States.

Goeschl and Swanson in Chapter 10 also address the prospects for the use of GURTs in developing countries, relying on a comparison with hybrid crop technologies to simulate GURT impacts on developing countries. The authors compare the yield history of eight crops for a 39-year panel of countries for hybrid crops (maize and sorghum) and non-hybrid crops (barley, cotton, millet, rice). They find that the diffusion of innovations (as measured by a yield increase in one country being transferred to another country) is highest in non-hybrid crops. Simulations of yields lead to differences between countries, and the authors conclude that the replacement of conventional varieties by IPR protected varieties will have negative impacts on the diffusion of innovations from one country to another.

These two chapters raise a number of interesting questions regarding GURTs and ag biotech generally. One of the questions is the role of competition from other private firms and more importantly from public sector NARS and their associated seed production systems (Certified and Registered growers). The hybrid technology in maize, initially realized in the United States in Iowa, was diffused to Alabama with a 20-year time lag. The time lag from Iowa to sub-Saharan Africa was 75 years. This was because the necessary platform varieties for the technology had to be developed before investments were justified (Griliches, 1957).

Once hybrid maize varieties were developed for a location, the hybrid methodology made all or almost all hybrids inherently higher yielding than open pollinated varieties. This, however, is not the case with GURTs. The installation of GURTs on varietal platforms does not necessarily make them higher yielding as with hybrids. This means that GURTs will be unlikely to be widely adopted unless farmers already buy new seed each year (for reasons of better quality and higher germination rates) or if a regulation is implemented requiring GURTs for environmental and other reasons. Competition will ensure this. The competition effect is also important for public policy. Farmers with access to technology from strong public programmes can always rely on this technology. Anything the private sector offers that is better in their view is a bonus for them.

Chapter 11 reports a study into the use of proprietary process or tool technology in public sector IARCs and NARS programmes. Many of these proprietary technologies were protected by patents, and in most cases, materials transfer agreements and licenses provided the legal basis for use of proprietary technology. However, for some technologies, no clear agreements for use were recorded. A second study for Latin America is also summarized. At this point, it appears that public sector IARCs and NARS programmes have not fully addressed policy options regarding proprietary technology.

In virtually all economic studies of technology, the R&D process is treated as a kind of 'black box.' R&D funds are spent, inventions are made, some are given IPRs, inventions are then commercialized and diffused to buyers. The actual technology of the research process is seldom identified. Chapter 12 in this volume is an exception to this rule, as it compares alternative research techniques (marker assisted selection, a biotechnology technique and conventional breeding) to achieve a research objective (also see Chapter 1 for a research technique study). The research objective is to transfer the quality protein maize phenotype from one elite inbred line to another elite inbred line. This problem is approached from a cost minimization perspective, through careful measurement of costs, including equipment and materials costs. The procedures are carefully described. Economies of scale are considered. In many ways, this study is the research counterpart to a good farm cost or farm management study.

The study concludes that from a cost perspective, the 'modern' biotechnology method is not necessarily superior to the older conventional breeding (back-crossing) method. But it does find that the research objective is achieved earlier with marker-assisted selection techniques, a fact that has considerable value because it moves the ultimate benefit stream associated with the adoption of the technology forward in time, relative to costs.

Part IV of the volume includes seven chapters dealing directly with developing country experience to date with ag biotech developments. Actually, very few developing countries have ag biotech products in the field, due in some cases to regulations and in other cases to limited domestic capacity to conduct biotech research. In addition, the state of IPR development is limiting the sale of 'genes for rent' products in many countries.

Chapter 13 (Tripp) discusses the problems of bringing ag biotech to smallholder farmers (the poor). The chapter implicitly makes a distinction between commercial (presumably non-poor) farmers and asks whether ag biotech products can be expected to be 'smallholder friendly'. The distinction between commercial and smallholder farmers may be meaningful in some regions and for some crops but, as Barker and Herdt showed a number of years ago, virtually all of Asian rice production is by smallholders and Green Revolution access was not a matter of smallholder vs. commercial farm organization, but rather a matter of production conditions such as water supply and soil conditions.

Chapter 13 reviews a number of studies of farmer knowledge and the role of information in technology adoption. The role of seed systems is also discussed. The factors that affect the extent and spread of MVs were noted to have important welfare effects in Chapter 1, where the congruity between the Green and Gene Revolutions was argued to be a factor to Gene Revolution expansion beyond the limits of the Green Revolution.

Chapter 14 describes a research strategy for engineering virus resistance in crop plants in developing countries. This is a collaborative research effort by a private firm (Monsanto), The International Service for the Acquisition of Agro-biotech Application (ISAAA) and five developing country NARS in Southeast Asia. The initial focus of the research programme is on papaya ringspot virus. While this research programme is in its early stages, it represents an important institutional mechanism for the exchange of basic scientific information between private sector and public sector programmes, and between developed and developing countries.

Chapter 15 reports a case study of ag biotech institutions in India, one of the developing countries with significant capacity development. The study uses an institutional development framework. India has long had well developed science and applied research institutions. The agricultural research system (i.e. the NARS) is one of the largest in the developing world (rivalled only by China) and incorporates federal and state as well as public–private interfaces. India has utilized a project grant fund as a vehicle for supporting ag biotech projects.

Indian policy regarding scientific research and technological research has generally been to encourage the application of scientific knowledge to the development of technology. Many research programmes in India contributed to the Green Revolution in India, and the productivity performance of the agricultural sector has been impressive. Indian policy regarding international trade and multi-national enterprises has varied over time. Since the early 1990s, India has been engaged in a programme of economic reform leading to less regulation and more openness to trade. This has produced a favourable economic climate for ag biotech research in both public and private firms. India may be poised to become a leader in biotechnology (especially genomic) research as it has become in software development.

Brazil has also developed significant ag biotech research capacity. Chapter 16 describes this development. Brazil is also struggling with political issues and is presently prohibited from actual ag biotech production, partly in the interest of being able to assure non-GMO products for international markets. It appears, however, that farmers have found GMO products (soybeans and maize) to merit adoption and that a considerable degree of GMO seed is being planted in Brazil in spite of the injunction prohibiting it.

The Brazil chapter reports several *ex ante* research evaluations for ag biotech products. Brazil, as with other countries, is addressing biosafety concerns.

Chapter 17 reports a case study of ag biotech research incentives, particularly of IPRs, in Argentina. The chapter utilizes the business case method and provides methodological insights as well as specific conclusions. The context of the chapter is that of a net technology-exporting North and a net technology-importing South. This dichotomy has long been recognized as critical to international IPR systems, as noted in earlier chapters. The case study compares research investment in a crop with natural IPR protection, maize hybrids, and a crop with weak IPR protection, soybeans. The findings suggest that IPRs will be important to future ag biotech development.

Chapter 18 reports a study of *Bt* cotton in South Africa. South Africa implemented regulations allowing GM crops to be planted in 1998. This chapter presents results from a survey of 100 smallholder farmers (40 non-*Bt*, 60 *Bt* growers) and reports substantial gains to them from this ag biotech product.

Chapter 19 presents an economic assessment of the potential income and employment consequences that could be expected for transgenic herbicide-resistant cassava varieties in Colombia. Cassava is an important staple food crop for many poor families. Weed control for cassava has been a major part of production costs. This study reports a cost analysis for six regions for four technologies including transgenic varieties. The transgenic varieties produce the greatest economic surplus in this analysis, but also reduce later demand.

Section V on International Models addresses how new GMO technology and policies may impact trade and welfare. The underlying methodology used in the first of the three papers, computable general equilibrium (CGE) modelling, is well suited to this task because it allows for important linkages between international markets and trade policies.

There is widespread opposition to GMO production and trade due to environmental and food safety concerns. Anderson *et al.* (Chapter 20), argue that the economic effects of GMO-related policies should be assessed so that decision makers and the public can be better informed. They use two global computable general equilibrium (CGE) models to examine the effects of different policy and consumer preference responses to GMOs. They find that the government-implemented import ban causes a more severe reduction in the global benefit from new GM technology than when consumers express their preferences through market mechanisms. The cost of imposing the ban in Western Europe falls primarily on the region itself. In addition, these results suggest that developing countries that do not gain access to GM technology may be slightly worse off if they cannot guarantee that their exports to Western Europe are GMO-free.

They also use CGE models to explore the impact of increased preferences for GM-free food and find that price differentials between the segregated markets to GM and non-GM soybeans are significant but tempered by commodity arbitrage. Successful redirection of trade flows requires that the relative price premium on non-GM products be sufficient to outweigh forgone productivity from not adopting GM products plus costs of compliance.

Chapter 21 discusses experience with transgenic cotton varieties in China. This is one ot the more important cases of adoption and diffusion of biotechnology products in a developing country. The study finds that planting of *Bt* cotton reduces pesticide use and improves production efficiency, a finding that brings the issue of access to this technology into sharp focus. Cotton farmers with access to *Bt* cotton varieties realize cost reductions: cotton farmers without access to *Bt* cotton varieties do not realize cost reductions but lost income is inferred because of price reductions induced by increased supply from farmers with access to *Bt* cotton varieties.

References

Santaniello, V., Evenson, R.E., Zilberman, D. and Carlson, G.A. (eds) (2000) *Agriculture and Intellectual Property Rights: Economic, Institutional and Implementation Issues in Biotechnology*. CAB International, Wallingford, UK.

Barton, J.H. (1998) The impact of contemporary patent law on plant biotechnology research. In: Ederhard, S.A., Shands, H.L., Collins, W. and Lowert, R.L. (eds) *Intellectual Property Rights III: Global Genetic Resorces: Access and Property Rights*. Crop Science Society of America, Madison, Wisconsin, pp. 85–97.

Bennett, J. (1995) Biotechnology and the future of rice production. *GeoJournal* 35, 333–335.

Evenson, R.E., Gollin, D. and Santaniello, V. (eds) (1998) *Agricultural Values of Plant Genetic Resources*. CAB International, Wallingford, UK.

Graff, G.D., Rausser, G.C. and Small, A.A. (1999) Agricultural biotechnology's complementary intellectual assets. Paper presented at The Shape of the Coming Agriculttural Biotechnology Transformation: Strategic Investment and Policy Approaches from an Economic Perspective. ICABR, Rome and Ravello, Italy, June 1999.

Griliches, Z. (1958) Research costs and social returns: hybrid corn and related innovations. *Journal of Political Economy* 66, 419–431.

Jarvis, L.S. (1999) The potential effects of recombinant somatotropin (rbST) on world dairying. Paper presented at The Shape of the Coming Agricultural Biotechnology Transformation: Strategic Investment and Policy Approaches from an Economic Perspective. ICABR, Rome and Ravello, Italy, June 1999.

Johnson, D.K.N. and Evenson, R.E. (1999) R&D spillovers to agriculture: measurement and application. *Contemporary Economic Policy* 17, 432–456.

Lin, W., Amurtha, C.S., Datta, K., Potrykus, I., Muthukrishnan, S. and Datta, S.K. (1995) Genetic engineering of rice for resistance to sheath blight. *Bio/technology* 13, 686–691.

Ruttan, V. (2001) *Technology, Growth and Development: an Induced Innovation Perspective*. Oxford University Press, Oxford, UK.

Qaim, M., Krattiger, A.F. and von Braun, J. (eds) *Agricultural Biotechnology in Developing Countries: Toward Optimizing the Benefits for the Poor*. Kluwer Academic Publisher, Dordrecht.

Santaniello, V., Evenson, R.E. and Zilberman, D. (eds) (2002) *Market Development for Genetically Modified Foods*. CAB International, Wallingford, UK.

Chapter 1
From the Green Revolution to the Gene Revolution

R.E. Evenson
Economic Growth Center, Department of Economics, Yale University, New Haven, CT 06520, USA

The second half of the 20th century was witness to the Green Revolution. This revolution was characterized by the development of 'modern' crop varieties (MVs) in National Agricultural Research System Crop Genetic Improvement (NARS-CGI) programmes in developing countries. These NARS-CGI programmes were supported by a system of international agricultural research centres (IARCs) built in the 1960s and 1970s. The scientific foundations for the Green Revolution were established late in the 19th century and were further developed during the early decades of the 20th century. Plant breeding improvements became the province of specialized plant breeders trained in the biological sciences in the 20th century. Most developed countries achieved a Green Revolution in the first half of the 20th century. Developing countries achieved a partial Green Revolution in the second half of the century.[1]

The scientific foundations for the Gene Revolution were established roughly a century after the comparable foundations for the Green Revolution. Gene Revolution products use 'genetic engineering' techniques. We are now witnessing the early stages of the Gene Revolution. These early stages have been characterized by considerable controversy over food safety and consumer and political movement resistance to genetically modified organisms (GMOs). There are important differences between the Green and Gene revolutions. These are particularly pronounced regarding the role of private firms. Private plant-breeding firms played a minor role in the Green Revolution but have taken the lead in developing Gene Revolution products. However, there are similarities as well, and these are not well recognized in the current controversy over GMOs. The chief similarity is that Gene Revolution techniques are presently best suited to qualitative trait breeding, a major strategy used in the Green Revolution. Much of the spread and extent of the delivery of MVs to farmers has been governed by the incorporation of qualitative traits such as host-plant resistance to insect pests and disease, and host-plant tolerance to abiotic stresses in crop varieties.

A further similarity between the Green and Gene Revolutions is that the economic

[1] Hayami and Ruttan (1985), provide a basic overview of the early Green Revolution. Dalrymple (1986a,b) and Timothy *et al.* (1988) provide early data on the adoption of Green Revolution varieties.

penalties and rewards associated with differential access to MVs have been a major feature of the Green Revolution and will be for the Gene Revolution.

In Part I of this book, a recent study of the Green Revolution is reviewed and assessed. In Part II, the Gene Revolution (for plants) as it has unfolded to date, is reviewed. Part III utilizes data from the International Rice Biotechnology Programme (IRBP) supported by the Rockefeller Foundation to assess potential impacts of the Gene Revolution for rice farmers. A scientist survey undertaken as part of the IRBP is used in this assessment. Part IV discusses economic penalties and rewards associated with differential access to MVs.

The Green Revolution: an Overview

This section is based on a study of IARC and NARS-CGI programmes sponsored by the Standing Panel on Impact Assessment – Technical Advisory Committee (SPIA-TAC) unit of the Consultative Group on International Agricultural Research.[2,3]

Production of modern varieties

Figure 1.1 shows annual releases of MVs for 11 food crops. The figure shows high rates of variety production in rice and wheat, the crops most closely associated with the Green Revolution. Figure 1.1 also shows that, across all crops, regions and time periods, a substantial amount of breeding success has been achieved.[4]

There are, however, important disparities in the production of MVs. For cereals grown primarily under semi-arid and dry land conditions, there was relatively little improved germplasm available until the 1980s. The same was true for the major pulses and for root crops – especially cassava. The production of MVs in these crops has lagged behind production in other crops. This effect has been particularly pronounced for the Middle East–North Africa and for sub-Saharan African countries.

Figure 1.1 can be summarized as follows:

1. MVs have been produced for all 11 food crops.
2. MV production rates differ by crop, reflecting the state of breeding experience and genetic resource development. By the late 1990s, varietal production was roughly proportional to planted area of the crop.
3. Varietal production for all crops doubled from the 1960s to the late 1970s, then doubled again by the 1990s.
4. Varietal production rates for wheat and rice have been roughly stable for the past 15 years and have been rising for all other crops.

Adoption of MVs

In most cases, the MVs[5] produced by national and international research institutions have been thoroughly evaluated by farmers. When a farmer chooses to adopt a new variety in place of an older variety, it reflects the farmer's judgement that the new variety offers some net benefit or advantage. Thus, farmer adoption of MVs is a good indication that research is generating appreciable benefits.[6] (Dalrymple, 1986a,b, and Timothy *et al.*, 1988 provide early data; Evenson *et al.*, 1991, review extension roles in adoption.)

Figure 1.2 depicts adoption rates by region and decade. The figure shows that, for most crops, in most regions, MV adoption has been significant. There are, how-

[2] See Evenson (2001) for a review of the early Gene Revolution production. Ruttan (2001) provides a useful review.
[3] This study is reported in full in Evenson and Gollin (2001).
[4] Stakman *et al.* (1967) provide a review of the foundations of the Green Revolution.
[5] The basic economic model of farmer adoptions of crop varieties is provided in Griliches (1957).
[6] Dalrymple (1986a,b) and Timothy *et al.* (1988) provide data on early adoption. More complete data are reported in Evenson and Gollin (2001).

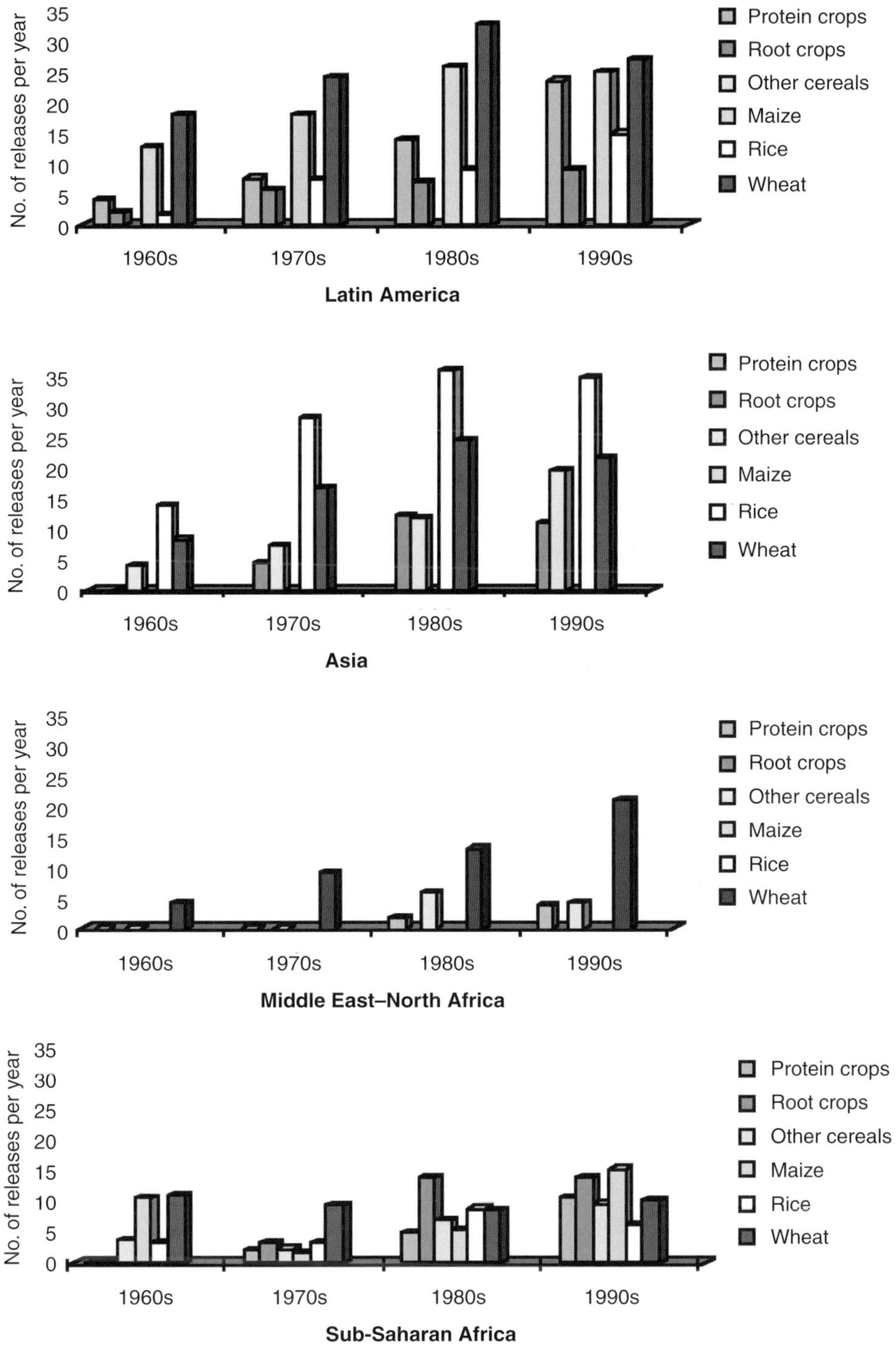

Fig. 1.1. Modern variety production by decade and region (annual MV releases).

ever, important differences across crops and regions in the date of first adoption and in the subsequent rates of increase in adoption. The date of first adoption is not always closely associated with the date of first production of MVs.

Particularly striking are the data on MV adoption from sub-Saharan Africa. Although significant numbers of MVs were produced in this region in the 1960s and 1970s, there was little adoption by farmers except for wheat. Why was sub-Saharan Africa different? In large measure, the data suggest that, in the 1960s and 1970s, national and international programmes sought to 'short-cut' the varietal improvement process in sub-Saharan Africa by introducing improved varieties from Asia and Latin America, rather than engaging in the time-consuming work of identifying locally adapted germplasm and using it as the basis for breeding new varieties. The introduced varieties released during this period were not useful beyond some limited domains and, as a result, they had relatively little impact. The pattern remained until the 1980s, when more suitable varieties finally became available.

The contributions of international research

How much of a role did international institutions play in the development of modern varieties? Figure 1.3 documents the contributions of IARCs to the MVs released over the 1965–1998 period. It also shows IARC contributions to the varieties actually grown by farmers in 1998. Two indexes of IARC content by crop and crop group are shown in Fig. 1.3. The first is the *direct IARC* contribution. It is the proportion of all released varieties that were based on a breeding cross made in an IARC programme. The second is the *indirect IARC* contribution. It is the proportion of all released varieties based on a NARS programme cross that have at least one ancestor based on an IARC cross.

There are two points of note regarding these data. The first is that IARC content is impressively high in released varieties. For all crops combined, 36% of MVs were based on an IARC cross and 22% were based on a NARS cross with at least one IARC crossed ancestor. Fifteen per cent of these ancestors were parents, 7% grandparents and other ancestors.

To put these figures in perspective, note that the IARCs account for only small fractions of the scientists working in crop-improvement programmes in developing countries – roughly 3% of the developing world's maize researchers; about 4% of the developing world's wheat researchers, and no more than 15% of the rice scientists in South and Southeast Asia, excluding China. The fractions of expenditures on crop improvement in developing countries are somewhat higher, since IARCs spend more money per researcher. Even by this measure, however, the IARC shares of research input are not large. In wheat, for example, international centres probably account for little more than 10% of the amount spent by developing countries' NARS, excluding China (Heisey *et al.*, 2001). Thus, the direct IARC contributions reported above suggest that the IARCs are contributing to MVs far out of proportion to those institutions' shares of scientific manpower or spending.

The second point to note is that there were very few contributions, direct or indirect, from research organizations in developed countries. The agricultural research system in the USA, for example, made almost no direct contributions and very few indirect contributions to the Green Revolution. At an aggregate level, the same can be said for research programmes in Japan, France and other European countries. This is in itself quite remarkable, given the perceived strength of agricultural research in these countries.

Figure 1.3 also reports a comparison of IARC content in MVs actually adopted by farmers in 1998. This comparison shows that for most crops (rice, minor cereals, protein crops and root crops) the IARC-crossed varieties account for a disproportionately large share of area planted. The sum of direct and indirect IARC contributions is

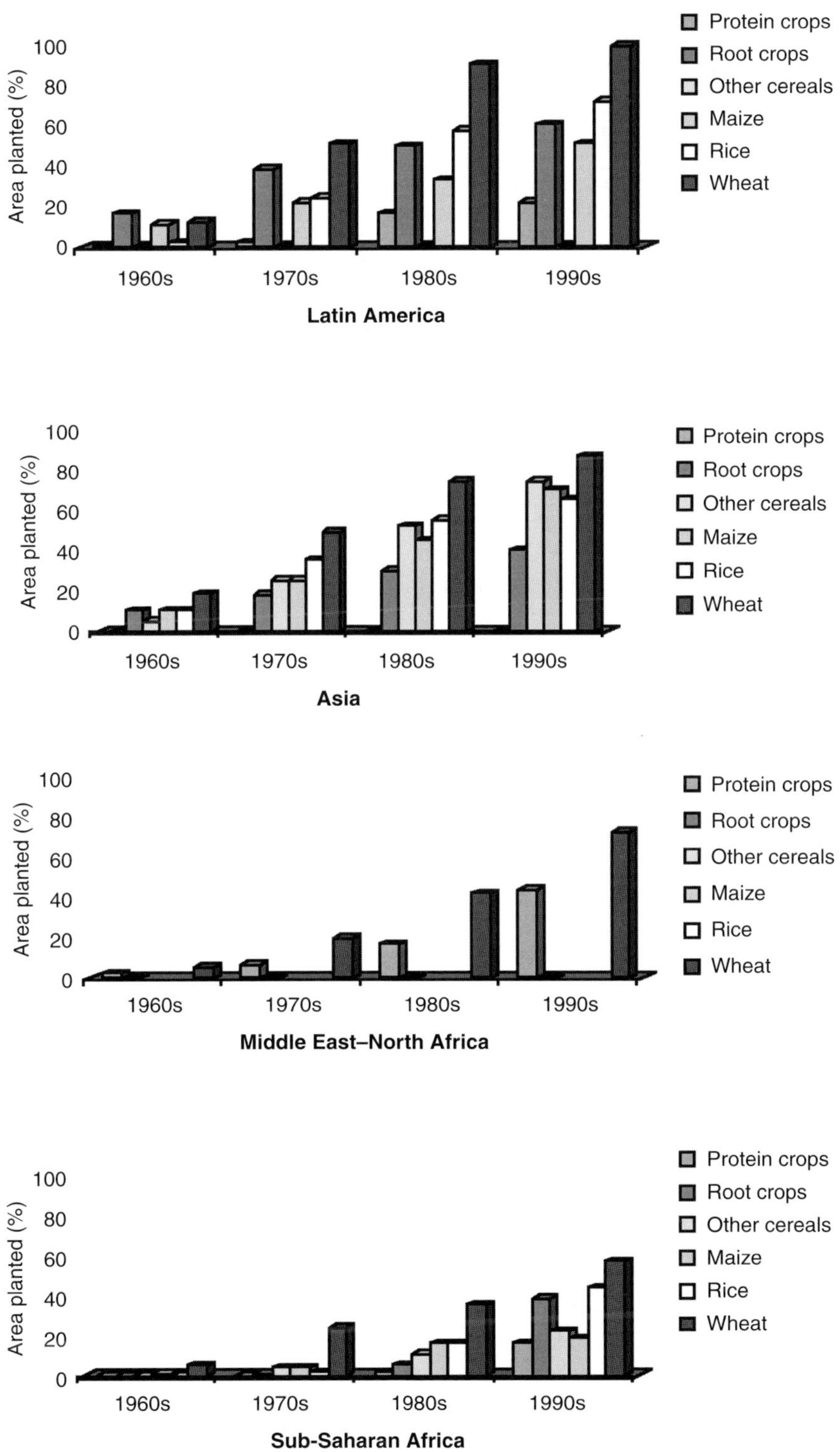

Fig. 1.2. Modern variety diffusion by decade and region (% area planted to MVs).

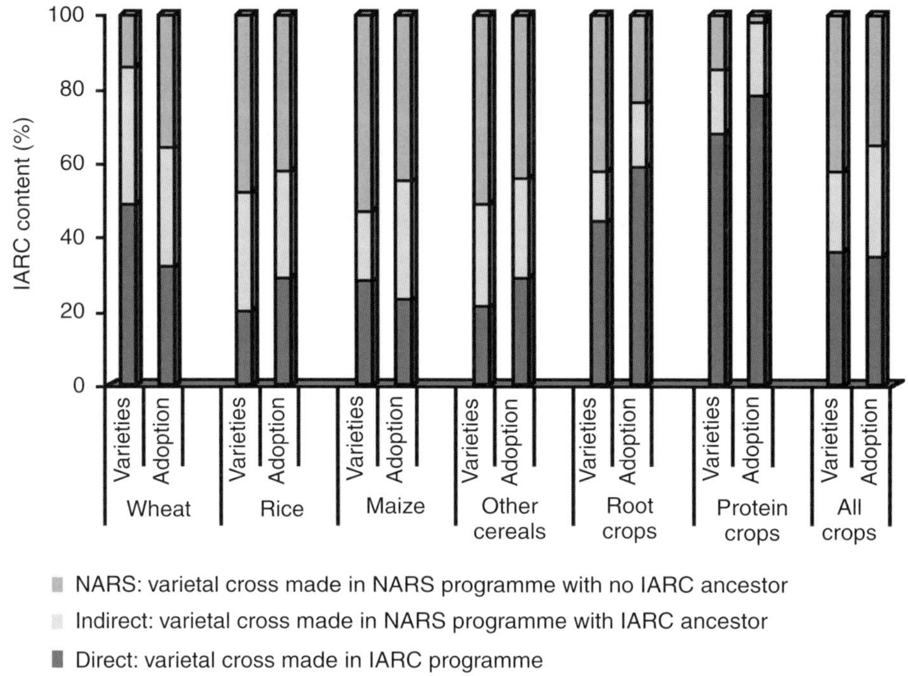

Fig. 1.3. IARC content in MVs (%).

higher for adopted varieties than for released varieties, for all crops except for wheat. Somewhat surprisingly, the highest IARC contributions to adoption are in protein crops and root crops. Across regions, IARC contributions are highest in the Middle East–North Africa and sub-Saharan Africa, where national breeding programmes are generally weak.

Germplasm effects

The data on indirect contributions can be used to estimate the contributions of IARCs to the development of MVs in NARS research programmes. Two channels through which IARC research affects national programme releases are relevant.

First, IARCs provide useful germplasm for national programmes. A statistical analysis indicates that the doubling of resources in NARS crop-breeding programmes between 1975 and 1995 would have produced 70% more varieties, even in the absence of IARC-developed germplasm. In fact, with these genetic resources available to NARS breeders, varietal production more than doubled over this period. This indicates that IARC germplasm made possible an expansion of the pool of released varieties.

Second, the presence of the IARCs has implications for investment in national crop-improvement programmes. Theory does not tell us whether international research should 'crowd out' national research investments or whether the two are complementary. The SPIA-TAC study concluded that the complementary effect dominated the competitive effect. NARS investments have been stimulated by the availability of relevant international research.

These germplams impacts, it should be noted, will be relevant in the Gene Revolution, because many genetically engineered traits can be incorporated into crop varieties using conventional breeding methods, after incorporation by genetic engineering methods.

Productivity growth impacts of CGI programmes

The study summarized here included estimates for each crop of the productivity advantages of converting crop acreage from traditional varieties to MVs. In some cases, estimates of productivity advantages of converting from early generation MVs to later generation MVs were also reported. Because the crop-by-crop approach may obscure effects across crops, the study also included three country studies, for India, China and Brazil. For example, one accomplishment of rice breeding has been to develop varieties with shorter duration to maturity. Such varieties might actually yield less than the longer-duration varieties that they replace, making it appear that productivity has fallen. However, if these varieties enable farmers to double-crop rice and wheat on land where this was previously not possible, the benefits of rice research might show up as increases in wheat production. The three country studies were designed to pick up some such effects. All three approaches to measuring productivity gains reported similar estimates of impact.

Figure 1.4 depicts annual crop-breeding contributions to growth by crop by decade. These contributions were obtained by multiplying MV adoption rates by the productivity estimates and converting these to 10-year growth rates. These calculations were then compared to actual yield growth over the periods. The calculated growth rates were highly correlated with actual yield changes by crop and region. It was estimated that more than half of the real productivity growth in developing country agriculture can be attributed to crop breeding.

Regional differences in the effectiveness of varietal improvement reflect differences in crop mix and in rates of adoption. Figure 1.4 thus goes a long way towards explaining one of the puzzles of the agricultural development literature. Observers have noted that sub-Saharan Africa and the Middle East–North Africa regions have had relatively high 'investment intensities' in agricultural research and extension. Yet the productivity performance of these two regions has not matched Asia's record. Figure 1.4 indicates that in the 1960s, Asian agriculture was already realizing significant growth from varietal improvement, while sub-Saharan Africa was realizing none. In the 1970s and 1980s, Asian agriculture was realizing roughly 1% per year from crop breeding alone, while sub-Saharan Africa was realizing only one-quarter as much. Even in the 1990s, sub-Saharan Africa was realizing only about half the growth of other regions. For all crops in all regions, the gains from breeding were highest in the 1980s and 1990s. Popular perceptions suggest that the Green Revolution was effectively over by this time but, in fact, as Fig. 1.4 shows, plant-breeding contributions were highest for the 1980s, even for rice and wheat. This was particularly important because, in most developing countries, the 1980s and 1990s saw the largest increases ever recorded in human populations.

Diffusion effects of MVs

As MVs are adopted, they create a kind of 'Kuznets curve' on incomes of farmers. When the level of MV adoption is low and MVs are available only to farmers in favourable environments, this creates income inequalities. Then, as MVs are diffused to all production environments, an equalization of income takes place.

In this section, two points regarding diffusion are emphasized. The first is that the Green Revolution was characterized by a two-stage breeding strategy in most crops. The first stage entailed the building of high-yielding 'platform' plant types based on multiple gene quantitative traits. This stage is a complex stage and requires the building of platform genetic resource stocks suited to the broad ecosystem in

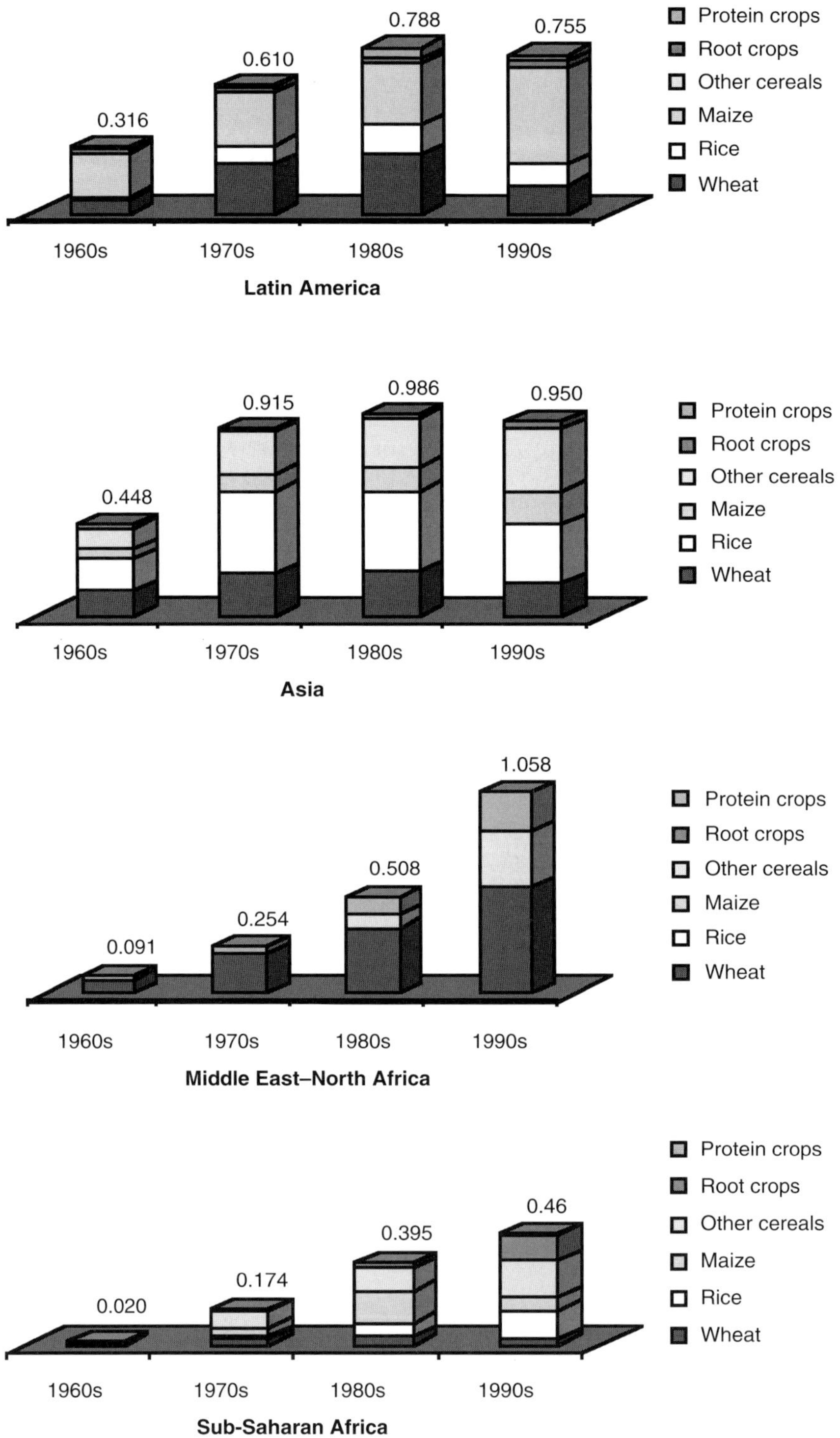

Fig. 1.4. Annual growth contributions of modern varieties.

question. The second stage entails the installation of qualitative traits – host-plant resistance to insect pests and to plant diseases, and host-plant tolerance of abiotic stresses (drought, submergence, etc.) on platform varieties. The incorporation of these qualitative traits is an important determinant of MV diffusion, once the first stage is achieved. As the next section will show, first-generation Gene Revolution methods are well suited to qualitative trait breeding.

The second point of this section is that the MV diffusion effects have important welfare implications.

Table 1.1 reports estimates of the relationship between MV production and MV diffusion at the international level. These estimates show that the elasticity of MV adoption (evaluated at the mean hectare levels) range from 0.25 to 0.5, indicating strong diminishing returns to MV extension beyond initial favourable ecosystem conditions.

Table 1.2 supports the first point, i.e. that the incorporation of qualitative traits on to MV platforms is an important determinant of MV diffusion. The table reports estimates from the India study in the SPIA-TAC Green Revolution study. The India study treated MV adoption or diffusion rates as an endogenous variable. (Area planted, irrigated area and yields are also treated as endogenous variables in the model.)

The determinants of rice MV adoption include variables measuring qualitative traits incorporation on the semi-dwarf high-yield platform plant types. Rice breeding in India (and Asia generally) is characterized by stages. Stage 1 is the achievement of the HYV platform variety. Stage 2 is the emergency-driven incorporation of qualitative traits to respond to field problems with insect pests and diseases (where insects are often the vectors of diseases) (HPR traits). Stage 3 is the systematic incorporation of qualitative traits with more emphasis on abiotic stress traits (HPT traits). Stage 4 is the incorporation of traits to better meet local tastes and preferences.

Table 1.2 from the India study reports estimates of the determinants of MV adoption for five crops in India. The rice diffusion estimates include variables measuring the number of varieties available to a state with host plant resistance (HPR) and host plant tolerance (HPT). The implications of the estimates are that MV area, as a proportion of total rice area, would have reached around 35% if only the Stage I – platform varieties (IR5–IR22) – were available to farmers. The Stage II varieties (IR26–IR36) raised adoption levels to the 50–60% range. Stage III trait incorporation led to more than 85% of production environments in India being planted to MVs.

The welfare implications of MV adoption rates are illustrated in Table 1.3. In this table, determinants of 'wasting' in

Table 1.1. Estimated relationship: MV production and MV adoption: international data for three periods, 1965–1975, 1975–1985, 1985–1995 (OLS Regression–weighted by area planted).

Independent variables	Rice	Wheat	Maize	Sorghum-millets	Cereals	Potatoes, cassava and beans
	–1.296	–2.809	–3.608	–0.778	–2.730	–0.812
Ln (CMVR)	(2.17)	(4.62)	(4.76)	(3.49)	(8.86)	(1.51)
Ln (CMVR) × Ln (HA)	0.1130 (3.92)	0.2041 (7.42)	0.2663 (6.30)	0.2461 (11.52)	0.1926 (12.50)	0.1239 (1.51)
No. obs.	51	66	43	27	187	64
R^2	0.904	0.904	0.912	0.984	0.834	0.981

MV, modern variety; CMVA, cumulated area planted to MVs; CMVR, cumulated numbers of MVs released; HA, hectares planted to the crop.
All regressions included: crop dummy variables, climate zone dummy variables.
Ratios in parentheses.

Table 1.2. Estimates of the determinants of MV adoption.

	Rice	Wheat	Maize	Sorghum	Pearl millet
	0.3689755***	1.347849***	0.5113548***	0.949578***	3.340208***
IRR	(0.0494216)	(0.0391511)	(0.0785826)	(0.2030388)	(0.4236795)
	0.0022844***	0.0007315***	−0.0021518***	−0.0006521*	0.0013057*
EXT	(0.0004075)	(0.0001765)	(0.0004165)	(0.0003925)	(0.0007925)
	0.0000819	−0.0000121	0.0003355***	0.0003952***	0.0002293*
STRESS	(0.0000567)	(0.0000295)	(0.0000736)	(0.0000451)	(0.0001200)
	0.0008705***	0.000055	0.0007506***	0.0015738***	−0.0008357***
PRCHEM	(0.0001426)	(0.0000656)	(0.0001695)	(0.0001238)	(0.0002990)
	0.4118516***	0.0412663	−0.3652459***	0.5044381***	0.2056575*
STORIE	(.0595180)	(0.0291028)	(0.0799168)	(0.0422108)	(20.1157527)
	1.395171***	0.3871841***	3.558084***	1.556***	23.344414***
MARKET	(0.1443209)	(0.1126576)	(0.2531967)	(0.2206623)	(0.8782209)
	−0.001838				
AVGNOLR	(.0014809)				
	0.000259				
AVGXOLR	(0.0035152)				
	0.3806231***				
HPR-disease	(0.1113970)				
	1.095059***				
HPR-insects	(0.0798016)				
HPT-abiotic	1.367718***				
stresses	(0.1482571)				
RMSE	0.3198495	0.333575	0.249133	0.2198743	0.5482644
R^2	0.4481	0.1971	0.4985	0.4337	−0.0026

children aged 0–6 years old in rural and urban locations are analysed. The measure of wasting is the percentage of children (0–6) two or more standard deviations below regional norms of weight per unit of height. These data are available for a number of countries from the World Health Organization. Data on the percentage of cropped area planted to MVs and on income distribution have been added to the database. Observations are for 1990 and 1998.

Specification (1) shows that calorie availability (from FAO sources) per capita is an important determinant of wasting. As calories available increase, wasting decreases. The schooling levels of adult females also reduces wasting, while the schooling levels of adult males does not.

Specification (2) adds interactions of schooling with calorie availability to specification (1). This extension indicates that, when calorie availability improves, the positive effect of adult male schooling on wasting is converted to a negative effect.

Specification (3) adds World Bank data on income distribution to the data set. The variable B20 is the estimated proportion of total consumer expenditures by the bottom two income deciles. Thus a higher B20 indicates a more equal distribution of income. Specification (3) indicates that, as B20 increases, the role of adult schooling in reducing wasting becomes more important. It also shows that calorie-availability effects on wasting are strengthened as income distributions are more equal.

Specification (4) incorporates the MV variable and its interaction with calories available. This specification also shows a powerful MV interaction with calorie availability. When MV levels are low, wasting increases for a given level of calorie availability. As MV levels rise, the impact of MVs is converted from increasing wasting

Table 1.3. Wasting determinants: international data-dependent variable; proportion of children (aged 0–6) 2 × SD below mean.

	Rural children				Urban children			
Independent variables	(1)	(2)	(3)	(4)	(1)	(2)	(3)	(4)
Calories available	−0.0093	0.0150	0.0036	0.0139	−0.79	0.0163	0.0044	0.0148
	(4.16)	(0.258)	(0.65)	(3.50)	(3.69)	(3.06)	(0.83)	(4.04)
Schooling: adult	−2.879	−19.343	−0.191	−2.89	−2.720	−20.2	−0.237	−2.835
females	(3.18)	(4.86)	(0.7)	(3.96)	(5.12)	(4.15)	(0.09)	(4.12)
Schooling: adult	1.456	23.355	5.30	0.734	1.31	24.07	5.25	0.770
males	(1.76)	(4.51)	(2.1)	(0.95)	(1.65)	(5.36)	(2.13)	(1.11)
Proportion rural	−0.0259	0.123	0.113	0.028	−0.035	0.117	0.078	0.056
	(0.49)	(2.31)	(1.56)	(0.67)	(0.70)	(2.21)	(1.10)	(0.39)
D98	4.296	5.066	4.449	4.533	3.63	4.59	3.309	3.908
	(4.48)	(5.87)	(4.55)	(6.02)	(3.95)	(5.56)	(3.87)	(5.46)
CAL × school F		0.0070				0.0074		
		(3.31)				(3.72)		
CAL × school M		−0.0089				−0.0092		
		(4.62)				(5.14)		
B20			6.182				6.816	
			(3.24)				(3.49)	
CAL × B20			−0.511				−5.28	
			(1.16)				(1.25)	
School F × B20			−0.399				(1.20)	
			−0.378				(1.19)	
School M × B20			−0.0013				−0.0012	
			(1.67)				(0.69)	
MV				94.85				92.46
				(6.95)				(7.78)
CAL × MV				−0.0349				−0.0345
				(6.52)				(6.93)
Constant	83.89	−37.65	−30.90	−27.41	30.012	−43.29	-30.93	−30.24
	(4.89)	(2.73)	(1.04)	(10.68)	(4.53)	(2.76)	(5.05)	(6.31)
R^2	0.566	0.669	0.753	0.740	0.535	0.608	0.737	0.435

CAL, calorie availability; MV, modern variety; T ratios in parentheses.

to decreasing wasting. When calorie availability improves (partly due to MVs), wasting is decreased and this effect is stronger, the higher the MV levels.

These estimates are similar for rural and urban children. They attest to a strong welfare impact of MVs and particularly to the expansion of MV area beyond the original favourable ecosystem conditions.

The Gene Revolution: Early Stages

The Gene Revolution is based on technology made possible by 'enabling science'. The scientific foundations underlying the invention of biotechnology or genetically engineered products in health and agriculture is impressive.

Biotechnology firms were very quick to recognize the commercial potential for genetically engineered products in both health (including animal health) products and in crop varieties. Many companies were formed, often by scientists or with scientist consultations, as early as the 1970s. By the end of the 1980s, large investments by many companies had been made and new products were being tested and approved for commercialization.[7]

Experience with health products has generally been consistent with investor expectations in situations of high risk. Much of the early innovation has taken

[7] See Evenson (2001) and Ruttan (2001) for reviews of the early Gene Revolution.

place in smaller, newly formed companies. Product introductions have had success ratios comparable to those in other new fields. Consumer resistance has not been a serious problem for Gene Revolution health products.

This is not the case for genetically modified (GM) crop products. For these products, large investments were made in the early stages, not by the traditional seed companies engaged in plant breeding (Pioneer, DeKalb) but by chemical firms (Monsanto, Ciba-Geigy, Novartis, DuPont). Investment was guided by the 'Life Science' model, where GM crops would serve to complement chemical-industry products. Herbicide-tolerant GM crops were designed to complement broader use of herbicides for weed control. *Bacillus thuringiensis* (*Bt*) insect-resistant GM crops were designed to control insects not easily controlled by insecticides. They were also designed to reduce insecticide use.

Both herbicide tolerant and *Bt* GM crops were put on the market in 1996. Farmer acceptance was very good and adoption rapid. However, the consumer resistance to unlabelled GM products has proven to be very strong. GMOs have become a controversial element of political movements, especially in the 'globalization backlash' movement.

At this point, the ag biotech industry is in turmoil. This is partially the result of consumer response fuelled to some extent by the incorporation of GMOs into broader political movements. However, more fundamentally, it is determined by investor dissatisfaction with Gene Revolution products. A number of large firms have been pressed by investors to divest of their ag biotech divisions.

The Green Revolution: congruity with the Gene Revolution

It is important to note first that the methods and techniques of biotechnology are far more subtle and fundamental to plant breeding than implied by the contemporary controversies over GMOs. The sciences underpinning genetic or biological invention have truly changed. The current controversy stresses 'products' but, over the longer run, it is the 'processes' of biotechnology that will be important. It is also important to note that the applications of this new science will require capacity building and training. The Green Revolution required NARS-CGI programmes with a sustained long-term commitment to the development of germplasm stock suited to regional ecosystems. While there are possibilities for 'gene-renting' with Gene Revolution products, it will continue to be the case that the 'platform' varieties must be developed in NARS programmes, and this will require training and institutional capacity.

The distinction between single gene-based qualitative traits and platform quantitative traits was important in the Green Revolution and will be even more important in the early decades of the Gene Revolution. With the development of complete genome maps for most crop species, the stage is now set for genomic analysis. At present, however, the basic techniques are best suited to single gene-trait breeding for host-plant resistance to insect pests and to diseases, and for host-plant tolerance to abiotic stresses. Prospects are also good for some quantitative trait breeding.

The International Programme on Rice Biotechnology

The Rockefeller Foundation recognized the complexities and difficulties associated with achieving Gene Revolution capabilities in developing countries in the mid-1980s. It established the IPRB to support the use of biotech methods for rice improvement and to assist in bringing biotech capacity to developing countries. This programme supports research projects and programmes in both developed and developing countries. (Some 65 research programmes were included in the programme including the three IARCs with mandates for rice improvement in developing countries, International Rice Research

Institute (IRRI) for Asia, the International Center for Tropical Agriculture (CIAT) for Latin America and the West Africa Rice Development Association (WARDA) for Africa.)

A major economic study of the IRBP was undertaken in the late 1990s. This study utilized a research project classification with two dimensions. Research problem areas (RPAs) were defined and linked to crop losses. Research techniques (RTs) were defined for managerial, conventional breeding (including wide crossing) and biotechnology techniques. For each RPA and RT, scientist evaluations of research potential (RP), research achievement to date (RA) and years to 25% (Y25) and 75% (Y75) attainment were obtained.[8]

The question of relevance for this chapter is the estimation of the added contribution that research programmes can make when they have biotech capabilities.

To proceed to answer this question, three steps are required. The first is to model the discovery process for a given RPA–RT research programme. The second is to model the contributions of different RTs to achieving progress in the RPA objective (e.g. reducing crop losses from disease). The third is to combine RPAs for the commodity.

Consider the first step. Suppose we have three RTs, management research (M), conventional breeding research – including wide crossing (CB) and biotech research – marker-aided breeding and transgenetics (BT). The search model applied to research programmes (Evenson and Kislev, 1978) can be used to specify an invention function for each RT. This model implies:

$$I_M = a_M + b_M \, 1n\,(M)$$

$$I_{CB} = a_{CB} + b_{CB} \, 1n\,(CB)$$

$$I_{BT} = a_{BT} + b_{BT} \, 1n\,(BT)$$

Suppose now that the units over which M, CB and BT apply are the same. Then we will want to maximize the number of inventions for a given level of cost. Further, suppose that the marginal cost of each type of research is equal, $MC_M = MC_{CB} = MC_{BT}$.

The maximizing invention requires

$$\frac{dI_M}{d_M} = \frac{b_M}{M} = \frac{dI_{CB}}{d_{CB}} = \frac{b_{CB}}{CB} = \frac{I_{BT}}{dBT} = \frac{b_{BT}}{BT} = MC$$

Then for simplicity set $MC = 1$.

This implies that for cost minimization:

$M = b_M$, $CB = b_{CB}$ and $BT = b_{BT}$.

The scientists' estimates of RP in the IPRB study can be considered to be estimates of b_M, b_{CB} and b_{BT}.

The second step in the procedure is to model the relative contributions of I_M, I_{CB} and I_{BT} to productivity improvement for the RPA. Clearly, this contribution is not one of perfect substitutions, where one would choose the inventions with the highest RP rating. All three types of invention contribute to productivity. Lacking a theoretical alternative specification, suppose that the Cobb–Douglas specification holds:

$$P_{RPA} = A\, I_M^{\varepsilon_M}\, I_{CB}^{\varepsilon_{CB}}\, I_{BT}^{\varepsilon_{BT}} \qquad (1)$$

Now suppose that $\varepsilon_M + \varepsilon_{CB} + \varepsilon_{BT} = 1$

Further suppose that ε_M, ε_{CB} and ε_{BT} are proportional to RP_M, RP_{CB} and RP_{BT}.

Then for a research programme with BT capacity:

$$P_{RPA} = A(RP_M)^{S_M}(RP_{CB})^{S_{CB}}(RP_{BT})^{S_{BT}} \qquad (2)$$

where $S_M = RP_M \big/ (RP_M + RP_{CB} + RP_{BT})$

(This specification is consistent with the first step in the process).

When a research programme does not have BT capacity, but still manages constant returns to scale:

$$P^1_{RPA} = A^1 RP_M^{\,S_M} RP_{CB}^{\,S^1_{CB}} \qquad (3)$$

where $S^1_M = RP_M \big/ (RP_M + RP_{CB})$

[8] See Evenson (1998) for review of the IRPB.

Equations (2) and (3) can be evaluated with Scientist Survey data for each RPA, to obtain a with-BT capacity and without-BT capacity comparison.

The third step in the process is to weight RPAs by units. The IPBR study also undertook an extensive survey of crop losses. These loss estimates were used as weights.

Table 1.4 reports estimates of the biotech advantage (BTADV) that rice research systems can achieve if they have a true biotech research capability. These are presented by RPA group and aggregated by region. Alternative 1 is based on the average RPA estimate in the scientist survey. Alternative 2 is based on the maximum RPA in the group.

These estimates of biotech technique advantage are based to a considerable degree on research processes and require training and skills of a high magnitude. In general, it is not possible to upgrade the skills of older plant scientists to obtain these skills. Long-term investments in training a new generation of scientists is ultimately required to achieve these capacities.

Economic Implications: Penalties and Rewards

The penalties and rewards to farmers, associated with access to the Green Revolution MVs were considerable. Farmers with access to MVs, where MVs of real value to them were actually delivered to them, adopted the MVs rapidly because the real costs of producing a unit of the crop went down. Because those farmers supplied more of the crop to the market, prices also went down. Most farmers with earliest access to MVs gained income because their costs decreased by more than prices decreased.

Consumers of these commodities also benefited from lower prices.

However, farmers without access or with delayed access suffered a significant economic penalty. Their costs did not decline, but the prices that they received did.

This penalty–reward system has now been globalized and whole regions are now impacted. The penalty to farmers in sub-Saharan Africa, shown in Fig. 1.4, was huge. Even though consumers do

Table 1.4. Estimates of biotechnology advantage.

	RP estimates			
	M	CB	BT	BTADV
HPR (insects)				
AIT (1)	2.76	2.56	3.88	1.177
AIT (2)	2.65	2.62	3.96	1.268
HPR (disease)				
AIT (1)	1.71	2.62	3.96	1.668
AIT (2)	1.71	2.52	3.36	1.513
HPT (abiotic stresses)				
AIT (1)	1.73	2.39	3.44	1.747
AIT (2)	1.59	2.28	3.29	1.802
Pests				
AIT (1)	3.50		4.00	1.107
AIT (2)	3.12		3.64	1.315
Bioefficiency				
AIT (1)		3.41	3.64	1.124
AIT (2)		3.12	3.61	1.304

BT ADV ALLRPAS	South Asia	East Asia	Southwest Asia
AIT (1)	1.35	1.40	1.19
AIT (2)	1.45	1.51	1.34

benefit from lower food prices, countries with large agricultural sectors were hugely penalized by delayed access and this was a major factor impeding more general economic development. Most of the countries without access to Green Revolution technology were also excluded from access to industrial technology. Much of this exclusion was related to wage levels (innovations in harvesting equipment were of little value in low-wage economies) but much of it is related to failures of industrial development policies in these countries.

Policy Implications for the Gene Revolution

Developing countries will have to address two important issues if they are to avoid the delay penalties that are associated with the Gene Revolution. The first issue is the intellectual property issue. The second is the issue of capacity building in public-sector NARS programmes. The two issues are related and the relationship is complementary – at least in the short run.

Consider the issue of intellectual property rights (IPRs). The strategy of maintaining relatively weak IPRs for the purpose of avoiding high licensing payments for foreign origin technology is an old one. The basic idea behind the strategy is to 'tilt' the IPR system toward domestic inventors and against foreign inventors. If skillfully done, this strategy produces a form of 'legal (or quasi-legal) IPR piracy'. The USA utilized this strategy in the 19th and early 20th centuries. Japan used the strategy in the early part of the 20th century and in the early post-World War II period. The Asian Tigers – South Korea, Taiwan, Singapore and Hong Kong – made at least partial use of the strategy.

Most developing countries have misused the strategy by tilting the system so far against foreign inventors that they lose the advantage of IPRs for domestic inventors. This is a fundamental flaw in the development policies of many developing countries. Domestic invention capability is absolutely essential to growth and development, and the achievement of this capacity requires IPR systems designed to stimulate domestic invention and a willingness to invest in public-sector research (as in the NARS programmes) when IPR systems are inadequate for the task. Domestic invention capability is critical because without this capacity it is very difficult to engage in technology piracy, legal or illegal.

The penalties for piracy or at least for perceived piracy are much higher as a result of the Trade Related Intellectual Property (TRIPs) component of the General Agreement on Tariffs and Trade (GATT) negotiations. Quasi-legal piracy now provokes such a 'steering clear' reaction by technology suppliers that it is no longer a viable development strategy. Quasi-piracy of biotech products is not going to be a viable strategy for developing countries.

Even legal piracy practices were not sufficient to produce the Green Revolution in developing countries. Countries without long-term commitment to and support for NARS programmes did not have a Green Revolution, and countries without a long-term commitment to NARS programmes with biotech capacity will not have a Gene Revolution.

The complementarity between IPR policy and NARS policy is that strong NARS programmes provide alternatives to farmers that can alleviate much of the hostile reactions to IPR payments to foreigners. If farmers have NARS MVs, the option to purchase private-sector MVs, whether from domestic or multinational companies, provides expanded opportunities. However, there is also a further complementarity in terms of benefits of competition provided to NARS programmes.

References

Agcaoili, M.C., Oga, K. and Rosegrant, M.W. (1993), Structure and operation of the International Food Policy and Trade Simulation (IFPTSIM) Model. Paper presented in the Second Workshop of the Research Project on Projections and Policy Implications of Medium- and Long-Term Rice Supply and Demand, IRRI, Los Banos, Philippines.

Dalrymple, D. (1986a) *Development and Spread of High Yielding Rice Varieties in Developing Countries.* Bureau for Science and Technology, Agency for International Development, Washington, DC.

Dalrymple, D. (1986b) *Development and Spread of High Yielding Wheat Varieties in Developing Countries.* Bureau for Science and Technology, Agency for International Development, Washington, DC.

Evenson, R.E. (1998) Biotechnology and genetic resources. In: Evenson, R.E., Gollin, D. and Santaniello, V. (eds) *Agricultural Values of Plant Genetic Resources.* CAB International, Wallingford, UK.

Evenson, R.E. (2001) IARC 'Germplasm' effects on NARS breeding programmes. In: Evenson, R. and Gollin, D. (eds) *Crop Variety Improvement and Its Effect on Productivity: the Impact of International Agricultural Research*, chapter 21. CAB International, Wallingford, UK.

Evenson, R.E. and Gollin, D. (eds) (2001) *Crop Variety Improvement and Its Effect on Productivity: The Impact of International Agricultural Research.* CAB International, Wallingford, UK.

Evenson, R.E. and Kislev, Y. (1975) *Agricultural Research and Productivity.* Yale University Press, New Haven, Connecticut.

Evenson, R.E. and Westphal, L. (1994) Technological change and technology strategy. In: Srinivasan, T.N. and Behrman, J. (eds) *The Handbook of Development Economics*, Vol. 3. North Holland, Amsterdam.

Evenson, R.E., Birkhauser, D. and Feder, G. (1991) The economic impact of agricultural extension: a review. *Economic Development and Cultural Change* 39, No. 3.

Evenson, R.E., Herdt, W. and Hossain, M. (1996) *Rice Research in Asia: Progress and Priorities.* CAB International, Wallingford, UK.

Gollin, D. and Evenson, R.E (1990) Genetic resources and rice varietal improvement in India. Economic Growth Center, Yale University, unpublished manuscript.

Gollin, D. and Evenson, R.E. (1998) Breeding values of rice genetic resources. In: Evenson, R.E., Gollin, D. and Santaniello, V. (eds) *Agricultural Values of Plant Genetic Resources.* CAB International, Wallingford, UK.

Griliches, Z. (1957) Hybrid maize: an exploration in the economics of technical change. *Econometrica* 25, 501–522.

Hayami, Y. and Ruttan, V.W. (1985) *Agricultural Development: an International Perspective.* Johns Hopkins University Press, Baltimore, Maryland.

Heisey, P.W., Lantican, M.A. and Dubin, H.J. (2001) Wheat. In: Evenson, R. and Gollin, D. (eds) *Crop Variety Improvement and Its Effect on Productivity: the Impact of International Agricultural Research*, chapter 4. CAB International, Wallingford, UK.

Pardey, P.G., Alston, J.M., Christian, J.E. and Fan, S. (1996) *Hidden Harvest: U.S. Benefits from International Research Aid.* International Food Policy Research Institute, Washington, DC.

Rosegrant, M.W. and Evenson, R.E. (1993) Determinants of productivity growth in Asian agriculture: past and future. Paper presented at the 1993 AAEA International Pre-Conference on Post-Green Revolution Agricultural Development Strategies in the Third World: What Next, Orlando, Florida.

Ruttan, V. (2001) *Technology, Growth, and Development: an Induced Innovation Perspective.* Oxford University Press, Oxford, UK.

Stakman, E.C., Bradfield, R. and Mangelsdorf, P.C. (1967) *Campaigns Against Hunger.* Belknap Press, Cambridge, Massachusetts.

Timothy, D.H., Harvey, P.H. and Doswell, C.R. (1988) *The Development and Spread of Improved Maize Varieties and Hybrids in Developing Countries.* Bureau for Science and Technology, Agency for International Development, Washington, DC.

Chapter 2

Conflicts in Intellectual Property Rights of Genetic Resources: Implications for Agricultural Biotechnology[1]

L.J. Butler

Department of Agricultural and Resource Economics, University of California, One Shields Avenue, Davis, CA 95616, USA

Abstract

Governments of developing countries are presently confronted with various, and directly conflicting, international obligations that regulate the use of genetic resources. They must recognize intellectual property protection of plant material and Farmers' Rights, and, they must take into account rules for access to plant genetic resources and related technology, and for sharing benefits of the use of genetic resources.

All these obligations may have an impact on plant-breeding activities and on the diffusion of new plant varieties and biotechnological innovations. This impact, however, may differ by region and by country, because there are large differences among countries in their plant-breeding capability, the share of farmers that use modern varieties, and the value of plant genetic resources available. These differences have caused an international controversy on the interpretation and operationalization of the international obligations.

This chapter discusses some effects of the plant variety protection (PVP) system. It is argued that the currently favoured International Union for the Protection of New Varieties of Plants (UPOV)-PVP system may have negative effects for the majority of small farmers in developing countries, and that these negative effects, and the absence of positive effects, have led to the demand for Farmers' Rights. It is suggested that developing countries should be allowed to introduce modified PVP systems. These systems should: (i) allow farmers to save and swap seed from protected varieties; and (ii) encourage breeding approaches which result in less uniform varieties which can be adapted by farming communities. Three policy options are discussed that could combine these requirements.

[1] The essential elements of this chapter were originally conceived by Jeroen van Wijk and Robin Pistorius, Department of Political Science, University of Amsterdam, Niels Louwaars, Centre for Plant Breeding and Reproduction Research, Wageningen, The Netherlands, and the author, in October 1996.

Introduction

Agricultural biotechnology involves, in large part, the use of genetic resources (or information), to transform plants and animals into marketable, industrial elements, or to impart desirable characteristics that enhance the plant or animal value.

In order to protect the investment in these new innovations, and/or to ensure compensation for the innovation, many plant breeders and western biotechnology firms have made extensive use of the various forms of intellectual property rights (IPRs) available to them. These include patents and plant variety protection (PVP), and in some cases, copyright and trademarks.

These forms of IPRs are, however, not necessarily available in all the markets to which these firms may consider exporting their new technologies. In fact, in most of the developing world, IPRs in any form are relatively rare.

In 1994, signatories to the Uruguay Round of the General Agreement on Tariffs and Trade (GATT)[2] also agreed to the Trade Related Intellectual Property (TRIPs) obligations as part of the entire GATT agreement. The industry and trade departments of industrialized countries initiated the negotiations on TRIPs in order to protect exported advanced technology from being copied in other parts of the world. The TRIPs agreement requires countries, which do not already have an IPR system, to introduce either a patent or an effective *sui generis* system for the protection of plant varieties and agricultural biotechnological innovations.

At the same time, IPRs in plant breeding were negotiated at a *second* international forum: the Food and Agriculture Organization of the United Nations (FAO). In this case, the agricultural departments of a number of developing countries, supported by international agricultural nongovernmental organizations (NGOs), debated the virtues of privatization of plant genetic resources (PGRs) through IPRs for the farming community in developing countries. Beginning in 1989, the FAO Undertaking on Plant Genetic Resources has sought an international consensus on ways to encourage the conservation of biological diversity, compensate countries and communities for the use of genetic resources, and introduced the concept of Farmers' Rights.

In 1992, the objectives of Farmers' Rights, i.e. compensation for the use of genetic resources and encouragement of biodiversity conservation, were incorporated into yet a *third* international negotiation forum, that of the Convention on Biological Diversity (CBD). The CBD objectives did not encounter much opposition because the delegates of the industrialized countries (mainly represented by the environmental departments of their respective governments), were not involved in IPR matters. For several years, the TRIPs and CBD negotiations were like two trains running in different directions. It was only a short time before the CBD was to be signed that the two agreements were related to each other. Even though some parts of the CBD text were adjusted to make it more compatible with the draft text of the TRIPs agreement, the adoption of the CBD also marked the birth of an international agreement, which, in a modest way, challenged the principles of IPR protection which were agreed upon in the GATT.

A similar dualism can be detected regarding the relationship between the FAO Undertaking and the TRIPs agreement. Although the FAO Undertaking recognized one important element of IPR (PVP), simultaneous acceptance of Farmers' Rights marked the FAO's opposition to IPR protection. Also, the relationship between the FAO and the CBD is not unambiguous. Currently, proposals are being negotiated in which the FAO Undertaking is linked to the CBD as a 'protocol'. This proposal, however, has provoked resistance from developing countries that consider the FAO Undertaking as the most suitable agreement to serve their interests. For developing countries depending on imports of plant genetic resources

[2] In 1994, the GATT secretariat was renamed the World Trade Organization, WTO.

(PGRs), the CBD is too broad in scope and its implementation too dependent on bilateral negotiations.

In summary, countries that are implementing the CBD, which also have signed the FAO Undertaking, and that meanwhile support the results of the TRIPs negotiations have to deal with genetic resources issues overlapping three political fora related to industry, agriculture and environment. The combined and conflicting interests in all three fora have triggered unexpected 'chemical reactions' and unwanted controversies.

This chapter examines these three international agreements which are in direct conflict with each other with respect to genetic resources, plant-breeders rights, conservation of biological diversity and property rights in genetic resources.

The TRIPs Agreement

By signing the new TRIPs agreement in 1994, all member states accepted a new international standard on the protection of intellectual property, which was significantly higher than earlier standards. Legal protection of innovations, particularly those involving living material, had only occurred in some industrialized countries prior to 1994. For most of the rest of the world, and particularly the developing world, IPR did not, and still does not, exist.

From the TRIPs agreement it can roughly be concluded that:

1. All signatories are obliged to grant intellectual property protection for all plant material that meets the requirements.
2. Plant varieties and inbred lines must be protected either under patent or an effective *sui generis* system. Parties are, in principle, free to design such a system. There is little doubt, however, that for industrialized countries, the wording '*adequate sui generis*' means accession to an International Union for the Protection of New Varieties of Plants (UPOV)-PVP system.
3. In principle, all types of genes that have been identified, isolated, modified and/or transferred must be patentable. This may result in the patentability of plants, their offspring and the entire transgenic species. The only exceptions to legal patent protection are plant-related inventions that are considered to be against the moral order.
4. Landraces and plants found in the wild do not fall within the scope of IPR. These types of plant material do not meet the requirements for a patent or PVP and cannot be protected.

The reason that many developing countries accepted the TRIPs agreement had less to do with a sincere desire to embrace a new universal framework for IPRs for biological innovations, but was rather dictated by a concern for hampered access to advanced foreign technology or for losing trade benefits. Moreover, the delegates in GATT were predominantly trade economists, representing the industrial and commercial sectors of their countries, and had little knowledge of the effects of the TRIPs agreement on agriculture.

One of the options for a *sui generis* system for the protection of plant varieties, as required under the TRIPs agreement is the PVP system, which is administered by UPOV. It is likely that new member countries can only accede to the 1991 act of UPOV and not to earlier acts of this organization. The discussion of PVP in this chapter will therefore solely refer to the effects of the 1991 act of UPOV.

Implications of the TRIPs Agreement

There are both costs and benefits to adopting a PVP system. Developing countries who are considering adopting a UPOV-PVP system need to consider carefully the relative costs and benefits associated with IPRs in PGRs.

IPRs may help breeders to get a better return on investments in plant breeding

One purpose of IPRs is to promote commercial plant breeding. Patents and PVP bestow a plant breeder with the right to prevent others from using his/her protected plant variety for commercial purposes. This right enables the breeder to get a better return on investments than would be the case if no protection were available and everyone could simply propagate and sell a variety bred by someone else. The rationale of patents or PVP, however, is not to make seed companies wealthier but to promote plant-breeding activities. The question, therefore, is whether or not a better return on investments results in more R&D expenditure.

A convincing answer to this question is not easy to find. The commercial seed industry attributes the development of the private seed industry to a great extent to PVP, but clear evidence for this assumption is absent. The conclusions of the very few studies that have been carried out on the effects of PVP on R&D investment are modest. An American study concluded that the US PVP Act has stimulated the development of new varieties of a few crops, such as wheat and soybean, but there is little evidence that it has affected the R&D output for most self-pollinating crops (Butler and Marion, 1985, p. 74; Butler, 1996, p. 28). Also the US Plant Patent Act (which provides a PVP type of protection for non-sexually propagating plants) had little impact on private investment in fruit breeding because the Act allows only a very narrow product space, and because enforcement costs are high (Stallmann and Schmid, 1987). A recent study carried out in Argentina, where PVP has been enforced since 1990, concluded the system has presumably played a role in preventing a reduction in plant breeding in soybean and wheat, rather than stimulating additional R&D expenditure (Jaffé and van Wijk, 1995).

It is clear that the temporary monopoly granted to a successful patentee is considered to be a 'suitable', even appropriate, reward for innovation and inventiveness. In fact it is painfully clear that many private research firms have built it into their portfolio, and any revocation of such rights would substantially change their justification for doing business. As Barton (1995) asserts, 'one of the most certain economic characteristics of patent law is that patents are more effective in encouraging the innovation and investment involved in bringing a product to market, than in encouraging research itself.'

The importance of PVP for plant breeding has probably been limited because, historically, the seed industry in OECD (Organization for Economic Co-operation and Development) countries has expanded by developing and selling hybrids. Nearly 40% of the total global commercial seed business of about US$15 billion is accounted for by hybrid sales in various crops. Hybrids are available for many important commercial grains, such as maize, sunflower, sorghum, rice, oilseed rape and various vegetables. Most monocotyledons, however, have escaped the ability to be hybridized. One of the characteristics of hybrids is they do not breed true to type, which makes them unattractive for seed saving. This is one of the reasons that the private sector has developed hybrids: IPR legislation is neither necessary nor desirable to develop the hybrid seed industry (Seghal, 1994). Only recently is the seed industry appreciating PVP for hybrids protection, because it is a means of preventing the use of inbred lines that have been acquired (stolen) by competing firms.

In summary, if PVP has a function in breeding, it is in the breeding of self-pollinating crops. In this sense, the PVP system may be relevant for developing countries, where breeding is predominantly directed to self-pollinators and where breeding of hybrids is rare.

IPRs may hamper diffusion of modern plant varieties (MVs)

An important problem in agriculture in developing countries is the slow diffusion of new varieties. To a certain extent, this situation can be traced back to an overly restrictive regulatory system that may prevent the quick release and distribution of newly bred plant varieties. These regulations have also been used as a means of protecting the national seed industry against imported seeds of foreign seed companies. Trade liberalization obligations under the new GATT agreement, and the need to import foreign breeding technology to boost food production, have encouraged most developing countries to lift import restrictions and regulations for seeds. Foreign companies are now allowed to market their varieties in most developing countries.

This is the background against which PVP is being introduced. PVP is supposed to bolster trade liberalization, as it would encourage innovators to export their technology. In the theory of free trade the absence of intellectual property protection is even considered to be an obstacle to trade. In reality, however, an IPR is inherently dualistic. An IPR sustains trade of technology *because* it restricts its use. It is this latter feature of IPR that may hamper the diffusion of new varieties.

PVP is likely to interfere with traditional seed distribution systems. Generally in developing countries, 60–90% of the seed is supplied outside of official markets. Farmers save their own seed or swap their grain for seed with other farmers or seed dealers, grain elevators and the like. When these farmers start to use the advanced protected varieties, this major distribution system is pushed into illegality. Under the 1991 Act of UPOV, PVP does not allow on-farm seed saving without compensating the breeder. Swapping seed is forbidden unless authorized by the breeder.

In other words, while PVP may bring advanced foreign plant varieties upon the market, the diffusion of these varieties depends on the official distribution circuits, which, in many developing countries, only reach a very small part of the farming community.

IPRs may improve access to foreign private varieties and inbred lines

In many countries, the adoption of PVP is an important consideration because it is hoped that PVP will stimulate the transfer of foreign varieties (e.g. of fruits and cutflowers), and inbred lines of hybrids (e.g. of maize, sunflower, sorghum). PVP may partially fulfil these expectations. Breeding in fruit and cutflowers is taking place mainly in OECD countries and all these varieties enjoy protection under PVP. Inbred lines of hybrids are increasingly being protected as well. The introduction of PVP, however, is likely to improve the access to foreign varieties and breeders' lines. A study in Latin America found that PVP will probably impact the licensing conditions and the type of varieties that can be obtained. Access to new varieties may be quicker, the quality of the germplasm higher, royalties lower, and the annexes to licensing contracts shorter (Jaffé and van Wijk, 1995).

The positive impact of PVP on access to foreign material does not apply to all crops, however. A great deal of foreign germplasm and related information of most staple crops is generally freely available at the international agricultural research centres (IARCs). In order to get better access to material from these centres, PVP protection is not necessary. On the contrary, most IARCs provide their germplasm under the condition that the material remains publicly available.

IPRs may restrict agricultural exports

For many developing countries it important to boost their agricultural exports in order to obtain foreign currency and to improve their balance of trade. For this reason several countries are encouraging the production of fruit and cutflower production for export. PVP plays a specific role in this policy. On one hand, PVP may facilitate access to foreign cutflower and fruit varieties, as has been mentioned earlier. This presumable effect is important because the fruit and cutflower production in developing countries depends almost entirely on foreign varieties, varieties that have been bred by Western companies. On the other hand, having a better access to foreign varieties does not mean that the exploitation opportunities improve as well.

Foreign breeders usually export their varieties under specific conditions, specifying, for example, how many plants may be produced from the variety and in which country these plants may be marketed. These conditions are very relevant, as fruit and cutflowers are already produced in the world's main export markets, Europe and the USA. Import of these crops from developing countries into these markets, also the home markets of the breeders, may result in a competition with local production. In order to prevent such competition, two scenarios can be expected.

In the first scenario, the breeder stipulates in the licensing contract that the produce may only be exported to new markets that have not yet been supplied. For example, Chilean producers of an American nectarine variety may export the fruit to other Latin American markets but not to the USA. In the second scenario, the import of the produce into the main markets is blocked because the PVP right-holder can show that only he is licensed to commercially use the variety in the market. For example, a European firm has been licensed by the breeder to grow a cutflower variety in glasshouses. The same variety is being produced in Colombia, also under a licence from the breeder. As the Colombian market for cutflowers is non-existent, all flowers are exported, partly to Europe. The European licensee therefore has the opportunity to prevent the import of Colombian flowers, even though these flowers have been produced with the consent of the breeder, and despite the payment of royalties.

In summary, even though PVP may improve the access to private foreign varieties, PVP may, at the same time, prevent the international trade of the produce of these varieties especially in case this competes with existing production elsewhere.

IPRs may hamper participatory plant breeding

The objective of conventional plant breeding is the release of uniform varieties with a wide adaptation so that they can be grown in large areas. This often results in varieties that perform well in large areas under relatively favourable ecological conditions but do not perform well elsewhere. Apart from their performance, their shape and taste is often adjusted to the demand of the average consumer, but not the specific demands of local communities. For these reasons, alternative breeding methods are being developed in which farmers play a more prominent role. This is known as participatory plant breeding (PPB). Although PPB involves a variety of methods (Eyzaguirre and Iwanaga, 1995), these methods often share a number of characteristics that make them distinct from conventional breeding:

1. The knowledge of farmers is being used in priority setting of breeding goals and in the selection of plants.
2. The resulting material is often genetically highly diverse among and within 'varieties'.
3. The use of a wide variation of germplasm in widely differing conditions requires an unrestricted use of material. The diffusion of material is uncontrolled after its initial release.

The Consultative Group on International Agricultural Research (CGIAR) centres are presently playing an important role in developing PPB in developing countries, and the question is whether or not the introduction of PVP legislation in these countries hampers or stimulates this breeding approach.

Farmer's participation in breeding is not necessarily hindered by PVP. For example, potato breeders in the Netherlands make use of the experienced 'eye' of experienced farmers to increase selection efficiency. The breeding company produces the true seeds through hybridization of promising parents and in some cases the first few generations of clones. These materials are then handed to so-called 'hobby breeders', who select among clones in the field. The company subsequently screens the selected clones in its laboratories for disease resistance and quality factors, after which the clone can be released. This form of cooperation has been very successful, such as in the case of the popular potato variety Mona Lisa. This variety met the requirements of PVP and the collected royalties are shared between breeder and farmer.

In other cases PVP may be at odds with PPB. Often in developing countries, PPB results in large numbers of genetically heterogeneous varieties. As these varieties do not meet the PVP requirements of uniformity and stability, protection is not possible and a financial incentive from the PVP system is thus absent. Furthermore, PPB results in a 'fuzzy breeding system' in which it is impossible to follow up the use of particular breeding lines or to trace specific contributors.

The success of conventional breeders can be measured by the number of releases or the popularity of their varieties (e.g. based on the quantities of seed sold). In PPB, the breeder loses control over the use of his variety, and also a remuneration which is related to the success of his contribution. The breeder may therefore prevent the use of his materials in PPB programmes. The unlikelihood of royalty income in the PPB approach may be a reason for public research centres, in need of new financial resources, to choose for the conventional methods.

IPRs are part of a modernization paradigm that affects preservation of biological diversity

The recognition of IPRs to plant breeders has raised opposition because the requirements for protection, such as uniformity and stability of the variety, are at odds with the objective of enhancing genetic diversity. Seed industry sources have opposed this argument by pointing out the distinction requirement, through which PVP would result in more genetic variability among varieties. Moreover, the seed industry says that many of the (traditional) varieties which farmers grow cannot be protected and fall necessarily outside the scope of PVP.

While there is truth in the arguments of both the advocates and opponents of PVP, it seems that the most important impact of PVP on biodiversity cannot be assigned to PVP legislation as such. PVP is just an element of a common sincere desire to 'modernize' the agricultural sector. This generally means to incorporate the national agricultural sector better in the international division of labour in the world's agricultural production. This policy entails the import of foreign technology, including seeds and chemicals. In this policy, the existence of farmers who are still growing traditional varieties is seen as a sign of backwardness.

The very reason for the adoption of PVP is to offer alternative MVs by stimulating domestic and foreign seed firms to breed more productive varieties. In conjunction with PVP, a combination of additional seed policies intends to encourage farmers to shift from traditional varieties to the MVs. In most developing countries, the use of certified seed of MVs is either recommended by extension services, linked to credit facilities or is obliged by the processing industry. The marketing of seed of traditional varieties may also be restricted by mandatory registration of varieties.

In short, PVP has a negative impact on the creation of biological diversity. Not so much because of its requirements, but more because it forms part of a deliberate attempt to separate the production fields, where genetic variability should be minimized, from genebanks – places especially designed to conserve genetic variability.

Summary

In summary, it should be clear to most readers that, from the point of view of many developing countries, the costs of an IPR or UPOV-based PVP system may outweigh the benefits of it. That is, while an IPR system may help breeders get a better return on investments in plant breeding, and may stimulate access to foreign varieties and inbred lines, IPRs may also hamper the diffusion of MVs, restrict agricultural exports and access to PGR in national genebanks, and discourage the conservation of biological diversity. In addition, there is no certainty that a PVP system would be compatible with PPB, or what its impact would be on public plant breeding. Clearly then, accession to a UPOV-based PVP system may not be acceptable to developing countries who wish to recognize Farmers' Rights.

TRIPs, Farmers' Rights and the Convention

Political controversies over IPRs, Farmers' Rights and the implementation of the CBD all have one thing in common: they attempt to regulate the allocation of economic benefits derived from genetic information in PGRs. For the majority of farmers in the world *any kind of regulation* on genetic resources is new. Distrust of it has taken two distinct forms: (i) opposition against IPRs, and (ii) support for Farmers' Rights in the FAO Undertaking, and claims of sovereign rights over genetic resources and related technology through the CBD.

Opposition to IPRs

Opposition to IPRs can be understood against the background of a general disapproval of IPR on living material. It is especially the agricultural community in developing countries that has opposed the TRIPs agreement. This reaction is understandable if we take into account that the concept of IPR in agriculture has emerged and developed in what now are advanced industrialized countries. The agricultural sector in these countries underwent a process of industrialization in which many traditional farm functions were taken over by specialized companies. This development also took place in plant breeding. Initially, during the first half of the 20th century, the first plant breeders were specialized farmers. PVP legislation was developed to support these breeders who were still so much related to, and appreciated by, the farming community. Due to the political power of the agricultural sector, on-farm seed saving had been allowed ever since the establishment of UPOV.

Towards the end of the 20th century, however, the plant-breeding sector, and its relationship with farmers, changed significantly. Plant breeding has become a fully commercialized and internationally operating business with close links to the biotechnology and chemical industry. These changes in the breeding sector were accompanied by an increasing demand for stronger intellectual property protection of innovation in plant breeding. This eventually has led to the granting of (industrial) patents for plant material and a widening of the scope of PVP (through revision of UPOV in 1991). The wider scope of protection inherently entailed a debate on seed saving, a debate which farmers, who represent a small percentage of the population, could not win.

The UPOV-PVP system is the result of a gradual and pervasive industrialization of the plant-breeding and agricultural sectors in a number of industrialized countries. Nevertheless, it is this system which, through the TRIPs agreement, is being integrally transferred to developing countries whose societies, in most cases, are still agricultural in nature. In these countries, the majority of whose population is still working in the agricultural sector, the seed supply takes place predominantly outside the official distribution channels, and a private seed industry is in its infancy or nonexistent.

The distrust against IPR is augmented because this legislation attributes private property to intellectual contributions with respect to living material. This is often difficult to understand for cultures in which the transition from public physical property to private property has not taken place. Because PVP is strongly related to an industrial and capital-intensive agriculture, proposals have been brought forward for alternative regulation, which is intended to support the type of agriculture in which the vast majority of farmers are involved. This is the background against which the demand for Farmers' Rights and regulated access to genetic resources and related technology through the CBD must be placed.

Farmers' Rights

Farmers' Rights came into being through two resolutions adopted in the FAO in 1989. In general terms, Farmers' Rights should, on the one hand, counterbalance the operative IPR systems under which plant material was protected. Farmers' Rights would be the justification for a compensation for the use of germplasm, provided by farming communities in developing countries, by the seed industry or industrialized countries. On the other hand, the Farmers' Rights concept should provide incentives for farmers, especially in the centres of origin, to conserve, improve and make available PGRs. Farmers' Rights are considered as a means of stimulating the use of local landraces which in turn may support agrobiodiversity.

The operationalization of the Farmers' Rights concept faces a number of difficulties:

1. Most industrialized countries do not want to donate additional funds to support the concept and make it operational. This refusal can partly be explained by policies that reject subsidizing agriculture. On the other hand, industrialized countries are not eager to erect an international fund under FAO auspices.
2. Farmers' Rights are not related to any international legal framework. The concept is difficult to judge from the perspective of IPR theory, because it is not based on a fundamental principle of IPR, the private exploitation right.
3. Like the CBD provisions on the access to technology and the sharing of benefits, FAO's Farmers' Rights have been made subordinated to the TRIPs agreement. Farmers' Rights is not acceptable for industrialized countries if it alters the IPR regime as agreed upon in the Uruguay Round.
4. Farmers' Rights provokes technical problems, as it is extremely difficult to measure the specific genetic contribution of a landrace to a finished commercial variety. Commercial varieties usually consist of a large blend of landraces and of commercial varieties. The costs for checking the respective genetic contributions are likely to be high.

The CBD

Regulation of access to genetic resources and related technology has also been one of the objectives of the CBD, which entered into force in December 1993. Rather than speaking of Farmers' Rights, the CBD states that countries have the sovereign right over their own biological resources. This implies that countries can determine under what conditions access to genetic resources is regulated. In addition, the CBD provides that source countries of genetic resources shall receive a fair share in the benefits obtained from those resources, and requires parties to facilitate the transfer of relevant technologies to other parties.

Attempts to put the CBD into effect have encountered a number of problems:

1. Many developing countries do not have the financial resources to set up a control mechanism to effectively regulate access and use of their genetic resources.
2. Unlike governments, private parties are not directly bound by the CBD. The effect of the CBD on private parties will depend on the legal and technical infrastructure a country develops to implement the CBD.
3. In the absence of clear national legislation, the implementation of the CBD will be executed on an *ad hoc* basis. Private parties (industry, genebanks, etc.) will often dominate these negotiations.
4. The CBD has become subordinated to other treaties. For example, the CBD calls for sharing the benefits of the use of PGRs and for a better access to advanced technology, but the CBD also states that '... access and transfer shall be provided on terms which recognise and are consistent with the adequate and effective protection of intellectual property rights' (Art. 16-2).

This provision is an obvious reference to the TRIPs agreement.

The need for an alternative PVP

Multilateral negotiations in the late 1990s indicated that support for Farmers' Rights and the CBD will not lead to adaptations of PVP. This was not only due to the incompatibility between the three types of regulation on genetic resources involved, but also to the bargaining power of the respective interest groups involved. This situation is not expected to change in the near future. Nevertheless, concerns about PVP remain.

A PVP regulation more compatible with the interests of small- and medium-scale farmer/breeders in developing countries should therefore contain the following elements:

1. It should encourage private breeders to invest in advanced plant breeding.
2. It should encourage farmer-breeders to improve landraces and to collaborate in PPB programmes which may or may not result in less uniform varieties, but attempt to preserve available biodiversity.
3. It should not hamper the rapid diffusion of new varieties and landraces through on-farm saving and swapping of seed.

Although the 1991 UPOV-PVP act is not entirely incompatible with these objectives, it does not serve these three objectives. The following section therefore is an attempt to draw alternative scenarios to solve this problem.

PVP and Other Options

PVP and accession to UPOV

For some countries with relatively strong and modernized farming sectors and existing private breeding sectors, the UPOV model of PVP may be a viable option. Domestic breeders may increase their income by collecting royalties and reducing seed sales on the 'black market', while they may have a better starting position for collaboration with foreign breeding companies.

Governments could allow farmers to save seed. The 1978 act of UPOV provides room for this, while the 1991 act includes an optional clause in which members may restrict the breeders' right in respect of seed saving. The part of the clause that specifies the conditions for such a restriction 'within reasonable limits and subject to the safeguarding of the legitimate interest of the breeder' (Art. 15-2) is vague enough to allow on-farm seed saving. Swapping or trading saved seed is illegal under both acts. In many countries, however, it may be impossible for breeders to exercise their rights adequately, so that the unofficial trade in saved seed may continue unhindered.

Countries that opt for the 1991 act of UPOV have the opportunity to operate the dependency principle (referred to as essentially derived variety or EDV[3]) not only with respect to protected original varieties, but also public original varieties. They could provide that varieties essentially derived from landraces may not be eligible for protection in order to avoid cosmetic breeding of landraces (Ghijsen, 1996). The situation that the introduction of PVP stimulates private companies to file protection for simple selections of landraces is one of the points of concern for opponents of PVP.

PVP but early exhaustion of the right

Countries who have the opinion that the UPOV-PVP system is too restrictive as it comes to the diffusion of new varieties may introduce a PVP system with a limited scope. Such a system could only pro-

[3] An essentially derived variety (EDV) is defined as: 'a variety that is predominantly derived from another variety, or from a variety that is predominantly derived from the initial variety'; that is 'clearly distinguishable from the initial variety'; and 'conforms to the initial variety in the expression of the essential characteristics that result from the genotype, or combination of genotypes of the initial variety'.

tect the certified seed of a variety. In this option, rights of breeders are exhausted as soon as the certified seed has been sold on the market. Farmers may use the seed to save, swap and sell it. In this option, the protection for breeders is much weaker than under UPOV-PVP. On the other hand, breeders could protect their market share by making the difference between the quality of certified and of on-farm saved seeds greater. At present, in many developing countries the quality of certified seed is hardly better than that of saved seed. By improving the quality controls of certification, sales of fresh seed, as well as the remuneration for the breeder, may increase.

Breeders' Fund

An alternative system for both the previous options may be the establishment of a Breeders' Fund. This Fund is not an IPR system as such; as no property rights on exploitation of varieties are assigned. That such a system, nevertheless, may be worthwhile is shown by the experience in the Netherlands where a Breeders' Fund was operational between 1941 and 1966. In the Dutch system, all farmers were levied a small tax on each crop based on the number of hectares that they had planted to that crop. The fund was used to pay plant breeders a remuneration for breeding new varieties. The payment was based on the proportion of total hectares planted to the new variety each year. Remuneration continued for 15 years. After the seed was released commercially, it could be grown, swapped, multiplied and sold by anyone.

The major advantages of a Breeders' Fund are that farmers maintain their basic freedom, i.e. to transfer grain into seed, thereby quickly diffusing new varieties. Plant breeders receive remuneration for their innovative efforts through the Fund, and the incentives to continue innovative plant breeding efforts remains. The Fund should consist of tax contribution from large farmers who are identifiable and who are the main users of fresh, certified seed.

In the meantime, the conflicting nature of the three international agreements that have been discussed may make it impossible for those developing countries that do not have an IPR system, or any sort of effective intellectual property protection system, to implement *any* such *sui generis* system. Intellectual property protection of new biotechnological innovations developed in industrialized countries will continue to be unprotected in developing nations to which they are exported. This may result in decreased innovation in industrialized countries, and slower development in developing countries.

Conclusions

In order to protect investments in new biotechnological innovations, many plant breeders and western biotechnology firms have made extensive use of the various forms of IPRs available to them. These forms of IPRs are, however, not available in all the markets to which these firms may consider exporting their new technologies.

In 1994, signatories to the Uruguay Round of the GATT also agreed to the TRIPs obligations as part of the entire GATT agreement. The TRIPs agreement requires countries, which do not already have an IPR system, to introduce either a patent or an effective *sui generis* system for the protection of plant varieties and agricultural biotechnological innovations.

At the same time, IPRs in genetic resources were negotiated in *two* other fora: (i) the FAO Undertaking on Plant Genetic Resources, which seeks international consensus on ways to encourage the conservation of biological diversity, compensate countries and communities for the use of genetic resources, and introduced the concept of Farmers' Rights; and (ii) the CBD, which aims to encourage biological diversity and states that countries have the sovereign right over their own biological resources.

Countries that are implementing the CBD, which also have signed the FAO Undertaking, and that meanwhile support the results of the TRIPs negotiations have to deal with genetic resources issues overlapping three political fora related to industry, agriculture and environment. The combined and conflicting interests in all three fora have triggered unexpected 'chemical reactions' and unwanted controversies.

By signing the new TRIPs agreement in 1994, all member states accepted a new international standard on the protection of intellectual property, which was significantly higher than earlier standards. Legal protection of innovations, particularly those involving living material, had only occurred in some industrialized countries prior to 1994. For most of the rest of the world, and particularly the developing world, IPR did not, and still does not, exist.

One of the options for a *sui generis* system for the protection of plant varieties, as required under the TRIPs agreement is the PVP system which is administered by UPOV. There are both costs and benefits to adopting a PVP system. Developing countries who are considering adopting a UPOV-PVP system need to carefully consider the relative costs and benefits associated with IPRs in PGRs.

The costs of an IPR or UPOV-based PVP system may outweigh the benefits of it. An IPR system may help breeders get a better return on investments in plant breeding, and may stimulate access to foreign varieties and inbred lines. However, IPRs may also hamper the diffusion of MVs, restrict agricultural exports and access to genetic resources in national genebanks, and discourage the conservation of biological diversity. In addition, there is no certainty that a PVP system would be compatible with PPB, or what its impact would be on public plant breeding. Clearly then, accession to a UPOV-based PVP system may not be acceptable to many developing countries.

Political controversies over IPRs, Farmers' Rights and the implementation of the CBD all have one thing in common: they attempt to regulate the allocation of economic benefits derived from genetic information in PGRs. For the majority of farmers in the world *any kind of regulation* on genetic resources is new. Distrust of it has taken two distinct forms: (i) opposition to IPR, and (ii) support for Farmers' Rights in the FAO Undertaking, and claims of sovereign rights over genetic resources and related technology through the CBD.

Multilateral negotiations since the late 1990s have indicated that support for Farmers' Rights and the CBD will not lead to adaptations of PVP. This is not only due to the incompatibility between the three types of regulation on genetic resources involved, but also to the bargaining power of the respective interest groups involved.

A PVP regulation more compatible to the interests of small- and medium-scale farmer/breeders in developing countries should: (i) encourage private breeders to invest in advanced plant breeding; (ii) encourage farmer-breeders to improve landraces and to collaborate in PPB programmes to preserve available biodiversity; and (iii) not hamper the rapid diffusion of new varieties and landraces by allowing on-farm saving and swapping of seed.

It is suggested that some form of modified PVP system or the establishment of a breeder's fund as *sui generis* systems could help developing nations achieve these objectives and overcome some of the incompatibilities between the three international agreements. In the meantime, intellectual property protection of new biotechnological innovations developed in industrialized countries will continue to be unprotected in developing nations to which they are exported. This may result in decreased innovation in industrialized countries, and slower development in developing countries.

References

Barton, J.H. (1995) 'Patent scope in biotechnology'. *International Review of Intellectual Property*, Draft, 6 February.

Butler, L.J. (1996) Plant breeders rights in the U.S.: update of a 1983 study. In: van Wijk, J. and Jaffé, W. (eds) *Intellectual Property Rights and Agriculture in Developing Countries.* University of Amsterdam, Amsterdam.

Butler, L.J. and Marion, B.W. (1985) *Impact of Patent Protection in the U.S. Seeds Industry and Public Plant Breeding.* NC-117 Monograph 16, University of Wisconsin, Madison.

Butler, L.J., Louwaars, N., Pistorius, R. and van Wijk, J. (1996) Intellectual property protection in plant breeding: plant variety protection and other options. Discussion paper, Department of Political Science, University of Amsterdam, The Netherlands, October.

Eyzaguirre, P. and Iwanaga, M. (eds) (1995) *Participatory Plant Breeding: Proceedings of a workshop on Participatory Plant Breeding*, held in Wageningen, 26–29 July. IPGRI, CGN, IDRC.

Ghijsen, H. (1996) Discussion paper concerning plant variety protection (PVP). Background information. Paper for the seminar 'Asian Seed' of the Asian and Pacific Seed Association, 25–27 September, Jakarta.

Jaffé, W. and van Wijk, J. (1995) *The Impact of Plant Breeders' Rights in Developing Countries: Debate and Experience in Argentina, Chile, Colombia, Mexico, and Uruguay.* DGIS, The Netherlands Ministry of Foreign Affairs, The Hague.

Seghal, S. (1996) IPR driven restructuring of the seed industry. *Biotechnology and Development Monitor* 29, 19–21.

Stallmann, J.I. and Schmid, A.A. (1987) Property rights in plants: implications for biotechnology research and extension *American Journal of Agricultural Economics* 69, 423–437.

Chapter 3

Sui generis Protection of Plant Varieties in Asian Agriculture: a Regional Regime in the Making?

Henrik Egelyng*

International Rice Research Institute (IRRI), PO Box 3127 (MCPO), 1271 Makati City, Philippines

Abstract

This chapter examines the plant variety protection (PVP) regimes emerging throughout Asia. The overview provided includes China, Indonesia, Korea, Bangladesh, Malaysia, Pakistan, Thailand and Vietnam. With particular emphasis on the Philippines, it analyses the introduction of PVP from an institutional development perspective. Focusing on the challenge for the provision of international rice research public goods, questions are further raised as to how implementation of PVP legislation will affect existing networks for the exchange of rice varieties and improved rice germplasm. The emerging PVP regimes are also investigated from a perspective of regulation and potential for economies of scale and other possible benefits from a regional division of labour in implementing plant variety protection. Finally, conclusions are made, pointing to the needs for both researchers and policy decision makers to apply new perspectives to this emerging regime of intellectual property as it expands into existing public domains and economies of natural capital.

Introduction

To comply with Article 27.3(b) of the World Trade Organization (WTO) agreement on Trade Related Aspects of Intellectual Property Rights (TRIPs), member states were given a choice to protect plant varieties through the patent system – or by a *sui generis* system or a combination of the two. Most Asian nations will meet these requirements by introducing a *sui generis* regime for the 'protection' of plant varieties. Only recently have the contours of these emerging individual national plant variety protection (PVP) regimes become clear. Less clear is the extent to which the countries in question will establish technical and institutional capacity to 'ground' solidly the PVP bills being passed and transform these into well-working institutions and successful avenues of agricultural development.

* Current contact e-mail address: he@kvistcon.dk

Whether the resulting developments will follow a path toward sustainability is an even more open question.

PVP – a Regime for All Crops

It has long been clear that a broad range of possible *sui generis* systems will be compatible with TRIPs (Leskien and Flitner, 1997). It is also clear that decision makers face complex issues (e.g. IPGRI, 1999). The early members of the International Union for the Protection of New Varieties of Plants (UPOV) allowed themselves significant time to put in place the institutions, infrastructure and instruments required to provide for PVP and the extension of such to progressively higher numbers of genera. The 1978 act of the UPOV required members, on accession, to protect only the largest possible number of species. The 1991 version requires members to protect at least 15 plant genera and species immediately, and all plant genera and species after 10 years. Although there may be little attention to the particularities of each individual crop in the legal framework for PVP, even from a narrow technocratic perspective, a host of crop-specific measures need to be taken regarding the technical examination of, for example, rice varieties. That is required in order to set up an appropriate technical examination system adapted to the nature, special characteristics and special growing conditions of each particular crop (Yasuoka, 2000).

National Asian Models: an Overview

China and Japan are already members of the PVP system of UPOV. India and the Republic of Korea have officially applied for UPOV membership. Thailand has its own law on PVP. Indonesia, Pakistan, the Philippines and Vietnam, as well as Bangladesh, Malaysia and Sri Lanka, are in varying stages of enacting PVP laws.

In what follows, the results of a preliminary analysis of the emerging systems are presented. All of the Asian *sui generis* laws are based on UPOV and therefore have much in common. Worded differently, in each national PVP law, the farmer's privilege is a common feature. However, they also display significant qualitative differences. Some laws explicitly recognize published descriptions of new lines such as by public research organizations, as sufficient to block anyone else from claiming discovery and ownership. Others explicitly refuse granting PVP to old or other public varieties or landraces.

While all regimes put a time limit on PVP, so that eventually all protected material goes into the public domain, the duration of protection for annuals such as rice varies from 7 years in the Bangladeshi draft to 12 years in Thailand, and to 20 years in Indonesia. The first suggestion following our analysis is therefore that from a regional perspective the emerging PVP regime in Asia is not too homogeneous and does not seem overly coordinated across national boundaries.

The operational complexity of effective PVP is high. Multiple ministries may have operational responsibilities for implementing or managing the system, through various bodies and authorities, including both legal and technical institutions. In addition to the PVP issues 'proper', other complexities are added such as biosafety, food safety of genetically modified organisms (GMOs) and, last but not least, bioprospecting. To the extent Convention on Biodiversity (CBD) issues (Farmers' Rights, farmer's privilege, landraces, wild varieties, community rights, etc.) are included, the complexity is further elevated. The following few country sections do not aim to rigidly compare countries with a view to listing differences and similarities. Instead they report some salient features of each national system, simply to illustrate and support the conclusion made above on the level of homogeneity and regional coordination.

Among the group of countries considered here, The Plant Varieties Act of Bangladesh is the one most deserving of the description *sui generis*. While applying the distinct, uniform, stable (DUS) criteria, the draft speaks about the economic values and possible economic, social and environmental impacts and national needs as additional criteria in need of consideration prior to registering a plant variety. While it is not clear how these 'social good' criteria will be operationalized and enacted, the Bangladeshi Act is unique in explicitly recognizing and addressing these aspects of PVP. Also, the Bangladeshi act is unique in stressing explicitly that public breeders will not be allowed to own varieties – instead, public-sector breeders will receive a citation award acknowledging their varieties. In addition to the PVP Act itself, Bangladesh is further introducing a Biodiversity and Community Knowledge Protection Act that will affect PVP (NCPGR, 1998a,b; Razzaque, 2000).

Enacting its PVP regulation in 1997 and acceding to the UPOV (1978 version) in 1999, China has published its first catalogues of PVP, featuring 18 genera, including rice. An administrative office for PVP has been set up, 15 DUS testing centres responsible for testing different agricultural crops have been established and 30 agents for new agricultural plant varieties have been nominated. A storage centre for the conservation of new plant varieties has been established and a PVP 'Gazette' is now being published. By 1 November 1999, nine rice varieties were PVP-protected in China, five of which originated from one Hybrid Rice Research Centre. In China, the patent law was revised in 1993 and now allows for patent protection of GMOs. However, article 10 of the 'Regulations of the People's Republic of China on the Protection of New Varieties of Plants' states that use of a protected variety for propagating purposes by farmers, on their own holdings, is allowed without farmers having to pay any royalties (Shumin and Lijun, 2000).

The Indonesian draft PVP law is based on the UPOV 1991 Act. Given to the People's Representative Body for approval in 1999, the draft law awaited hearing and debate by February 2000. The draft planned for registered varieties to be approved by a national seed board and planned the creation of a PVP office within the Agriculture Department of the Republic of Indonesia. GMOs are treated in a manner similar to other varieties, but require a biosafety and food safety assessment. A special feature of the Indonesian PVP draft law (Chapter XI) deals with 'Criminal Stipulation'. Imprisonment of up to 7 years and fines of up to Rp 500,000,000 are laid out, making any intentional breach of the PVP law a felony (Government of Indonesia, 1999; Sumarno, 2000).

Under the Korean Seed Industry Law, plant variety protection is allowed for, consistent with the UPOV 1991 Act. The implementing organization is the National Seed Management Office (NSMO). A Variety Protection Appeal Committee handles appeals on decisions from the NSMO. By 1999, the number of rice varieties protected under the Seed Industry Law in Korea was almost 100, the great majority of which originated from public breeders. In general, intellectual property protection for plants is on the rise in Korea. The number of plant patents granted by the Korean Intellectual Property Office, covering asexually reproduced varieties, grew from nine in 1995 to 46 in 1999 (Choi and Ryu, 2000).

A final draft of the Plant Variety Protection act was submitted to the Malaysian Parliament in 1999. A new division for Plant Variety Protection and Registration will be set up in the Department of Agriculture. Already, a Genetic Modification Advisory Committee (GMAC) and Institutional Biosafety Committees have been established to deal with GMOs. At present, Malaysia has one rice-breeding organization, at the Malaysian Agricultural and Development Institute (MARDI). It is owned and funded by the government, and releases all its varieties for free. MARDI varieties cover 90% of Peninsular Malaysia, while Sabah and Sarawak are dominated by traditional varieties. Under the new agricultural policy, however, the government envisions involving private breeders in rice breeding (Othman, 2000).

In Pakistan, the draft Plant Breeder's Rights Act is awaiting clearance from the cabinet. With the PVP regime in force, the government envisions private-sector rice breeding being established. A particular feature of the Pakistani act is a strong emphasis on the requirement that varieties shall be 'designated by an acceptable denomination'. This emphasis is probably based on the concern that the famous 'Basmati' rice and its derivatives may be used in ways that could be considered inappropriate. Another particular feature is the creation of a national gene fund. The Pakistani draft explicitly states that PVP shall not affect farmers' traditional rights to save, use, exchange, share, or sell their farm produce of a protected variety, 'except where a sale is for the purpose of reproduction under a branded marketing arrangement' (Malik and Hussain, 2000).

The Philippine Draft Plant Variety Protection Act of 2000 is now in both houses of Congress (House Bill 1070, Senate Bill 108). Varieties are considered distinct when distinguishable from 'any publicly or commonly known variety' and new only if they have not been sold or otherwise disposed of to others with the consent of the breeder. Consistent with the UPOV 1991 version, protection comprises essentially derived varieties. The farmers' right portion of the Philippine (draft) bill explicitly excludes from protection plant varieties used for non-commercial purposes, including the use of propagating materials by small farmers on their own landholdings as a product of their harvests and exchange of seeds among and between small farmers. The broader implications of the act for biodiversity, bioprospecting and indigenous communities have been much debated. Following the 'one bill–one subject matter' principle, a consensus was reached to defer such issues to other laws such as the Indigenous Peoples Rights Act and the existing Executive Order No. 247 for bioprospecting. This principle also applies to the issue of possible conflict between the PVP and patent regime, since the bill will grant protection regardless of whether a variety is obtained through conventional breeding or genetic engineering (Faylon *et al.*, 2000). The Intellectual Property Code of the Philippines, enacted in 1998, excludes plant varieties from patent protection.

The Thai PVP act was approved by the House of Representatives in November 1999. The scope of the Thai draft is wide, based on UPOV 1978, and includes community rights. Protection covers new varieties, local domestic varieties, general domestic varieties and wild species. DUS criteria are required, except for the latter two categories, for which uniformity is not required and for which protection is automatic, without registration. The Thai act is the result of extensive public hearings conducted throughout the country in 1997 and 1998. Transgenic varieties may be protected once approved by the biosafety authorities. Other than breeders' rights, the draft allows for farmers' privilege and the rights of local governments to assert community rights. To institutionalize a sharing arrangement on derivatives, the draft envisions establishing a PVP fund (Chitrakon and Thitiprasert, 2000).

The Vietnamese Ministry of Agriculture and Rural Development (MARD) drafted a 'Decree on the Protection of New Plant Varieties' in 1999, which has not yet been approved. The draft is based on the UPOV 1991 convention, however, and establishes a National Plant Variety Protection Office under the MARD to receive, process and publish all PVP applications (Nghia and Quang, 2000).

Implications for Institution Building: the Philippine Case

PVP bills, proposed or enacted as described, imply a picture of the legal and highly technical institution-building task that most developing countries are facing once the bills have passed their national parliaments, as the first step on a long journey. New challenges surface as the various new bodies, committees and expert groups are formed, technical centres of expertise are appointed, and testing facilities are required. It becomes apparent that making a PVP regime operational and effective will require significant resources, new skills, new equipment, new patterns of behaviour in both the public and private sectors, new ethics, and efforts of good governance and best practices.

The Philippine PVP bills will establish several new institutions. These include a National Plant Variety Protection (NPVP) Board, an NPVP Registrar (manned by examiners) and the publication of an official *Journal of Plant Variety Protection*. They also include various verifying institutions for each crop, such as centres for genetic identity or variety evaluation as well as two funds to be set up based on PVP royalties. For rice, the national Philippine rice research institute, PhilRice, is to be verified as an official centre for verification of rice varieties proposed for registration[1]. The law will stipulate that at least two verification institutions per crop be used as a basis for issuance of a licence, aiming to avoid that a registrant institution is designated as a verifier of one of its own varieties. In the Philippines, the NPVP Board will consist of representatives from academia, the seed industry, farmer organizations, indigenous organizations, environmental organizations, consumer organizations, and the secretaries of the Departments of (i) Agriculture, (ii) Science and Technology, and (iii) Environment and Natural Resources. The functions of the board will include establishing and supervising the NPVP Registrar, the centres for variety evaluation and the Community Gene and Biodiversity Rehabilitation Trust Funds created under the act. The act defines the structures and procedures, function and financing of these new institutions in significant detail. The Philippine bills therefore invite a study of the institutional and technical capability-building challenges that face countries seeking seriously to abide with TRIPs demands.

Social and Biophysical Costs

In any analysis of PVP, the resources needed to operate the system effectively and efficiently must be taken into account. When implementing a national PVP regime, diverse direct costs become clearly visible. In a classical economic sense, these include salaries and operating expenses of the following:

- NPVP Board secretariat and the Registrar's office;
- initial costs involved in establishing additional technical equipment and extra skills necessary for verification purposes;
- the cost of establishing and maintaining a data bank with all descriptions of variety characteristics;
- the cost of establishing and maintaining reference collections (variety banks);
- initial and ongoing training costs for examiners and technical personnel;
- cost of shipping materials to verifier institutions;
- costs of verification at verifying institutions, including laboratory costs, and labour and land costs. The latter vary depending on whether the variety is a perennial or annual crop – for rice, probably 4–6 months are required for field-testing, per variety.

On top of the immediate direct costs are social costs, which are still difficult to stipulate. These will include, for instance:

[1] One can get a flavour of the technical challenges facing the verifying centre, by referring to the relevant UPOV Test Guidelines which are the protocol for standardization of the testing of plant varieties, providing yardsticks for parameters such as DUS of a given species. A plant variety may be defined as a plant grouping within a single botanical taxon of the lowest known rank, characterized by its morphological features, growth, leaf, flower, seed and other distinct characteristics. The classical methods of DUS testing are based on morphological and physiological characteristics, such as disease resistance and grain length, shape, colour, stem length, time of heading, leaf colour, number of panicles, and time of maturity. These tests are labour- and land-intensive to carry out. Increasingly advanced methods such as electrophoresis have been applied and use of DNA profiling techniques may be demanded in the future. The cost implications of these prospects are presently unclear.

- arbitration (settlement of disputes);
- socio-economic impacts on farmers and consumers;
- environmental effects on ecosystems, their natural resources and biodiversity.

While the government treasury can cover some of the direct costs through user fees, the indirect costs must be included in any social cost–benefit analyses trying to measure the national and regional transaction costs of a PVP regime. In the real world, of course, cost–benefit equations are often complicated and go beyond direct and indirect costs. An intellectual property regime such as PVP creates economic distortions, or incentives, favouring *ceteris paribus*, human-made capital (registered varieties) over natural capital (old varieties, landraces, wild germplasm and other public varieties). A possible cost from liquidation of natural capital (perhaps by genetic erosion brought about by economic disincentives against public-domain rice biodiversity) does concern some analysts.[2] More fundamentally, one might wish to examine the extent to which assigning PVP creates additional value in terms of gains to the ecological and social economy of the biosphere, and to what extent it simply privatizes some existing values and perhaps indirectly cause others to be liquidated. There is a particular need to begin such studies in rice because rice is a unique crop – a major food crop, little commercialized, hardly traded across borders and is still underpinned by a unique research system.

The number of patents recently taken out in rice is growing and recent socio-economic and technological developments such as functional genomics have triggered more private-sector interest in the crop. Most rice grown in the world today is not 'protected' by PVP. As of December 1999, however, UPOV listed 12 member countries as reporting PVP extended to rice. Two countries accounted for the major share of rice plant variety certificates issued: Japan (342 varieties protected) and the USA (55 varieties protected). The Community Plant Variety Office of the European Union had registered 22 varieties of rice as protected. At the other end of the spectrum, Australia, Chile, Hungary and Portugal each had a single rice variety protected. These figures demonstrate not only that private rights to rice varieties are currently distributed asymmetrically. They also show that a beginning has been made with a critical mass of countries that transfer some rice varieties from the public to the private domain. A middle group is slowly gaining momentum in this process: Argentina (16 rice varieties with PVP), Colombia (six), France (ten), Italy (six), Spain (11) and Uruguay (six) (Yasuoka, 2000). Although Korea and China were mentioned above as having PVP protected rice recently, no Asian country is on the UPOV list. This article argues that any unilinear interpretation of these figures as 'Asia lagging behind' in PVP would not only be simplistic, but miss the point about rice as a crop.

Economies of Scale – a Regional Option

In the academic field of agricultural input regulation studies, a range of questions have been asked to assess whether a particular design of instruments is efficient and effective in achieving set goals. Such issues can be analysed by responding to the basic questions of what should be regulated, how and why. The regulatory 'objective' of PVP law is often defined as assigning ownership of new varieties.[3] To the extent that such an intermediate

[2] A theoretical parallel may be made to the Asian mangroves, where the introduction of proprietary rights triggered liquidation of natural capital (biodiverse biomass) in order to produce smaller amounts of human-made capital (shrimp) for a limited period.

[3] In the absence of PVP legislation, but with compulsory seed certification some governments have also met the same regulatory objective (promoting privatization and quality standards) by refusing to certify seed for anyone except the company that registered the variety.

instrumental objective can serve as a development policy objective, one can distinguish three main strategies for achieving that objective: one using government tests, the other relying on company data and the third a combination of the two[4] (Gisselquist *et al.*, 2000).

The most interesting aspect of the input regulation discussion is perhaps the factors affecting the potential for achieving economy-of-scale benefits through regional collaboration. A simple technocratic yet ambitious objective of international collaboration in rice PVP would include the following basic objectives:

- to avoid duplication of effort where test results are already available;
- to ensure that a variety having been considered distinct, uniform and stable in country A is not rejected in country B.

Now, however, it is not clear to what extent the Asia-Pacific Economic Co-operation Forum (APEC) or Association of Southeast Asian Nations (ASEAN) is taking a leading role in any harmonized or coordinated regional regime for plant variety protection. As an example of why such cooperation could be significant, one can refer to the fact that the present UPOV test guidelines from 1985 were prepared without much concern for applicability to tropical and subtropical rice varieties (Yasuoka, 2000), or perhaps, for their agroecological, technical and institutional environments. Those guidelines are being revised at UPOV, but it is unclear to what extent experts from tropical and subtropical rice producing countries are involved.

TRIPs and CBD Perspectives Converging

The UPOV PVP regime allows for exemptions, including the research exemption and farmers' rights or seed saving exemptions. The TRIPs regime has no such exemptions. Recent attempts to introduce, into national *sui generis* legislation, certificates of geographic origin and proof of bioprospecting collectors having secured prior informed consent, seek to integrate the UPOV exemptions into the TRIPs regime. These attempts have not been very successful and a tension remain between an intellectual property or TRIPs axis versus a CBD (Blakeney, 2000). One may ask why this is important and what the relevance is to researchers and decision makers, in the field of international agricultural research and PVP. One answer is that in the USA, the combination of PVP and patents has already changed the dynamics of semi-public agricultural research institutions. Erbisch (2000) describes how this resulted in costs to society in terms of lost opportunities caused by institutional deadlocks, obstructing public–private research collaboration[5]. For developing countries introducing PVP, similar institutional complexities can be readily envisioned. Even in technically advanced countries, there is not yet a consensus about what biotechnology products and processes can be patented. Principles are evolving through (expensive) court cases, patent are and international discussions. In reality, therefore, most biotech patents are therefore best perceived as hypotheses for a court to test, validate or reject.

[4] The US–Canadian approach bases decisions to assign ownership on DUS data that companies submit from their own tests. Regulators in these countries perceive this model as effective in time and cost, and though falsification of data is possible, disputes can be referred to courts for resolution. The European strategy bases decisions to assign ownership on DUS data from official (government) tests, while accepting tests from other governments. This is seen as workable if cooperating countries include some countries with large seed markets. The third strategy bases decisions to assign ownership on DUS data from official (government) tests only, and requires in-country tests. Costly in time and paperwork, this latter model may be suited only to large countries and markets (Gisselquist *et al.*, 2000).

[5] The Erbisch example is about a superior turfgrass variety developed by Michigan State University (MSU). The variety had to be destroyed because it contained a gene patented by one company and a promoter patented by another company and neither company wanted to license its gene or promoter to MSU.

The European Patent Board of Review now indicates that a plant defined by a single recombinant DNA sequence may be patentable in Europe also, in contrast to plant varieties *per se*, which are not (Blakeney, 2000).[6] In tandem with technological developments, these current changes could mean that patents become a relatively more important instrument of IP protection, even for the seed industry. To the extent PVP loses relevance from the point of view of bio-industrial interests pushing the IP-biotech agenda in the WTO, it might therefore be relevant to re-examine the rationale for developing nations committing many resources to fully implement their PVP (*sui generis*) for all crops. Such full and across the board implementation is considered unfeasible and uneconomical even for industrialized countries (Gisselquist *et al.*, 2000). The new PVP entrants might find themselves investing in a legal and technical infrastructure that might become economically and politically less relevant, even before it has a chance of becoming functional. One could argue that developing countries were bound to introduce PVP anyway to spur their domestic seed industries. Whether that argument holds in a globalizing and highly interdependent reality is questionable. In industrialized countries, PVP, trade secrets and trademarks are currently mixed and matched according to their relative strengths and weaknesses to provide IP protection for major commercial plant cultivars and their propagating material. In the USA, utility patents of individual genes have been added to this IP toolbox and these patents are used to protect aspects in the cultivars that could otherwise not be protected.

Implications for International Exchange of Germplasm Through Networks Sharing Crop Varieties and Improved Germplasm

This section examines the impact of the emerging *sui generis* regimes on networks for international germplasm exchange. One such system is the International Network for Genetic Evaluation of Rice (INGER).[7] So far, INGER has been able to facilitate free sharing of rice germplasm and related information across international borders and has evaluated more than 20,000 rice breeding lines and varieties for yield performance and/or resistance to abiotic and biotic stress. These evaluations resulted in the release of 525 rice varieties in 62 countries, with each variety valued at US$2.5 million (Chaudhary *et al.*, 1998). The following sections analyse the impacts on INGER of an 'all members adopt PVP' scenario. The INGER collection is based on contributions of varieties and germplasm mainly from the International Rice Research Institute (IRRI), but also with significant national contributions, historically including high numbers from individual nations such as India and Sri Lanka. Qualitatively, it is difficult to estimate which of these contributions have been and will be the most valuable, and quantitative trends do not seem to have been subject to policy analysis in the past, perhaps because relevant data have not been available. Today, however, some such data are becoming available and some are presented here. The total number of breeding lines/varieties contributed to INGER by national agricultural research systems (NARSs) declined from more than 350 in 1991 to less than 100 in 1999 (INGER reg-

[6] In December 1999, the Enlarged Board of Appeal (the highest court) at the European Patent Office, in decision 'G01/98', clarifies the law on patenting plants (e.g. GM plants), saying basically, that if not protectable by means of plant variety rights, plant innovations are patentable. This will from now on allow patents on GM plants at the European Patent Office (a full text of the decision is at http://www.european-patentoffice.org/dg3/biblio/g980001ex1.htm).

[7] The global INGER has three regional targets: Asia, Africa, and Latin America and the Caribbean (LAC). The global INGER (IRRI-based) coordinates the whole programme and focuses on Asia. INGER-Africa was coordinated by the International Institute of Tropical Agriculture in Nigeria in the past and is now under the coordination of the West Africa Rice Development Association (WARDA). INGER-LAC is being coordinated by the Fund for Latin American Irrigated Rice (FLAR) and the International Center for Tropical Agriculture (CIAT).

istry figures, 2000). This general trend can be detected by looking at some individual countries as well (Figs 3.1 and 3.2).

Figure 3.1 shows that in the first 3-year period from 1991 to 1993, INGER registered 112 breeding lines and/or varieties contributed by Bangladesh. In the next 3-year period (from 1994 to 1996), this figure was 53 and in the following 3 years (1997 to 1999) only five Bangladeshi contributions were registered. Figure 3.2 shows similar figures for China, where the number of contributions dropped from 130 in the first period to 68 in the second and 37 in the final 3-year period. Similar trends are clear for India and Korea. For other individual countries, the picture is less pronounced, but the overall trend remains clear.

To understand this trend, one possible explanation may well be that following the discussions prior to the United Nations Conference on Environment and Development (UNCED) in 1992, national contributors to began consider their rice varieties as a national heritage. With the entry into force of the CBD and subsequently the TRIPs negotiations with PVP regimes emerging on the horizon, contributors probably began to consider whether they should begin to use PVP for the varieties they send to INGER or whether to share these at all. Contributors of such rice germplasm may want to prevent their rice varieties distributed through INGER from being 'appropriated' or protected by third parties at the national level. (PVP can only be taken out at the national level.)

In summary, the prospects INGER member countries have for introducing PVP significantly change the rules of the game and the situation that INGER is in. The new *sui generis* laws in combination with patents (possibly) and new administrative management regimes (certainly), have begun to affect the international flow of rice germplasm and INGER's time as a rather loose network for the free exchange of rice germplasm seems to be over. INGER was founded at a time when genetic resources were perceived of as common property.

To survive, INGER may need a formal constitution to set the scene for the exchange of improved germplasm and varieties for which PVP rights can be asserted. Already, some contributors of material to INGER have reportedly started to require the use of material transfer agreements, perhaps attempting to secure some rights over the national heritage being shared

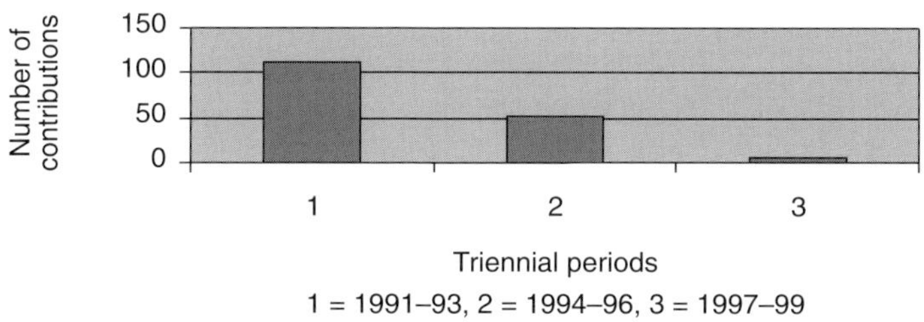

Fig. 3.1. Breeding lines and/or varieties contributed by Bangladesh.

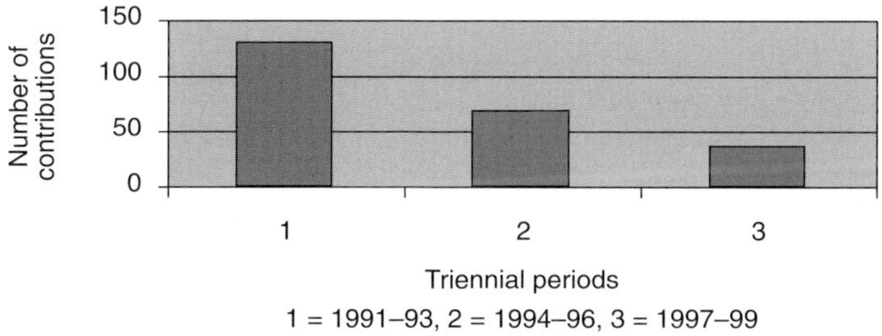

Fig. 3.2. Breeding lines and/or varieties contributed by China.

through INGER.[8] For its functionality, INGER is closely dependent on and institutionally linked with the IRRI.[9] Whether a continued exchange of rice varieties (and related information) through INGER can be ensured has become a highly relevant question. For a new INGER regime, taking into account the new reality of PVP across the region, it will only be an additional challenge that the details of PVP regulations will differ in each member country. Finished varieties are what PVP is designed for. To prevent the complications of a future (Asian) regional PVP regime from affecting INGER, one possible future scenario could be to keep finished varieties out of INGER. That, however, could hurt fragile NARS and INGER donors might not like that idea. In the future, INGER will have to address other similar policy questions.[10]

Conclusions

National PVP regimes are emerging and maturing throughout the developing world. In many Asian countries, such regimes have been transformed into law already or are quite mature and ready to become law within a short time. This article presents seven preliminary conclusions. The first is that from a regional perspective the emerging PVP regime in Asia features a set of PVP laws that, despite common reference to UPOV, remain non-homogenous and do not always appear fully coordinated. The second is that the step from law to making PVP a reality for all crop species, including rice, represents a major challenge, in terms of technical, legal *and* administrative institution building. The third point suggests that, for nations meeting these challenges, major direct costs as well as indirect costs and externalities (social and environmental costs) may be involved. To what extent such costs are justifiable to the average Asian nation is not discussed. Fourth, PVP introduces distortions (or perhaps incentives) favouring human-made and privately owned varieties over natural capital and public-domain varieties. In other areas of the biosphere and earth's ecological economy, the introduction of such distortions through the promotion of intellectual property rights regimes, has caused a loss of biodiversity and environmental service functions. The fifth point is about the potential for economies of scale through regional collaboration. Even if one is not convinced whether implementing PVP in rice throughout Asia is right or 'experimental', one may have to admit that economies of scale in doing policy experiments together may be preferable to each nation doing such experiments on its own. Significant potential exists for PVP regimes to become more efficient and effective through regional collaboration. A social force to bring about such regional collaboration is not clearly visible. Harmonization of the various national laws and integration of the various legal bodies and technical instruments envisioned to implement the laws are therefore not on the immediate agenda. Sixth, the possibility of PVP becoming economically and politically less relevant for Asian nations, even before it has a chance of becoming functional, was reflected on. Seventh, the CBD already

[8] The question of who will be responsible for honouring and enforcing these agreements may have to be clarified, however. IRRI, acting to date as the *de facto* institutional 'embodiment' of INGER, may not wish to accept responsibility for infringement of rights to any material distributed through INGER.

[9] That is, with the International Rice Genebank Central Information System (IRGCIS), the International Rice Information System (IRIS), and the System-Wide Information Network for Genetic Evaluation of Rice (SINGER).

[10] As far as the kinds of materials are concerned, one should distinguish varieties and 'fixed' inbred lines from hybrids. At the moment, neither private nor public hybrids are tested or shared through INGER. If that were to happen, a major question would be: under which terms? Would national public breeders wish to PVP protect inbred lines disseminated through INGER? If INGER were to become testing facility for private hybrids, would INGER charge a fee? Would INGER, in general, devise a mechanism so that contributors of biological assets would be able to benefit in more tangible ways? Would the materials shared through INGER be of sufficiently high value to warrant such discussion?

seems to have severely impacted international germplasm networks, such as INGER, in terms of the number of contributions of rice varieties made by member countries to INGER for testing and dissemination. With a regional Asian PVP regime in place, it seems likely that the exchange of germplasm through INGER will, at the very least, be slowed down, unless new rules of the game are invented. To continue to be able to provide international public goods of quantity, quality and relevance to its current beneficiaries, INGER may need a new constitution and perhaps a more independent, transparent and formal organization than it currently has.

Note Added in Proof

Parts of this chapter have been previously published in the article 'Plant variety protection in Asia: some issues of implementation and implications for germplasm exchange networks, in *Bio-Science Law Review* (2000), 6, 187–193. They are republished here with permission of the publishers, Lawtext Publishing Ltd (www.lawtext.com).

References

Blakeney, M. (2000) Protecting plant varieties: patents or *sui generis* protection. Paper presented at the Workshop on the Impact on Research and Development of *sui generis* Approaches to Plant Variety Protection of Rice in Developing Countries, 16–18 February. IRRI, Los Baños, Philippines.

Chaudhary, R.C., Sesha, D.V., Alluri, K., Cuevas-Perez, F., Lopes, V.C and Khush, G.S. (1998) *INGER-derived Rice Varieties Directly Released in Various Countries.* IRRI, Los Baños, Philippines.

Chitrakon, S. and Thitiprasert, W. (2000) Plant variety protection system in Thailand. Paper presented at the Workshop on the Impact on Research and Development of *sui generis* Approaches to Plant Variety Protection of Rice in Developing Countries, 16–18 February. IRRI, Los Baños, Philippines.

Choi, K.-J. and Ryu, H.-Y. (2000) Plant variety protection and its implications on rice in the Republic of Korea. Presentation made at the Workshop on the Impact on Research and Development of *sui generis* Approaches to Plant Variety Protection of Rice in Developing Countries, 16–18 February. IRRI, Los Baños, Philippines.

Erbisch, F.H. (2000) Challenges of plant protection: how a semi-public agricultural research institution protects its new plant varieties and markets them. Presentation made at the Workshop on the Impact on Research and Development of *sui generis* Approaches to Plant Variety Protection of Rice in Developing Countries, 16–18 February. IRRI, Los Baños, Philippines.

Faylon, P.S., Crisanto, R.E., Sonny, P.T. and Conception, E.M. (2000) Status of PVP in the Philippines. Presentation made at the Workshop on the Impact on Research and Development of *sui generis* Approaches to Plant Variety Protection of Rice in Developing Countries, 16–18 February. IRRI, Los Baños, Philippines.

Gisselquist, D., Nash, J. and Van Der Meer, C. (2000) *Agricultural Inputs Regulation: Issues and Options for Sustainable Growth.* World Bank Ad Hoc Inputs Committee, Washington, DC.

Government of Indonesia (1999) Draft of Act of the Republic of Indonesia, #_, Year_, on Plant Variety Protection. Djakarta.

IPGRI (1999) *Key Questions for Decision-Makers: Protection of Plant Varieties under the WTO Agreement on Trade-Related Aspects of Intellectual Property Rights.* International Plant Genetic Resources Institute, Rome, Italy.

Leskien, D. and Flitner, M. (1997) Intellectual property rights and plant genetic resources: options for a *sui generis* system. *Issues in Genetic Resources,* No. 6, June. International Plant Genetic Resources Institute, Rome, Italy.

Malik, K.A and Hussain, A. (2000) The plant variety protection system in Pakistan. Presentation made at the Workshop on the Impact on Research and Development of *sui generis* Approaches to Plant Variety Protection of Rice in Developing Countries, 16–18 February. IRRI, Los Baños, Philippines.

NCPGR (1998a) Plant Varieties Act of Bangladesh, Draft only. National Committee on Plant Genetic Resources, Dhaka.

NCPGR (1998b) Biodiversity and Community Knowledge Protection Act of Bangladesh, Draft only. National Committee on Plant Genetic Resources, Dhaka.

Nghia, N.H. and Quang, P.D. (2000) Country report prepared by Prof. Dr Nguyen Huu Nghia, Director General, Vietnam Agricultural Science Institute, and Dr Pham Dom Quang, Director, National Centre for Variety Evaluation and Seed Certification. Presentation made at the Workshop on the Impact on Research and Development of *sui generis* Approaches to Plant Variety Protection of Rice in Developing Countries, 16–18 February. IRRI, Los Baños, Philippines.

Othman, O. (2000) The *sui generis* plant variety protection system in Malaysia. Presentation made to the Workshop on the Impact on Research and Development of *sui generis* Approaches to Plant Variety Protection of Rice in Developing Countries, 16–18 February. IRRI, Los Baños, Philippines.

Razzaque, A. (2000) IP legislation and its impact on rice research. Presentation made to the Workshop on the Impact on Research and Development of *sui generis* Approaches to Plant Variety Protection of Rice in Developing Countries, 16–18 February. IRRI, Los Baños, Philippines.

Shumin, W. and Lijun, L. (2000) The emerging *sui generis* PVP system in China. Presentation made at the Workshop on the Impact on Research and Development of *sui generis* Approaches to Plant Variety Protection of Rice in Developing Countries 16–18 February. IRRI, Los Baños, Philippines.

Sumarno (2000) Plant variety protection in Indonesia. Presentation made at the Workshop on the Impact on Research and Development of *sui generis* Approaches to Plant Variety Protection of Rice in Developing Countries, 16–18 February. IRRI, Los Baños, Philippines.

Yasuoka, S. (2000) Protection of rice varieties under the UPOV system: overview and technical aspects. Presentation made at the Workshop on the Impact on Research and Development of *sui generis* Approaches to Plant Variety Protection of Rice in Developing Countries, 16–18 February. IRRI, Los Baños, Philippines.

Chapter 4

Intellectual Property Aspects of Traditional Agricultural Knowledge

Michael Blakeney

Queen Mary Intellectual Property Research Institute, CCLS, University of London, Mile End Road, London E1 4NS, UK

Role of Indigenous and Traditional Communities in Biodiversity Conservation and Innovation

The traditional knowledge of indigenous peoples throughout the world has played an important role in identifying biological resources worthy of commercial exploitation. For example, the search for new pharmaceuticals from naturally occurring biological material has been guided by ethnobiological data (McChesney, 1996). The recent passion for environmental sensitivity in Western countries has resulted in a heightened interest in natural products. Research into these products has been guided by the knowledge of indigenous peoples. In 1991, Merck, a multinational pharmaceutical company, entered into a bioprospecting agreement with the Costa Rican Association Instituto Nacional de Biodiversidad (INBio) a non-profit organization. Under the agreement, over a 2-year period, Merck received 10,000 plant samples (IPBN, 1995). The samples were supplied with information about their traditional use. Merck has paid a reported US$1.35 million for the 10,000 samples, and has agreed to pay a royalty of between 2% and 3% (IPBN, 1995). Currently, three of the drugs that Merck sells earn over US$1 billion each. If one of the 10,000 samples becomes a billion-dollar drug then Merck has agreed to pay US$20–30 million in royalties. Conceivably, the royalties from the 10,000 samples could earn Costa Rica well in excess of US$100 million per annum. This is clear evidence of the commercial value that the pharmaceutical industry places on indigenous peoples' intellectual property. In 1995 the estimated market value of pharmaceutical derivatives from indigenous peoples' traditional medicine was US$43 billion (IPBN, 1995) worldwide. Under current intellectual property law, there is no obligation for companies that utilize the traditional knowledge of indigenous peoples to provide any compensation to recognize their equity in the commercial application of this knowledge.

A similarly significant contribution has been made by the knowledge of indigenous peoples and traditional farmers in the development of new crop types and biodiversity conservation. These groups have been an important agency in the conservation of plant genetic resources (PGRs) and the transmission of these resources to seed companies, plant breeders and research institutions. They have not typically been paid for the

(eds R.E. Evenson, V. Santaniello and D. Zilberman)

value they have delivered, whereas breeders and seed companies have resorted to intellectual property rights (IPRs) to recover their development expenditures. On the other hand, farmers who utilize improved varieties are obliged to pay for them.

The economic value of biological diversity conserved by traditional farmers for agriculture is difficult to quantify (e.g. Brush, 1994). It has recently been suggested that 'the value of farmers' varieties is not directly dependent on their current use in conventional breeding, since the gene flow from landraces to privately marketed cultivars of major crops is very modest' (Correa, 2000), because 'conventional breeding increasingly focuses on crosses among elite materials from the breeders own collections and advanced lines developed in public institutions.' On the other hand, those collections and advanced breeding lines are often derived from germplasm contributed by traditional groups.

An increasingly significant economic value of biodiversity is the extent to which it provides a reservoir of species available for domestication, as well as genetic resources available for the enhancement of domestic species. The modern biotechnological revolution has enabled the engineering of desirable genetic traits from useful local species. It is estimated that about 6.5% of all genetic research undertaken in agriculture is focused upon germplasm derived from wild species and landraces (McNeely, 2001).

This paper examines a number of recent cases in which traditional agricultural knowledge has been converted into IPRs and explores the possibilities for the protection of this knowledge within the context of the Convention on Biological Diversity (CBD) and the World Trade Organization (WTO) agreement on Trade Related Aspects of Intellectual Property Rights (TRIPs) and the various national implementations of these international obligations. Finally, the paper considers the possibilities for the protection of traditional agricultural knowledge within the context of the International Undertaking on Plant Genetic Resources (IUPGR).

The Consultative Group on International Agricultural Research (CGIAR)

Each of the case studies discussed below concerns the seeking of IPRs in relation to germplasm deposited with or developed by a number of the agricultural research centres comprising the CGIAR. The CGIAR, founded in 1971, is an informal association of public and private donors that supports an international network of 16 international agricultural research centres (IARCs), each with its own governing body. The major sponsors are the Food and Agriculture Organization of the United Nations (FAO), the World Bank, the Rockefeller and Ford Foundations, the United Nations Development Programme (UNDP), the United Nations Environment Programme (UNEP) and the aid programmes of the EU and a number of individual countries. With a budget of some US$340 million per annum, the CGIAR oversees the largest agricultural research effort in the developing world.

This agricultural research commenced with the work of Norman Borlaug, an American plant breeder, who won the Nobel Prize in 1970 for his work in developing high-yielding wheat varieties for Mexico. Borlaug was the founding father of the Centro Internacional de Mejoramiento de Maiz y Trigo (CIMMYT), which became the first of the CGIAR centres. Following the establishment of CIMMYT, 15 other IARCs have been established, each focusing on crops and materials of interest to developing countries.[1] In addition to conducting

[1] These centres are: the Centro Internacional de Agricultura Tropical (CIAT), Center for International Forestry Research (CIFOR), Centro Internacional de la Papa (CIP), International Center for Agricultural Research in the Dry Areas (ICARDA), International Center for Living Aquatic Resources Management (ICLARM), International Center for Research in Agroforestry (ICRAF), International Crop Research Institute for the Semi-Arid Tropics (ICRISAT), International Livestock Research Institute (ILRI), International Institute of Tropical Agriculture (IITA), International Plant Genetic Resources Institute (IPGRI) International Rice Research Institute (IRRI) and the West Africa Rice Development Association (WARDA).

research, the CGIAR supports a collection of germplasm, which currently comprises over 600,000 accessions of more than 3000 crop, forage and pasture species which are held at the research centres. These germplasm collections are held under the auspices of the FAO 'in trust for the benefit of the international community, in particular the developing countries', and include up to 40% of all unique samples of major food crops held by genebanks worldwide. The FAO Commission on Genetic Resources for Food and Agriculture determines the policy under which the network of *ex situ* collections operates.

In addition to the so-called 'designated germplasm', which is held under the trust relationship with the FAO, the various CGIAR centres have developed 'elite germplasm' and biological tools, such as isogenic lines, mutants and mapping populations, from the materials which have been deposited with them.

Since the late 1990s, difficult questions have been raised concerning the legal status of the germplasm collections of the agricultural research institutes that are members of the CGIAR. At the time of their establishment, the questions of ownership and IPRs in the collections were very much subordinated to the mission to increase crop yields to feed a burgeoning world population. It has only been in recent years that ownership issues have become important, either as a bargaining counter in North–South negotiations or as a source of revenue (Blakeney, 1997).

Case Studies

The status of the CGIAR collections was questioned in a series of so-called biopiracy episodes in which IPRs were sought in relation to germplasm obtained from a number of CGIAR centres.

1998 Australian 'biopiracy' episode

Germplasm ownership concerns were raised in 1998 as a consequence of Plant Breeder's Rights applications made in Australia by a number of agricultural research institutes in relation to a peavine and a lentil which had been bred from genetic stock obtained from the CGIAR genebank: the ICARDA, located in Aleppo, Syria (Editorial, 1998; Edwards and Anderson, 1998). A charter for ICARDA had been established in November 1975 on the basis of an agreement between the World Bank, FAO and UNDP, and the Canadian International Development Research Centre (IDRC), as the executing agency. Subsequently, establishment agreements were negotiated by IDRC with Syria (28 June 1976), Lebanon (6 July 1977) and Iran (20 July 1976). These parallel agreements provide for the establishment of ICARDA in 'the region' defined as the 'Near East, North Africa and the Mediterranean region'. ICARDA's headquarters were established in Aleppo, Syria. The agreements also provided for the Chairman of CGIAR to declare that ICARDA has been established as a legal entity allowing IDRC's role to lapse. This formal step does not appear to have been taken. The question of ownership of the ICARDA collection was raised in the context of whether its Director-General acted in breach of trust obligations, which he owed, in relation to ICARDA genetic material, in permitting the Australian agricultural research institutes to seek IPRs in applications of that material (Blakeney, 1998).

The 14 February 1998 issue of *New Scientist* contained an editorial and leading article on the alleged biopiracy of two Australian agricultural agencies. The two agencies: Agriculture Western Australia and the Grains Research and Development Corporation (GRDC) had apparently applied for Plant Breeder's Rights (PBRs) under the Australian Plant Breeder's Rights Act, 1994, in relation to two species of chickpea, which had been bred from material that had been provided by ICRISAT. These PBR applications had to meet the statutory tests prescribed in section 43 of the Act – that the new variety has a breeder, that it is distinct, uniform and stable and that it has not been or has only recently been exploited. The Australian Plant Breeder's Rights Office did not have an opportunity to make a determination on these matters because the furore caused by these applications led to their withdrawal, prior to determination.

The *New Scientist* editorialized that 'it was hard to imagine what two Australian government agricultural agencies thought that they were up to when they applied for property rights on chickpeas grown by subsistence farmers in India and Iran' (Editorial, 1998). A feature article in *New Scientist* carried an accusation from a spokesperson from the South Asian Network on Food, Ecology and Culture which described the PBR applications as 'blatant biopiracy' by 'privatising seeds that belong to our farmers and selling them back to us' (Edwards and Anderson, 1998).

Research by RAFI suggested that there were numerous other instances of 'biopiracy' by other Australian agricultural research institutes. Reacting to the biopiracy controversy, CGIAR called for a moratorium on the granting of IPRs over plant germplasm held in its centres. The CGIAR Chairman, Dr Ismail Serageldin, explained the call for a moratorium as 'the strongest signal the CGIAR can send governments to ensure that these issues be resolved and the materials in the CGIAR remain in the public domain' (CGIAR Press Release, 1998). In Australia, serious concerns were expressed about the implications such a moratorium would have, particularly for its cultivation of cereals. Consequently, to prevent a recurrence of this incident, the operating regulations of the Australian Plant Breeders Rights Office were amended to oblige applicants for PBRs in relation to varieties derived from germplasm obtained from CGIAR centres, to document that such applications were made with the permission of the relevant centre.

Responding to concerns about the impact of IPRs upon the operation of the CGIAR, it commissioned a report on the use of proprietary technologies by CGIAR centres by the International Service for National Agriculture Research (ISNAR), which operates as its legal advisory body (Cohen *et al.*, 1998). The report noted the burgeoning use of proprietary technologies by the centres and recommended that they undertake audits of their intellectual property management policies. ISNAR established a Central Advisory Service to provide legal counsel for the centres on intellectual property matters.

An issue that has not yet been addressed by the CGIAR or the FAO is the question of the rights, if any, of the indigenous and traditional communities from which seeds may have been collected by the various CGIAR institutes – that collection may have been informed by the knowledge of those communities, or may have occurred without the communication by the collector to those communities of the implications of the act of collection.

Blight-resistant rice

In the late 1970s a strain of rice from Mali, *Oryza longistaminata*, was identified by a researcher, working in Cuttack, North India, as being resistant to bacterial blight, a disease which particularly afflicts rice. In 1978 this resistant sample was taken to the IRRI in Los Baños, Philippines for further investigation. Over a 15-year period, IRRI researchers developed, through conventional breeding, a high-yielding, blight-resistant strain of rice. The IRRI researchers identified that this resistance was contributed by a single locus called *Xa21*. A post-doctoral research fellow, Dr Ronald, from the University of California at Davis, who was working at IRRI, was permitted with co-workers at Stanford University to map, sequence and clone the *Xa21* gene. The molecular mapping process was facilitated by the construction of a BAC library utilizing a biological tool provided by IRRI.

On 7 June 1995, the Regents of the University of California filed a patent application for 'Nucleic acids, from *Oryza sativa*, which encode leucine-rich repeat polypeptides and enhance *Xanthomonas* resistance in plants'. The inventors named in the application were Dr Ronald and her co-workers. The patent was granted by the US Patents and Trademark Office (USPTO) on 12 January 1999 (US patent 5,859,339).

This patent generated some controversy in CGIAR circles because it was perceived to compromise IRRI's research efforts and those of its clients in the rice-producing regions of Asia. Bacterial blight is not a particular problem for US rice producers and a primary effect of the patent was to prevent the export of bacterial blight-resistant rice, utilizing the patent to the USA. This patent also raised the question of equitable compensation, at least for the traditional farmers of Mali who had conserved *O. longistaminata*. The UC Davis dealt with the issue of compensation by establishing a Genetic Resources Recognition Fund (GRRF) as a mechanism to share benefits arising from the commercial utilization of its patent. It was also acknowledged that in the absence of this sort of mechanism, it would have been 'more difficult for the university in the future to obtain research access to developing countries' national genetic materials' (Ronald, 2001, p. 13).

UC Davis also agreed to allow non-commercial researchers access to the gene, provided they did not develop commercial products based on that gene. The University also agreed that 'IRRI would have full rights to develop new rice varieties incorporating cloned *Xa21* and distribute this material as well as the clone to developing countries' (Ronald, 2001, p. 14).

The attempts of UC Davis to provide a mechanism for the sharing of benefits with traditional peoples to be derived from its patent on *Xa21* were analysed in a case study by the World Intellectual Property Organization and the UNEP (Ronald, 2001, p. 14). This study found that with one exception, the traditional peoples of Mali regarded *O. longistaminata* as a weed with a very low grain yield. It concluded that in designing benefit sharing arrangements:

- the stakeholders are not limited to the formal scientific research institutions of the country of origin of the genetic resource;
- the ethnobotanical knowledge of local PGRFA is not necessarily and only held by local land-owning farmers, but can also be held by local communities that are landless and subsist on mixed modes of income...;
- the ethnobiological knowledge is not limited to indigenous communities.... (Ronald, 2001, p. 24)

This case study illustrates a number of the practical problems with proposals for benefit sharing under the CBD. Firstly, the research into the blight-resistant properties of *O. longistaminata* commenced prior to the CBD coming into force; secondly, the local farming and indigenous communities were largely unaware of these properties; and thirdly, the itinerant poor community that made use of the rice would be unlikely to enjoy any of the benefits that UC Davis could make available. It is for this reason that benefit-sharing norms are being developed by the international community.

Enola bean

On 13 April 1999, the US Patent and Trademarks Office granted patent no. 5,894,079 to Larry Proctor for an invention described in the patent grant as relating to 'a new field bean variety that produces distinctly colored yellow seed which remain relatively unchanged by season.' On 28 May 1999, Larry Proctor also obtained a US Plant Variety Protection Certificate on the bean variety.

Larry Proctor was the president of a Colorado-based seed company, POD-NERS. Upon the grant of the patent, this company was reported to have written to all the importers of Mexican beans in the USA, requiring the payment of a royalty of 6 cents per pound (Biotechnology Noticeboard, 2000). According to Miguel Tachna Felix, of the Agricultural Association of Rio Fuerte, this would have meant an immediate drop in export sales, over 90%. POD-NERS was reported to have brought infringement actions against two companies that were selling Mexican yellow beans in the USA. In January 2000, the Mexican government announced that it would challenge the US patent. José Antonio Mendoza Zazueta, Under-Secretary of Mexican Rural Development declared that 'We will do everything necessary, anything it takes, because the defense of our beans is a matter of national interest' (quoted, Biotechnology Noticeboard, 2000). RAFI denounced the yellow bean patent as 'Mexican bean biopiracy' and demanded that the patent be legally challenged and revoked. RAFI formally requested that FAO and the CGIAR investigate the patent as a likely violation of their 1994 Trust agreement (RAFI Geno-Types, 2000).

On 20 December 2000, the CIAT filed a formal request for re-examination of the US patent concerning the yellow bean, which was alleged to be the Mexican Enola bean (RAFI, 2001). CIAT's official request for re-examination of the patent stated that the claims for inventiveness contained in the patent failed to meet the statutory requirements of novelty and non-obviousness, and ignored the prior art widely available in the literature (RAFI, 2001). The challenge was particularly critical of the patent's claim of exclusive monopoly on any *Phaseolus vulgaris* (dry bean) having a seed colour of a particular shade of yellow, pointing out that 'it will make a mockery of the patent system to allow statutory protection of a color per se.' Although there was no evidence that the patent owner obtained his yellow beans from CIAT's genebank, the patent challenge noted that CIAT maintained some 260 bean samples with yellow seeds, and six of the accessions were 'substantially identical' to claims made in the patent (RAFI, 2001).

CIAT's patent challenge also asserted that the yellow bean was 'misappropriated' from Mexico, and that this was in breach of Mexico's sovereign rights over its genetic resources, as recognized by the CBD. The USPTO is currently determining this challenge.

Yacon

In November 1999, five traditional Peruvian varieties of yacon held in the genebank at the CIP in Peru were distributed to the Peruvian Ministry of Agriculture, which passed them to researchers in Japan. Yacon (*Smallantus sonchifolius*), an ancient Andean crop, is eaten raw as a fruit in the Andes. It has a high fructose content with a high percentage of insulin and leaves reported to have anti-diabetic properties (National Research Council, 1989). CIP's Potato Germplasm Curator, Dr Zozimo Huaman, alleged that this distribution of yacon by CIP was in breach of its trust obligations, particularly because the biosafety requirements of the Centre were apparently not followed (e.g. Huaman, 2001). Japanese researchers, in a seminar at CIP in September 2000, indicated that the area cultivated with yacon in Japan had been greatly increased in recent years. Yacon was utilized as a vegetable, pickles and juices. They also reported that the National Shikoku Agriculture Experiment Station had released the first commercial variety named 'Sarada-Otome' on 25 August 2000. Then Dr Huaman expressed concern that, apparently because of plant breeders' rights, the Japanese researchers were not prepared to send germplasm of Sarada-Otome to be tested in Peruvian farmers' fields. He questioned the equity of denying new derivatives of deposited germplasm to a source country (Huaman, 2001).

Upon learning of Dr Huaman's allegation, CIP requested its Genetic Resources Policy Committee (GRPC) to determine if any violation of the FAO agreement had occurred. The GRPC, chaired by Dr M.S. Swaminathan, was established by CIP as an independent advisory committee made up of internationally known scientists, as well as representatives of the non-governmental organization (NGO) community, private sector, and developing and developed country governments. The committee concluded that CIP had no right to interfere in Peru's sovereign decision to send the germplasm to Japan and commended CIP for its proper management of its germplasm held 'in trust' (Zandstra, 2001).

As discussed below, one impact of intellectual property issues upon the CGIAR has been calls for the limitation of the availability of IPRs in relation to PGRs. Another approach has been to embrace the new legal reality. Some CGIAR centres perceive that CGIAR-generated intellectual property might be used as a bargaining chip, to be traded for biological tools patented by the private sector. For example, the *Policy on Intellectual Property* of CIMMYT envisages that intellectual property protection may be sought 'to facilitate the negotiation and conclusion of agreements for access to proprietary technologies of use to CIMMYT's research and in furtherance of its mission (CIMMYT, no date). This proprietization of public sector agriculture research is questioned, particularly by those NGOs opposed to patenting in the life sciences (RAFI, 2000).

Nuna beans

On 21 March 2000, a patent was granted to a US corporation (US Patent No. 6,040,503) in relation to a 'bean-nut popping bean' apparently derived from crosses involving at least 33 Andean nuna bean varieties from Peru, Bolivia, Ecuador and Colombia. RAFI reported that a meeting of a tribunal of indigenous elders from six Andean communities that grow nuna beans met in February 2001 and condemned the patenting as biopiracy of their Andean heritage and demanded that CIAT – the CGIAR centre based in Cali, Colombia – uphold its obligation under a United Nations 'trust agreement' to keep farmer-bred bean varieties in the public domain (RAFI).

RAFI reported the views of a number of scientists working with nuna who were concerned that the patent would limit improvements in the crop to the prejudice of the traditional Andean peoples. In particular it was observed that toasting nunas used less fuel than boiling beans, a feature important to economic and environmental conditions in areas of the world where fuel is scarce. Additionally, RAFI reported that the US patent also prejudiced a stratagem to use popping beans grown in the Andes as a substitute for illicit crops.

The patent was described as 'particularly offensive to Andean farmers and indigenous people' because it extended to crosses involving at least 33 Andean nuna varieties traditionally bred and developed over centuries in Peru, Bolivia and Ecuador, all of which were freely provided by Andean farming communities 'who allowed their bean varieties to be put into the public realm in order to ensure the continued maintenance of the world's seed biodiversity.' Nine of these varieties were held in CIAT's international bean collection as designated in-trust accessions, all being farmers' varieties collected in Peru.

This case has been adopted by a number of NGOs as a test of the international intellectual property community's resolve to support the conservation and development of indigenous knowledge.

The CBD

These case studies will be analysed in the context of the international legal environment that bears on IPRs and the protection of biological diversity.

The Rio Earth Summit, which was convened in June 1992, promulgated the CBD, The Rio Declaration on Environment and Development, and Agenda 21. The CBD represented an attempt to establish an international programme for the conservation and utilization of the world's biological resources (McConnell, 1996) and for the 'fair and equitable sharing' of the benefits arising from the utilization of genetic resources (Art. 1). 'The single most divisive issue in the negotiations was the relationship between intellectual property rights and access to genetic resources' (Chandler, 1993, p. 161). The developing countries of the South, generally speaking those with the most substantial sources of genetic resources, sought to use the CBD as a means of bargaining access to those resources for royalties, technology and research data. Thus the CBD contains articles on access to genetic resources (Art. 15), access to and the transfer of technology (Art. 16), and informed consent and the distribution of benefits of biotechnological innovations (Art. 19). The industrialized group of countries, obviously the principal source of biotechnological innovation, insisted that the CBD did not conflict with IPRs. Thus for example, Art. 16(2) contains the statement that 'In the case of technology subject to patents and other intellectual property rights, such access and transfer shall be provided on terms which recognize and are consistent with the adequate and effective protection of intellectual property rights'.

Reflecting the uncomfortable political deal that was struck in bringing the CBD to conclusion, the language of the Convention is unfortunately vague. The positive affirmation of principles in a number of areas is qualified by vague transcendental values. Thus the respect for intellectual property affirmed by Art. 16(2) is counterbalanced by the phrase in the same provision that 'access to and the transfer of technology...shall be provided and/or facilitated under fair and most favourable terms...'. Similarly, Art. 15(4) provides that 'access (to genetic resources) where granted shall be upon mutually agreed terms'. Article 19(2) provides that 'access...to the results and benefits arising from biotechnologies ... shall be on mutually agreed terms'. Since mutuality is a precondition for an agreement of any sort, these provisions may be mere rhetoric. On the other hand, they may be a guarantee against unilateral expropriation.

Scope of the CBD access regime

Article 1 of the CBD envisages 'appropriate access to genetic resources' and 'the fair and equitable sharing of benefits arising out of the utilization of genetic resources'. 'Genetic resources' are defined in Art. 2 as meaning 'genetic material of actual or potential value'. The term 'genetic material' is then defined in Art. 2 to mean 'any material of plant, animal, microbiological or other origin containing functional units of heredity'. On a strict analysis of this definition, it is suggested that biochemical extracts that do not contain DNA or RNA would be outside the scope of the CBD (Glowka *et al.*, 1994). Thus the Convention would apply to seeds, cuttings and DNA extracted from a plant, such as a chromosome, gene, plasmid or any part of these such as the promoter part of a gene (Glowka, 1998).

Article 9 deals with 'the conservation of components of biological diversity outside their natural habitats', for example, in germplasm and seed banks, botanical gardens, museums, laboratories and agricultural research institutions. This article calls for national legislation to provide for the acquisition, conservation, storage and management of these *ex situ* collections. The access and benefit-sharing provisions of the CBD do not apply to the genetic resources of a country which were collected prior to the entry of the CBD into force in that country (Art. 15(3); Yusuf, 1995). Thus a country with a pre-existing collection of genetic material has the sovereign right to control access to that collection, but has no legal right to insist upon a share of any benefits derived from the use of that collection. Furthermore, the CBD applies to those genetic resources which originate in the country of a contracting party (Art. 15(3); Yusuf, 1995).

Sovereign rights over genetic resources (Art. 15 (1))

Article 15(1) of the CBD affirms 'the sovereign rights of States over their natural resources' and provides that 'the authority to determine access to genetic resources rests with the national governments and is subject to national legislation'. This provision, dealing as it does with access to genetic resources, does not refer to the question of the ownership of genetic resources. This leaves unanswered the ownership issues raised by the creation of the CGIAR germplasm collections.

Mutually agreed terms, prior informed consent and benefit sharing

Article 15(4) of the CBD envisages that where access is granted it will be subject to mutually agreed terms. Currently the conventional form of access agreement is the Material Transfer Agreement (MTA) (Cohen *et al.*, 1998). A number of the provisions of the CBD refer to the equitable sharing of benefits arising from the utilization of the genetic resources of a signatory. Article 15(7) requires each Contracting Party to 'take legislative, administrative or policy measures, as appropriate' and in accordance with a number of specified provisions of the Convention, 'with the aim of sharing in a fair and equitable way, the results of research and development and the benefits arising from the commercial and other utilization of genetic resources with the Contracting Party providing such resources'. Article 8(j) envisages the 'equitable sharing' of benefits with indigenous and local communities, arising out of the use of the traditional knowledge, innovations and practices of those communities. Article 21 provides for the establishment of a 'mechanism' for the provision of financial resources to developing country parties to the CBD.

Complementary to the equitable sharing of benefits, the CBD provides for the access of developing country signatories to technologies that may result from the utilization of the genetic resources they may provide. Article 16(1) recites the importance of access to biotechnologies to attain the objectives of the CBD and Art. 16(2) provides for the access to technologies by developing countries on 'fair and equitable terms, including on concessional and preferential terms'. Article 19(1) requires parties to take appropriate measures to 'provide for the effective participation in biotechnological research activities by those Contracting Parties, especially developing countries, which provide the genetic resources for such research'. Article 19(2) requires parties to 'take all practicable measures to promote and advance priority access on a fair and equitable basis..., especially developing countries, to the results and benefits arising from biotechnologies based upon genetic resources provided by those Contacting Parties' on mutually agreed terms.

Indigenous and local communities

The Rio Declaration in Principle 22 stated that 'Indigenous peoples and their communities ... have a vital role in environmental management and development because of their knowledge and traditional practices'. Chapter 26 of Agenda 21 detailed the relationship which conference participants recognized between indigenous peoples and their lands. The Agenda, at para. 26.3(a), required governments:

> to establish a process to empower indigenous peoples and their communities' through measures that include:
>
> - recognition of their values, traditional knowledge and resource management practices with a view to promoting environmentally sound and sustainable development;
> - enhancement of capacity-building for indigenous communities based on the adaptation and exchange of traditional experience, knowledge and resource-management practices, to ensure their sustainable development;
> - establishment, where appropriate, of arrangements to strengthen the active participation of indigenous peoples and their communities in the national formulation of policies, laws and programs relating to resource management and other development processes that may affect them.

The Preamble to the CBD recognized the

> ... close and traditional dependence of many indigenous and local communities embodying traditional lifestyles on biological resources, and the desirability of sharing equitably arising from the use of traditional knowledge, innovations and practices relevant to the conservation of biological diversity and sustainable use of its components.

Article 8(j) of the Convention required each signatory

> ... subject to its national legislation, respect, preserve and maintain knowledge, innovations and practices of indigenous and local communities embodying traditional lifestyles relevant for the conservation and sustainable use of biological diversity and promote their wider application with the

> approval and involvement of the holders of such knowledge, innovations and practices and encourage the equitable sharing of the benefits arising from the utilization of such knowledge, innovations and practices.

The provisions of Art. 8(j) require implementation through national legislation. It is expressed to be subject to national legislation, in order to preserve legislation on this subject which predates the CBD (Chandler, 1993).

The discussion, in the context of the CBD, of the IPRs of traditional and local communities has not tended to focus upon the rights of traditional farming communities. This subject has been taken up as an aspect of the negotiations for an IUPGR.

Farmers' Rights and the IUPGR

In 1983 the Conference of the FAO adopted the IUPGR as a non-legally binding instrument. The IUPGR provides for the exploration and collection of genetic resources (Art. 3), for conservation *in situ* and *ex situ* (Art. 4), for the availability of PGRs (Art. 5), for international cooperation in conservation, exchange and plant breeding (Art. 6), for international coordination of genebank collections and information systems (Art. 7) and for funding (Art. 8). Some 113 countries have subscribed to the IUPGR, excluding the USA (FAO, 1996).

The IUPGR was originally predicated on the principle that PGRs should be freely exchanged as a 'heritage of mankind' and should be preserved through international conservation efforts. In subsequent years the principle of free exchange was gradually narrowed. In November 1989, the 25th Session of the FAO Conference in Rome adopted two resolutions providing an 'agreed interpretation' that plant breeders' rights were not incompatible with the IUPGR (Resolutions 4/89 and 5/89). The acknowledgement of plant variety rights obviously benefited industrialized countries, which were active in seed production. The contribution made by traditional farmers, who conserved and improved the breeding materials from which new varieties are derived, was not recognized in the International Convention for the Protection of Plant Varieties (UPOV) scheme.

The concept of Farmers' Rights was developed as 'a counterbalance to intellectual property rights' (FAO, 1994) In exchange for this concession, developing countries won endorsement of the concept of Farmers' Rights. This was a moral commitment by the industrialized countries to reward 'the past present and future contributions of farmers in conserving, improving and making available PGRs particularly those in centres of origin/diversity. Farmers' Rights were defined in a Resolution of the 1989 FAO Conference (Annex II, Resolution 5/89) as:

> ... rights arising from the past, present and future contribution of farmers in conserving, improving and making available plant genetic resources, particularly those in centres of origin/diversity. These rights are vested in the International Community, as trustee for present and future generations of farmers, for the purpose of ensuring full benefits to farmers, and supporting the continuation of their contributions.

Farmers' Rights were intended to promote a more equitable relation between the providers and users of germplasm by creating a basis for farmers to share in the benefits derived from the germplasm which they had developed and conserved over time (Glowka, 1998, p. 6). An International Fund for Plant Genetic Resources was proposed in a Resolution of 1991 as a means of implementing Farmers' Rights. This Fund will support plant genetic conservation and utilization programmes, particularly in the developing countries.

Farmers' Rights are conceived of as a 'retrospective equity' (Brush, 1996) primarily as the recognition of the moral obligation, rather than an economic incentive. Its implementation is uncertain, although suggestions have been made in India for a seed tax, where the revenue yield will be distributed through a Community Gene Fund (Swaminathan and Hoon, 1994).

At its eighth session in 1999 the members of CGRFA agreed the following article:

Article 15 – Farmers' Rights

15.1 The Parties recognize the enormous contribution that the local and indigenous communities and farmers of all regions of the world, particularly those in the centres of origin and crop diversity, have made and will continue to make for the conservation and development of plant genetic resources which constitute the basis of food and agriculture production throughout the world.
15.2 The Parties agree that the responsibility for realizing Farmers' Rights, as they relate to Plant Genetic Resources for Food and Agriculture, rests with national governments. In accordance with their needs and priorities, each Party should, as appropriate, and subject to its national legislation, take measures to protect and promote Farmers' Rights, including:

(a) protection of traditional knowledge relevant to plant genetic resources for food and agriculture;
(b) the right to equitably participate in sharing benefits arising from the utilization of plant genetic resources for food and agriculture;
(c) the right to participate in making decisions, at the national level, on matters related to the conservation and sustainable use of plant genetic resources for food and agriculture.

15.3 Nothing in this Article shall be interpreted to limit any rights that farmers have to save, use, exchange and sell farm-saved seed/propagating material, subject to national law and as appropriate.

The debate on Farmers' Rights in the context of the FAO has subsumed some of the content of the contemporaneous debate on the necessity to protect the IPRs of indigenous communities. However, it is not always the situation that the indigenous communities are engaged in farming. Furthermore, as is illustrated by the case study concerning bacterial blight-immune rice from Mali, a genetically interesting landrace might be of no interest either to the farming or to the indigenous communities.

WTO Agreement on TRIPs

Paralleling the formulation of the CBD, were the negotiations of the Uruguay Round of the General Agreement on Tariffs and Trade (GATT). Attempts by the World Intellectual Property Organization to revise the Paris Convention on Industrial Property, 1883, which deals with the international patents, industrial designs and trade marks regime, had foundered on the irreconcilability of the position of developing countries and industrialized countries on the compulsory licensing of patents (Blakeney, 1989). For this and other reasons, the US proposed that the GATT formulate legislative norms for intellectual property protection and that it require the introduction of a range of mechanisms for the enforcement of IPRs (Blakeney, 1995). The resultant Agreement on TRIPs was annexed as a condition of membership to the Agreement Establishing the WTO (Blakeney, 1996). Article 27.3 of the TRIPs Agreement permits signatories to exclude from patentability 'plants and animals other than microorganisms, and essentially biological processes for the production of plants or animals, other than non-biological and microbiological processes'. However, the provision requires that 'Members shall provide for the protection of plant varieties either by patents or by an effective *sui generis* system or by any combination thereof'.

Sui generis protection of plant varieties

There is a vigorous debate on the sorts of *sui generis* systems that might comply with Art. 27.3(b). The TRIPs provision makes no reference to UPOV. This is considered to provide some leeway in the formulation of *sui generis* systems (e.g. *Biotechnology and Development Monitor*, 1998). Furthermore, key elements for the shaping of *sui generis* systems are either unclear or not defined. First, there could be several ways to define the term 'plant variety'. For granting protection under the traditional PBR system, plant varieties must meet the criteria of being distinct, uniform and stable (DUS). It has been suggested that 'uniformity' and 'stability' could be replaced by the criterion of identifiability, allowing the inclusion of plant populations which are more heterogeneous, thus taking into account the interests of local communities (Seiler, 1998). The scope of protection could be limited to cover only the reproductive parts of plants, or could be extended to include also harvested plant materials.

Secondly, the TRIPs Agreement does not prohibit the development of additional protection systems, nor does it prohibit the protection of additional subject matter to safeguard local knowledge systems and informal innovations as well as to prevent their illegal appropriation.

The original formulation and promulgation of the TRIPs Agreement had occurred largely without the active participation of developing countries. Their principal negotiating position during the Uruguay Round had been to question the relevance of intellectual property for the GATT, particularly as WIPO had already been established as the United Nations' specialized agency for intellectual property matters. The failure of developing countries to address the substance of TRIPs during the Uruguay Round was sought to be remedied by their active participation in the review procedure. The various regional groupings of developing countries held meetings to agree a common negotiating position for the TRIPs and also the CBD reviews.

A Communication to the WTO from Kenya, on behalf of the African Group, to assist the Preparations for the 1999 Ministerial Conference, pointed out that as the deadline for implementation of the obligations by developing countries of the TRIPs Agreement was January 2000, the review would precede the implementation of obligations undertaken by developing countries. As developing countries would have had insufficient experience with the operation of the Agreement, they would have had no prior opportunity to conduct impact assessment studies of implications resulting therefrom.

Furthermore, the Communication pointed out that the review would preempt the outcome of deliberations in other related fora such as CBD, UPOV, FAO, IUPGR and the development of an OAU model law on Community Rights and Control of Access to Biological Resources. They proposed that an additional 5 years be allowed, prior to the review of Art. 27.3(b).

The African group proposed that:

> after the sentence on plant variety protection in Article 27.3(b), a footnote should be inserted stating that any *sui generis* law for plant variety protection can provide for:
>
> (i) the protection of the innovations of indigenous and local farming communities in developing countries, consistent with the Convention on Biological Diversity and the International Undertaking on Plant Genetic Resources;
> (ii) the continuation of the traditional farming practices including the right to save, exchange and save seeds, and sell their harvest;
> (iii) preventing anti-competitive rights or practices which will threaten food sovereignty of people in developing countries, as is permitted by Article 31 of the TRIPs Agreement.

On 25 July 1999, a federation of indigenous peoples groups issued a statement for the purposes of the TRIPs review. The Statement commences with the observation that 'Humankind is part of Mother Nature, we have created nothing and so we can in no way claim to be owners of what does not belong to us. But time and again, western legal property regimes have been imposed on us, contradicting our own cosmologies and values.' It expresses concern that Article 27.3(b) 'will further denigrate and undermine our rights to our cultural and intellectual heritage, our plant, animal, and even human genetic resources and discriminate against our indigenous ways of thinking and behaving.'

The Statement drew the distinction between private proprietorial rights and 'Indigenous knowledge and cultural heritage (which) are collectively and accretionally evolved through generations...The inherent conflict between these two knowledge systems and the manner in which they are protected and used will cause further disintegration of our communal values and practices'.

The Statement pleaded for a legislative structure which 'Builds upon the indigenous methods and customary laws protecting knowledge and heritage and biological resources' and which prevents the appropriation of traditional knowledge and integrates 'the principle and practice of prior informed consent, of indigenous peoples as communities or as collectivities'.

This Statement was picked up by a submission of Cuba, Honduras, Paraguay and Venezuela to the TRIPs Council (*Proposal*, 1999), which stated that these countries 'consider it fair to recognize the specific contribution of indigenous and tribal peoples and local communities to the cultural diversity and social and ecological harmony of mankind'.

Responding to these developing country initiatives, the USA has urged that an effective *sui generis* system should clearly identify: (i) the subject matter of protection; (ii) any limitations to the rights which will be granted under such a system; and (iii) the legal remedies available to rights holders (WTO, 2000). In relation to a *sui generis* system for the protection of plant varieties, the US submission was that all plant varieties should be covered, with the objective of encouraging the development of new varieties from the widest possible range of genera and species (WTO, 2000, pp. 2–3). This submission also recommended confining this system of protection only to breeders or others specifically entitled through contract law or succession. The US submission was unsympathetic to the claims of indigenous people for the protection of oral knowledge and practices, because of the inaccessibility of this information beyond the relevant indigenous community (WTO, 2000, p. 5).

The US submission was hostile to suggestions to facilitate benefit sharing by requiring the identification of the source of genetic materials and traditional knowledge in patent applications. Its preferred approach was to oblige parties to negotiate benefit-sharing arrangements as a condition of the grant of access.

WIPO and access to genetic resources

WIPO's involvement with the issue of access to genetic resources commenced in 1999 with a study, commissioned jointly with the UNEP, on the role of IPRs in the sharing of benefits arising from the use of biological resources and associated traditional knowledge. These matters were taken up at the third session of the Standing Committee on the Law of Patents (SCP) in September 1999. The SCP requested the International Bureau to include the issue of protection of biological and genetic resources on the agenda of a Working Group on Biotechnological Inventions, to be convened at WIPO in November 1999. The Working Group, at its meeting the following month, recommended the establishment of nine projects related to the protection of inventions in the field of biotechnology. The Working Group decided to establish a questionnaire for the purpose of gathering information about the protection of biotechnological inventions, including certain aspects regarding intellectual property and genetic resources, in the Member States of WIPO.

In response to the invitation issued by the SCP, WIPO organized a Meeting on Intellectual Property and Genetic Resources on 17–18 April 2000. The Meeting addressed issues that generally are raised in the context of access to, and *in situ* preservation of, genetic resources in their direct or indirect relationship with intellectual property. The Chairman's conclusions from the Meeting state that a consensus was reached that:

> WIPO should facilitate the continuation of consultations among Member States in coordination with the other concerned international organizations, through the conduct of appropriate legal and technical studies, and through the setting up of an appropriate forum within WIPO for future work.

At the third session of the WIPO Standing Committee on the Law of Patents in September 1999, the delegation from Colombia proposed the introduction into the Patent Law Treaty, as a means of achieving some global harmonization of patent registration procedures, an article which provided that:

> **1.** All industrial protection shall guarantee the protection of the country's biological and genetic heritage. Consequently, the grant of patents or registrations that relate to elements of that heritage shall be subject to their having been acquired made legally.
> **2.** Every document shall specify the registration number of the contract affording access to genetic resources and a copy thereof whereby the products or processes for which protection is sought have been manufactured or developed from genetic resources, or products thereof, of which one of the member countries is the country of origin.

The Diplomatic Conference, which commenced on 11 May 2000, became bogged down on the question of obliging the identification of source countries in biotechnological patent applications. To facilitate progress on the procedural aspects, the source country question was referred to an expert group for further consideration. In a press release issued on 1 June 2000, WIPO reported that it had also received a mandate to discuss this issue from the COP 5 meeting in Nairobi and that this request would be referred to its General Assembly in September 2000.

In a note dated 14 September 2000, the Permanent Mission of the Dominican Republic to the United Nations in Geneva submitted two documents on behalf of the Group of Countries of Latin America and the Caribbean (GRULAC) as part of the debate on in the WIPO General Assembly on 'Matters Concerning Intellectual Property and Genetic Resources, Traditional Knowledge and Folklore' (WIPO Doc. WO/GA/26/9). The central thrust of these documents was a request for the creation of a Standing Committee on access to the genetic resources and traditional knowledge of local and indigenous communities:

> The work of that Standing Committee would have to be directed towards defining internationally recognized practical methods of securing adequate protection for the intellectual property rights in traditional knowledge. (WIPO Doc. WO/GA/26/9, Annex I, 10)

The GRULAC documents suggested that questions concerning the use and exploitation of genetic resources and biodiversity and also traditional knowledge, could be divided into two groups depending on whether they are currently recognized or being addressed by intellectual property in the international environment. The first group includes problems whose solutions could in principle be found in known intellectual property regimes. In this group, the preferred remedy would be the broadening or clarification of existing intellectual property remedies.

It was suggested that the second group comprised those aspects, questions and problems the settlement of which calls for recognition and acceptance of the values and interests whose protection is sought, and the creation of new disciplines and provisions so that their protection may be established at the international level. The GRULAC documents envisaged that the Committee would examine the protection needs and expectations of sectors that possess traditional knowledge and to determine the manner in which they require an adjustment of existing intellectual property regimes or the creation of new ones. The Committee might also consider it necessary to ascertain whether some of the protection claims were not completely outside the present or prospective framework of intellectual property.

In order to clarify the future application of intellectual property to the use and exploitation of genetic resources and biodiversity and also traditional knowledge, it was suggested that the Committee could clarify: (i) the notions of public domain and private domain; (ii) the appropriateness and feasibility of recognizing rights in traditional works and knowledge currently in the public domain, and investigating machinery to limit and control certain kinds of unauthorized exploitation; (iii) recognition of collective rights; (iv) model provisions and model contracts with which to control the use and exploitation of genetic and biological resources, and machinery for the equitable distribution of profits in the event of a patentable product or process being developed from a given resource embodying the principles of prior informed consent and equitable distribution of profits in connection with the use, development and commercial exploitation of the material transferred and the inventions and technology resulting from it; and (v) the protection of undisclosed traditional knowledge.

Finally, it was suggested that in concert with the secretariat of UPOV, the Committee could embark on the exploration of possible options for defining *sui generis* systems for the protection of genetic resources and biodiversity.

At the WIPO General Assembly the Member States agreed the establishment of an Intergovernmental Committee on Intellectual Property and Genetic Resources, Traditional Knowledge and Folklore. Three interrelated themes were identified to inform the deliberations of the Committee: intellectual property issues that arise in the context of: (i) access to genetic resources and benefit sharing; (ii) protection of traditional knowledge, whether or not associated with those resources; and (iii) the protection of expressions of folklore (WIPO, 2000).

In November 2000, WIPO convened an Inter-regional Meeting on Intellectual Property and Traditional Knowledge at Chiangrai, Thailand, to provide an opportunity to developing countries to contribute to the activities of the Intergovernmental Committee. The meeting recommended that governments identify, catalogue, record and document genetic resources and traditional knowledge and that national mechanisms be established to regulate access to and benefit sharing in genetic resources and the protection of traditional knowledge and folklore.

Treaty to Share the Genetic Commons

The difficulties which have attended the appropriation of traditional knowledge, particularly by the life sciences companies, explain in part a proposal, by a group of NGOs, in anticipation of the preparatory meeting in New York in April 2001 for a conference to be convened on the tenth anniversary on the Rio Earth Summit, for a Treaty to Share the Genetic Commons. This treaty requires that

> ... the nations of the world declare the Earth's gene pool, in all of its biological forms and manifestations, to be a global commons, to be protected and nurtured by all peoples and further declare that genes and the products they code for, in their natural, purified or synthesized form as well as chromosomes, cells, tissue, organs and organisms, including cloned, transgenic and chimeric organisms, will not be allowed to be claimed as commercially negotiable genetic information or intellectual property by governments, commercial enterprises, other institutions or individuals.
>
> The Parties to the treaty – to include signatory nation states and Indigenous Peoples – further agree to administer the gene pool as a trust. The signatories acknowledge the sovereign right and responsibility of every nation and homeland to oversee the biological resources within their borders and determine how they are managed and shared. However, because the gene pool, in all of its biological forms and manifestations, is a global commons, it cannot be sold by any institution or individual as genetic information. Nor can any institution or individual, in turn, lay claim to the genetic information as intellectual property.

References

Biotechnology and Development Monitor 3 (1998) Various systems for sui generis rights systems, No. 38.

Biotechnology Notice Board (2000) Mexican bean biopiracy, posted by PANUPS; panupdates@panna.org, 24 January.

Blakeney, M. (1989) *Legal Aspects of the Transfer of Technology to Developing Countries.* ESC, Oxford.

Blakeney, M. (1995) Intellectual property in world trade. *International Trade Law and Regulation* 76.

Blakeney, M. (1996) *Trade Related Aspects of Intellectual Property Rights: a Concise Guide to the TRIPs Agreement.* Sweet & Maxwell, London.

Blakeney, M. (1997) Access to genetic resources: the view from the south. *Bioscience Law Review* 97–103.

Blakeney, M. (1998) Intellectual property rights in the genetic resources of international agricultural research institutes – some recent problems. *Bioscience Law Review* 3–11.

Brush, S. (1994) *Providing Farmers' Rights Through In Situ Conservation of Crop Genetic Resources.* University of California, Berkeley.

Brush, S. (1996) Whose knowledge, whose genes, whose rights? In: Brush, S.B. and Stabinsky, D. (eds) *Valuing Indigenous Knowledge: Indigenous Peoples and Intellectual Property Rights.* Island Press, Washington, DC, p. 12.

CGIAR Press Release (1998) CGIAR urges halt to granting of intellectual property rights for designated plant germplasm. 11 February, http://www.cgiar.org:80/germrel.htm

Chandler (1993) The biodiversity convention: selected issues of interest to the international lawyer. *Colorado Journal of International Environmental Law and Policy* 141, 161.

CIMMYT, *Policy on Intellectual Property*, Article III.4.v, <www.cimmyt.org/resources/obtaining/seed/ip_policy/htm/ip-policy.htm>.

Cohen, J., Falconi, C., Komen, J. and Blakeney, M. (1988) *The Use of Proprietary Biotechnology Research Inputs at Selected CGIAR Centers.* International Service for National Agricultural Research (ISNAR)/CGIAR, The Hague.

Correa, C. (2000) *Options For The Implementation of Farmers' Rights at The National Level*, South Centre, Trade-Related Agenda, Development And Equity Working Papers, No. 8, December 2000, citing Wright, Intellectual property and Farmers' Rights. In: Evenson, R., Gollin, D. and Santaniello, V. (eds) *Agricultural Values of Plant Genetic Resources.* CAB International, Wallingford, UK, p. 228.

Editorial (1998) Lest we starve. *New Scientist*, No. 2121, 14 February, 3.

Edwards and Anderson (1998) Seeds of wrath. *New Scientist*, No. 2121, 14 February, 14.

FAO (1994) Revision of the International Undertaking. Issues for consideration in stage II: access to plant genetic resources and Farmers' Rights. CPGR-Ex1/94/5, Rome.

FAO (1996) Global participation in the development of major components of the global system for the conservation and utilization of plant genetic resources, September.

Glowka, L. (1998) *A Guide to Designing Legal Frameworks to Determine Access to Genetic Resources.* IUCN, Gland, p. 4.

Glowka, L., Burhenne-Guilmin, F. and Synge, H. (1994) *A Guide to the Convention on Biological Diversity.* IUCN, Gland, p. 3.

Huaman, Z. (2001) Unethical distribution to Japan of Yacon held in trust by CIP. Circulated on the biodiv-conv listserver run by BIONET (http://www.bionet-us.org), 7 April.

IPBN (1995) *Indigenous People, Biodiversity, and Health.* COURTS Canada IPBN Factsheet, November.

McChesney (1996) Biological diversity, chemical diversity and the search for new pharmaceuticals. In: Balick, M., Elisabetsky, E. and Laird, S. (eds) *Medicinal Resources of the Tropical Forest: Biodiversity and Its Importance to Human Health.* University of Columbia Press, p. 12.

McConnell, F. (1996) *The Biodiversity Convention. A Negotiating History.* Kluwer, London.

McNeely, J.A. (2001) Biodiversity and agricultural development: the crucial institutional issues. In: Lee, D.R. and Barrett, C.B. (eds) *Tradeoffs or Synergies? Agricultural Intensification, Economic Development and the Environment.* CAB International, Wallingford, UK, pp. 399–404.

National Research Council (1989) *Lost Crops of the Incas: Little Known Plants of the Andes with Promise for Worldwide Cultivation.* National Academy Press, Washington, DC.

Proposal on the Protection of the Intellectual Property Rights of the Traditional Knowledge of Local and Indigenous Communities, WT/GC/W/362, 12 October 1999.

RAFI (2000) *In Search of Higher Ground. The Intellectual Property Challenge to Public Agricultural Research and Human Rights and 28 Alternative Initiatives.* The Occasional Paper Series, Vol. 6, No. 1, September.

RAFI (2001) Enola bean patent challenged. News Release, 5 January; www.rafi.org

RAFI Geno-Types (2000) Mexican bean biopiracy, 15 January.

Ronald, P. (2001) Quoted in WIPO/UNEP, *The Role of Intellectual Property Rights in the Sharing of Benefits Arising from the Use of Biological Resources and Associated Traditional Knowledge. Selected Case Studies.* WIPO, Geneva, p. 13.

Seiler (1998) Sui generis systems: obligations and options for developing countries. *Biotechnology and Development Monitor*, No. 34.

Swaminathan, M.S. and Hoon, V. (1994) *Methodologies for Recognizing the Role of Informed Innovation in the Conservation and Utilization of Plant Genetic Resources.* CRSARD Proceedings, Madras, no. 9.

US Patent No. 6,040,503; Patent Cooperation Treaty patent no.WO99/11115.

WIPO (2000) Matters concerning intellectual property genetic resources traditional knowledge and folklore. WIPO Doc. WO/GA/26/6, 25 August.

WTO (2000) Review of the Provisions of Article 27.3(B). Further views of the United States of America. WTO Doc. IP/C/W/209, 20 September.

Yusuf, A.A. (1995) International law and sustainable development: the Convention on Biological Diversity. In: Yusuf, A.A. (ed.) *African Yearbook of International Law*, Vol. 2. Kluwer, London, p. 109.

Zandstra (2001) Potato Center upholds letter and spirit of FAO Agreement. (http://www.bionet-us.org, 9 April.)

Chapter 5

Farmers' Rights and Intellectual Property Rights – Reconciling Conflicting Concepts

Daniel Alker and Franz Heidhues

University of Hohenheim, Institut 490A, D-70593, Stuttgart, Germany

Abstract

The chapter discusses the relation of Farmers' Rights and intellectual property rights (IPRs) on plant genetic resources (PGRs). It describes the nature and purpose of both concepts and the relevant national and international institutional frameworks. Farmers' Rights can be regarded as a counter-concept to IPRs, which advocates the interests of developing countries and their traditional farmers and tries to remunerate conservation and informal innovation efforts with regard to PGRs. It also aims at the conservation of PGRs. Possible conflicts between both concepts are depicted and options for reconciliation through the parallel implementation on the national and international level are shown. It is argued that a successful conclusion of the revision of the International Undertaking on Plant Genetic Resources (IUPGR) is crucial to reconciliation on the international level. On the national level, many developing countries do already enact plant variety protection legislations with various provisions for the implementation of Farmers' Rights but it is still not decided if these provisions are in accordance with their obligations under the World Trade Organization (WTO).

Introduction

Since the 1980s rapid developments in the field of agricultural biotechnology have resulted in a private sector-driven push to strengthen intellectual property rights (IPRs) on genetic resources worldwide. The culmination of this development has been the linkage of intellectual property issues to trade issues through the integration of the Agreement on Trade-related Aspects of Intellectual Property Rights (TRIPs) as one of the three constituting pillars in the World Trade Organization (WTO). All WTO member countries are now obliged to enact patent protection for plants or an 'effective *sui generis*[1] system' of plant variety protection (PVP) or a 'combination thereof'.[2] PVP legislation rewards the efforts of formal agricultural innovators, which is expected to lead to increased research and development (R&D) invest-

[1] *Sui generis* means unique or of its own kind. It is customarily used in law if a special circumstance is not covered by existing laws.
[2] See Annex 3: TRIPs Agreement Article 27.3(b).

ments in the seed sector and consequently to enhanced varieties. Yet, the effects of PVP on agricultural productivity, agrobiodiversity and food security in developing countries (DCs) are at present far from clear.

On the other hand, farmers see their traditional practices of replanting and exchanging seeds endangered through modern PVP. Thus Farmers' Rights were originally conceived as a counter-concept to IPRs and to mirror the concerns of the developing world about the effects of globally strengthened IPRs. Today they include various moral and functional aspects: they aim at balancing the potentially negative impacts of IPRs on traditional farmers in developing countries, at remunerating their past efforts to conserve and improve germplasm and they shall function as an instrument continuously to entice these efforts. Traditional farmers' conservation efforts have constantly provided the basic resources for modern plant breeding and thus have been, and will continue to be, essential for global productivity increases in agriculture. However, since traditional varieties were seen as a 'common heritage of mankind' (FAO Resolution 8/83, reaffirmed in FAO Resolution 4/89 and 5/89) and thus as a global public good, traditional farmers have not been able to share the benefits of their use.

The entry into force of the Convention on Biological Diversity (CBD) in 1993 has added new momentum to the discussion about Farmers' Rights and today they are at the threshold of implementation. On the international level they are likely to be incorporated in the revised International Undertaking on Plant Genetic Resources (IUPGR) and on the national level they form part of various new *sui generis* PVP systems in developing countries.

The paper discusses the purposes of IPRs and Farmers' Rights, their effects and their relation. It describes the current status of implementation and depicts possible options for reconciliation through parallel implementation on the national and international level.

IPRs

The concept of IPRs

IPRs are intended to prevent the commercial exploitation of intellectual goods, that is, ideas and inventions, without compensating their originators. Like other forms of property rights, IPRs grant their holders a defensive right, which allows them to exclude others from using the protected intellectual good. IPRs thus implicitly confer a monopoly right to their holders and thereby entice the production of new knowledge. Contrary to conventional property rights, IPRs are temporary rights. IPRs are grouped into copyrights (literary and artistic work) and industrial property (patents, plant breeders' rights, industrial designs, trademarks and geographic indications of source). As an instrument of economic policy, IPRs are used to direct R&D investments to knowledge-creating sectors. In the case of inventions, for which patent protection can be sought, the right holder is obliged publicly to disclose his work in return for the temporary monopoly right. In doing so, new knowledge enters the public domain and allows subsequent innovators to use this knowledge for new inventions, which in turn have to meet the criteria of protection.

From a static point of view, the dissemination of new knowledge at the marginal costs of transmitting this knowledge leads to a maximization of welfare, because knowledge is non-rival in nature. From a dynamic point of view, incentives for the creation of new knowledge have to be given by granting a temporary monopoly, because without the prospect of adequate returns, risky R&D investments that produce new knowledge will not be undertaken. The World Bank (1998, p. 33) concludes, 'IPRs are a compromise between preserving the incentive to create knowledge and the desirability of disseminating knowledge at little or no cost'.

IPRs are national in scope and extend only to the territory of the state in which the application was filed. However, various international treaties have sought to harmo-

Box 5.1. PGRFA and related classifications.

Plant Genetic Resources for Food and Agriculture (PGRFA) are the genetic material of food and agricultural plants of actual or potential value (FAO, 1997). They are arguably the most important of the earth's biological resources for humans since they are the foundation of all food production and key to further productivity increases, half of which are commonly attributed to genetic improvement (Koo and Wright, 1999). The value of crop germplasm is vastly increased by the rapid growth of the human population and the limited amount of new agricultural land. Advancements in biotechnology will help to improve PGRFA more effectively, which further increases their value. Future human welfare thus depends on improved crop conservation and breeding. Though they embody very different intentions and approaches, IPRs and Farmers' Rights are central institutions for these endeavours.

Various partly overlapping classifications of PGRFA are used subject to the goal of analysis:

- **In situ *and* ex situ *PGRFA.*** The material growing in farmers' fields and its wild and weed-like relatives is termed *in situ* PGRFA, the material stored in genebanks is referred to as *ex situ* PGRFA.
- ***Modern and traditional PGRFA.*** Modern PGRFA (MPGRFA) are the result of formal plant breeding activities whereas traditional PGRFA (TPGRFA) and their wild and weed-like relatives are the plants conserved and developed by the informal plant-breeding activities of traditional farmers. Landraces and traditional varieties are synonyms for TPGRFA, commercial varieties for MPGRFA. Two forms of IPRs can protect modern PGRFA: patents and plant breeders' rights (PBRs). Traditional PGRFA currently belong to the public domain. They are the subject matter of Farmers' Rights.
- ***Varieties vs. genetically coded information.*** Another subdivision of PGRFA is needed to determine the diversity of PGRFA and its economic values. This is complex, because the diversity of PGRFA cannot simply be derived from the sole number of plant varieties. Smale (1997, p. 1259) found that 'phenotypically similar populations of plant varieties may contain a very different set of genes while phenotypically distinct varieties may contain a very similar set of genes'. Likewise, Hoisington *et al.* (1999, p. 5942) state: 'Molecular dissection is much more powerful for determining the usefulness of a species than casual analysis at the morphological or physiological level. Useful alleles exist in both the related and unrelated species of all crop plants'. Therefore Virchow (1999, p. 19) proposes the term 'genetically coded information' (GCI) as the most suitable unit for economic and institutional analysis since GCI is the ultimate determinant of plant characteristics and the most precise unit of PGRFA diversity analysis. Due to the current costliness of technologies to identify GCI, varietal diversity is, however, still being used as the most practical indicator for PGRFA diversity.

nize IPR practices internationally and TRIPs has finally integrated most of these efforts by setting minimal protection criteria for WTO members. For formal innovations in relation to PGRFA, two forms of IPR are currently granted in most industrialized countries: patents and/or PBRs. Developing countries have only recently begun to enact PVP laws in order to comply with their obligations under TRIPs and some of them attempt to extend IPRs to landraces.

Trade and welfare effects of internationally strengthened IPR

IPRs affect international trade flows in several ways. A firm may be deterred to export its IPR-protected good into a foreign market, if potential copiers can diminish the profitability of the firm's activity in that market because of a weak IPR regime. Accordingly, a strengthening of a country's IPR regime would tend to increase imports, as foreign firms would face increasing net demand for their products, reflecting the displacement of copiers. On the other hand, a firm may choose to reduce its sales in a foreign market as a response to stronger IPR protection because of its greater market power in an imitation safe environment. These opposing market-expansion and market-power effects imply that the overall effect of IPR protection on bilateral trade flows is theoretically ambiguous (Maskus and Penubarti, 1995).

The implications of tighter IPRs on economic welfare are also highly complex and involve both static and dynamic effects. In a two-country model, from a static, partial-equilibrium point of view, the source country of the trade flow is likely to gain from tighter protection, because it can capture increased monopoly profits from the sale of its goods abroad. In contrast, the static effects on the welfare of the destination country are likely to be negative because increased market power by foreign title holders leads to deadweight losses (Deardorff, 1992). Thus, many small, innovation-consuming developing countries fear that increased patent protection will only lead to a rent transfer to industrialized, innovation-producing countries.

From a static, general-equilibrium point of view, tighter IPRs tend to be further detrimental to the destination country of the trade flow because the reallocation of production from the previously copying destination country to the source country worsens the terms of trade in favour of the source country. From a static welfare point of view, IPRs can be viewed as a rent-transfer mechanism, which deteriorates the international allocation of production. Most studies conclude that the destination country loses from tighter protection whereas the source country is usually better off (Deardorff, 1992).

From a dynamic point of view, the introduction of IPRs stimulates innovation in the source country and thus increases future trade flows, which is beneficial to both trading partners. Through IPRs, innovation-producing countries have an incentive to develop new technologies, which in their next generation are manufactured by follower countries. This mechanism thus leads to continued technological progress and economic growth and from a dynamic point of view is beneficial for both leaders and followers (Maskus and Penubarti, 1995). The international recognition of IPRs can be seen as a mechanism, which guarantees the functioning of dynamic competition between countries. Although benefits of a dynamic nature can be identified for both trading partners, on average, it is unlikely that these dynamic benefits can compensate for the static losses in the innovation-consuming developing countries with strengthened IPR systems and it is also unclear whether tighter IPRs improve world economic welfare via their impact on trade flows.

Trebilcock and Howse (1995, p. 251) found that '(a) country where innovation is not a major source of economic activity and growth is likely to choose, on balance, a less stringent intellectual property regime than would a country whose economy is highly dependent on innovations.' The scope and duration of protection required for under TRIPs is oriented towards the standards of the mostly innovation-producing industrialized countries and therefore hardly optimal for the mostly innovation-consuming developing countries.

Thus there is a strong case for leaving the system flexible, for requiring only minimal global standards and for allowing developing countries a learning-by-doing approach, but the obligations under TRIPs and the International Union for the Protection of New Varieties of Plants (UPOV) are restricting this freedom to choose for most developing countries. Political economy considerations may explain better than economic theory why innovation-consuming developing countries are opting in favour of strengthened IPRs (Primo Braga, 1996).

Effects of IPRs on investments in the improvement of PGRFA

Strong IPRs are expected to increase investment in the protected sector, but empirical evidence could not yet prove convincingly that this classical justification for IPR holds true for the modern seed sector as well (World Bank, 1998; Alston and Venner, 2000). Even if a correlation between the level of investment and the strength of IPRs can be found, causality is difficult to establish because it is questionable whether strong IPRs attract additional investments or whether powerful and knowledge intensive industries, such as the modern, highly integrated breeding companies, lobby and push for strong IPRs. The welfare effects are consequently equally difficult to determine. Other factors such as the political and economic stability of a country, the size and dynamics of the relevant market, the resource endowment and the physical and legal infrastructure often seem to have a much stronger influence on investment decisions than IPR (Maskus and Penubarti, 1995). Thus, it is still unanswered, whether strong IPRs for PGRFA actually entice investment in crop improvement or whether they are just an instrument of marketing, advocated and employed by powerful seed companies (Fowler, 1994; Alston and Venner, 2000).

Besides, in evaluating options for IPR protection in developing countries, it must be recognized that virtually no empirical analyses have been conducted on the welfare impact of IPR on food and agriculture, especially in developing countries (Blakeney *et al.*, 1999).

Institutional framework for IPR on PGRFA

The global framework: TRIPs

WTO members are required to provide for patents 'for all inventions, whether products or processes, in all fields of technology' (TRIPs Article 27.1). They may, however, exclude from patentability 'plants and animals other than micro-organisms' if they 'provide for the protection of plant varieties either by patents or by an effective *sui generis* system or by any combination thereof' (TRIPs Article 27.3(b)). The criteria for the patent protection of a plant variety under a TRIPs compliant law are: (i) novelty; (ii) non-obviousness; and (iii) usefulness. The duration of protection is a minimum of 20 years. Only few countries like the USA, Korea and Guatemala currently allow patent protection for plant varieties.

Under a *sui generis* PVP system, different criteria for and periods of protection can be established but the TRIPs agreement itself does not provide any definitions thereto. Since no dispute over a *sui generis* system has been brought to the dispute-settlement body of the WTO yet, it remains unclear what the protection criteria, scope and duration of protection of an effective *sui generis* system would be. Industrialized countries deem a UPOV-style protection system (see below) as effective and advocate its implementation for developing countries (CPGR, 1994).

Developing countries were required to implement the TRIPs provisions by 1 January 2000, least-developed countries (LDC) have to comply by 1 January 2006. However, Article 27.3(b) (see Annex 5.3) has a 'built-in review', which required a revision in 1999 – prior to the implementation. Due to the current political stalemate at the WTO, this revision has not taken place yet and is even unlikely to conclude in 2001. Developing countries therefore argue that they are only obliged to implement PVP laws after the review – which could change the provisions of the Article substantially – will have taken place (WTO, 1999a).

The national framework: UPOV

PBRs are IPRs specifically designed to protect modern varieties. Of primary importance on the international level is the PBR system established by UPOV. UPOV seeks to harmonize legislation and to simplify the application procedure for plant breeders by requiring all member states to accept the test results produced by others as the basis for their decision on the granting of a PBR. The goals of these harmonization and simplification efforts are to increase the trade in modern varieties among member states and to encourage investments in the modern commercial seed sector. Different acts of this convention are in force (1972, 1978 and 1991), but new members may only accede to the 1991 act.[3] As of April

[3] The deadline for joining the UPOV 1978 act was 24 April 1999.

2001, 47 states are members of the convention, 21 of which are developing countries.

The protection criteria that have to be met under UPOV 1991 for the protection of a variety are: (i) novelty; (ii) distinctness; (iii) uniformity; and (iv) stability. UPOV 1991 has strengthened the PBR as compared to the 1978 version. It now requires farmers to pay a license fee for planting back their own harvested material and does not allow them to exchange it with other farmers. An exemption provided for under Article 15(2) of UPOV 1991 states that 'each Contracting Party may, within reasonable limits and subject to the safeguarding of the legitimate interests of the breeder, restrict the breeders' right in relation to any variety in order to permit farmers to use for propagating purposes, on their own holdings, the product of the harvest which they have obtained by planting, on their own holdings, the protected variety.'

This potentially allows the exemption of small-scale and non-commercial farmers from the obligation to pay royalties, as shown, e.g., by the UPOV 1991-compliant PVP legislation of the EU (Council Regulation EC No. 2100/94). Besides, Article 15(1) of UPOV 1991 also provides that 'the breeder's right shall not extend to acts done privately and for non-commercial purposes', which can be interpreted as allowing subsistence and resource poor farmers to freely replant and exchange their harvest. Nevertheless, the 1991 act is restricting the so-called 'Farmers' Privilege' for commercial farmers, which comprises the right to plant back the harvested seeds and the right to exchange propagating material 'over the fence' from farmer to farmer.

RECENT DEVELOPMENTS WITH REGARD TO UPOV IN DEVELOPING COUNTRIES. The UPOV system is tailored to the needs of the commercial seed sector and, although commercial modern varieties are not expected to provide more than 15% of the total seed requirements in developing countries (Srivastava and Jaffe, 1993), many developing countries also regard an UPOV-style protection system as appropriate for their needs and have enacted UPOV-compliant PVP legislation. Twenty developing countries have already joined UPOV since 1995 and a further 63 countries are expected to follow or to enact UPOV-based *sui generis* PVP laws in the near future, which would raise the number of nations that have UPOV-style PVP laws in place to 110.[4] This is striking, especially since the newly acceding developing country have to comply with the strict UPOV 1991 convention, which limits the Farmers' Privilege and therefore the scope of action for the implementation of Farmers' Rights. These countries obviously ascribe a high value to UPOV's advantages of a simplified PBR application procedure, technical and legal cooperation and the expectation of investments in the national modern seed sector.

Prior to TRIPs, the majority of the developing country did not have any PVP legislation in place at all, because most new varieties originated from public breeding efforts and were made available as public goods. It remains to be seen if new PVP systems will attract private investments, improve the availability of enhanced varieties for traditional and modern farmers in developing country and facilitate public–private partnership arrangements or if they will severely interfere with farmers' traditional practices and serve as an excuse to further cut back public spending for agricultural R&D.

PBRS FOR LANDRACES: FARMERS' RIGHTS AS IRPS? Only modern varieties can meet the UPOV protection criteria whereas landraces fail in all of them, except for distinctness. Landraces are usually not new, but have been in use for generations. They are not uniform but show phenotypic and genotypic variability, which explains why they are not stable over time but evolve constantly. All of these characteristics are inherent of landraces and are valuable because they indicate a rich genepool,

[4] Personal communication by the UPOV secretariat, 30 November 2001.

which deserves conservation and which enables them to adapt successfully to changing environmental conditions.

An IPR protection for landraces under PVP legislation is sometimes discussed as a possible measure to advance Farmers' Rights. It is argued that traditional farmers could appropriate part of the benefits of the utilization of their landraces in cultivation and breeding, if the users had to pay a royalty. The TRIPs agreement possibly also allows parties such as indigenous and other rural communities or farmers associations to apply for intellectual property protection if the national legal system and practice does include this kind of rights.[5] Some countries already grant community IPR on biodiversity under their *sui generis* PVP systems (see Annex 5.1).

In order to facilitate such a protection of landraces, the conventional UPOV protection criteria would need to be broadened. The criteria of (i) distinctness; (ii) usefulness; and (iii) identifiability have been proposed, with usefulness replacing novelty and identifiability replacing uniformity and stability (IPGRI, 1999). Another approach is a 'Dual System', which would establish a protection system for landraces parallel to the UPOV-style PVP system (Correa, 2000). The protection requirements for landraces would be less strict and, accordingly, the scope of the rights conferred to the traditional farmers would be less extensive.

Several problems arise, however, with respect to the advancement of Farmers' Rights through the protection of traditional PGRFA under a conventional IPR system: (i) due to the nature of the innovation process for landraces, it is extremely complex to identify a single farming community as the informal innovator and therefore as the potential right holder (FAO, 1994b); (ii) other traditional farming communities could be required to pay royalties if they use a landrace on which another community holds the right. The resulting financial transfer from resource-poor farmers to resource-poor farmers is certainly not advancing Farmers' Rights; (iii) although landraces constitute a great potential value, their current use in modern plant breeding is quite marginal (Virchow, 1999). Thus, significant royalty payments cannot be expected at all.

It seems unlikely that the adjustment of conventional PVP systems to accommodate traditional PGRFA could confer significant benefits to traditional farmers and advance Farmers' Rights. Since it is not possible to reconcile IPRs and Farmers' Rights by merging them, non-IPR mechanisms for Farmers' Rights, which allow their parallel implementation with IPRs, have to be found. Moreover, it would seem illogical to make Farmers' Rights part of the IPR system because it is that very system that has created some of the problems that the concept of Farmers' Rights aims to solve.

Effects of PVP in developing countries

The importance of conventional PVP laws for food security in developing countries is disputed. UPOV argues that new plant varieties are an essential tool for the sustainable development of agriculture and the achievement of a country's food security (Heitz, 1998). This assertion is reaffirmed in an official communication sent to the governments of the 'Organisation Africaine de la Propriété Intellectuelle' (OAPI) in June 1999 for the revision of their common intellectual property law for the adoption of UPOV 1991 (UPOV, 1999). The principal advantages of introducing plant variety protection in Africa are listed as follows:

1. *Food security* by the increase in quantity, quality and diversity of foodstuffs.
2. *Sustainable agriculture*, e.g. by a more efficient use of available resources and inputs or by the use of pest- and disease-resistant varieties.

[5] Article 1(1) of the TRIPs agreement states: 'Members shall be free to determine the appropriate method of implementing the provisions of this Agreement within their own legal system and practice.' According to Girsberger (1999), the interpretation of this Article could be 'permissive'.

3. *Protection of the environment and of biodiversity*, e.g. by reducing pressure on natural ecosystems through better productivity of cultivated lands, increase in species- and varietal-diversity and increase in the interest in conservation and use of genetic resources for food and agriculture

Opponents see in the UPOV 1991 act a potential threat to food security (GRAIN, 1999a) because it restricts the Farmers' Privilege by making the free planting back of seeds contingent upon an explicit exception, because it entirely prohibits the exchange of seed 'over the fence' (UPOV 1991, Article 15.2) and because it requires that old commercial varieties, which do not meet the protection criteria must be taken from the market, whether they are useful to farmers or not. Besides, opponents regard UPOV 1991 as a threat to agricultural biodiversity since genetic extinction has, until today mainly occurred in the form of the replacement of traditional varieties through genetically uniform modern varieties, the use of which is promoted by UPOV-compliant laws (FAO, 1997).

UPOV itself considers a size of the agricultural sector that justifies investments in plant breeding as the essential basis for a successful development of a private seed sector (Heitz, 1998). It therefore remains doubtful, whether a PVP system can attract investments for food crops in countries where the potential market is small in size or purchasing power, where the physical infrastructure to distribute the seeds is inadequate and where the legal and institutional infrastructure to enforce PBR is weak. In these countries, UPOV-style PVP is likely to attract investment to the seed industry for the export-oriented and capital-intensive branches of the agricultural sector. This is valuable in terms of elevated foreign direct investment (FDI) and export value, but may have ambiguous effects in terms of food security. A WTO case study for Kenya and Argentina underlines this: it judges UPOV-style PVP as conducive to the development of the Argentine agricultural sector during the last 25 years but found for Kenya that 'the implementation of PBR resulted in some hardships to small scale farmers who depended on old varieties' (WTO, 2000a,b). Moreover, UPOV 1991 has to be enacted in a way that the cost-free planting back of protected varieties is ensured for traditional farmers, if the act is not to impact negatively on food security.

Box 5.2. PVP in developing countries – some evidence from Africa.

The following experiences stem from the only three African countries that have PVP in place for some time (Cullet, 2001):

- Kenya adopted its PVP law in 1975. By May 1999, of the 140 PVP applications approved, only one was on a food crop: a variety of green bean, which Kenya predominantly exports to Europe. More than 90% of the PVP certificates were for flowers, while the rest went to export crops such as coffee, sugarcane and to barley for the beer industry. Though PVP seems to have benefited the capital-intensive sectors of the Kenyan agriculture, it is evident that the law did not have any effect on food security; 90% of the applications came from foreign breeders.
- In Zimbabwe, the Plant Breeders' Rights act was enacted in 1973. As of 1999, over 70% of all applications were on cash crops: ornamentals, fibres, oilseeds and tobacco. 30% of the applications were on food crops. Over two-thirds of the applications came from foreign breeders.
- In South Africa, the PVP system became operational in 1977. As of the end of 1998, a total of 1435 PVP grants had been made. Half of them were for cash crops.

In all three countries, it seems that the capital-intensive and export-oriented agriculture is the main beneficiary of PVP and the chief of FAO's Seed and Plant Genetic Resource Service acknowledges that 'these (efforts to strengthen the seed industry in SSA) have had relatively little impact on the majority of resource-poor farmers' (Menini, 1998).

Given these assessments, public sector breeding will remain essential for the development of seeds geared to the needs of resource-poor farmers and to distribute them as a public good. The public breeding sector should not use UPOV laws as a justification for its retreat but explore the potential for ventures with the private sector to develop better plants. Countries with a large number of resource-poor farmers and an underdeveloped legal and physical infrastructure have good reasons not to enact UPOV-compliant PVP laws but to deliberately draft *sui generis* PVP laws, which can be better suited to the needs of their farmers and seed sector and facilitate the advancement of Farmers' Rights.

Farmers' Rights

The concept of Farmers' Rights

Farmers' Rights are not conventional rights, such as property rights or IPRs. They have not been conceived as such and the content of IPRs, that is, the protected subject matter, the protection criteria, the right holders and the rights and obligations of these holders, can only partially capture the nature and purpose of Farmers' Rights. Farmers' Rights are rather a political concept, though by no means a homogeneous or a consensual one. It includes certain conventional rights but its overall nature and purpose is more comprehensive. Some parts of this concept – the issues of benefit-sharing, participation and technology transfer – can rather be promoted and advanced, another can be legally protected: the Farmers' Privilege as the right of traditional farmers freely to replant and exchange farm-saved seeds.

For almost two decades of discussions on issues relating to plant genetic resources (PGRs) within the FAO, this concept has been the basis for recognition and remuneration of important contributions that traditional farmers have made and continue to make for the conservation and development of PGRs. Though it has been interpreted by many as nothing more than a vague moral appreciation of these efforts, various versions of Farmers' Rights are presently at the threshold of being implemented at the national and international level: they are recognized in various international agreements and integrated in many newly drafted PVP laws of developing countries. Most importantly, they are recognized and made operable in a new major, legally binding international agreement, the revised International Undertaking on Plant Genetic Resources (IUPGR), which is expected to enter into force in 2002.

Thus the transition of Farmers' Rights from a political basis of discussion to an operable political and legal concept is currently in progress and the various forms of Farmers' Rights will have to prove their practicability and success in the years to come. Farmers' Rights complement existing forms of IPRs. They are not, however, intended to compete with or replace, existing IPRs (Girsberger, 1999).

Origin and evolution of Farmers' Rights

Concerns of the developing world and their advocates have been growing that strengthened IPRs in agriculture are harmful to small-scale farmers and accelerate the erosion of agricultural biodiversity through the replacement of genetically diverse landraces by uniform modern varieties. Moreover, the perceived inequality in the distribution of benefits between farmers as suppliers of TPGRFA and the producers of commercial varieties that ultimately rely on such germplasm, have resulted in the quest for a counter-concept to IPRs. The term 'Farmers' Rights' came up in the early 1980s (Fowler, 1994) and was featured in the debates held within FAO on the inequality in the distribution of the benefits of PGRFA use: while a commercial variety generates returns to the breeder on the basis of PBR, no parallel appropriation mechanism to act as an incentive for the providers of germplasm to continue to maintain and make available these resources existed (Esquinas-Alcazár, 1998).

The debates at FAO finally led to a negotiated compromise: the simultaneous and parallel international recognition of PBRs and Farmers' Rights. This recognition is embodied in the parallel FAO Conference Resolutions 4/89 (Recognition of PBR) and 5/89 (Recognition of Farmers' Rights), which were unanimously adopted by more than 160 countries in 1989 and annexed to the current, legally non-binding IUPGR. Until the revised IUPGR will enter into force as a legally binding instrument and provide a new, enforceable definition of Farmers' Rights, the definition in FAO resolution 5/89 remains the only, yet unenforceable, definition in an international agreement until today (see Annex 5.2). Its central features are the international community's appreciation of farmers' past, present and future contributions to the conservation and provision of PGRFA and the acknowledgment of the need for conservation and benefit-sharing of PGRFA. Although FAO Resolution 3/91 further elaborates on the financial and institutional aspects of Farmers' Rights, the implementation, especially the provision of the financial means to realize Farmers' Rights, has proven extremely difficult and has still not been accomplished.

In the 1990s, the political and economic environment has changed significantly. The breeding sector and the biotechnology industry have undergone an unprecedented process of concentration. Meanwhile the scientific evidence about the loss of agrobiodiversity has grown (FAO, 1997) and the commercialization of the first genetically modified plants sparked a public debate over corporate control over genetic resources, putting genetic resource policies under greater public scrutiny. The entry into force of the legally binding CBD in December 1993 reflects this development. It has significantly changed the global legal status of genetic resources by specifying that they be under the sovereignty of the government of the state in which they developed their distinctive properties. This provision has considerably strengthened the bargaining position of gene-rich developing countries on the emerging markets for genetic resources. It has also posed new questions to the global exchange of *in situ* and *ex situ* PGRFA, which previously had been regarded as the common heritage of mankind and the exchange of which was regulated in an open access regime under the IUPGR. Furthermore, the CBD has introduced the concept of benefit sharing for the use of genetic resources and established rules in relation to the access to genetic resources.

As a consequence of these changes, the IUPGR went under revision in 1994, in order to bring it in line with the CBD. This revision has since then been the forum for the debate over Farmers' Rights. It is expected to conclude in 2001, resulting in a new major, legally binding instrument with the objectives of 'the conservation and sustainable use of plant genetic resources for food and agriculture and the fair and equitable sharing of the benefits arising out of their use, in harmony with the Convention on Biological Diversity' (Article 1). The draft Article 9 (see Annex 5.6) of the IUPGR explicitly deals with Farmers' Rights. It states that the responsibility for realizing Farmers' Rights rests with national governments, which is seemingly a clear departure from the current vesting of the rights in the international community. However, the draft IUPGR further defines the right to 'equitably participate in sharing benefits arising from the utilization of PGRFA' as a 'measure to protect and promote Farmers' Rights. Since some of the main provisions of the draft IUPGR (Article 8, 11, 13, 16) elaborate on the concrete realization of benefit-sharing at the international level, a successful IUPGR can be interpreted as protecting and promoting Farmers' Rights also at the international level. The legal status of the revised IUPGR is likely to be a protocol to the CBD or a stand-alone international agreement.

Box 5.3. Recognition of Farmers' Rights in other international agreements.

Besides the original recognition in the IUPGR, Farmers' Rights have been recognized in the following international instruments (Correa, 2000):

- Chapter 14.60(a) of the United Nations Conference on Environment and Development (UNCED) Agenda 21 states that the appropriate UN agencies and regional organizations should 'strengthen the Global System on the Conservation and Sustainable Use of PGRFA by ... taking further steps to realize Farmers' Rights'.
- The Global Plan of Action for the Conservation and Sustainable Utilization of Plant Genetic Resources for Food and Agriculture (GPA) included the realization of Farmers' Rights at the national, regional and international level, as one of the long-term objectives of the Plan, in the context of *in situ* conservation (para. 32).
- Resolution 3 of the Nairobi Conference for the Adoption of an Agreed Text of the UN Convention on Biological Diversity identified the realization of Farmers' Rights as one of the 'outstanding issues' for further negotiation.
- A June 1999 study by the UN Economic and Social Council (ECOSOC) on the Right to Food, submitted to the Commission on Human Rights, urged that Farmers' Rights be promoted as part of the 'Right to Food', especially since 'our future food supply and its sustainability may depend on such rights being established on a firm footing' (UN Commission on Human Rights, 1999).

Nature and purpose of Farmers' Rights

The nature and purpose of Farmers' Rights is usually derived from three lines of reasoning, which are not always clear-cut but often interdependent (Girsberger, 1999; Correa, 2000):

1. *Equity* reflected in the 'right to equitably participate in sharing the benefits arising from the utilization of PGRFA' (Draft Article 9 of the revised IUPGR).
2. *Protection* of traditional farmers from potentially restrictive effects of IPRs by ensuring the Farmers' Privilege to save, exchange and sell seeds.
3. *Conservation* of traditional PGRFA.

The protection of the Farmers' Privilege is the part of Farmers' Rights that can be protected as a right in the conventional sense, whereas equity and conservation goals can rather be promoted or advanced.

While conservation is a functional objective of Farmers' Rights to the benefit of all humans, equity considerations are based on moral considerations, which largely derive their legitimation from traditional farmers' past contributions in conserving and making available PGRFA. Equity and conservation goals are highly interlinked: the implementation of benefit-sharing mechanisms to advance equity goals can be achieved in a way that conservation goals are also reached, e.g. through planned *in situ* conservation.

The protection of traditional farmers from potentially harmful effects of IPR has both moral and functional aspects.

Further elements of Farmers' Rights as defined by the draft Article 9 of the IUPGR are the protection of traditional knowledge and the right to participate in decisions in relation to PGRFA.

Equity

Equity can be defined as consideration of fairness, reasonableness and good faith and is as such used in international law (Girsberger, 1999). Equity considerations are mirrored in the IUPGR's recognition of farmers' past contributions in the conservation and development of PGRFA as a central legitimation of Farmers' Rights. The question of equity has also gained strong momentum in the debate over genetic resources and Farmers' Rights with the coming into force of the CBD in 1993, which for the first time introduced the concept of 'benefit-sharing' (Article 8(j)) in a legally binding international instrument as a means of promoting equity. Equity is also referred to in other international instruments, *inter alia* in the Preamble of the Rio Declaration (UNCED), in the Agenda 21 (Chapter 15.5e) and in the CBD itself (Article 1). However, since none of these instruments defines the term equity and all make its implementation contingent upon 'necessity'[6] and 'appropriateness',[7] Girsberger (1999) concludes that states are not required to take specific legal action. Nevertheless, equity can serve as a moral basis for the realization of Farmers' Rights.

Another dimension of equity is the question of intergenerational equity, which focuses on the relation of present and future generations in relation to the use of the world's natural and cultural resources. In the context of PGRFA, intergenerational equity can be achieved through conservation and Farmers' Rights can be employed as an instrument to facilitate this.

Protection of the Farmers' Privilege

Protecting traditional farmers from potentially restrictive effects of IPR by ensuring the Farmers' Privilege is another element of Farmers' Rights. Its moral basis is the consideration that any form of IPR imposed restriction on resource-poor farmers would mean a further hardship and a danger to food security. Its functional aspect is that *in situ* conservation and informal breeding is only feasible if the Farmers' Privilege is protected, because the swapping of PGRFA in between farming communities is essential for its *in situ* conservation and development. A granting of IPR for landraces possibly also conflicts with this goal (see The national framework, above).

The right of farmers to save, use, exchange and sell their seeds is a part of the proposed version of Farmers' Rights in the revised IUPGR (see Annex 5.6). This comprehensive protection of farmers' practices exceeds the Farmers' Privilege because it also extends to the right to sell propagating material. Yet, this extended Farmers' Privilege has also already been incorporated in various national PVP laws. Although this principally conflicts with UPOV 1991, exemptions may be granted under such laws to ensure this extended Farmers' Privilege only for resource poor farmers and thus to make them compatible with Farmers' Rights.

Conservation of PGRFA, its economics and institutions

Conservation of PGRFA is the most consensual legitimation of Farmers' Rights, since conservation activities benefit all humans and their implementation can be designed in a manner that the development of traditional farmers as the key actors of *in situ* conservation is furthered.

A rapid loss of PGRFA and the consequent need to conserve it by means of complementary *in situ* and *ex situ* conservation strategies are widely acknowledged (Brush, 1994; FAO, 1997). This loss is to a large extent caused by the replacement of traditional by modern varieties. The discussions circle around the extent of conservation in general, the emphasis on each strategy and the methods for *in situ* conservation.

Traditional farmers are the actors of *in situ* conservation. Their past, present and future efforts in *in situ* conservation are recognized through Farmers' Rights but the question today is: how they can be encouraged to conserve the global socially optimal amount of PGRFA and how Farmers' Rights can be implemented to achieve this.

ECONOMIC VALUATION OF PGRFA. Traditional PGRFA have to be economically valued in order to determine the global socially optimal amount of conservation and to compensate adequately traditional farmers as the suppliers. Yet, no market mechanism exists today to accomplish this and it is unlikely that this will change in the foreseeable future because the transaction costs of a market solution in the form of institutional and informational hindrances are

[6] Article 15.7 of the CBD.
[7] Chapter 15.5e of Agenda, Article 8(j) of the CBD.

very high and so are the opportunity costs: the global benefits of a multilateral system of open or 'facilitated' access and benefit-sharing (MUSE). Thus the economic value of PGRFA can only be estimated.

Since PGRFA loss is irreversible, losing PGRFA always implies losing future options. On the other hand, the conservation of PGRFA requires resources. Consequently, a rational decision about the right amount of conservation needs to analyse the costs and benefits of conservation as far as possible (Evenson *et al.*, 1998). The costs of *in situ* and *ex situ* conservation can be quantified quite reliably (Virchow, 1999), but the estimation of the benefits of PGRFA conservation involves considerable uncertainty:

The benefits of PGRFA conservation can be estimated using the total economic value (TEV), which is composed of direct and non-direct use values (Virchow, 1999). The direct use value of PGRFA diversity is a static value. It can be determined with some certainty by quantifying its contribution to crop improvements and its current insurance function against yield fluctuations and unforeseeable events (breeding and insurance value). Evenson *et al.* (1998) introduced techniques to estimate the breeding value.

The estimation of PGRFA's non-direct use value, which comprises the existence and the heritable value, is a dynamic task, which is severely hindered by incomplete information about the future: (i) the existence value is the intrinsic value of life and as such subject to changing ethical assessments; (ii) the heritable value is the value of the known and unknown GCI for future utilization. In order to estimate this heritable value, presumptions on future technologies to use the presently known and unknown GCI as well as on future environmental and market conditions, in which GCI could be of value, have to be made. This is highly speculative. Since the heritable value of PGRFA constitutes a large share of the TEV, the estimation of the latter is thus severely handicapped.

CURRENT OBSTACLES TO A MARKET SOLUTION FOR PGRFA EXCHANGE. The demand for traditional PGRFA is quite marginal today and it is not expected to increase significantly in the future, because conventional breeding increasingly focuses on crosses among elite materials from the breeders' own collections and advanced lines developed in public institutions (Wright, 1998). Therefore it would be unrealistic to think that substantial value may be derived from gene flows of landraces held in *in situ* conditions (Gollin, 1998). None the less, with the advancement of biotechnology, traditional varieties can be screened more efficiently for agronomically interesting traits, which is likely to increase their value for agronomic improvement, especially in the regions of their predominant origin, where this improvement is most needed. Today, however, any economic measure directly linked to such gene flows, which reflect the current demand, would grossly underestimate the global values generated by the conservation of traditional varieties over time.

On the supply side, traditional PGRFA is provided as a positive externality of the low-input farming systems of traditional farmers. Until today, no mechanism has been established to enable farmers to appropriate a part of these benefits. This and the public good characteristics of PGRFA accelerate the loss and cause a potential future undersupply of PGRFA. PGRFA have been regarded as a 'common heritage of mankind', which legitimates their status as a global public good. A public good is characterized as follows:

- No legal or technical possibility to exclude others from the utilization of the good. Hence the provision of the good cannot be made contingent on payments of the users.
- No use-rivalry, which means that the use of the good by one individual does not affect the use of the good by any other individual. Due to the characteristics of hereditary information, PGRFA can be reproduced infinitely without depleting the genetic substance.

PGRFA also show the characteristics of an environmental good such as intergenerational existence and irreversibility of extinction.

All these characteristics prevent an efficient allocation of PGRFA by the market mechanism because they enable today's users to 'free ride' on PGRFA at the expense of today's suppliers and of future generations: users benefit from PGRFA without having to pay the costs of its supply. Thus, as with all public goods, a collective decision about the right amount of supply is required. The benefits of PGRFA are of concern to all humans and this might explain why it is particularly difficult to come to a collective decision over the desired amount of conservation. Until today, conservation is conducted almost exclusively *ex situ* by more or less coordinated efforts in national and international genebanks (Virchow, 1999). Since virtually no planned activities for *in situ* conservation exist, it is presently only carried out as a positive external effect of low-input farming.

INTERNALIZATION OR COMPENSATION? FINDING THE RIGHT MECHANISM FOR THE OPTIMAL LONG TERM SUPPLY OF TPGRFA AND THE PROMOTION OF FARMERS' RIGHTS. Basically two options exist to prevent the further erosion of traditional *in situ* PGRFA and to facilitate its conservation and development: a collective political decision over the right level of conservation by compensating traditional farmers as the suppliers of TPGRFA ('compensation solution') or the creation of a market mechanism that enables traditional farmers to directly appropriate the benefits derived from its use ('internalization solution').

Some argue that the lack of market institutions is the cause for the decline of PGRFA and argue for an internalization solution (Virchow, 1999). The rationale behind this argument is that if farmers can appropriate the external benefits they create through TPGRFA conservation, they will have an incentive to keep traditional varieties instead of replacing them by modern ones. Others regard the open access to PGRFA and its status as a global public good combined with a politically defined level of compensation for farmers as essential for the conservation and ongoing development of traditional PGRFA (Brush, 1992). The opportunities of and hindrances to both approaches will be shortly outlined below.

Internalization: transforming the public good traditional PGRFA into a private good. A public good can be transformed into a private good if one or both of the above-mentioned characteristics of public goods are changed. While the characteristic of 'no use-rivalry', which is intrinsic of the hereditary information of TPGRFA, cannot be altered,[8] a legal option to exclude others from the use of PGRFA has been opened through FAO Resolution 3/91 and the CBD. Both acknowledge the sovereignty of states over their genetic resources. TPGRFA could now theoretically be turned into a private good, with states, communities or individuals as the holders of the property rights on these resources.

In theory, a complete internalization of the external benefits of conservation and production of TPGRFA would lead to a global socially optimal amount of PGRFA supply and would eliminate the current equity distortions which affect the suppliers of TPGRFA. The obstacles that a market solution with such an internalization encounters are to a great extent owed to the environmental good characteristics of PGRFA (Hampicke, 1991) and the difficulties to assign a total economic value to an environmental good have been described.

[8] There are two exceptions in which the genetic information in a plant loses its agronomic value when it is reproduced: (i) hybrid seeds and (ii) genetically modified organisms (GMOs) with genetic use restriction technology (GURT). The former lose general agronomic value when being replanted, whereas GURTs are genetic technologies that either render the harvested seeds sterile ('varietal GURTs', popularly dubbed 'terminator technology') or that 'turn off' certain agronomically valuable traits in a plant when it is replanted ('trait-related GURTs'). As of April 2001, GMOs that contain GURTs were not yet on the market and several large seed companies had agreed not to commercialize seed-sterilizing GURTs. However, GURTs could be a powerful tool for companies to appropriate value in environments without enforceable IPRs.

A large share of the value of PGRFA lies in the potential use of its largely unknown genetic information (heritable value), which is obviously of no direct use to today's individual demander. The private market demand will thus not reflect the intertemporal global socially optimal demand. Conventional market mechanisms are not able to incorporate intergenerational aspects of PGRFA conservation and the irreversibility of extinction. Therefore, the state or the global community has to appear as a demander and the amount of its demand should be guided by the estimation of the TEV. The level of public demand is necessarily a political decision, and the estimation of TEV of different environmental goods or of different ways to conserve PGRFA can help to direct public financial resources to the most efficient allocation.

Besides these difficulties to internalize the external benefits of PGRFA conservation, a further fundamental problem in a market solution is to determine the property right holder on PGRFA. While the CBD confers states the right over their genetic resources, indigenous communities and individual farmers have been brought up in the discussion about Farmers' Rights as potential right holders as well (Girsberger, 1999). Since an intrinsic feature of landraces is their development in a communal and intertemporal effort, it seems inappropriate to confer certain individuals the property right on a specific landrace. Moreover, the market value of GCI cannot be determined *a priori* but only be observed *a posteriori* as a result of their performance on the market (Virchow, 1999). Due to the intergenerative structure of GCI benefits, the accruement of the total benefits may often exceed the life span of the individual farmers, which speaks in favour of communities or states as the potential property right holders.

It seems appropriate to try to identify certain communities as the 'inventors' of landraces, or at least as the place where landraces have developed their distinctive characteristics. However, it could be shown that the high interdependence of PGRFA makes the pedigree of landraces almost impossible to track (Gollin, 1998). A single landrace has typically been developed over centuries in various communities and often in various countries so that granting a specific community the ownership right on a landrace would be highly arbitrary. Defining nation states as the owners of PGRFA decreases this problem to a certain extent, but states have also been highly interdependent on PGRFA and it is by no means easy to identify the territories in which PGRFA have developed their distinctive properties, as demanded by the CBD. The global interdependence of PGRFA is much higher than for other genetic resources because plants of agricultural value have been traded, exchanged and bred globally for centuries.

Although states seem to be the most apt property right holders in a market solution, they may not ensure that the benefits of PGRFA are passed on to the supplying farming communities. In many countries, the farmers who are involved in the conservation of particularly diverse PGRFA are economically and technologically isolated and members of marginalized ethnic minorities. The national government is unlikely to be a strong advocate of these groups. Examples include the Kurds of Southwest Asia (wheat), the Quechua speaking indigenous groups in Peru (potatoes), Mayans in Mexico (maize), the Naga of India, the Ifugao of the Phillipines and the Karen of Thailand in the case of rice (Brush, 1992).

If states use their newly awarded market power on the emerging markets for genetic resources – the market for PGRFA potentially being one of them – for rent-seeking activities, neither conservation nor equity goals will be promoted.

Internalization and the advancement of Farmers' Rights. Leaving the allocation of PGRFA solely up to the market will neither lead to a global socially optimal amount of conservation and thus supply of PGRFA nor will it allow traditional farmers equitably to share the benefits of its use because these benefits are to a great extent intertemporal (Virchow, 1999). In an internalization solution, states are best qualified as potential property right holders and to advance Farmers' Rights, but only if they act in the interest of traditional farmers and only if they complement the market demand with an additional demand that takes into account the intertemporal benefits of PGRFA. Benefit sharing as an element of equity in Farmers' Rights includes the participation of traditional farmers in the benefits derived from the R&D activities in relation to PGRFA, such as the engineering of genetically modified plants to their needs as well as other forms of technology transfer, information transfer and capacity building. This is a development task, which is largely political, and the market alone will not be capable of achieving this. Additionally, Farmers' Rights have a historical dimension since they partly arise from farmers' past efforts to conserve and make available PGRFA (see Annexes 5.2 and 5.6). A market solution cannot accomplish a compensation of these past efforts.

In summary, a market solution, which assigns property rights on traditional PGRFA to states, communities or farmers, is presently not able to contribute significantly to the objectives of Farmers' Rights. In the future, this situation could change, however, because the advancement of biotechnology will allow identifying, utilizing and assessing the potential value of the GCI of landraces more efficiently, thereby lowering the transaction costs in a potential market of PGRFA. Enhanced information combined with enforceable property rights are prerequisites for a functioning of the market mechanism, which would enable the suppliers to internalize a larger share of the currently external benefits of traditional PGRFA. Likewise, the incentives to conserve traditional PGRFA would be improved. Conservation and equity as two main purposes of Farmers' Rights could therefore in the future possibly be achieved by a market solution with enforceable property rights on traditional PGRFA. Today, however, extremely high transaction costs make it necessary to find a political solution to the tasks of conservation and equity. Farmers' Rights as envisioned under the IUPGR are such a political solution and can therefore be regarded as a temporary political instrument against market failure for PGRFA exchange and conservation (see below).

Compensation: establishing a multilateral system of access and benefit sharing. The alternative to the market solution for the exchange and conservation of PGRFA is the compensation solution. A possible institutional framework for this is a multilateral system of access, exchange and benefit sharing of PGRFA (MUSE), which is currently practised under the current IUPGR, although due to the voluntary nature of this agreement, enforceable benefit sharing provisions could not be implemented yet. A compensation approach tries to accomplish a politically defined level of conservation of PGRFA and a politically defined level of equity through benefit-sharing.

Unlike a market solution, a compensation solution does not assign property rights on PGRFA but leaves them in the public domain by not restricting the access. The rationale for this before the background of the high interdependence of PGRFA is that each member of such an open access regime gains access to more genetic resources then he himself contributes and is therefore a net beneficiary (Crucible II Group, 2000). Transaction costs are greatly reduced in comparison to a market system because the informational deficits, which prevent the finding of an adequate price, do not matter. However, if no price is paid for the use of PGRFA, no incentives exist for their conservation. Thus conservation activities have to be politically devised. Although principally all members gain from such an open access regime, technology-rich countries usually have a stronger breeding sector than others and will therefore demand considerably more PGRFA and benefit more from the open access.

Yet, PGRFA and technology-rich countries are especially suspicious towards a free-access regime since they give away their resources for free without being able to derive much benefit from the access to the resources and to the technology of other countries. They are highly independent, because they dispose over the technological capacities to conserve and improve PGRFA for their needs within their borders and they could reap additional benefits by selling these resources on a potential market. Thus, if the benefits are not satisfactorily shared within the MUSE, technology- and gene-rich countries are most likely to be the first to exit the system and opt for a market solution, in the hope of being able to appropriate a larger share of the value of their PGRFA. However, the often-cited divide between the technology-rich and gene-poor North and the gene-rich and technology-poor South seems to oversimplify the situation since '(d)eveloping countries are by no means a homogeneous group when it comes to the fundamental controversy over internalisation versus compensation' (von Braun and Virchow, 1997, p. 24). PGRFA- and technology-rich countries like Brazil, China or India are examples of such potentially weak advocates of a MUSE. PGRFA- and technology-poor countries like the Central African Republic are likely to be the strongest advocates of a MUSE because they can expect to 'free ride' on access to technology and PGRFA in a MUSE. Technology-rich and gene-poor countries and gene-rich but technology-poor countries have a reasonably strong interest in a MUSE.

In the long run, further technological development will improve the assessment of the value of PGRFA, lower the transaction costs and facilitate the assignment of property rights on PGRFA, which will increase the attractiveness of a market solution. Countries will then reconsider the decision for an internalization or a compensation solution. In the medium run, however, in order to encourage the sustainable participation of as many countries as possible, mechanisms have to be established that oblige the countries that benefit most from a MUSE to share these benefits. A concrete measure of benefit sharing is the financing of *in situ* conservation activities in PGRFA-rich countries. Others include the exchange of information on and the transfer of technology relevant to the use and development of PGRFA, as well as capacity building. Until today, benefit-sharing mechanisms have been envisioned but not been implemented due to a lack of political will. The current revision of the IUPGR includes all of the above-mentioned forms of benefit sharing (FAO, 2001) and a successful conclusion of the negotiations would result in a legally binding and enforceable multilateral system of access, exchange and benefit sharing.

In a MUSE, the global socially optimal amount of traditional PGRFA supply has to be determined, using the valuation techniques for PGRFA mentioned above. The demand side then has to provide the necessary financial resources to facilitate the conservation of this level of TPGRFA. Various options to provide these financial resources have been discussed during the revision of the IUPGR: (i) state contributions (in accordance to the UN scale of assessment or depending on the area planted with IPR protected crops in its territory); (ii) the private seed industry should pay 'an equitable royalty in line with commercial practice' (Article 13.2(d)(ii); see Annex 5.5) for each crop that is developed by the use of PGRFA accessed under the MUSE and for which IPR are granted that restrict the further free use in research and breeding; or (iii) a tax on consumers as the ultimate beneficiaries of PGRFA conservation.

Since the benefits of such a system are a public good and of a very long-term nature, strong tendencies to 'free ride' for all sectors will always be inherent to the system and a continuous political effort to reach consensus over its goals and benefits will be necessary in order to overcome these instabilities.

Compensation and the advancement of Farmers' Rights. It is argued that a compensation solution should provide financial means for the *in situ* conservation of traditional PGRFA. Only *in situ* conservation can maintain the dynamic adoption of landraces to changing agroecological conditions (Brush, 1994). Yet, the effect of planned *in situ* conservation on traditional farmers is disputed, because if they are encouraged to keep their landraces, they will not be able to participate in agricultural development through the use of modern varieties. *In situ* conservation is therefore potentially conflicting with development and productivity goals. Farmers who are encouraged to continue to grow traditional varieties must benefit from this decision at least as much as they would if they had chosen to grow modern varieties. This can be achieved through complementary measures to improve the livelihood of traditional farmers.

Today, large areas are still grown under traditional varieties and agricultural conservation and development policies must simultaneously strive to replace a large share of this area with modern varieties while conserving all currently existing traditional varieties on a much smaller area. If *in situ* conservation strategies can be found – and are sufficiently financed – that reconcile conservation with development goals, Farmers' Rights are advanced. Theoretically, a mere 1% of the 1.4 billion ha of the world's arable land would suffice to conserve today's 3 million varieties in *in situ* conditions (Virchow, 1999), while the rest of the land could be used for agronomic goals other than conservation. Only a globally planned and coordinated conservation effort could reach such a minimization of the conservation area without sacrificing diversity and thereby largely resolve the tension between conservation and development goals.

The Leipzig Global Plan of Action (GPA) introduces such a global strategy to implement *in situ* conservation and is generally regarded as the most appropriate instrument to allocate the financial resources of the MUSE and also as a concrete means of realizing Farmers' Rights (Girsberger, 1999). Likewise, the draft Article 8 of the revised IUPGR states, '(t)he implementation of the GPA contributes to the realization of Farmers' Rights'. Accordingly, the protection and promotion of Farmers' Rights and the *in situ* conservation of PGRFA could be designed in a mutually supportive way.

If the different forms of benefit sharing are implemented in a way that they are conducive to the development of traditional farmers, they can also be interpreted as an ex-post compensation of traditional farmers' past efforts to conserve and make available PGRFA. This can also be interpreted as a promotion of Farmers' Rights.

In conclusion, a multilateral system of access and benefit sharing for PGRFA with a sustainable funding strategy is currently more apt to achieve the global socially optimal amount of conservation than an internalization solution, which entails the assignment of property rights on traditional PGRFA. Additionally, if a MUSE is designed in a way that the livelihoods of traditional farmers are improved, Farmers' Rights are also better advanced by the MUSE than by the alternative assignment of property rights on traditional PGRFA.

Farmers' Rights: a temporary instrument against market failure?

The revised IUPGR intends a multilateral approach to benefit sharing, in which the distribution of the resources of the common fund is not linked to the amount of genetic resources provided by a country, but implemented through plans and programmes in areas of high conservation priority in accordance with the GPA. A country, which provides profitable germplasm, does thus not necessarily participate in its benefits. Nevertheless, as laid out above, the benefits of a MUSE for any country are higher than under a market arrangement. In the future, however, this situation could change substantially:[9] some countries will reach a high degree of

[9] For genetic resources other than PGRFA, markets are already in the making. Again, the starting point for this development was the granting of national sovereignty over genetic resources by the CBD in 1993. A prominent example for this development is the agreement of the pharmaceutical company Merck and the Costa Rican conservation agency InBio over the prospecting, use and benefit sharing of Costa Rican Biodiversity.

national independence in the supply and development of PGRFA and will encounter an elevated international demand for PGRFA. They will tend towards a market solution. As a preparation for this development, Virchow (1999, p. 175) proposes the establishment of national 'Conservation and Service Centers' to coordinate the conservation and prospecting of GCI and to act as a supplier of GCI on the national and international market for PGRFA. Once such a development has started and with the further advancement of biotechnology, more and more countries can be expected to leave the multilateral system of access and exchange and to opt for a market solution instead, depending on their national cost–benefit analysis.

These dynamics will put a constant pressure on the IUPGR, which is therefore not an inherently stable institutional arrangement. A transitory step from a compensation to a market solution would be to guide the benefit-sharing mechanism in a more bilateral direction and to include more market-based elements of PGRFA exchange in the MUSE.

As discussed above, a market solution cannot compensate for the past efforts of farmers to conserve PGRFA and therefore not promote Farmers' Rights in a way that a compensation solution can. This justification of Farmers' Rights will, however, lose its legitimation after a compensation for these past efforts will have been provided for for a lapse of time because it is not reasonable that these past efforts justify an eternal compensation but rather one for a limited time.

It could thus be argued that Farmers' Rights as discussed today are a transitory solution to conservation and equity questions until the informational and institutional hindrances, which lead to market failure, are eliminated. It should be stressed again that only states qualify as actors on these international markets since other possible actors have no facilities to internalize the intertemporal benefits of PGRFA diversity. A market that is capable of internalizing most of these external benefits promotes conservation and equity and renders the concept of Farmers' Rights needless.

Options for Reconciliation of Farmers' Rights and IPRs at the International Level

IUPGR

The IUPGR came into force in 1983 and since then it regulates the access to *ex situ* and *in situ* germplasm, which is *de facto* still free today because more than 160 countries adhere to the IUPGR's principle of multilateral and open access to PGRFA. As a legally non-binding agreement, the current IUPGR has failed in implementing Farmers' Rights and benefit-sharing provisions, but the revised IUPGR is expected to become a legally binding agreement and thus to be more in this regard. Historically, the IUPGR has provided the ground for reconciliation by respecting the parallel legitimacy of IPRs and Farmers' Rights in the FAO resolutions 4/89 and 5/89 (Cooper, 1994). Since then, however, no agreement has been reached to transform this acknowledgement into concrete policies. Yet, with the revision of the IUPGR to bring it in line with the CBD, which has been underway since 1994 and is expected to conclude in 2001, globally enforceable commitments relating to the conservation of PGRFA, Farmers' Rights and IPRs are being drafted (FAO, 1994b).

The conceptualization of the MUSE in the draft revised IUPGR does not restrict the rights of states to grant IPR on modern PGRFA in accordance with their national PVP law, but it links the issue of IPR directly to the issue of Farmers' Rights because it requires right holders of IPR on modern varieties to pay royalties into an international fund. These royalties are used to support the implementation of Farmers' Rights worldwide. Thus, a successful conclusion of the revision of the IUPGR with such a provision, as envisioned in Article 13.2(d)(ii) (see Annex 5.5), would clearly reconcile Farmers' Rights and IPRs at the global level (Crucible II Group, 2000).

Since the royalty payments of the seed industry would be used for the conservation of traditional PGRFA and related benefit-sharing measures, not only would traditional farmers benefit, but ultimately also the breeding sector itself because such a mechanism ensures the long-term availability of its resources. In the negotiations of the revised IUPGR, the representatives of the breeding sector also partly acknowledged this and did not generally object to such a benefit-sharing provision, although it involves additional taxation. The seeming paradox of financing a fund for Farmers' Rights by taxing the seed sector is that ultimately farmers themselves finance the fund as seed producers will pass on those taxes to the demanders of seed via the pricing mechanism. However, firstly it will depend on the market structure, how much of these additional costs will be passed on to farmers and secondly, such a mechanism redistributes funds from modern to traditional farmers and does therefore not infringe Farmers' Rights, if they are interpreted as only pertaining to traditional farmers.

On the side of developing country governments as the advocates of traditional farmers, the choice not to expand the current IPR system to accommodate landraces but to leave them in the public domain and share the benefits of the use rather through multilateral benefit-sharing arrangements than through internalization is also vital for the reconciliation of Farmers' Rights and IPRs. However, in order to comply with the CBD's recognition of national sovereignty over genetic resources, the revised IUPGR abolishes the current practice of open access and replaces it through 'facilitated' access in compliance with national access legislation (see Annex 5.4). This could impose certain access restrictions on the demanders of PGRFA in comparison to the open access case and lead to the establishment of bilateral and domestic benefit-sharing provisions – a step towards internalization. National PVP and access legislation could therefore be in accordance with the IUPGR but not promote its goal to establish an 'efficient, effective and transparent' multilateral system and 'to minimize transaction costs, obviate the need to track individual accessions, and ensure expeditious access' (FAO, 2001).

The revised IUPGR does not explicitly define the right holder of Farmers' Rights, but it recognizes 'the enormous contribution that the local and indigenous communities and farmers of all regions of the world, particularly those in the centres of origin and crop diversity, have made and will continue to make for the conservation and development of plant genetic resources' (Article 9). Yet, it leaves the responsibility for realizing Farmers' Rights with national governments, 'in accordance with their needs and priorities' and 'subject to its national legislation'. Governments could therefore interpret Farmers' Rights as only belonging to resource-poor farmers.

The IUPGR is the political forum where all multilateral policies and commitments will be made and through which they will be enforced and implemented. As a consequence, a legally binding IUPGR that obliges the beneficiaries of traditional PGRFA to provide the financing for the implementation of the GPA is the principal instrument for the reconciliation of IPR and Farmers' Rights at the international level.

Agreement on TRIPs

The TRIPs agreement aims at the global promotion and harmonization of IPR. Consequently, it has no provisions for the advancement of Farmers' Rights. Girsberger (1999) has suggested using the revision of Article 27.3(b) to include Farmers' Rights in TRIPs and thereby to bring it in explicit harmony with the IUPGR and the CBD. This would oblige all WTO members to implement Farmers' Rights and non-compliance could be sanctioned by retaliation measures. This position is also adopted by the African Group in the WTO (WTO, 1999a). India argues that the TRIPs agreement conflicts with the CBD, and that the two must be reconciled before they can be properly implemented at the national level. This position is widely supported by governments across the South (WTO, 1999b). The CBD promotes the objectives of equity, benefit sharing and conservation in relation to biodiversity in general and has mandated the IUPGR to solve these problems for the subgroup of PGRFA, including the question of Farmers' Rights. The USA argue that TRIPs and the CBD are sufficiently flexible to carry out their parallel implementation on the national level in a non-conflicting manner and that an explicit harmonization is therefore unnecessary (WTO, 2000b).

Developing countries had to implement the provisions of Article 27.3(b) by 1 January 2000, least-developed countries have to implement them by 1 January 2006. Only 21 of the 68 developing country members of the WTO had complied with this obligation, not counting the 29 least-developed country members (GRAIN, 2000). This is not surprising, since the 'built-in review' of the article, which was scheduled prior to implementation, had not taken place yet. This review could bring a substantial change to the provisions and a definition of an 'effective *sui generis*' system. It seems, however, that some developing countries find Article 27.3(b) not particularly limiting and already make use of the *sui generis* provision to draft PVP laws that include Farmers' Rights and special provisions for access to their PGRFA (see Annex 5.1). For these countries, PVP is apparently rather a national objective than an international obligation. However, since an 'effective *sui generis*' system is not defined by TRIPs, the compliance of many of these laws with TRIPs is not yet decided upon. If the review does not bring a clarification to this question, it will, according to WTO rules, ultimately be defined by the rulings of the WTO dispute settlement body. Only then it can finally be judged, if TRIPs obstructs a reconciliation of Farmers' Rights and IPR, for example by abolishing the *sui generis* option (as proposed by the USA) or by restricting the scope of action for *sui generis* laws in a way that only UPOV-style laws are judged to be effective protection systems. Yet, the growing awareness of developing countries about the issues of conservation and valuation of their genetic resources and the legal and institutional support for these tasks through the CBD and the IUPGR make it improbable that the WTO would pose a hindrance to the reconciliation of Farmers' Rights and IPRs through parallel implementation in a *sui generis* PVP system.

Options for Reconciliation of Farmers' Rights and IPR at the National Level Through *sui generis* PVP Legislation

The national *sui generis* PVP and access legislation is the key instrument for the reconciliation of Farmers' Rights and IPR at the national level. The time pressure exercised upon developing countries by the TRIPs agreement has brought these issues to an elevated position in the national policy agenda so that many developing countries have recently drafted and enacted new PVP laws (see Annex 5.1).

PVP laws primarily aim at creating conditions of intellectual property protection with the objective to attract investments in the breeding sector and to facilitate national and international seed trade. However, developing countries also seek to include elements of Farmers' Rights in these laws such as the conservation of PGRFA, the equitable sharing of benefits from PGRFA use and the protection of traditional farmers' practices. If these goals can be reached in a non-conflicting manner through parallel implementation, Farmers' Rights and IPRs are reconciled. The integration of PGRFA conservation provisions and Farmers' Rights in PVP systems is an unprecedented undertaking because the PVP laws prior to TRIPs were in force almost exclusively in technology-rich but biodiversity-poor industrialized countries in which the issues of PGRFA conservation and Farmers' Rights are of minor importance. While countries gain experiences with different approaches, it is desirable that the international framework allows some flexibility to adjust the national laws to a changing socio-economic and natural environment.

Farmers' Privilege

A suitable instrument to reconcile Farmers' Rights and IPRs on the national level is the granting of the Farmers' Privilege only for certain disadvantaged groups of farmers in a *sui generis* system. Arrangements with plant breeders are conceivable that exempt farmers below certain prosperity levels – determined, e.g., on the basis of income, volume of output, size of landholdings, species planted, etc. – from the requirement to pay a licence fee for a UPOV 1991 or a patent-protected variety. Plant breeders who make available their varieties for free to these resource-poor farmers could be compensated in turn by reduced tax payments to the global and national compensation funds. This kind of market segmentation could either be achieved through legislation to protect disadvantaged farmers or through voluntary cooperation arrangements between the state, the public and the private breeding sector. It is, however, difficult to control and enforce market segmentation for a homogeneous good such as seeds. Severe leakage problems may arise, depending on the legal and physical infrastructure of a country.

In addition to the royalty exemption, further thinkable measures to protect the Farmers' Privilege for resource-poor farmers are to exempt exchanges of seed that take place within the same community or with neighbours, and between farming communities and to allow certain sales of seeds as propagating materials, for instance, those that take place within the farmers' customary market area (Correa, 2000). Such legislation conflicts, however, with UPOV 1991 if it extends to acts other than those done 'privately and for non-commercial purposes' (Article 14.1).

A concrete example of a voluntary public–private cooperation arrangement is currently under discussion in the case of the so-called 'Golden Rice', a GMO with high beta-carotene content. 'Golden Rice' was developed by public breeders, who used privately owned and patent protected technologies. The private-sector right holders now propose that Asian rice farmers who earn less than US$10,000 year^{-1} will be exempted from paying a licence fee (Potrykus, 2000).

Legislative provisions to explicitly protect the Farmers' Privilege were also chosen already by various developing countries, e.g. India and Nicaragua (see Annex 5.1).

National benefit sharing and access legislation

In addition to the benefit-sharing mechanism of the IUPGR, national governments may implement benefit sharing through a variety of modalities to promote Farmers' Rights. A concrete approach is the creation of a national conservation fund, which directs a share of the benefits of PGRFA use to traditional farmers via *in situ* conservation plans and programmes. The financing of this fund may arise from sources similar to those discussed under the revised IUPGR, or from remuneration payments in accordance to the use of TPGRFA by breeders.

For this purpose, national PVP laws may establish the obligation to reveal the source of genetic material used for the creation of a new variety and, if appropriate in the particular case, to prove that the applicant has complied with rules relating to access and sharing of benefits, e.g. through a 'certificate of origin' (Correa, 2000). They could further be obliged to reach 'prior informed consent' with farming communities when collecting *in situ* PGRFA. This type of 'facilitated access' as envisioned under the IUPGR, would not be inconsistent with the TRIPs Agreement, which does not limit the states' rights to make the granting of intellectual property protection conditional on complying with certain obligations. In contrast to the multilateral benefit sharing of the IUPGR, these elements intend to share the benefits bilaterally, which is also in the spirit of the CBD, but which necessarily imposes access restrictions on the demanders in order to identify the origin of the genetic material.

Additionally, breeders could be doubly taxed by the IUPGR and by national PVP legislation. Though this would signify a comprehensive benefit-sharing, it would also deter investments and slow down the rate of innovation in the breeding sector. Consequently, if these financial or administrative obligations burden the breeders inappropriately, they have to be interpreted as obstructing the reconciliation of Farmers' Rights and IPRs.

Examples of benefit sharing through a national conservation fund in PVP laws of developing countries include Thailand, India, Bangladesh and Pakistan. PVP laws which require the breeders to disclose the origin of the PGRFA used in breeding and to share the benefits with the providers are being drafted or in place, for example in Nicaragua, Bangladesh, Thailand and India (where the proof rests with the claimant). Various developing countries, e.g. Thailand and Costa Rica, also seek to enable their traditional farmers a sharing in the benefits of PGRFA by granting various forms of community IPR on traditional PGRFA and the related traditional knowledge (see Annex 5.1).

Acknowledgement

We gratefully acknowledge the financial support of the Eiselen Foundation, Ulm, which has made possible the research for this chapter.

References

Alston, J.M. and Venner, R.J. (2000) *The Effects of the U.S. Plant Variety Protection Act on Wheat Genetic Improvement.* Environment and Production Technology Division Discussion Paper No. 62. International Food Policy Research Institute, Washington, DC.

Blakeney, M., Cohen, J.I. and Crespi, S. (1999) Intellectual property rights and agricultural biotechnology. In: Cohen, J.I. (ed.) *Managing Agricultural Biotechnology – Addressing Research Program Needs and Policy Implications.* CAB International, Wallingford, UK.

Braun, J. von and Virchow, D. (1997) Conflict prone formation of markets for genetic resources: institutional and economic implications for developing countries. *Quarterly Journal for International Agriculture* 36, 6–38.

Brush, S. (1992) Farmers' Rights and genetic resource conservation in traditional farming systems. *World Development* 20, 1617–1630.

Brush, S. (1994) *Providing Farmers' Rights Through* In Situ *Conservation of Crop Genetic Resources.* CPGRFA Background Study Paper No. 3. FAO, Rome.

Cooper, D. (1994) *A Multilateral System for Plant Genetic Resources: Imperatives, Achievements and Challenges.* Issues in Genetic Resources No. 2. FAO, Rome.

Correa, C.M. (2000) *Options for the Implementation of Farmers' Rights at the National Level.* Trade Related Agenda, Development and Equity (TRADE) Working Paper No.8. South Centre.

CPGR (1994) *Revision of the International Undertaking: Analysis of some Technical, Economic and Legal Aspects for Consideration in Stage II.* FAO Doc. CPGR-Ex1/94/5 Supp. FAO, Rome.

Crucible II Group (2000) *Seeding Solutions,* Vol.1, *Policy Options for Genetic Resources.* IDRC, IPGRI, Dag Hammarskjöld Foundation, Rome.

Cullet, P. (2001) Plant Variety Protection in Africa. Towards compliance with the TRIPs Agreement. *Journal of African Law* 45, 97–122.

Deardorff, A.V. (1992) Welfare effects of global patent protection. *Economica* 59, 35–51.

Esquinas-Alcazár, J. (1998) Farmers' Rights. In: Evenson, R.E., Gollin, D. and. Santaniello, V. (eds) *Agricultural Values of Plant Genetic Resources.* CAB International, Wallingford, UK.

Evenson, R.E., Gollin, D. and Santaniello, V. (1998) Introduction and overview: agricultural values of plant genetic resources. In: Evenson, R.E., Gollin, D. and Santaniello, V. (eds) *Agricultural Values of Plant Genetic Resources.* CAB International, Wallingford, UK.

FAO (1994a) *International Undertaking on Plant Genetic Resources (FAO Conference Resolution 8/83).* Document CPGR-Ex1/94/Inf. 1. FAO, Rome.

FAO (1994b) *Revision of the International Undertaking: Issues for Consideration in Stage II: Access to PGRFA and Farmers' Rights.* CPGR/94/WG9/4. FAO, Rome.
FAO (1997) *The State of the World's Plant Genetic Resources for Food and Agriculture.* FAO, Rome.
FAO (2001) *Composite Draft Text of the International Undertaking on Plant Genetic Resource.* Document CGRFA/EX-6/01/2. FAO, Rome.
Fowler, C. (1994) *Unnatural Selection: Technology, Politics, and Plant Evolution.* Gordon and Breach, Yverdon, Switzerland.
Girsberger, M.A. (1999) *Biodiversity and the Concept of Farmers' Rights in International Law: Factual Background and Legal Analysis.* Lang, Berne.
Gollin, D. (1998) Valuing Farmers' Rights. In: Evenson, R.E., Gollin, D. and Santaniello, V. (eds) *Agricultural Values of Plant Genetic Resources.* CAB International, Wallingford, UK.
GRAIN (Genetic Resource Action International) (1999a) Ensnaring the South in UPOV net, 1999. In: *Seedling* 16(2). GRAIN, Barcelona.
GRAIN (1999b) Beyond UPOV. Examples of developing countries preparing non-UPOV sui generis plant variety protection schemes for compliance with TRIPs. Last visited 20 April 2001: www.grain.org/publications/reports/nonupov.htm
GRAIN (2000) For a full review of TRIPs. GRAIN Publications. Last visited 20 April 2001: www.grain.org/publications/reports/tripsfeb00.htm
Hampicke, U. (1991) *Naturschutzökonomie.* UTB, Stuttgart.
Heitz, A. (1998) Intellectual property rights and plant variety protection in relation to demands and farmers in Sub-Saharan Africa. In: *Seed Policy and Programmes for Sub-Saharan Africa. Proceedings of the Regional Technical Meeting on Seed Policy and Programmes for Sub-Saharan Africa.* Abidjan, Côte d'Ivoire, 23–27 November 1998. FAO, Rome.
Hoisington, D., Khairallah, M., Reeves, T., Ribaut, J.-M., Skovmand, B., Taba, S.and Warburton, M. (1999) Plant genetic resources: what can they contribute toward increased crop productivity? *Proceedings of the National Academy of Sciences, USA* 96, 5937–5943.
IPGRI (1999) *Key Questions for Decision-Makers. Protection of Plant Varieties under the WTO Agreement on TRIPs.* IPGRI, Rome.
Koo, B. and Wright, B.D. (1999) *Dynamic Implications of Patenting for Crop Genetic Resources.* Environment and Production Technology Division Discussion Paper No. 51. IFPRI, Washington, DC.
Kuyek, D. (2001) Intellectual property rights: ultimate control of R&D in Asia. GRAIN Publications. Last visited 20 April 2001: www.grain.org/publications/reports/asiaipr.htm
Maskus, K.E. and Penubarti, M. (1995) How trade-related are intellectual property rights? *Journal of International Economics* 39, 227–248.
Menini, U.G. (1998) Preface. In: *Seed Policy and Programmes for Sub-Saharan Africa. Proceedings of the Regional Technical Meeting on Seed Policy and Programmes for Sub-Saharan Africa.* Abidjan, Côte d'Ivoire, 23–27 November 1998. FAO, Rome.
Potrykus, I. (2000) Last visited 24 April 2001: http://www.biotech-info.net/GR_tale.html
Primo Braga, C.A. (1996) Trade-related intellectual property issues: The Uruguay Round Agreement and its economic implications. In: Martin, W. and Winters, L.A. (eds) *The Uruguay Round and the Developing Economies.* World Bank Discussion Paper No. 307. The World Bank, Washington, DC.
Smale, M. (1997) The Green Revolution and wheat genetic diversity: some unfounded assumptions. *World Development* 25, 1257–1269.
Srivastava, J. and Jaffe, S. (1993) *Best Practices for Moving Seed Technology.* World Bank Technical Paper No. 213. The World Bank, Washington, DC.
Trebilock, M.J. and Howse, R. (1995) Trade related intellectual property. In: *The Regulation of International Trade.* Routledge, London.
UN Commission on Human Rights (1999) *Updated Study on the Right to Food.* UN Document no. E/CN.4/Sub.2/1999/12. UNHCHR, Geneva.
UPOV (1999) *Aide Mémoire pour la Ratification du Nouvel Accord de Bangui et l'Adhésion à l'Union Internationale pour la Protection des Obtentions Végétales (UPOV).* UPOV, Geneva.
Virchow, D. (1999) *Conservation of Genetic Resources: Cost and Implications for a Sustainable Utilization of Plant Genetic Resources for Food and Agriculture.* Springer, Berlin.
World Bank (1998) *World Development Report 1998/1999: Knowledge for Development.* Oxford University Press, Washington, DC.

Wright, B.D. (1998) Intellectual property and Farmers' Rights. In: Evenson, R.E., Gollin, D. and Santaniello, V. (eds) *Agricultural Values of Plant Genetic Resources.* CAB International, Wallingford, UK.

WTO (1999a) *Review of the Provisions of Article 27.3(b). Communication from Kenya on behalf of the African Group.* WTO Doc. IP/C/W/163. WTO, Geneva.

WTO (1999b) *Review of the Provisions of Article 27.3(b). Communication from India.* WTO Doc. IP/C/W/161. WTO, Geneva.

WTO (2000a) *The Relationship between the Convention on Biological Diversity (CBD) and the Agreement on the Trade Related Aspects of Intellectual Property Rights (TRIPs); with a Focus on Article 27.3 (b).* WTO Doc. IP/C/W/175. WTO, Geneva.

WTO (2000b) *Review of the provisions of Article 27.3(b). Further Views of the United States.* WTO Doc. IP/C/W/209. WTO, Geneva.

Annex 5.1. Status of selected PVP Laws and Drafts (January 2001)

Sources: Kuyek (2001) and GRAIN (1999b)

Bangladesh

Title: Plant Variety Protection Act of Bangladesh

Status: Draft. Has been approved by relevant ministries and is under public discussion.

IPR:

- Based on UPOV 1978, Bangladesh is not member and has not formally approached UPOV. However, in early 2001, the European Union approved a development cooperation package for Bangladesh under which the country must accede to UPOV (1991) by 2006.[10]
- Criteria for protection: novelty, consistency, distinctness and stability. In addition, varieties must demonstrate 'immediate, direct and substantial benefit to the people of Bangladesh'. Hybrids only protected if parents are available as public domain.
- Short duration of breeders' right (e.g. 7 years for annuals).
- GMOs can be protected subject to further legislation.

Farmers' Rights:

- Country of origin of materials used to develop protected varieties shall be disclosed.
- Where community varieties, wild materials or indigenous varieties are used in the development of a protected variety, 25% of the revenue from its commercialization will be redistributed.
- Any variety that may lead to genetic or cultural erosion shall not be protected.
- Any variety developed by public institutes, or by farmers non-governmental organizations (NGOs) using public funds, shall be considered common property of the people of Bangladesh and shall receive Citation of Award rather than PVP certificate.
- Strong provisions for community rights and farmers' rights, which will be supported through a Plant Variety Development Fund.
- Protection is restricted to nationals of CBD member states, which have to obey to the principles of national sovereignty over genetic resources and benefit-sharing. Thus the obligations under the CBD are given priority over the obligations under TRIPs.

Costa Rica

Farmers' Rights:

- PVP law will be subordinate to the country's compliance with the CBD, which was formalized through the enactment of Law No. 7788 entitled 'Biodiversity Law' in May 1998.
- Protection of traditional knowledge via a *sui generis* system of community intellectual property rights, which extend to 'the knowledge, practices and innovations of the indigenous peoples and the local communities, related to the use of the components of biodiversity and associated knowledge'.

[10] 'Cooperation Agreement between the European Community and the People's Republic of Bangladesh in Partnership and Development', *Official Journal of the European Communities,* Luxembourg, C143/9, 21 May 1999, approved by the European Parliament under Consultation Procedure on 17 January 2001.

- The community intellectual property rights shall not be affected by PBRs, patents or any other form of intellectual property applied to biodiversity and associated knowledge.
- Any application for PBR in Costa Rica must receive clearance from the Technical Office of the Commission administering the Biodiversity Law to ensure that the application does not contravene community intellectual rights, even though these need not be formally registered.
- The recognition of community intellectual rights in Costa Rica 'oblige(s) the Technical Office to answer negatively any consultation related to the recognition of intellectual or industrial rights over the same component (of biodiversity) or knowledge' (Article 84).

India

Title: The Protection of Plant Varieties and Farmers' Rights Bill. Bill No. 123 of 1999.

Status: Draft. Undergoing parliamentary examination.

IPR:
- Based on UPOV 1978 and 1991. India has initiated the accession procedure.
- GMOs can be protected by PVP subject to further legislation.

Farmers' Rights:
- Farmers may sell the harvest of any protected variety, but not as reproductive material under commercial marketing arrangements.
- Foresees benefit-sharing arrangements between breeders and those, including farmers and communities, who claim to have contributed genetic material to a protected variety. The burden of proof rests with the claimant, not with the holder of the PVP certificate.
- A National Gene Fund will be built up with royalty fees from plant variety right holders, national and international contributions, etc., meant to be used for benefit sharing and compensation to farming communities, and for conservation and sustainable use of genetic resources.
- Specific and detailed provisions for communities to register collective rights.
- Farmers' Rights are formalized in the following terms (Article 31): 'Nothing contained in this act shall affect a farmer's traditional right to save, use, exchange, share or sell his farm produce of a variety protected under this act except where a sale is for the purpose of reproduction under a commercial marketing arrangement.'

Nicaragua

Status: Draft

IPR:
- Discoveries may not be protected.
- A plant variety shall be eligible for protection if it differs from another variety in several characteristics (not just one).
- Transgenic material shall be subject to separate biosafety legislation.
- It sets PVP apart from industrial property and therefore seeks to comply with UPOV 1978 which expressly prohibits double protection.

- Protection extends to the following acts: direct sowing, preparation for reproduction or multiplication as certified seed, repetitive use for the production of another variety. It does not offer protection for marketing, import or export.
- Criteria for protection are: novelty, distinction, uniformity or variability, stability or evolutionary capacity, plus the variety must carry a denomination.
- The provision regarding essential derivation shall be applied in cases whether the 'new' variety is at least 20% dependent on an earlier variety.

Farmers' Rights:

- The definition of breeder and breeding is wide in scope: it covers anyone making use of techniques of crop improvement.
- Plant breeders' rights shall not extend to the variety when it is used for consumption or sowing directly by farmers or when it is used by tenants, cooperatives or other non-landholding entities.
- A variety shall be deemed variable if its characteristics are adapted to different climatic and soil conditions of the country.
- A variety shall be deemed to have evolutionary capacity if it contains genes or genetic complexes which are expressed under environmental change.
- Registration requires: proof of compliance with CBD Art 8j and 15 (especially compensation to countries and communities of origin) and scientific proof of the variety's superiority to cultivars grown in the country through at least two production cycles of comparative tests.
- Wide compulsory licensing.
- The law is subordinate to the rights and obligations acquired through the Convention on Biological Diversity.

Pakistan

UPOV 1991 with Farmers' Privilege. Portion of royalties will flow to National gene fund for genetic conservation. Draft.

Thailand

Title: Plant Varieties Protection Act, B.E. 2542 (1999)

Status: Adopted by parliament but not yet in force. National PVP Committee is now being established.

IPR:

- Based on UPOV 78. Thailand has consulted UPOV on the conformity of its act with the UPOV Convention.
- Covers four kinds of plants: new varieties, local domestic varieties, general domestic varieties and wild species.
- Rights will be granted for 12 years in the case of registered annual species.

Farmers' Rights:

- Wild species need not express uniformity to be protected.
- General domestic plant varieties and wild species shall be protected automatically, without registration. There are special provisions for farmer's and community rights over local domestic plant varieties, which must be unique to a particular locality within the Kingdom.

- Revenue accruing from the procurement and use of general domestic varieties and wild species will be on a profit-sharing basis through a Plant Variety Protection Fund. The Fund will benefit local communities and government units involved in conservation, research and development of plant varieties.

Zambia

The Zambian government has made it clear that in order to fulfil its rights and obligations under CBD, its *sui generis* PVP system must recognize and reward the innovation of indigenous peoples and local communities. For this, their law, which is being drawn up with full stakeholder participation, defines innovation to include 'any inventive input done collectively, accretionary, inter-generationally and over a period of time, in relation to genetic resources.'

Africa/OAPI

The 15 francophone member states of the OAPI revised the Bangui Agreement which governs their common intellectual property regime. The new Agreement establishes, in Annex X, a common PVP system and foresees that the OAPI member states will join UPOV by depositing an instrument of accession to the 1991 act.

Africa/SADC

The Southern African Development Community, with the support of the International Plant Genetic Resources Institute, has examined whether alignment with UPOV would be appropriate for compliance with the *sui generis* principle of TRIPs. The conclusion was that UPOV is mainly appropriate to protect the interests of exporters of horticultural and ornamental varieties, but not for southern Africa. As a result, SADC is currently drafting a common legislative framework for *sui generis* rights that protects the gamut of plant biodiversity as well as traditional knowledge of the local communities, in cooperation with the OAU.

Annex 5.2. FAO Resolutions 5/89 on Farmers' Rights

The Conference,

Recognizing that:

(a) Plant genetic resources are a common heritage of mankind to be preserved, and to be freely available for use, for the benefit of present and future generations,

(b) Full advantage can be derived from plant genetic resources through an effective programme of plant breeding, and that, while most such resources, in the form of wild plants and old landraces, are to be found in developing countries, training and facilities for plant survey and identification, and plant breeding, are insufficient, or even not available in many of those countries,

(c) Plant genetic resources are indispensable for the genetic improvement of cultivated plants, but have been insufficiently explored, and in danger of erosion and loss,

Considering that:

(a) In the history of mankind, unnumbered generations of farmers have conserved, improved and made available plant genetic resources,
(b) The majority of these plant genetic resources come from developing countries, the contribution of whose farmers has not been sufficiently recognized or rewarded,
(c) The farmers, especially those in developing countries, should benefit fully from the improved and increased use of the natural resources they have preserved.
(d) There is a need to continue the conservation (*in situ* and *ex situ*), development and use of the plant genetic resources in all countries, and to strengthen the capabilities of developing countries in these areas.

Endorses the concept of Farmers' Rights (Farmers' Rights mean rights arising from the past, present and future contributions of farmers in conserving, improving, and making available plant genetic resources, particularly those in the centres of origin/diversity. These rights are vested in the International Community, as trustee for present and future generations of farmers, for the purpose of ensuring full benefits to farmers, and supporting the continuation of their contributions, as well as the attainment of the overall purposes of the International Undertaking) in order to:

(a) ensure that the need for conservation is globally recognized and that sufficient funds for these purposes will be available;
(b) assist farmers and farming communities, in all regions of the world, but especially in the areas of origin/diversity of plant genetic resources, in the protection and conservation of their plant genetic resources, and of the natural biosphere;
(c) allow farmers, their communities, and countries in all regions, to participate fully in the benefits derived, at present and in the future, from the improved use of plant genetic resources, through plant breeding and other scientific methods.

(Adopted on 29 November 1989)

Annex 5.3. TRIPs Article 27.3 (b)

Article 27: Patentable Subject Matter

3. Members may also exclude from patentability: (b) plants and animals other than micro-organisms, and essentially biological processes for the production of plants or animals other than non-biological and microbiological processes. However, Members shall provide for the protection of plant varieties either by patents or by an effective *sui generis* system or by any combination thereof. The provisions of this subparagraph shall be reviewed 4 years after the date of entry into force of the WTO Agreement.

Annex 5.4. Article 12.2 of the draft revised International Undertaking (May 2001)

Article 12 – Facilitated access to plant genetic resources for food and agriculture within the Multilateral System

12.2 Parties agree to provide such access to other Parties, in accordance with the conditions below:

(a) Access shall be provided solely for the purpose of and utilization in research, breeding and training for food and agriculture, provided that such purpose does not include chemical, pharmaceutical and/or other non-food/feed industrial uses. In the case of multiple-use crops (food and non-food), their importance for food security should be the determinant for their inclusion in the Multilateral System and availability for facilitated access;

(b) Access shall be accorded expeditiously, without the need to track individual accessions and free of charge, or, when a fee is charged, it shall not exceed the minimal cost involved;

(c) All available passport data and, subject to applicable law, any other associated available non-confidential descriptive information, shall be made available with the plant genetic resources for food and agriculture provided;

(d) Recipients shall not claim any intellectual property or other rights that limit the facilitated access to the plant genetic resources for food and agriculture, or their genetic parts or components, in the form received from the Multilateral System;

(e) Access to plant genetic resources for food and agriculture under development, including material being developed by farmers, shall be at the discretion of its developer, during the period of development;

(f) Access to plant genetic resources for food and agriculture protected by intellectual and other property rights shall be consistent with relevant international agreements, and subject to national legislation;

(g) Plant genetic resources for food and agriculture accessed under the Multilateral System and conserved shall continue to be available to the Multilateral System by the recipients of those plant genetic resources for food and agriculture, under the terms of this Undertaking;

(h) Without prejudice to the other provisions under this Article, the Contracting Parties agree that access to plant genetic resources for food and agriculture found in *in situ* conditions will be provided according to national legislation or, in the absence of such legislation, in accordance with such standards as may be set by the Governing Body.

FAO Document CGRFA/EX-6/01/2 (FAO, 2001)

Annex 5.5. Article 13.2(d)(ii),(iii),(iv) of the draft revised International Undertaking (May 2001).

Article 13: Benefit-Sharing in the Multilateral-System

13.2 (d): Sharing of monetary benefits on commercialization

(ii) Whenever the use of plant genetic resources for food and agriculture accessed under the Multilateral System results in a product that is a plant genetic resource covered by any form of intellectual property right that restricts utilization of the product for research and plant breeding, the rights-holder shall pay an equitable royalty in line with commercial practice on the commercial exploitation of the product into a mechanism referred to in Article 19.2(g), as a contribution to the implementation of agreed plans and programmes as established under this Undertaking.

(iii) Whenever the use of plant genetic resources for food and agriculture accessed under the Multilateral System results in a product that is a plant genetic resource covered by any form of intellectual property right that does not restrict utilization of that product for research and plant breeding, the Contracting Parties shall take measures, as appropriate, to encourage the rights-holder to pay into the above mechanism a royalty on the commercial exploitation of that product, taking into account the need to exempt farmers in developing countries, especially in least developed countries, from this provision.

(iv) The Governing Body shall review the provisions of Article 13.2(d)(ii) and

13.2 (d)(iii) within a period of 5 years of the entry into force of the International Undertaking, with a view to optimising benefits accruing from these provisions, and shall in particular assess the possibility of establishing a mandatory scheme in regard to the above paragraph. Following this review, any proposed amendment shall be addressed in accordance with Article 22.

FAO Document CGRFA/EX-6/01/2 (FAO, 2001)

Annex 5.6. Article 9 of the draft revised International Undertaking (May 2001)

Article 9: Farmers' Rights

(As negotiated during the Eighth Regular Session of the Commission on Plant Genetic Resources, April 1999)

9.1 The Parties recognize the enormous contribution that the local and indigenous communities and farmers of all regions of the world, particularly those in the centres of origin and crop diversity, have made and will continue to make for the conservation and development of plant genetic resources which constitute the basis of food and agriculture production throughout the world.

9.2 The Parties agree that the responsibility for realizing Farmers' Rights, as they relate to Plant Genetic Resources for Food and Agriculture, rests with national governments. In accordance with their needs and priorities, each Party should, as appropriate, and subject to its national legislation, take measures to protect and promote Farmers' Rights, including:

(a) Protection of traditional knowledge relevant to plant genetic resources for food and agriculture;
(b) The right to equitably participate in sharing benefits arising from the utilization of plant genetic resources for food and agriculture;
(c) The right to participate in making decisions, at the national level, on matters related to the conservation and sustainable use of plant genetic resources for food and agriculture.

9.3 Nothing in this Article shall be interpreted to limit any rights that farmers have to save, use, exchange and sell farm-saved seed/propagating material; subject to national law and as appropriate.

FAO Document CGRFA/EX-6/01/2 (FAO, 2001)

Chapter 6

Universities, Technology Transfer and Industrial R&D

Gregory Graff,[1] Amir Heiman,[2] David Zilberman,[1] Federico Castillo[3] and Douglas Parker[4]

[1] Department of Agricultural and Resource Economics, University of California, 201 Giannini Hall, Berkeley, CA 94720-3310, USA; [2] Department of Agricultural Economics and Management, Hebrew University, Rehovot, Israel; [3] Department of Environmental Science, Policy and Management, University of California, Berkeley, California, USA; [4] Department of Agricultural and Resource Economics, University of Maryland, College Park, Maryland, USA

Abstract

Some of the most important innovations to emerge in recent years – which have transformed whole industries, particularly in the areas of biotechnology (transforming medicine and agriculture) and information technology (transforming computing and communications) – were the result of research conducted in universities and public sector institutions. Many of these institutions have recently established Offices of Technology Transfer (OTTs) that aim to manage their intellectual property and to commercialize some of their research products. Technology transfer has meant more than just the licensing of patents: substantial numbers of university researchers have migrated to private industry, establishing start-up companies. This chapter provides a perspective on the economic forces at work in the transfer of technology from publicly funded research to commercial use. It provides an overview of the differences and the relationships between public and private research and then introduces, interprets and analyses results from several recent surveys of OTT operations and results. Finally, it draws implications for public policy makers, university administrators and company managers.

The University Enterprise and the 'Educational–Industrial Complex'

The 21st-century American university is an institution that has evolved to meet multiple objectives. Universities are expected to educate their student-customers, contribute works of culture, nurture the arts, host the intellectual conscience and discourse of society, and generate new scientific knowledge for the good of both the economy and society at large. These various outputs consist – in the parlance of economics – of both *public* and *private* goods. For example, through educational services universities provide individuals with skills that

(eds R.E. Evenson, V. Santaniello and D. Zilberman)

increase their private earning potential, generate human capital for firms to hire and to utilize, and, in general, make for a more educated and cultured citizenry. Through laboratory research, universities aim both to solve specific problems – thereby building specific intellectual capital that can be intentionally used by firms – and to create generalized, abstract forms of knowledge that are impossible to quantify or contain.

This multiplicity of objectives leads the university to draw upon a wide variety of funding sources to support its activities. General university revenues come from tuition and fees, from federal, state and local government grants and budget allocations (particularly for public universities), from endowment and investment income, and from private gifts and grants from industry, benefactors and foundations. In addition, individual programmes and research projects conducted at universities are supported by restricted (i.e. programme-designated) grants applied for and recieved from government and private sources. Figure 6.1 illustrates the variety of revenue sources for one of the largest public universities in the USA, the 10-campus system of the University of California. The pursuit of multiple objectives and the concommitant reliance on a variety of funding sources enable universities, when it comes to conducting scientific research, to establish research units that are quite unique in their capabilities and that have distinct relative advantages in terms of capacity and cost effectiveness in pursing certain kinds of research.

Taken together, this set of objectives of the 21st-century American university system and its economy-wide coalition of beneficiaries and benefactors make up what may be regarded as today's *educational–industrial complex*. This social and economic infrastructure may not be as menacing to many as that which supported the Cold War, but it similarly constitutes much of the very fabric of contemporary life and business that knits together the knowledge-based economy. The key element of the equation that has tipped the American research university from being 'provider of standardized educational ser-

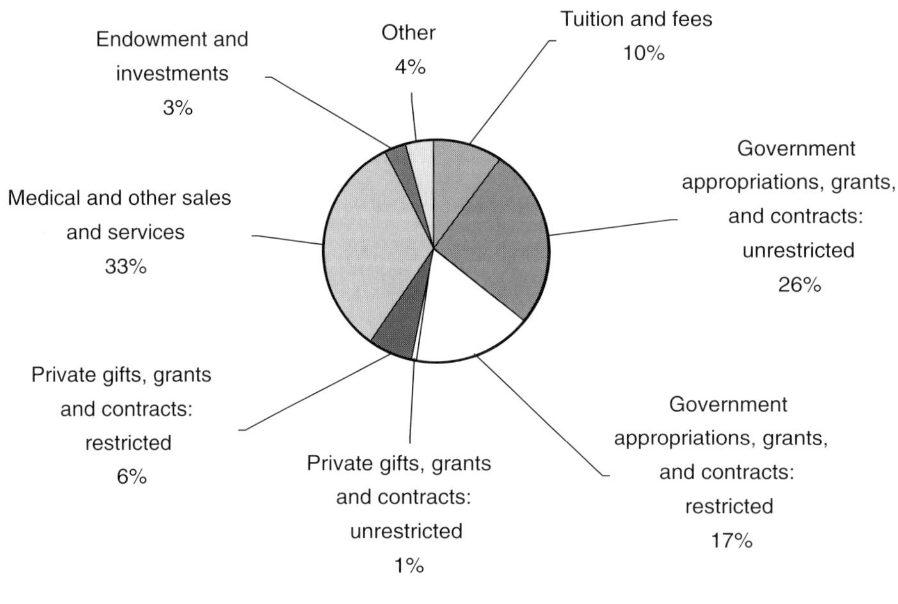

Fig. 6.1. Who pays the piper? The variety of funding sources for the University of California system in 2000. Source: The University of California, Annual Financial Report 1999–2000.

vices and public goods' to becoming 'key component in the national economic infrastructure' is the economic power wrought occasionally but repeatedly by new technologies that emerge from university research and move out into commerce, leading to the development of new products and processes. University research results have led to the creation of whole new firms and even industries, brought old ones down, and, in general, profoundly impacted rates of industrial innovation. University research is a source of competitiveness, effectively serving an anti-trust role when new technologies like young Davids rise up to challenge the technological base of oligopolistic Goliaths in established markets. For existing firms, the univeristy can be both a problem and a solution, a technological competitor and a technological saviour, and the entire difference turns on relationships formed and intellectual property rights won in an arcane sounding process known as *technology transfer*.

Is *Basic* Research Necessarily *Public* or *Academic* Research?

Since at least the 1940s a conventional distinction has been drawn between *basic* research and *applied* research. This distinction is enshrined in the annual national R&D statistics reported by the National Science Foundation (NSF), for whom 'basic' research is defined as that primarily intended 'to gain more comprehensive knowledge or understanding of the subject under study, without specific applications in mind' (National Science Board, 2000). Basic research is, economically speaking, assumed to have stronger *public goods* properties: to yield general knowledge and to generate benefits that are difficult for individual parties to appropriate as rents or as commercial profits. Thus, basic research has conventionally been assumed to be the province of universities while industry is left to concentrate where its interests and expertise lie, in applied research and development[1]. The prevailing paradigm holds that a mutually beneficial division of labour exists between the universities and industry. This division of labour is illustrated in Fig. 6.2 by the NSF categorizations of the character of the R&D undertaken in US universities and industry in 1998.

These NSF figures show that in 1998, 69% of the R&D undertaken at academic institutions was characterized as 'basic' research. Of course, the proportion of basic R&D is higher or lower than this in the different fields and departments across academia. Overall, however, it is arguably the primary focus of universitiy science: the academic sector is the single largest performer of basic research in the USA. Yet still, 24% of university research falls in the 'applied' category and 7% is considered 'development' work (National Science Board, 2000, see Fig. 6.1).

Rosenberg and Nelson (1994) challenge the convention of a black and white distinction between basic and applied research, particularly when it comes to describing the research undertaken at universities. They argue that in the USA, especially following World War II, much of the publicly funded university research was in fact targeted at solving practical, concrete problems. The lion's share of public research funding, they point out, was funneled through the Department of Defense (DOD), the Department of Energy (DOE), the National Institute of Health (NIH) and the Department of Agriculture (USDA), each with a general mandate to solve a specific class of problems faced by society at large. Even research projects supported by the NSF – an organization traditionally dedicated to basic science – have strong problem-solving orientations. Universities in America were among the first in the world to introduce industry-targeted scientific disciplines such as chemical engineering and computer science, aimed to

[1] We will follow NSF convention and use the term 'R&D' to refer to all of these categories collectively, unless otherwise specified (see National Science Board, 2000).

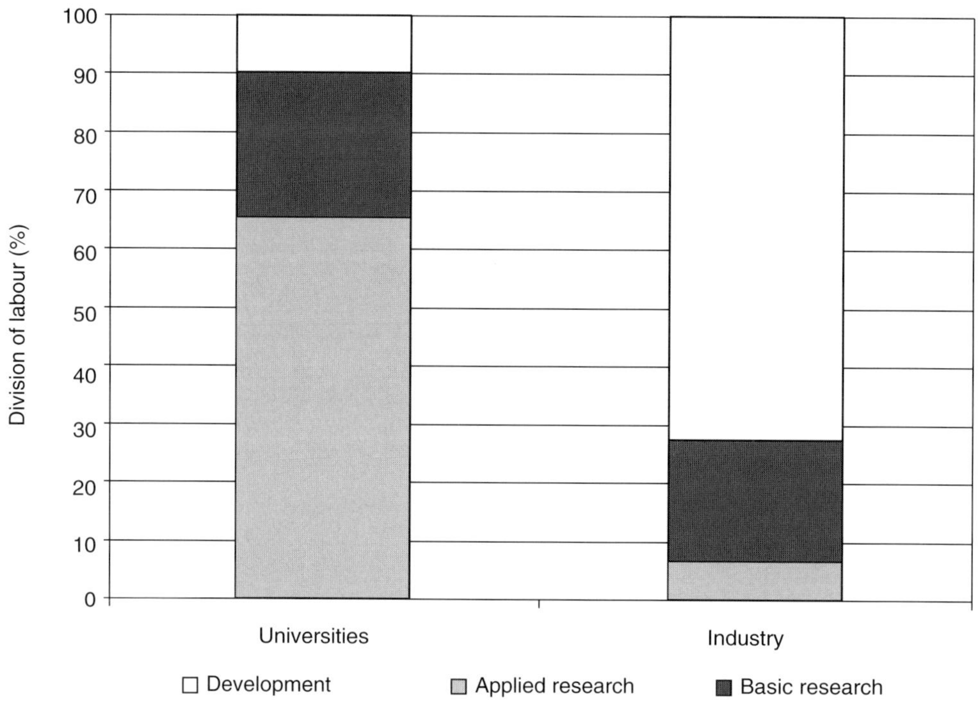

Fig. 6.2. The division of labour: categories of R&D conducted in US universities and industry in 1998. Source: National Science Board (2000).

provide solutions and to train scientists for industry's newly emerging problem areas. While basic research may concentrate on providing the fundamental knowledge and systematic methodologies of the scientific disciplines, it is ultimately justified as leading to the solutions of the practical problems facing government, industry and society at large.

This approach advocated by Rosenberg and Nelson views research projects as lying somewhere on a continuum stretching between the extremes of 'basic' and 'applied' – with particular pieces of research often encompassing both abstracts fundamental issues of human knowledge and targeted solutions to specific, concrete problems. While universities emphasize work that is in some respects closer to the 'basic' end of this continuum, their efforts are nevertheless intended to result in practical knowledge and technologies, which can and do lead to commercially viable inventions.

There are several fundamental reasons for which economists argue that public funds should be expended to support basic research, even despite its pragmatic and sometimes commercially profitable outcomes. The classic economic analysis maintains that basic research is far too uncertain and its return horizon too long to be left to private sector enterprises that need to meet Wall Street standards. Betting on the outcome of individual basic reaserch projects is risky, so much so in fact that, left to their own criteria, individual private companies simply cannot on their own invest nearly enough to supply the full gamut of basic research that society and even they themselves would need. Thus, basic research that is deemed valuable in broad social terms or, in medium to long run economic terms, must be funded by the public sector if it is to be done at all. The underlying economic reasons are, succinctly, as follows:

1. *Uncertainty*: The outcome of basic research is highly uncertain, and society as a whole is assumed to be less averse to shouldering the risk involved in the cost–benefit trade-off of basic research than are individual firms.
2. *Inappropriability or 'non-marketability'*: Some results from basic research, while valuable, are simply not appropriable, because they occur at such fundamental levels of scientific analysis. For example, the discovery of the double-helix structure of DNA was not patentable, nor could any product be made directly from that particular insight alone. However, Watson and Crick won the Nobel Prize for this contribution, and the entire biotechnology industry owes it existence, in part, to that fundamental knowledge.
3. *Spillovers*: Some results from basic research, while valuable, can spill over to competitors in the same line of business (that is, they can be learned and adopted by those competitors) so easily that the results may actually help the competitors more than they help the company that did the initial research.
4. *General purpose*: Results from basic research are often able to improve multiple lines of business, but individual private companies employ specialized assets and business methods to create, manufacture and market specific lines of products. They do not have a competitive advantage in the utilization of an innovation in lines of business where they are not already active, which means their expected return on investing in basic research is lower than the expected returns in the the economy as a whole. Thus, instead, most individual private companies conduct research that aims to improve their own product lines, to overcome quality or production problems, and to expand their offerings within their area of specialization.
5. *Competence destroying*: Occasionally, a *radical* innovation arises from basic research that renders the old way of doing business altogether obsolete. The tide let loose by such an innovation can be powerful and unpredictable enough that even industrial giants are challenged, become less cost-effective than new competitors, and lose market share. Technologically adept incumbent firms will eventually find that they have to manoeuvre to adopt such a radical innovation as it spreads and becomes standardized, but such established firms it is argued – despite the rhetoric of some of the more R&D savvy – do not face sufficient incentives to pursue the research that might lead to such radical innovation and thus undermine their existing market base (Henderson, 1993). There are no theoretical guarantees and scarce real-world precedents that the creators of radical innovations are able to control the new technology for long or to realize a net profit from its use.
6. *Indivisibilities and specificities*: And, finally, some basic research requires large dedicated investment in unique equipment and highly specialized skills of individual scientists.

Observing the Dynamic Nature of the Innovation Process: Technological Trajectories

Inventions of industrial importance are rarely one-off occurrences, regardless of where they originate. Instead, the innovation process typically involves quite a number of steps carried out over time, with each step leading to the next. The so-called 'linear hypothesis' of the relationship between basic science and applied technology claims the following: since, by definition, basic R&D precedes and enables applied R&D, the work at institutions specializing in basic R&D, largely universities, necessarily precedes and makes possible the work at institutions specializing in applied R&D, largely companies in the private sector.

There has been frustration, however, in trying to empirically demonstrate such a simple dichotomous causal relationship between public and private R&D work. In other work Rosenberg (1994) points to the bi-directional flow of knowledge between high-powered corporate research laboratories and universities, on what he calls the 'two-way street' of technology exchange. In pharmaceuticals, Cockburn and Henderson (1996) examine the immediate flow of knowledge between public and private researchers in a bibliometric-type analysis of co-authorship of research papers. They find a significant amount of co-authorship between public and private researchers and take that to imply that the 'simple linear model of the relationship between public and private research' is 'misleading'. In agricultural research Huffman (1998) dismisses a direct one-way relationship between basic and applied work, citing the long history of practical problem-solving on the farm driving the agenda for basic research at public land-grant universities.

Much of the literature in marketing emphasizes the importance of listening to customers as a source of new product innovation and product improvement ideas, and indeed, innovations can be the result of feedback from clients or from those marketing personnel who sell or service the product. Even in these cases, however, once the idea is introduced within a firm and committed to, a rather linear process of research, development, production and marketing will follow. A consensus holds that there is some kind of systematic and mutually beneficial cause-and-effect relationship between university research and industry R&D, but not one that is simple, nor necessarily linear.

More realistic views of the growth patterns followed by new technologies can be found in a general set of theories that describe whole families of new technological solutions being birthed by specific breakthrough discoveries that provide the technological or conceptual tools which allow companies both to focus on solving problems in their production technology (conducting supply-driven or 'technology-pushed' applied R&D) and to focus on meeting the demands of their customers in new ways (through demand-driven or 'demand-pulled' applied R&D). In a review of empirical economic studies of innovation, Cohen and Levin (1989) describe several economists' views of the interactions between basic and applied R&D along this kind of a causal sequence that evolves over time:

1. *Creation of technological opportunity*: A basic invention decreases the cost of seeking solutions to a practical problem, thus making the process of applied research more efficient by limiting the number of possibilities over which to search (Rosenberg, 1974; Evenson and Kislev, 1976; Nelson, 1982).
2. *Compulsive sequences*: A breakthrough in one area generates new problems or imbalances in the production system that require innovations in other areas (Rosenberg, 1969).
3. *Technology life cycles*: The nature of innovation changes in a predictable manner over the life span of its industrial application, beginning with a 'radical' innovation, proceeding through a phase of experimental 'product' innovations, and followed by a competitive phase of 'process' innovations as the industrial application becomes commodified (Abernathy and Utterback, 1978; Utterback, 1979).
4. *Technological paradigms and natural trajectories*: Technologies have a tendency to develop along a relatively clear path by repeatedly focusing on a particular class of problems or using a particular breakthrough discovery as a starting point. A cluster of closely related innovations form within a particular problem solving paradigm and it evolves with further research along a specific technological trajectory through time (Nelson and Winter, 1977; Sahal, 1981; Dosi, 1982, 1988).

The linear hypothesis, recast in terms of technological trajectories, describes basic research as the process of actively brainstorming and exploring the variety of possible problem-solving paradigms in a given field of human inquiry or technology. Several new technological paradigms might emerge from such basic work in a given field, but each is tested under the scrutiny of peer-reviewed science, the rigours of regulatory criteria, the selective forces of the technology market-place, and the popular voice of social acceptance or rejection. Occasionally, follow-on innovations, inspired by and clustering within the parameters defined by a successful new paradigm begin to take off, growing over time into a 'branch' on the technological 'family tree'. In the early 'basic' phases of such a natural trajectory, there is still much uncertainty about what direction the technology will take and whether it will be of any commercial value. Understandably, before profitable application has been reasonably proven, only those willing to shoulder a fair amount of investment risk or those who have a deep understanding or clear vision of the technology's potential can be found to financially back its further development. However, once the technology is proved, industry can be expected to become very interested in the technology and a broader set of investors can be found to back the incremental innovations necessary to adapt the technology to particular industrial products and processes, thus driving the growth of the trajectory through its middle and later 'applied' phases.

The Incentives and Constraints that Shape University Research

To really understand the differences between research in universities and private industry, one must compare the objectives and incentives faced by individual scientists located in both types of institution: 'the university' and 'the company'. The general objectives of virtually every scientist in the world can be summed up as the pursuit of (some combination of) the three 'F's – *fame, fortune* and *freedom*. What we mean by 'fame' and 'fortune' should be quite obvious; what we mean by 'freedom' is *self-determination*: a researcher's ability to select his or her own research objectives and strategies, the ability to 'be ones own boss'. However, constraints faced by researchers in the university and in industry differ significantly, resulting in different patterns of behaviour on the parts of individual researchers. At the end of the day, the collective effects of individual researchers' decisions made under the different systems of incentives and constraints determine the innovative results of the different sectors.

At universities, the three objectives of scientists are complementary. Academic fame generally results in individual fortune, in terms of higher salary, more lucrative opportunities for consulting and the like. Fame also gives rise to freedom. A well-regarded professor can generate more research funding and gain better access to research equipment and facilities, which in turn results in greater freedom to choose the path of research.

Furthermore, at universities, the objectives of research are at least broadly complementary to the other basic objectives of the university enterprise. Research that leads to innovations also helps to educate students and enhances the researcher's knowledge beyond the scope of previous training or research projects.

The output of university researchers is typically embodied in the publication of scientific papers and books, in the training of graduate students with knowledge and skills that they will carry on with them, and in the awarding of patents and other forms of intellectual property. Particularly for peer-reviewed, scholarly publications, the most important criteria by which success is judged are the significance of the problem and the originality and creativity of the solution posed. This pair of criteria – originality and creativity – drive university scientists to differentiate and diversify their projects, their careers and themselves, ultimately creating a broad portfolio of problem-solving paradigms and specific innovations, most of which may be merely 'interesting' to the layperson observing from outside the given scientific field and most of which may not be at all practical or cost-effective for a profit-seeking enterprise. However, occasionally – as this perspective suggests – the relentless drive at universities for diverse, creative research results in important breakthrough discoveries.

The Incentives and Constraints that Shape Industry R&D

In most cases, researchers working in industry are subject to a somewhat different set of constraints, which arise mainly from the need to show short-term profitability. While some research centres in industry operate like universities – as is the case with Bell Labs or Xerox PARC – these are the exception rather than the rule: most industry research efforts are quite targeted, and scientists in industry tend to have less freedom than those in academia. Weight is given to research that is expected to contribute directly to the firm's ability to create value by increasing revenues or reducing costs.

Much of the research and the development work in industry is managed sequentially, in phases leading up to the introduction of new production processes or new products. Millions of R&D dollars are spent, for example, on routinely pre-testing drugs, chemicals and other such products for regulatory approval. In making investments of this magnitude, private firms spend primarily on the improvement of specific proprietary industrial processes or products that are owned or licensed by the company and that are very certain to yield a positive rate of return within a reasonable time horizon, such as the term of a patent.

Research Specialization and the Division of Labour Between Universities and Industry

As a result of these differences, one expects university research to result more often in ideas and methodologies (radical innovations) that are original and differentiated from existing processes and products already found in industry, while research within industry is more likely to result in (incremental) innovations that are derived from and improve upon existing products and processes. Since much of the reward in academic circles is given to new ideas and creative concepts, with less attention for their ultimate implementation, research products of universities that do promise eventual commercial application tend to provide only a starting point that requires significant amounts of follow-on innovation, further development and refinement.

Such research results in the university can be patentable, even when the potential of their commercial application is subject to much uncertainty. In their embryonic stage university research products may not be interesting or promising to the established firms in an industry, especially when the research does not appear to have links to the firms' existing production processes, product lines or strategic plans. This suggests that some research products need first to pass through an incubation phase that will continue their development in directions that lead closer to the demand side.

Investors specializing in high-risk, high-return opportunities may be interested in developing some of the university's more promising research products. Indeed, a large number of start-up companies in information technology, biotechnology and other fields have been established through alliances of researchers, venture capitalists, and entrepreneurial business managers and salespeople, expressly to take a new innovation from its initial stage closer to a final product. Often these start-ups are then absorbed within large corporations who rely on the start-ups to complement their own research laboratories. For example, Cetus was purchased by Hoffmann-LaRoche, while Calgene was taken over by Monsanto. The major corporations have a much stronger apparatus for product testing, as well as marketing and production. These corporations rely on start-ups to complement their research laboratories and other capabilities, and in turn start-ups rely on univerity reasearch. These patterns of development, especially as manifested in biotechnology and information technology, thus suggest complementarily between corporate and university research.

The process of innovating and introducing new products can for most cases be generalized and simplified into a few basic stages. These include research, development, production and marketing. We consolidate the organizations involved in this process into three major categories according to our discussion: universities (U), entrepreneurial start-up companies (S) and established corporations (C). Table 6.1 presents patterns of the division of responsibility among these three kinds of organizations.

1. In some cases, all the innovation activities occur within a single established corporation (Pattern 1 in Table 6.1).
2. In other cases, university researchers come up with an initial discovery and license or sell it directly to an existing corporation (Pattern 2).
3. In another set of cases, university researchers develop an idea, sell it to a start-up, or obtain financing from venture capitalists to spin it out as a start-up company. Then the start-up company commercializes and markets products based on the new innovation[2] (Pattern 3).
4. A variant of Pattern 3 occurs when the start-up company is taken over by an existing major corporation which then commercializes and markets products based on the original university innovation (Pattern 4).
5. In a more rare variant of Pattern 3, university researchers come up with an idea, sell it or spin it out to a start-up company, which itself then grows to become a major corporation based on the new technology (Pattern 5). Examples include Genentech and Chiron.
6. The process of technology transfer is not limited to universities. A fairly recent phenomenon has been a surge in out-licensing of intellectual property by companies, and it has become an serious source of earnings for corporations with strong research capacities. For example, IBM earned close to US$1 billion in 1998 from royalties that are largely outcomes of its own in-house research (Valery, 1999). In some of those cases of corporate out-licensing, such as at Lucent, practitioners or researchers within the company may come up with an idea that does not fit closely within the firm's core specialization; they are encouraged to refine it, obtain financing from venture capital or other such sources, and start their own firm to commercialize the idea separately from the parent company (Buderi, 2000) (Pattern 6). It is not a brand new idea, however; Sony, for example, was founded by a Masaru Ibuka and Akio Morita when they obtained rights to the transistor by licensing patents from Bell Laboratories (Morita, 1988).
7. Finally, it has also been recently observed that, on occasion when company researchers have created patented technologies that do not fit closely enough with the firm's core business, the company will donate the patents to a university to further refine and manage its commercialization (Pattern 7). This can result in a significant tax write-off for the donor firm.

University *offices of technology transfer* (OTTs) are especially important in providing the link between university inventors and existing companies (in Pattern 2) as well as between university inventors and

Table 6.1. Common patterns of the division of labour of the innovation process.

Pattern	Research	Development	Production	Marketing
1	C	C	C	C
2	U	C	C	C
3	U	S	S	S
4	U	S	C + S	C + S
5	U	S	C = S	C = S
6	C	S	(then any of 3–5...)	
7	C → U	(then any of 2–5...)		

U, universities; S, start-up companies; C, corporations.

[2] It should be clear that technology transfers go well beyond the simple licensing of intellectual property. In fact, often, university professors and ex-graduate students may start new companies without any formal transfer of intellectual property rights.

investors in new start-up ventures (in Patterns 3–5). In the next section we discuss how OTTs are organized and how they perform in meeting these goals.

University OTTs: Objectives, Challenges and Operations

OTTs were established to facilitate the development and utilization of commercially viable innovations discovered by university and government scientists. The Research Corporation, founded in 1912 by a University of California-Berkeley faculty member to sell rights to use patents of several affiliated universities, was a predecessor to the contemporary OTT (Mowery *et al.*, 2001). Several leading research universities, including the University of California, Stanford, MIT and the University of Wisconsin, established their OTTs over 30 years ago. However, it was the passage of the Bayh–Dole Act in 1980 that catalysed the development of technology transfer programmes at most institutions. This Act of Congress created a uniform patenting policy among all the federal agencies that fund research, which allows research institutions to retain intellectual property title to the material and products invented by their employees doing research under federal funding. A survey of 106 university and government research laboratory OTTs conducted by several of the current authors (Castillo *et al.*, 2000) found that 78% of the OTTs surveyed were created after the passage of the Bayh–Dole Act. Overall, the Association of University Technology Managers (AUTM) reports that in 1980 only 25 offices existed, but a decade later, by 1990, 200 such offices existed (reported in Mowery *et al.*, 2001).

OTTs are subject to faculty, administration and state priorities

The initial impetus for establishing OTTs, even prior to the Bayh–Dole Act, was the concern by faculty and administrators that many of their most promising ideas were not being sufficiently utilized by the private sector. Universities came to recognize that private firms would not be interested in developing university-spawned technologies unless they could obtain (often exclusive) rights to them. Two other objectives of OTTs have grown more prominent over time: the provision of legal and intellectual property management services to university researchers and the collection of licensing royalty revenues for the university. When university researchers wish to commercialize their ideas, OTTs provide a formal, above-board and relatively effective mechanism. The revenues generated by OTT licensing, while still only a minor percentage of universities' operating budgets, have grown substantially. Beyond their monetary value the growth of such royalty revenues serves to demonstrate the success of OTTs in diffusing the fruit of the university's research.

In a study undertaken to review the economic effects of the Bayh–Dole Act, Jensen and Thursby (2001) surveyed 62 research universities concerning their technology transfer activities for the years 1991–1995. Questions were asked of the technology transfer offices, the faculty, and the administrations about their objectives for technology transfer. The different parties' responses demonstrate the differences that exist within the university over technology transfer. Administrators tend to consider technology-associated revenues most important, while the faculty see the ability to attract research sponsorship as paramount (Table 6.2). OTTs often operate under somewhat conflicting mandates from their administration and faculty and emphasize the more immediate and tangible outcomes of executed licenses and commercialized inventions.

Beyond what the various parties report to be priorities for technology transfer, actual investments of time by OTT staff in particular activities reveal how OTT staff respond to (and perhaps balance) the conflicting demands placed on them (Castillo *et al.*, 2000). OTT officials undertake multiple tasks: they scout and assess inventions at their institutions, make patent applica-

Table 6.2. Divergent priorities: ranking the importance of university technology transfer outcomes by OTT officers, administrators and faculty.

Outcome	Technology transfer officers' priority ranking	University administrations' priority ranking	University faculty members' priority ranking
Revenue	1	1	2
Inventions commercialized	2	3	3
Licences executed	3	4	5
Sponsored research	4	2	1
Patents granted	5	5	4

Source: Jensen and Thursby (2001).

tions, market patent rights to possible buyers, mediate contacts between their institution's personnel and outside investors, and monitor and enforce licensing and research contracts. The time spent on a successful invention by an OTT is typically split 50–50 between efforts expended prior to and efforts expended after patenting. Prepatenting activities involve solicitation of invention disclosures from faculty, evaluation of inventions, and assessment of their economic potential. Patent preparation involves, on average, only about 10% of the total time and efforts of OTT staff[3]. Commercialization efforts, including the negotiation of a license, involve about 25 to 30%, and follow-up activities of monitoring and enforcement involve only about 10% of an OTTs expended time (see Table 6.3).

Prior to the mid-1990s, private universities appeared to put more effort than public universities into securing earnings from technology transfer. Private universities emphasized market-related activities (evaluating inventions and assessing markets) and spent relatively more time enforcing and monitoring contracts[4].

OTT decisions of what to patent, and when

One of the greatest challenges faced by offices of technology transfer is identifying and separating out commercially promising

Table 6.3. Revealed priorities: percentage of time on the job that public and private university OTT officers spent on various activities in 1999.

Activity	Public universities (% time)	Private universities (% time)	Both (% time)
Soliciting ideas	11.3	7.8	10.0
Evaluating inventions	14.1	16.6	15.0
Assessing markets	12.5	13.6	12.9
Referring inventors	3.8	2.4	3.3
Preparing patent applications	8.6	10.1	9.1
Drafting licences	24.3	24.6	24.4
Enforcing patents	2.8	5.0	3.6
Monitoring contracts	7.2	11.5	8.7
Other	15.8	10.1	13.9

Source: Castillo *et al.* (2000).

[3] This figure is in accordance with the relatively low priority given to patenting by all of the parties in the university in the Jensen and Thursby survey (see Table 6.2).
[4] Spending more time on monitoring contracts also increases contact time with clients and builds and enhances the social networking that may lead to future transfers.

ideas from among the large volumes of publicly oriented knowledge constantly being created at the university. The probability that an OTT will patent a certain innovation is greater if:

- a technology is clearly patentable;
- a technology has favourable cost–benefit considerations (good commercial prospects or a potential buyer is already interested in the technology);
- a potential buyer offers to patent collaboratively; or
- the OTT is approached and encouraged by a faculty inventor.

If none of these conditions are met, the OTT may chose not to patent the invention but rather to give the the faculty inventor the right to utilize and pursue it. Some faculty inventors have used unpatented technology to start firms.

In survey interviews, most OTT officials agree that, despite the importance of revenue earnings from technology transfer, in the decision of when to patent, financial concerns do not dominate academic considerations. While a short delay in publication may be imposed for the practical purpose of allowing a priority date to a patent application, faculty members are permitted and encouraged to publish the results and to compete in the academic research race (Postlewait *et al.*, 1993).

Allocation of revenues within the university

There are a variety of different formulas used as university policies for the allocation of OTT revenues from licence royalties. The most common formula is equal sharing among the university (33%), the department (33%) and the employee inventor (33%), while another commonly reported alternative is a 50–50 sharing between the university and the inventor (Castillo *et al.*, 2000). Jensen and Thursby (2001) report average net revenue distributions going to university (35%), department (25%) and faculty inventor (40%). When a department shares in the royalties it is often justified as taking into account the research and collaboration efforts of the entire team that led to an invention or indirectly helped to make it possible. In this way the whole organization that contributed to the success, and not just the principal researcher, benefits from the proceeds. However, patterns of royalty sharing that dilute the inventor's share may increase their incentive to depart from the university.

Measuring the Performance and Results of OTTs

The results of OTT activities can be measured in several dimensions. OTTs issue patents and licences, help establish start-up companies and collect revenues in the forms of signing fees, royalties, penalties and equity in start-up companies. OTT activities also result directly or indirectly in research contracts, grants and donations to the university.

University technology licensing revenues: cash and equity

Figure 6.3a and b shows the distribution of technology licensing revenues for the top 30 universities in 1992 and 1995 out of a sample of 106 institutions compiled from AUTM data by our Castillo *et al.* survey[5]. In 1992, six of the institutions were big hitters, with total earnings each between $10 million and $30 million. In 1995, the top three universities had each exceeded $30 million in earnings with an additional four universities bringing in around $10 million. The overall earnings for the top 30 in 1995 was up roughly 80% from the overall earnings for the top 30 in 1992[6]. This may

[5] The distribution of earnings in Figs 6.3(a) and (b) do not show equity or other forms of technology revenue.

[6] The figures presented here are in nominal terms.

(a)

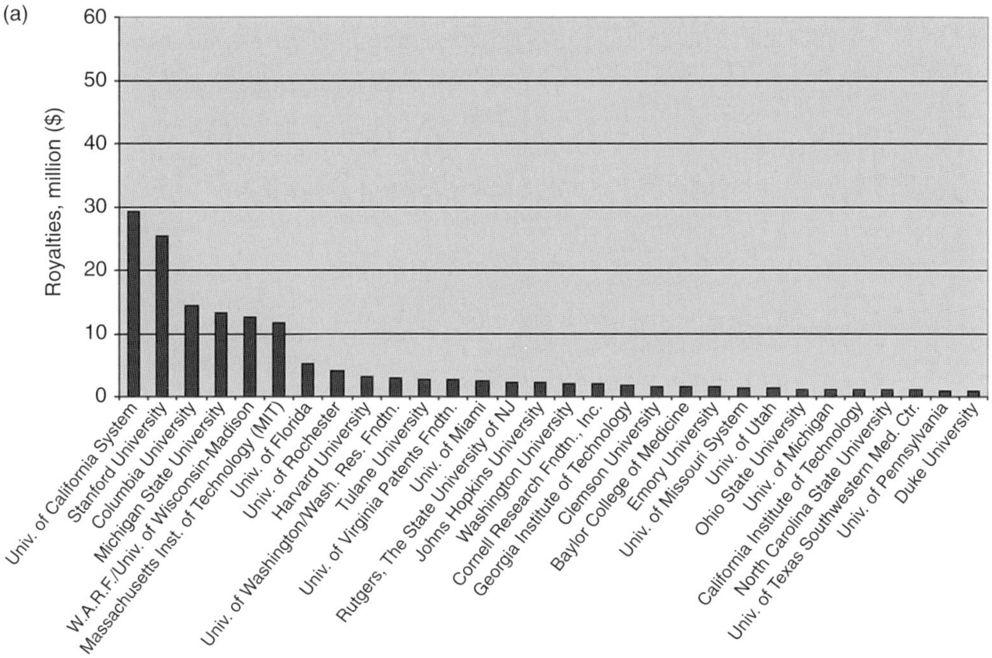

(b)

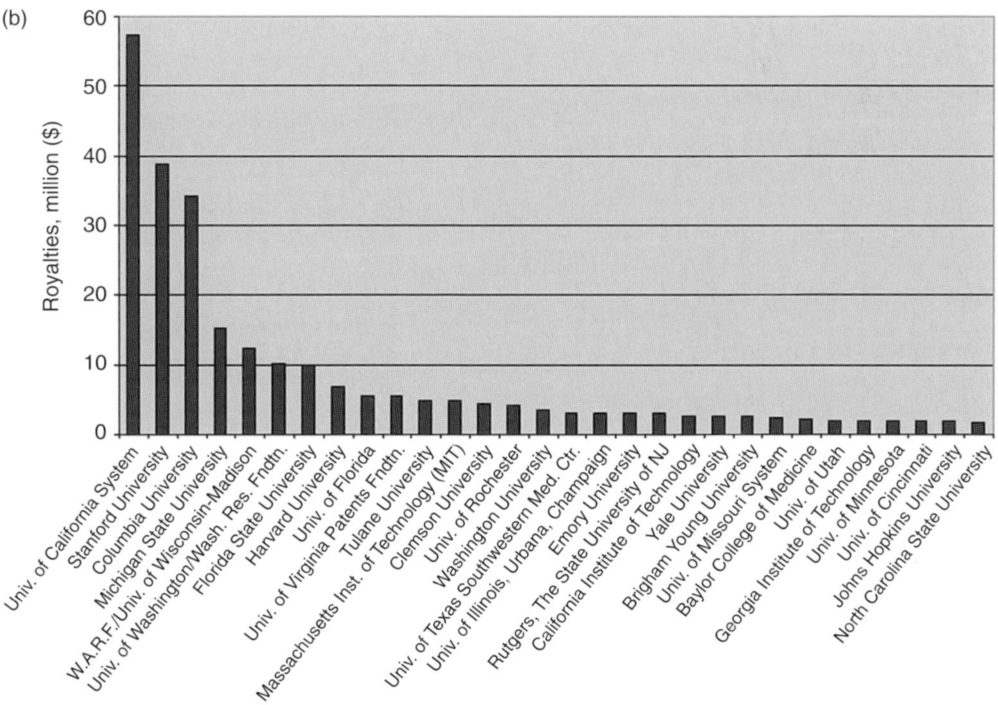

Fig. 6.3. (a) Distribution of royalties for the top 30 US universities (1992); (b) Distribution of royalties for the top 30 US universities (1995).

reflect royalties from new patents, higher royalties generated by products resulting from existing university patents and more aggressive commercialization on the part of the OTT in general. There were two turnovers in the top ten from 1992 to 1995, and eight turnovers in the top 30.

Table 6.4 presents the results of the authors' survey regarding preferences among OTTs for compensation in cash or in equity of the licensing firm. Forty per cent of private universities take the position that 'my institution believes equity and cash are of equal value', relative to just 27% of public institutions. These numbers coincide with the AUTM data indicating that a greater proportion of the top 20 institutions that took equity in companies were private as opposed to public. The rule is not absolute, however; while private universities seem generally more open to equity than pubic schools, the top three universities most often accepting equity are public: the University of Arizona, the University of North Carolina at Chapel Hill and Rutgers University (AUTM, 1997). The survey results in Table 6.4 indicate that many OTTs are open to accepting equity as compensation. These provide a single time-slice of an ongoing change in policy and practice at many universities in the 1990s: allowing the university to take ownership in private firms. Jensen and Thursby reported that, by 1995 for the 62 institutions they surveyed, 23% of the licence agreements included equity. Yet, they caution that the magnitude of university ownership of private equity should not be overstated, as only 3% of the total OTT revenues were in the form of equity.

Even for the top ten technology licensing universities in 1995 the total technology transfer earnings average just 2.5% of the university research budgets. Mid-range institutions generated revenues averaging just 1.5% of their research budgets, and the bottom 10% generated less than 1%. It is clear that technology transfer revenues do not pay for university research. Furthermore, much of this money does not return directly to the university's research programmes but instead goes toward administrative costs.

Licensing revenues, however, are only a small part of the full benefits that accrue to society from university innovations and technology transfer activities. These innovations in the end generate much more income than the OTTs are ever able to realize. If a university receives only 3–5% of the sales, that means 95–97% must go elsewhere. University innovations lead to spin-off products and, within the legal bounds, copycat or follow-on products. In most cases the returns to these accrue to the firms that shoulder the risk and make the necessary investments to develop processes and products from early stage university ideas and bring them to market. Consumers benefit handsomely from being offered new kinds of products that meet their needs in new and valuable ways. Thus, the revenue numbers presented here tell only a small part of the whole story.

Numbers of university patents

The magnitude of patenting by the university sector, while still small relative to industry, has increased substantially in the

Table 6.4. Equity preferences reported by OTT officials (1999).

Question	Public universities, %	Private universities, %
'My institution will not or is not allowed to consider equity'	10.0	13.3
'My institution prefers cash payments but will consider equity'	63.3	46.7
'My institution believes cash and equity are equally viable forms of payment'	26.7	40.0
'My institution prefers equity'	0.0	0.0
Total	100.0	100.0

Source: Castillo *et al.* (2000).

last 30 years in absolute numbers of patents and it has increased more quickly than the overall trend of patenting in the USA (see Fig. 6.4). The increase in patenting by universities began in the 1970s, even before the passage of the Bayh–Dole Act, but the pace quickened in the 1980s after the Act's changes were brought to bear (Mowery *et al.*, 2001), suggesting that the economic forces that brought that legislation about were perhaps already at work in the years prior to its enactment. Furthermore, the number of patents assigned to universities per research dollar spent at universities has more than tripled (Henderson *et al.*, 1998). This increase in universities' '*propensity to patent*' is due both to an increase in the number of patentable inventions resulting from changes in the orientation of underlying university research agendas and to a general decline in the threshold of university standards for patentability (Henderson *et al.*, 1995). The latter phenomenon likely reflects the large influx of patents from the many smaller institutions with generally weaker research programmes and less experience in patenting their inventions.

Figure 6.5 shows the distribution of the numbers of patents accumulated by the 30 top patenting US universities in 1999 according to the AUTM data. The University of California system dominates the field[7] followed closely by a strong cadre of six other big-hitting universities, each with a portfolio of over 80 patents by 1999. After these top seven, the rest of the top 30 universities levels off around an average of 50 patents.

It should be noted that the role of a patent at a university is not as straightforward as it may be at a company. Jensen and Thursby report that in just 28% of the cases is a patent issued at the time of a licence. In most of the other 72% of cases a patent application has been filed, but it is clear that much of the inventive work coming out

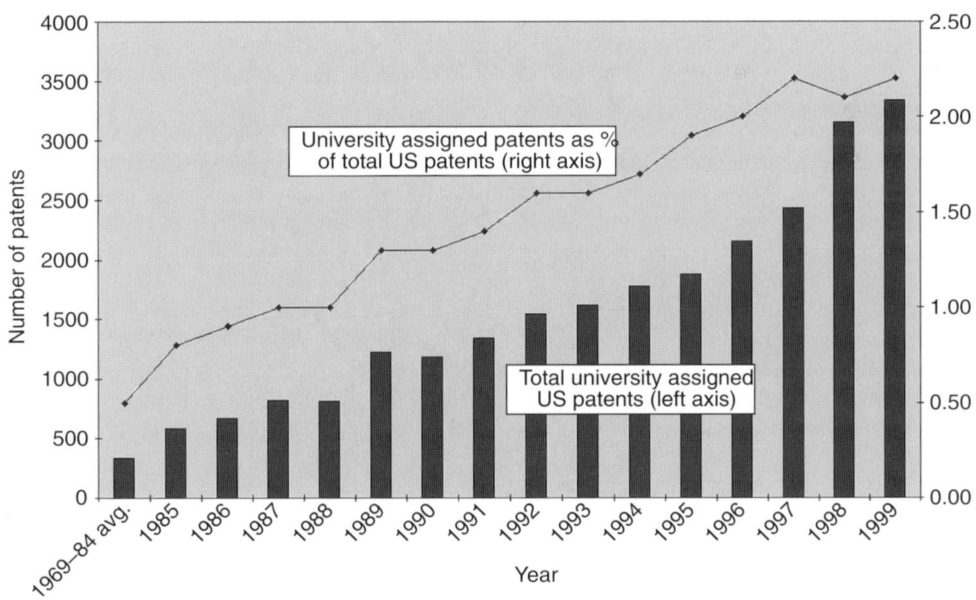

Fig. 6.4. Patents assigned to US universities: number and percentages of total US patents per year, 1969–1999. Data source: US Patent and Trademark Office.

[7] The UC system represents ten campuses, including several that would stand alone as prominent research universities – UC Berkeley, UC San Francisco, UC Davis, and UCLA. None the less, a single central office of technology transfer exists at the University of California Office of the President which coordinates the technology transfer activities of the entire system in cooperation with autonomous licensing offices at the largest campuses.

of the univeristies is transferred out at the pre-patenting stage. The number of granted patents is just the tip of the iceberg when it comes to indicating the value generated by university research.

Licensing revenues and numbers of patents in different fields of technology

Table 6.5 compares the distribution of the average income and number of patents by academic discipline according to the results of the Castillo *et al.* survey. Biotechnology looms largest in the universities, making up a large part of the fields of both medicine, which dominates in terms of income (earning 55% of the total) and patents (taking 46% of all patents), and agriculture, which generated about 9% of the income and 8% of the patents. Engineering and physics accounts for about 24% of the income and 34% of the patents. Computer sciences generated approximately 5% of the income and 4% of the patents. In an in-depth analysis of three universities, the University of California, Stanford and Columbia, Mowery and colleagues (Mowery *et al.*, 2001) found similar patterns. At Columbia, almost 75% of all inventions disclosed to the OTT between 1981 and 1995 were biomedical, the remaining 25% were mostly software and electronics. At the University of California, by 1990, roughly 65% of all inventions disclosed were biomedical. At Stanford, however, only about 20% of all invention disclosures by 1990 were biomedical, while about 30% were in software.

This distribution reflects, to some extent, the division of research labour between universities and industry as discussed earlier. Many of the technological trajectories played out in engineering and computer hardware are now squarely situated in the mature competitive phase characterized by 'process' developments with many of the industrial applications becoming commodified. Uncertainty is smaller and the markets are well defined. In contrast, biotechnology is still in its early phase, i.e. it is very risky and is proceeding without well-defined markets. In close relation to the uncertainty and the investment requirements of the stages in the life cycles or the points in the natural trajectories of innovations, profitability is a key factor in determining the allocation of research between private firms and universities in each of the disciplinary or industrial fields. This perhaps explains why much agricultural research has historically been done in universities, though it is mostly related to established mature innovations.

In addition, the data in Table 6.5 show that for public universities, relative to private universities, a greater proportion of patents and a higher percentage of technology income are in agriculture. Many of the largest public universities are Land Grant institutions with historically strong schools of agricultural science (such as UC Davis, Texas A&M, Cornell University, University of Wisconsin-Madison, Ohio State University, Florida State University, Iowa State University, Purdue University, etc.) In addition, while we do not present time-series data here, there is some evidence that, even for these public universities, the

Table 6.5. Average proportion of licensing revenues and patents by academic field (1999).

	Average revenues, as % of total			Average number of patents as % of total		
Academic field	Public	Private	Both	Public	Private	Both
Agriculture	10.3	6.9	9.1	9.5	5.5	8.1
Engineering and physics	19.8	32.3	24.1	32.2	38.7	34.1
Medicine	55.2	55.2	55.2	44.4	51.0	46.4
Computer science	5.5	5.2	5.1	3.9	4.2	4.0
Other (including chemistry)	10.3	0.5	6.6	11.2	0.5	7.4

Source: Castillo *et al.* (2000).

composition of patents and invention disclosures has shifted from the agricultural field to the more profitable biomedical field (Mowery *et al.*, 2001).

Logic suggests that university royalty rates should be higher in industry sectors with higher overall profit rates or where the relative contribution of research to profitability is higher. University royalties should be higher particularly in situations where the university has provided a more commercially developed product and thus has contributed more to reaching the market. Table 6.6 shows royalty rates in a variety of technology fields. The high royalty rates for medical research reflect the relatively high contribution of the university research to the value of marketed products. The average value of the minimum annual royalty payment for diagnostic and therapeutic innovations is much higher than other categories, especially agriculture and medical research. Medical therapeutic licence contracts have high up-front fixed fees. However, there does not appear to be a correlation between the up-front fee and the annual royalty payment. This suggests that there is independence between technology fields. We do not yet have enough data to test for correlation between up-front fees and royalty payments within technology fields.

Numbers of companies started

For a variety of reasons, corporations do not often give university innovations a very enthusiastic reception; to say it another way, Pattern 1 as presented in Table 6.1 is relatively scarce. Over the last two decades, recognition of this corporate reluctance has prompted OTTs to seek other avenues for the commercialization of promising university inventions. A common strategy that has emerged for OTTs is to facilitate negotiations and relationships between university researchers and venture capitalists in order to start up companies that can finance the further development and commercialization of university ideas. A shortlist of major companies spawned by university OTTs in the San Francisco Bay area includes Sun Microsystems, Cisco, Genentec, Chiron and Amgen, all of which followed Pattern 5 in Table 6.1. Table 6.7 reveals that across the USA the number of start-up companies associated with university innovations was increasing through the 1990s.

University-spawned start-ups play an important role in increasing competitiveness. For example, while the first computers were developed in universities and commercialized by giant companies such as IBM, some of the subsequent breakthroughs that made them more affordable and user-friendly resulted from university innovations made available to consumers via start-ups. Many foundational elements of the Internet were born this way as well; for one, Netscape came out of work at the University of Illinois. Likewise, most biotechnology-based medical treatments are the result of university research.

Table 6.6. Key indicators of earnings by field of technology.

Type of product	Average royalty as % of sales	Average value of up-front fixed fee (US$)	Average value of minimum annual royalty payment (US$)
Agricultural	3.9	20,105	6,928
Engineering	6.3	32,236	16,397
Medical (therapeutics)	6.3	98,437	83,010
Medical (diagnostics)	6.6	36,906	46,227
Medical (devices)	6.6	37,115	38,775
Medical (materials and reagents)	9.4	12,942	4,444
Other (includes chemicals)	7.6	78,583	42,687
All fields	6.6	45,189	34,066

Source: Castillo *et al.* (2000).

University innovations, once transferred to the private sector, accelerate the rate of product change and force established companies to take action and to adjust to a new reality. In industries such as pharmaceuticals and agricultural chemicals, there are large established players who, if left unchallenged, might choose to slow the rate of innovation and product development in order enjoy a longer time horizon to take advantage of their positions in certain market segments. When start-up companies introduce new products and innovations in those markets, they force the large corporations to respond by reducing prices, improving their own product mix, or perhaps taking over the start-up and marketing the new innovative products alongside their own old line of products.

Skewed distribution of outcomes from university research

The lopsidedness of OTT earnings, as seen in Fig. 6.3a and b, reflects a very important general fact about the odds on returns to investment in upstream research: it is quite rare to make a hit, but when it is made it can be very big. A handful of university patents, including the Cohen–Boyer patent for genetic recombination (Stanford and University of California-San Francisco), the Hepatitis B vaccine patent (University of California-San Francisco), the Taxol patent (Florida State University) and the Gatorade sports drink patent (University of Florida), have each generated tens of millions of dollars over their lifetimes. Most university patents, however, generate incomes ranging from zero to just a few tens of thousands of dollars.

Scherer and Harhoff (2000) have compiled extensive data on the financial returns to broad sets of patents, both industry and university, and find that this kind of skewness of returns is typical for outcomes from research. They find that the value distribution of patents can be so skewed that the average rate of return on investments to a portfolio of patents can be completely determined by the size of its few biggest outliers, which raises questions about the viability of typical portfolio-management strategies. A patent portfolio where more than half of the 'investments' fail to earn anything at all and where only one in 20 yields appreciable returns appears indeed to be a normal state of affairs. It is reasonable then to ask what the determinants of success are for university technology patenting and transfer, in order to help pick the winning technologies and design mutually beneficial policies for technology transfer.

Successful Technology Transfer is a Function of OTT Age, Research Quality and Inventor Involvement

Several general factors have been observed to play roles in the overall success of university efforts at technology transfer, and attention to these factors can be important for company managers looking to universities as a source of new technologies. First, the age and professional experience of a

Table 6.7. Spinning out: the number of start-ups based on university inventions, 1980–1999.

Year(s)	Number of institutions reporting	Start-ups formed
1980–1993	154	1169
1994	156	241
1995	172	223
1996	168	248
1997	171	333
1998	176	364
1999	188	344
Total		2922

Source: AUTM Licensing Survey, FY 1999.

university's OTT matters. It takes time to develop an expertise in intellectual property management, to build up a portfolio of patents, and to sucessfully sell and manage the many contingencies that can arise with technology licences. There is typically a 4- to 9-year lag between making an academic research discovery until the first introduction of a new commercial product or process based on that discovery (Mansfield, 1991). Even after the market debut of a new technology, slow rates of technology diffusion and adoption[8] can mean that royalties grow only gradually. As a result, younger OTTs tend to lag significantly in their earnings relative to older OTTs. Indeed, the top ten-ranked OTTs in terms of revenue are predominantly the older ones; however, over the last 10 years their share of the total number of university licences and total university royalties has decreased, reflecting the greater licensing and revenues of the younger OTTs as they mature (Table 6.8).

The fact that the dominance of the top ten OTTs has decreased more in total numbers of licences than in royalties results from the time lag for newer OTTs between licensing and royalty collection as more and more new licences have come on line against a backdrop of many older OTTs that were highly ranked because of older individual big hits (such as Taxol at Florida State and the Cohen–Boyer patent of UC San Francisco and Stanford). Since it is impossible to predict accurately where the next big research breakthrough will occur, there is an element of sheer chance in capturing those kinds of big research hits. For a university, the longer it has an OTT business, the greater is the statistical probability of licensing a breakthrough invention. For a company or investor seeking potentially lucrative undeveloped technologies, approaching older OTTs means working with more experienced professionals who have dealt with success in the past and know to handle it when it happens again.

Another identifiable force behind the success of a university's technology transfer is, not surprisingly, the university's academic standing together with the amount of research dollars spent on research in the fields that tend to matter commercially, such as biology, medicine, engineering and computer science (as indicated in Tables 6.5 and 6.6). Highly ranked research institutions such as the University of California, Stanford, Columbia and MIT are the leading technology-income earners among US universities (Fig. 6.3b). While many of the applied programmes at smaller universities can and do produce technologies that are highly valuable, all else being equal, one should look for cutting-edge technologies at leading institutions. As a rule of thumb, approach the OTTs at schools where you would want your children to go to college and which have respectable programmes in the 'majors' in which you are particularly interested.

Finally, ongoing inventor involvement is essential to successful technology transfer. Only the university inventor can communicate all the ins and outs of an invention to a

Table 6.8. Relative performance of the top ten university OTTs in 1992, 1995 and 1997.

	Top 10 in 1992, as % of 1992 totals	Top 10 in 1995, as % of 1995 totals	Top 10 in 1997, as % of 1997 totals
Licences	52.5	43.8	43.1
Royalties	71.9	67.5	64.1
Research expenditures	38.2	33.6	33.9

Source: AUTM (various years), reported in Castillo *et al.* (2000).

[8] The literature on diffusion suggests that the relationship between adoption (and thus earnings) and the age of an innovation is S-shaped, with a slow early build-up period, a subsequent rapid take-off, and then a slow-down approaching saturation (Griliches, 1957).

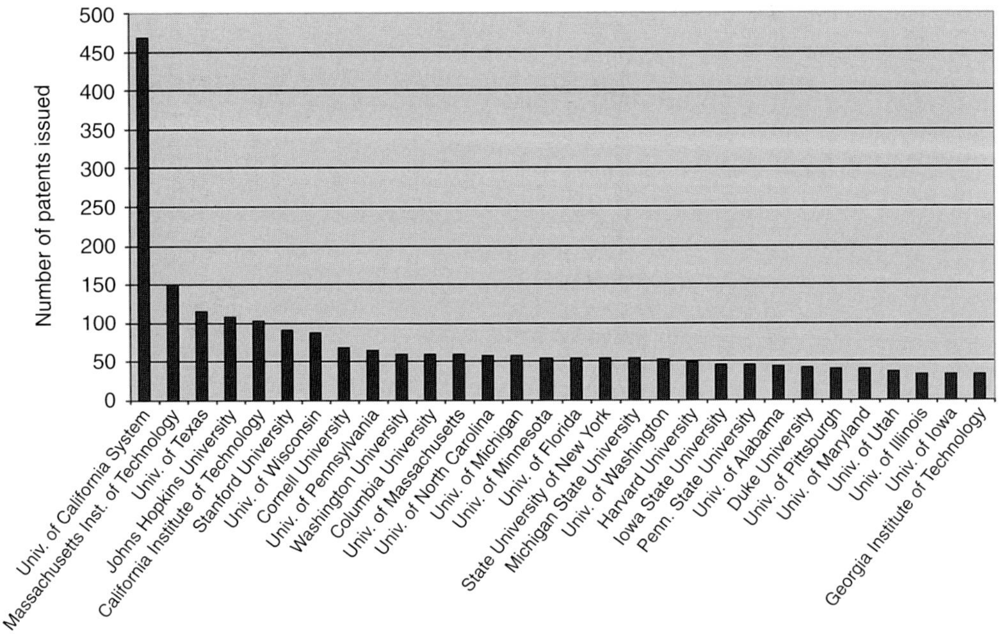

Fig. 6.5. Distribution of number of patents assigned to the top 30 US universities in 1999 (see Table 6.9).

licensee who will develop, produce and market a product based on it. A successful transfer requires knowledge and know-how at both the sending and receiving ends. Jensen and Thursby find in their survey that most university inventions at the time they are licensed are little more than 'proof of concept' disclosures (48%) or laboratory-scale prototypes (29%), and they emphasize that most licensed inventions (71%) require inventor cooperation for successful commercial development.

Summary and Conclusions

General summary

Today, universities pursue multiple objectives, both public and private, and in so doing rely on a variety of funding sources, also both public and private. This diversification allows university scientific research to emphasize work that is largely described as 'basic' but which occasionally and rather unpredictably results in practical technologies that can be very valuable when applied commercially.

Technologies have a tendency to develop along relatively clear paths or 'technological trajectories', resulting from research that repeatedly focuses on a particular class of problems or uses a particular breakthrough discovery as its starting point. In the early or 'basic' phase, research in a new paradigm tends to be more uncertain, and its return horizon is often quite long. If public support is not available it is not likely to be pursued. However, if and when more certain 'applied' innovations, promising sufficient economic returns, emerge within such a problem-solving paradigm, venture or corporate investors will begin to be able to afford to support further R&D in that technology. The work leading to an eventual product release will be observed to have passed through four stages: research, development, production and marketing.

The very different financial constraints of universities and companies translate into rather different sets of objectives and

Table 6.9. Number of patents assigned to US universities in 1999.

Univ. of California System	468
Massachusetts Institute of Technology	151
Univ. of Texas	115
Johns Hopkins University	108
California Institute of Technology	103
Stanford University	91
Univ. of Wisconsin	87
Cornell University	69
Univ. of Pennsylvania	64
Washington University	60
Columbia University	59
Univ. of Massachusetts	59
Univ. of North Carolina	58
Univ. of Michigan	58
Univ. of Minnesota	55
Univ. of Florida	55
State University of New York	54
Michigan State University	54
Univ. of Washington	53
Harvard University	49
Iowa State University	46
Penn. State University	45
Univ. of Alabama	44
Duke University	42
Univ. of Pittsburgh	41
Univ. of Maryland	40
Univ. of Utah	37
Univ. of Illinois	34
Univ. of Iowa	34
Georgia Institute of Technology	34

incentives for the individual researchers they employ which encourage them to specialize at different stages within these patterns of technological growth, in response to the varying probabilities of success. The private objectives of all scientists can be summed up as a pursuit of some combination the three 'F's – *fame*, *fortune* and *freedom*. In the university these are pursued by seeking original and creative solutions to a diverse array of significant problems, resulting in a broad portfolio of university innovations, of which a few turn out to have enormous value but most offer little or no economic value. In industry, researchers pursue the 'F's by focusing research and developing technologies to contribute to the firm's ability to create value. Industry scientists are largely limited to working on improvements in proprietary processes or products that are held by the company which are likely to yield sufficient returns within a reasonable time horizon.

The link between university inventors and industry is provided by university OTTs, which manage the intellectual property of universities with the goal of commercializing inventions and earning revenues. Commercialization is achieved either by licensing a patented invention to an existing company, or by helping the university inventor to obtain venture capital to start up a company that will commercialize the technology. The essence of the technology transfer process is to supply companies with new ways to meet customers' needs.

Conclusions for policy makers and university officials

Government policies and support have enabled universities to discover and develop breakthrough technologies that have changed the economy. A string of massive public support for research in defence, aerospace, medicine, agriculture and the environment, along with federal funds for more targeted problems and financing of higher education, has historically supplied the foundation on which universities have flourished. US patent law, together with the Bayh–Dole Act of 1980, provide a framework that enables universities to retain title to their new technologies and innovations. Offices of technology transfer have begun to capitalize on the new legal possibilities for transferring knowledge that arises in the public sector and now plays a crucial role in developing the networks for exchange between university researchers and potential users of their research results in the private sector. The result is the engine of the educational–industrial complex: university research plays essential roles setting the pace of innovation in the economy, creating new technological paradigms, and breaking old technological bottlenecks. A powerful combination is achieved at the university in the co-production of human capital and intellectual capital, with education and training complementing research and inventing. Investment in university research does not just result in research for its own sake, but serves to constantly maintain and upgrade the knowledge infrastructure of the economy.

University research also can be viewed as an important source of competitiveness in the economy and could be promoted as part of anti-trust policy. If universities are encouraged to continue the innovation and commercialization of new technologies, they can be expected to generate a constant flow of competitive new ideas and new entrants into the economy.

Finally, it is reasonable to assume that the marketing of university research would benefit from economics of scale, meaning that, at least to an extent, the larger the represented research base, and thus the larger the portfolio of new technologies made available to license, the lower the marketing costs and the higher the probability of successfully selling licences. These conditions could be achieved, for instance, if several universities used a single OTT to market their innovations, a move that would increase the variety and depth of technologies available and thus increase the attractiveness of the combined OTT to potential business customers. However, in actuality, universities appear to be seeking to market their research results independently. Each university's first goal in marketing its research appears to be demonstration of its research vitality to potential public and private research funding sources rather than simple maximization of earnings from licensing. Even so, it would be much more efficient for technology transfer if interested companies could go to one place that hosts a full variety of research results and new product innovation opportunities. A viable alternative may be the establishment of a middle-man organization which respresents and markets the products of multiple university OTTs.

Conclusions for company managers

Marketing and strategic management scholars argue that there are two sources of ideas for the innovation of new products. The first source, advocated by the customer orientated approach, is 'listening to your customer's voice'. A company first identifies the needs of customers (both potential and existing customers and both their stated and hidden needs) and develops a new product; identifying the gap between customers' demand and the current supply offered by you and your competitors opens up a strategic opportunity.

The second source, supply-driven innovations – development of new products based on new technological capabilities – can be more risky. Inventions that are the outcome of university research are by their very definition supply-driven innovations (although it is possible to argue that most basic research arises as a response to society's more generally articulated needs). They have potential to succeed, but, to the extent that they were developed with technological opportunities and not consumers' needs in mind, their rate of acceptance by the market and their final value is uncertain. In fact, according to classical marketing-management perspectives, the pursuit of product innovation from the supply side is considered by some to be downright myopic. In his paper 'Marketing Myopia', Theodor Levit wrote, 'Another big danger to a firm's continued growth arises when top management is wholly transfixed by the profit possibilities of technical research and development' (Levit, 1960). Kotler (1997) expounds on the theme: 'Under the (product) concept, managers assume that buyers admire well-made products and can apprise product quality and peformance. However these managers are sometimes caught up in a love affair with their product and do not realize that that the market is less turned on. The product concept leads to a kind of marketing Myopia ... ' One way to avoid this downside of technology driven innovation is to conduct market research, asking customers what they think, at the research or development stage, before moving into production.

These two sources of product ideas, i.e. customers' unfulfilled needs and university research, can be used simultaneously and made to complement each other. Customer's needs can be used for short-term planning since it requires product improvement and upgrading while the commercialization of new research products is a more long-term activity. Implementation of applied research is always easier and faster than the implementation of basic research, which needs higher capabilities and resources. However, the rare company that succeeds in adopting and implementing basic research gains huge competitive potential.

One important conclusion is that private companies should build mechanisms to help them identify research done in the universities and screen it according to its marketing potential. Once a systematic identification of research is implemented, a company has a constantly updated base of new ideas to work with. The next step is to screen these ideas regularly and select according to both market needs and probability for successful research completion.

In closing, a short list of suggestions for R&D managers to consider:

1. Look beyond just the technologies in the current product market. Do not regard university research as remote or irrelevant. You will miss great opportunities or may get undercut by new technologies you did not see coming. Instead, try to identify successful research teams at universities. View faculties as resources on which to draw. Establish mechanisms for detecting new technologies and new talent in the universities. Learn before others what kind of work is going on in universities.

2. Develop an R&D strategy that takes into account technologies available from universities. Seek cost effective options for licensing and technology transfer. Shop what is 'on the market' from universities. In some cases you will be able to avoid reinventing the wheel. It may cost less to license rather than to make something comparable from scratch. Diversify your R&D strategy:

(a) work in house, but,

(b) recognize that the best value may be to augment from outside. For example, consider how Microsoft and Cisco regularly buy up small firms with proven technologies from outside and integrate them into the company.

3. It is essential to capture technologies at the appropriate stage in their evolution: not too early, not too late. It is the age-old balance of exposure to potential losses versus potential benefits.

4. Establish a venture fund to sponsor specific university projects or teams. You may be able to negotiate 'first right of refusal' from a university. If you are not ready to shoulder the full investment by bringing a technology in house. Your venture fund could partner with other investors to create an independent firm to develop the technology. This limits your exposure if the technology does not work out.

5. Keep an eye on the start-ups coming out of universities. They have already passed through one stage of screening for commercial viability and are a good indication of what is available. It is possible to approach such start-ups to license the technology they bring from the university or to contract them to develop the technology to meet your company's needs. This again may be more cost effective or a better way to manage the risk than to internalize the technology directly.

6. Move technologies out from your company that do not belong in your portfolio. Do your own technology transfer: license to other firms; create your own start-ups with the help of venture capital; donate blocks of patents to universities, non-profits or the government.

References

Abernathy, W.J. and Utterback, J.M. (1978) Patterns of industrial innovation. *Technology Review*, 41–47.

AUTM (1997) *AUTM Licensing Survey, Various Years.* The Association of University Technology Managers, Norwalk, Connecticut.

Buderi, R. (2000) Lucent ventures into the future. *Technology Review* 103, 94–106.

Castillo, F., Parker, D. and Zilberman, D. (2000) *Offices of Technology Transfer and Privatization of University Discoveries*, Department of Agricultural and Resource Economics, University of California, Berkeley.

Cockburn, I. and Henderson, R. (1996) Public–private interaction in pharmaceutical research. *Proceedings of the National Academy of Sciences, USA* 93, 12725–12730.

Cohen, W.M. and Levin, R.C. (1989) Empirical studies of innovation and market structure. In: Schmalensee, R. and Willig, R.D. (eds) *Handbook of Industrial Organization.* Elsevier Science, Amsterdam, II: 1059–1107.

Dosi, G. (1982) Technological paradigms and technological trajectories: a suggested interpretation of the determinants and directions of technological change. *Research Policy* 11, 147–162.

Dosi, G. (1988) Sources, procedures, and microeconomic effects of innovation. *Journal of Economic Literature* 26, 1120–1171.

Evenson, R.E. and Kislev, Y. (1976) A stochastic model of applied research. *Journal of Political Economy* 84, 265–281.

Griliches, Z. (1957) Hybrid maize: an exploration in the economics of technical change. *Econometrica* 25, 501–522.

Henderson, R. (1993) Underinvestment and incompetence as responses to radical innovation: Evidence from the photolithographic alignment equipment industry. *RAND Journal of Economics* 24, 248–270.

Henderson, R., Jaffe, A.B. and Trajtenberg, M. (1995) *Universities as a Source of Commercial Technology: A Detailed Analysis of University Patenting 1965–1988.* National Bureau of Economic Research, Cambridge, Massachusetts.

Henderson, R., Jaffe, A.B. and Trajtenberg, M. (1998) Universities as a source of commercial technology: a detailed analysis of university patenting, 1965–1988. *Review of Economics and Statistics* 80, 119–127.

Huffman, W. (1998) Finance, organization, and impacts of U.S. agricultural research: future prospects. *Knowledge Generation and Transfer: Implications for Agriculture in the 21st Century*, University of California, Berkeley.

Jensen, R. and Thursby, M. (2001) Proofs and prototypes for sale: the licensing of university inventions. *American Economic Review* 91, 240–259.

Kotler, P. (1997) *Marketing Management.* Prentice-Hall, Englewoods Cliff, New Jersey.

Levit, T. (1960) Marketing Myopia. *Harvard Business Review* (July–August), 45–56.

Mansfield, E. (1991) Academic research and industrial innovation. *Research Policy* 20, 1–12.

Morita, A. (1988) *Made in Japan.* Penguin Books, Harmondsworth, UK.

Mowery, D.C., Nelson, R.R., Sampat, B.N. and Ziedonis, A.A. (2001) The growth of patenting and licensing by U.S. universities: an assessment of the effects of the Bayh–Dole act of 1980. *Research Policy* 30, 99–119.

National Science Board (2000) *Science and Engineering Indicators*. National Science Foundation, Arlington, Virginia.

Nelson, R.R. (1982) The role of knowledge in R&D efficiency. *Quarterly Journal of Economics* 97, 453–470.

Nelson, R.R. and Winter, S.G. (1977) In search of a useful theory of innovation. *Research Policy* 6, 36–76.

Postlewait, A., Parker, D. and Zilberman, D. (1993) The advent of biotechnology and technology transfer in agriculture. *Technological Forecasting and Social Change* 43, 271–287.

Rosenberg, N. (1969) The direction of technological change: inducement mechanisms and focusing devices. *Economic Development and Cultural Change* 18, 1–24.

Rosenberg, N. (1974) Science, invention, and economic growth. *Economic Journal* 84, 90–108.

Rosenberg, N. (1994) Science–technology–economy interactions. In: Granstrand, O. (ed.) *Economics of Technology*. North-Holland, Amsterdam, 323–338.

Rosenberg, N. and Nelson, R.R. (1994) American universities and technical advance in industry. *Research Policy* 23, 323–348.

Sahal, D. (1981) *Patterns of Technological Innovation*. Addison-Wesley, Reading, Massachusetts.

Scherer, F.M. and Harhoff, D. (2000) Technology policy for a world of skew-distributed outcomes. *Research Policy* 29, 559–566.

Utterback, J.M. (1979) The dynamics of product and process innovation in industry. In: Hill, C.T. and Utterback, J.M. (eds) *Technological Innovation for a Dynamic Economy*. Pergamon Press, New York.

Valery, N. (1999) Survey: innovation. *The Economist* 350, after 52.

Chapter 7

Mergers and Intellectual Property in Agricultural Biotechnology

Alan C. Marco[1] and Gordon C. Rausser[2]

[1] *Department of Economics, Vassar College, Poughkeepsie, NY 12604-0708;*
[2] *Department of Agricultural and Resource Economics, University of California at Berkeley, 201 Giannini Hall, Berkeley, CA 94720-3310, USA*

Introduction

Over the past decade the structure of the plant-breeding and agricultural biotechnology industries has been radically transformed. Through dozens of mergers, acquisitions and strategic alliances, there has been a rapid and dramatic concentration of control over value-generating assets. At the time of many of these acquisitions and mergers, the recorded valuations were surprising. In August of 1996, the announced purchase of Plant Genetics Systems (PGS) for (US)$730 million was made at the time PGS's market capitalization was $30 million. According to AgrEvo, $700 million of the purchase price was assigned to the valuation of the patent-protected trait technologies owned by PGS. The acquisition of Holden's Foundation Seeds by Monsanto may have been even more surprising. Here, a privately owned company, Holden's, with gross revenues of only $40 million, was acquired for a purchase price of $1.1 billion. A principal regulatory issue in this merger was the potential effect that might arise for germplasm access by Monsanto's competing trait developers. Holden's germplasm is widely disbursed throughout the industry and at least one of its elite lines is present in most commercial maize pedigrees. In the case of Monsanto's acquisition of DeKalb Genetics, Monsanto paid not only a control premium of 122% for the 60% of DeKalb that they did not already own, but also indemnified DeKalb against any disapproving regulatory action. Moreover, DuPont acquired 80% of Pioneer for $7.7 billion that it did not already own. In this instance, the control premium was only 14%, while the initial premium paid for 20% of Pioneer (purchase price of $1.7 billion) was significantly higher. More generally, the pattern of acquisition in agricultural biotechnology is presented in Fig. 7.1, over the period of 1984–2000.

As assets have been reshuffled, and in many instances newly created, much controversy has arisen. The content of the controversy has ranged from regulatory concerns about the exercise of market power, academic researchers' concerns about freedoms to operate, competitors' concerns about litigation threats, consumer concerns about genetically altered foods, and environmental concerns about insect resistance build-up. Our specific concern in this chapter is with the role that intellectual property has played, if any, in the

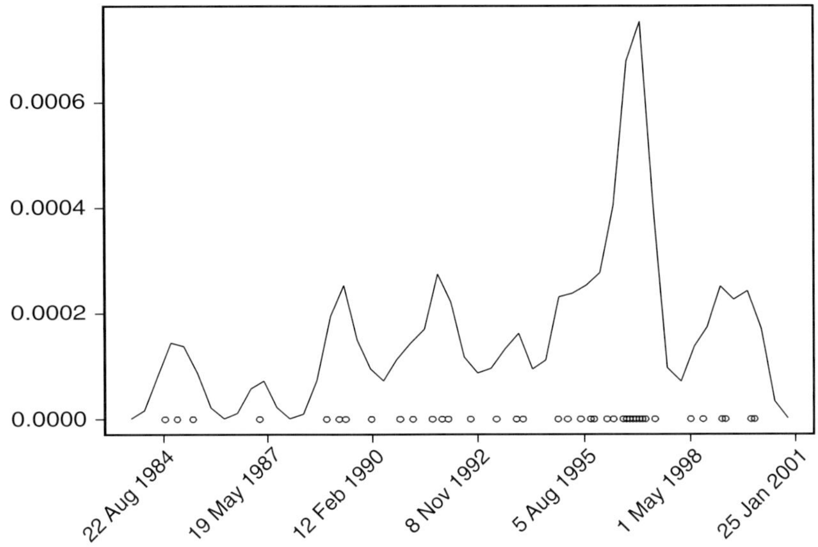

Fig. 7.1. Frequency and timing of acquisitions in agricultural biotechnology.

consolidation and restructuring of the US agricultural biotechnology industry.

The validity and scope of intellectual property in agriculture has a long and controversial history. During much of this century, conflicts and disputes have continuously arisen between farmers' and breeders' rights. Up until the 1930s, farmers' rights generally prevailed. Prior to the development of hybrid maize varieties, farmers had direct access to any germplasm that was developed. Attempts to secure premiums or differentiable pricing for new germplasm were generally unsuccessful. Annual crops had short reproductive cycles and their seeds could easily be saved by farmers. Until the introduction of hybrid maize, these saved seeds bred true and maintained their productivity. In the case of maize, however, the introduction in the 1930s of hybrid varieties meant that saved seed was no longer a viable option for farmers. As a result, the biological science of the time allowed private breeders to protect their discoveries and innovations through trade secrecy. Initially, private breeders used public 'inbred lines' to develop the parents of their proprietary hybrid varieties. As a result, hybrid maize was the first significant example of intellectual property in the agricultural industry.

The first legislation in the USA to protect the investment of plant breeders came through the Plant Patent Act of 1930. The act protects asexually reproduced varieties, i.e. those that are reproduced by cutting, layering, budding or grafting. The legislation specifies that in order to qualify as intellectual property, the variety must be 'distinct' and 'new'. These requirements are much weaker than those that apply to utility patents. Both kinds of patents are administered by the US Patent and Trademark Office.

It was not until 1970 that protection was provided to sexually reproduced varieties through the Plant Variety Protection (PVP) Act. This act is administered by the US Department of Agriculture (USDA), which offers certificates to breeders on the basis of distinctiveness, uniformity and stability. Infringement covers selling, importing or sexually multiplying a protected variety. There are, of course, several exemptions allowed under this act. In particular, there is a research exemption, which allows anyone to use a protected seed variety to breed a new variety. Moreover, another exemption allows farmers to save seed for reproductive purposes as well as to sell seed to other farmers whose primary occupation is growing crops for consumption or feed. The PVP Act was amended in 1994 to eliminate controversial provision allowing the sale of 'saved seed' to others for reproductive purposes. Under this amendment, a farmer can use saved seed only for his own replantings, and can sell his purchased seed only for purposes other than reproduction. The subsequent court decision, *Asgrow Seed Company vs. Winterboer* (115 USC 788, 1995), established that agents who sell any amount of seed for reproductive purposes now violate the rights of certificate holders.

With the introduction of modern biotechnology, utility patents have provided stronger intellectual property protection for plant-related innovations. A utility patent is a property right granted by the US government to inventors to exclude others from imitating, manufacturing, using or selling the invention over a specified period of time. In exchange for this exclusive right, the public receives a detailed description of the invention, so that others can use it after the patent-specified term has expired. To obtain a utility patent, the subject matter must meet certain criteria: it must be novel, non-obvious, useful and amenable to the descriptive requirements of the law.

Until the landmark Supreme Court ruling in the matter of *Diamond vs. Chakrabarty* (447 US 303, 1980), plant-related inventions based in genes or cells from nature or applied to living organisms were viewed as natural phenomena and were thus deemed unpatentable. In this case, however, the court held that 'anything under the sun that is made by man' is patentable subject matter. Specifically the court found that 'the patentee has produced a new bacterium with markedly different characteristics from any found in nature and one having the potential for significant utility. His discovery is not nature's handiwork, but his own; accordingly it is patentable subject matter under the section 101.' This decision broadened the narrow reach of utility patent laws to encompass living organisms. Accordingly, utility patents are now granted in the USA for genetically engineered organisms, for processes that transform cells and express proteins, and for the genes themselves.

With the current widespread issuance of utility patents, agricultural biotechnology suffers from the same problems as biotechnology intellectual property in general: (i) many layers of patented technology are necessary for production, and those layers may be owned by different firms; and (ii) new technologies embodied by biotechnology patents are frequently ill-defined, which leads to uncertainty over patent scope and validity.

The new patent structure is one possible causal source for the merger wave that has swept through the industry. Indeed, it may be that uncertainty in patent rights causes a breakdown in contracting which provides incentives for consolidation. However, to date, there is no hard evidence to support this hypothesis. Weak or uncertain property rights may lead firms to develop alternative organizational forms with which to manage their intellectual property. Certainly consolidation is one such alternative and may in part be explained by patent right uncertainty. In this context, one obvious question that arises is whether the evolution of patent ownership mirrors the changing structure of the industry. As shown in Figs 7.2 and 7.3, the concentration of patents holdings fell from the mid-1980s. There is a trough between 1995 and 1996 reflected by the Herfindahl–Hirschman index (HHI) of the patents in our sample.

To our knowledge, no empirical study has investigated the role of intellectual property holdings in merger decisions. Using patent data enables us to directly address the intellectual property issues that the industry raises in regard to consolidation, and affords us two methodological benefits. By using patent data, we are able to examine both public and private firms. Most merger studies focus exclusively on public firms because of the dearth of economic data on private firms. Since patent data are available for all patenting

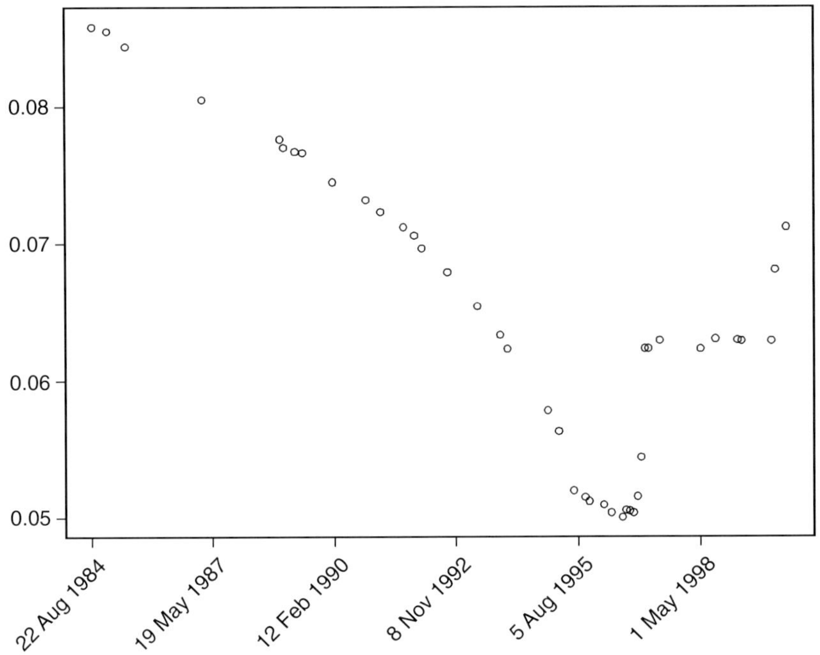

Fig. 7.2. Herfindahl–Hirschman index (HHI) of agricultural patents in sample.

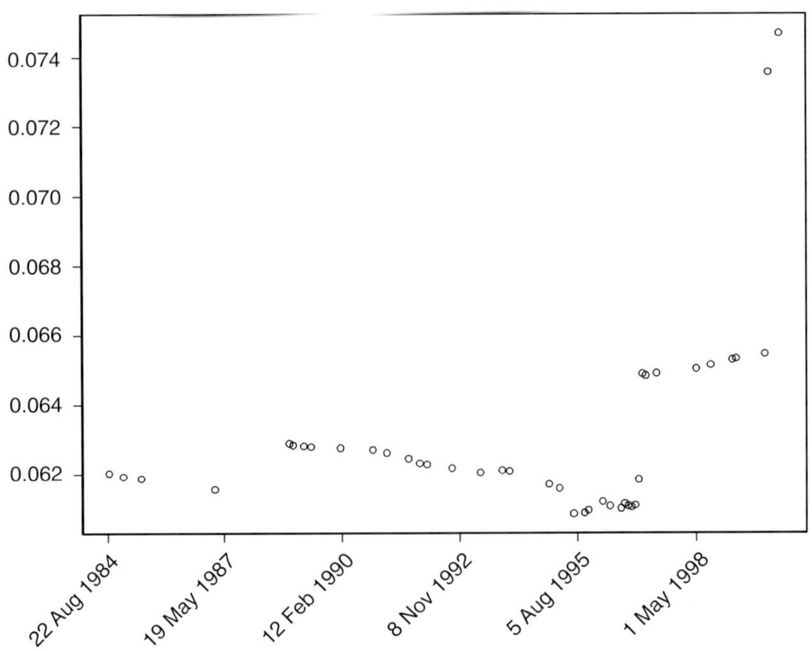

Fig. 7.3. Herfindahl–Hirschman index (HHI) of patents in sample.

firms, they enable us to use a wide variety of informative measures without relying on information available for only public firms. Moreover, by building on previous work of Marco (2000) we are able to investigate the consequences of intellectual property uncertainty in the consolidation of agricultural biotechnology.

Following a brief literature survey on causal analysis of mergers and other types of restructuring, the methodology is presented. To address the limitations of conventional methodologies discovered by the prior literature, we combine duration and logit models. By doing so, we hope to capture the strength of each approach. We use duration analysis to investigate the timing and factors influencing the merger decision. However, once that decision is made, we use a conditional logit model to estimate the probability of 'matching'. The conditional logit model is appropriate for the matching decision since – by conditioning on the acquisition decision – we have reduced the decision to a static one.

Literature Review

A variety of tools exist to examine the causes and consequences of mergers and other types of restructuring. Hall (1990) and Sinay (1998) investigate the consequences of mergers by comparing merged firms to non-merged firms. Hall examines the R&D behaviour of firms under different types of restructuring (including mergers and leveraged buyouts; LBOs). Sinay (1998) investigates the effects of mergers on hospital costs by examining pre- and post-merger cost function estimates.

Qualitative choice models have also been used to examine the determinants of mergers. Hall (1988) presents an excellent description of the econometric issues that arise in applying qualitative choice models to the market for corporate control. The paper discusses the problems of a market where the buyers and sellers are *ex ante* indistinguishable, and the problems involved in defining the choice set. In the merger market, the set of choices is equal to the number of possible participants in the market. In some cases, this calls for some simplification in order feasibly to analyse a given sample. Since Hall uses a large inter-industry sample, sampling is advanced in order to reduce the choice set for each firm.

Some authors have analysed industries in isolation. For instance, Bacon *et al.* (1994) use a logit analysis to predict whether firms belong to the merged or non-merged groups in the rural electricity market. Tremblay and Tremblay (1988) estimate the probability that beer manufacturers will be involved in mergers. The benefits of focusing on a particular industry are that the choice set becomes feasible, and also that the results do not suffer from potentially contaminating differences among industries. Of course, to apply the results then to other industries becomes problematic. Also, see Hall (1988) for problems that arise from assuming too narrow a choice set.

With regard to merger analysis, qualitative choice methods suffer from the problem that they are inherently static. When the analysed period of time is small, this may not be a problem. However, when the range of the study is large, static analysis does not seem appropriate. In the Tremblay and Tremblay study this problem is dealt with by estimating the probability of merger year by year. However, because of the dynamic nature of merger decisions, and the fact that they occur at different times within a sample, duration analysis has also been used.

Two examples of duration models applied to corporate structure are Ravenscraft and Scherer (1991) and Van de Gucht and Moore (1998). Van de Gucht and Moore use a duration model to estimate the factors that influence the survival of LBOs. Many LBOs revert to public forms of ownership, while some remain LBOs. Since these events happen over a length of time, and since some observations are truncated, duration analysis is appropriate.

Ravenscraft and Scherer (1991) investigate the probability that firms will sell off divisions (the flip side of the merger market).

They point out three benefits of using duration analysis: (i) when events occur at different times; (ii) when the probability of events may be changing over time; and (iii) when observations are censored. The intuition is that duration analysis uses valuable information about the timing of events that logit analysis is not able to capture. However, they also point out two problems with the approach: (i) it requires specification of a particular hazard function (at least for parametric approaches); and (ii) it is difficult to deal with time-varying covariates. Regarding (ii), they argue that while theoretically time-varying covariates can be incorporated, '(i)n practice, this step is plagued with computational complexities and collinearity'.

Methodology

Three models are estimated that identify the factors that increase the likelihood of consolidation. First, we estimate a duration model measuring the rate at which firms pursue acquisitions in agricultural biotechnology. Then, we estimate a second duration model – this one on the rate of being acquired. Last, conditional on a firm deciding to pursue an acquisition, we use a fixed effects model to investigate the probability that a given acquirer will match with a potential target.

Likelihood of making an acquisition

The first duration study examines the probability (the hazard rate) of making an acquisition in the agricultural biotechnology industry. The factual support for this study is event data. In particular, we identify a sample of agricultural biotechnology firms and track control changes in these firms in the post-1994 period (Graff *et al.*, 1999). Beginning from this sample, we obtained merger dates and other information from searching Lexis' Mergers and Acquisitions file. The sample was augmented with agricultural mergers going back to 1984, up to April 2000.

In total we have merger histories for 111 firms (see Appendix 7.1). Note that we include agricultural mergers only. So, while Dow is involved in agricultural chemicals, if it purchased an electronics firm, that firm and that merger were not included in the sample. We obtained data on the patent portfolios of these firms from Micropatent.

In the merger context, the duration – or 'spell' – refers to the length of time without making an acquisition. We tracked the merger history for all potential acquirers, where history refers to the number of previous acquisitions by the firm in the sample. With regard to tracking mergers, we needed to make several assumptions about who was buying whom. These assumptions are laid out below:

- The sample consists only of patent holders. Since we are interested in the consolidation in agricultural biotechnology, we do not examine mergers of non-patent holders.
- Parent firms are always the acquirers. That is, if a subsidiary makes an acquisition, we code that as an acquisition by the parent.
- Parents are assumed to have a patent portfolio consisting of the current patents of all their subsidiaries.
- Companies formed by the merger of equals are considered to be new entities, e.g. Novartis formed by the merger of Ciba Geigy and Sandoz. This makes a difference only in coding the merger history of the firm. However, a name change is not considered to be a new entity, e.g. ELM becomes Savia. So, Savia retains the merger history of ELM.
- The beginning of a firm's spell is assumed to be the month in which it applies for its first patent, or 1 January 1984, whichever is later. When a firm makes an acquisition, its spell has ended, and the following month it begins a new spell with its history augmented by one.
- A firm remains in the sample until the earliest of: (i) the date it is acquired;

(i) 10 years after the issuance of its last patent; or (iii) the end of the sample period (4 April 2000).
- Measuring time-varying covariates in duration models necessitates some simplification. In our data, time-varying explanatory variables for a given firm are measured at the end of the spell (at the event date, t_1). So, the probability that a firm will make an acquisition at any time $t < t_1$ is a function of the firm's characteristics at time t_1. For shorter spells this simplification is not troublesome, since patent portfolios change slowly over time.

Patent data were obtained from Micropatent by searching on company name. The data consist of 94,976 US patents issued by the 111 firms in the sample between the years of 1975 and 1998. For each firm, and for each event date (the date of a merger between any two firms in the sample) we constructed the variables defined in Table 7.1.

All of the explanatory variables listed above are calculated using firm's 'live' patent portfolio as of time t. In our analysis, a patent is alive from the application date until 17 years after the issue date. Because we use the application date, the portfolio includes patents that are 'in the pipeline', i.e. those that have been applied for, but not yet issued. This is a reasonable measure since firms will base their decisions on in-process technology as well as developed technology.

Once a firm acquires a target, the target's portfolio is absorbed by the parent. Since HHIs needed to be calculated at all event dates, it was necessary to calculate each firm's market share at all event dates. We tracked the portfolios of each firm over each event date, accounting for all consolidations of portfolios through mergers, and using only live patents.

Calculation of patent enforceability

Firms in agricultural biotechnology claim that one of the reasons that they engage in mergers is because of the difficulty in enforcing their property rights and the difficulty in producing where other firms are enforcing theirs. A patent is only enforceable if a court finds it both valid and infringed. Therefore, we interpret the predicted probability of validity and infringement (conditional upon being litigated) as 'enforceability'. A variable *prvi.ag* is constructed reflecting the average enforceability for a firm's agricultural patents.

The probability that a patent will be found valid if it is brought to court is estimated by:

$$\Pr(V = 1) = f(X\beta),$$

Table 7.1. Variable definitions for acquirer duration analysis.

$event_{jt}$	1 = firm j made an acquisition at time t. 0 = firm j exited the sample at time t
dur_{jt}	Duration of the spell before an acquisition or exit from the sample
$history_{jt}$	Number of previous agricultural mergers by firm j prior to the current spell
$countpat_{jt}$	Number of patents by firm by event date
$share.pat_{jt}$	Firm j's share of all patents issued by firms in the sample, by firm and event date. This variable is a proxy for firm size
$pct.ag_{jt}$	The proportion of firm j's patents that are agriculture related at time t. This variable measures firm intensity[a]
$pct.yng_{jt}$	The proportion of firm j's patents that are less than 4 years old at time t. This variable measures whether a firm is an active patenter or not
$prvi.ag_{jt}$	The estimated enforceability of firm j's agricultural patents. This is a derived covariate, which is calculated in section
$hhi.ag_t$	HHI of agricultural patents at time t

HHI, Herfindahl–Hirschman index.
[a] Agricultural patents are defined as those assigned to international patent classes A01, C07H, C07K, C12M, C12N or C12Q.

Table 7.2. Probability that a patent will be found to be valid.

Variable	Value	SE	*t* value
(Intercept)	−1.185266711	0.542101636	−2.18642895
age1	−0.033180446	0.042868631	−0.77400292
age2	0.096305935	0.048768442	1.97475931
iss8389	0.705183797	0.226737814	3.11012876
iss9098	−0.044040362	0.432098707	−0.10192200
forlif	0.032895578	0.051975455	0.63290601
selfor	−0.246259809	0.444770101	−0.55367888
numback	0.002974284	0.007923651	0.37536787
numicd4	0.325841631	0.165883991	1.96427412
tekag	−0.754300642	0.482221807	−1.56421927
tekmed	−0.084324084	0.353821260	−0.23832396
tekchem	−0.480499971	0.364076162	−1.31977872
tekelec	0.087339886	0.298389091	0.29270469
tekmech	−0.203172823	0.295179983	−0.68830149
patdelay	−0.002353177	0.058711893	−0.04008007

SE, standard error.

where X is a matrix of patent characteristics, including:

- age1: The age of the patent at the time of litigation;
- age2: The age of the patent at the time of adjudication;
- dummy variables for the year in which the patent was issued (1982 or before, 1983–1989 and 1990 or after);
- forlif: average annual forward citations to the patent;
- selfor: the proportion of forward citations that are self-citations;
- numback: The number of backward citations;
- numicd: the number of unique four-digit international patent classes to which the patent has been assigned;
- dummies for the technology field of the patent (agriculture, medicine, chemicals, electronics, mechanical);
- patdelay: the delay of the patent between application and issuance.

For validity, we estimate a probit model and obtain the parameter estimates found in Table 7.2. Using these parameters, we predict the probability that each patent in our sample would be found valid if litigated. We repeat this analysis for the probability of infringement to obtain the parameter estimates found in Table 7.3.

To create our enforceability measure, we construct an interaction term equal to the product of the predicted probability that a patent is valid and the predicted probability that it would be found infringed. Presuming independence, *prvi.ag* = Pr(patent is valid and infringed) = Pr(valid)-Pr(infringed).

Note that we could have used the litigation data to directly estimate a logit model of the joint probability of validity and infringement findings. However, some cases do not rule on both matters. So we are able to increase the sample size by estimating them separately[1].

Summary

The merger data yield 133 observations: 48 acquisitions, 31 observations are censored on the right because they are acquired and 54 observations are censored because the firms do not acquire anyone before they exit the sample. The means of the variables for the acquisition analysis are given in Table 7.4.

Note that the maximum *history* is seven. This firm is Monsanto, who acquires seven

[1] For a more detailed description of calculating the probability of validity and infringement, using both the specification here and using market reactions, see Marco (2000).

Table 7.3. Probability that a patent will be found to have been infringed.

Variable	Value	SE	*t* value
(Intercept)	−0.020983861	0.563869153	−0.03721406
age1	−0.152577136	0.053684553	−2.84210499
age2	0.075749491	0.060030993	1.26183970
iss8389	−0.372731833	0.215375029	−1.73061768
iss9098	−1.163235936	0.355732960	−3.26996952
forlif	0.027514074	0.036201606	0.76002356
selfor	−0.265419946	0.441443120	−0.60125514
numback	0.006851342	0.007798113	0.87858984
numicd4	0.094409502	0.178207595	0.52977260
tekag	−0.677229984	0.537134680	−1.26081970
tekmed	0.446340466	0.321154700	1.38979895
tekchem	0.255104302	0.352575424	0.72354533
tekelec	0.557850718	0.272965460	2.04366779
tekmech	0.529058984	0.293193278	1.80447174
patdelay	0.041218947	0.060843000	0.67746408

SE, standard error.

firms in the sample before it is acquired. After its seventh acquisition, its *history* is seven, at which point it exits the sample by being acquired. Also, the maximum *duration* is 16 years, which reflects firms who are in the sample for the entire sample period, but which never acquire. All the Japanese firms belong in this category.

Estimation

We specify a reduced form model for the probability of acquisition. Our model is motivated by the assumption that a firm will choose to make an acquisition in the next small interval of time when the value of doing so exceeds the reservation value (the value of no acquisition). Of course, the value to any particular firm of an acquisition is dependent upon the choice set of possible targets.

However, we are not (at this point) interested in which target will be chosen, but only whether a firm chooses to make an acquisition at all. Since the choice set is (almost) the same for all firms, the only distinguishing characteristics are the characteristics of the potential acquirer. The choice set is almost the same, because for any firm *j*, the set does not include the firm *j*. Or alternatively, the set includes *j*, but acquiring oneself is equivalent to making no acquisition. Because of this, the probability that a given firm *j* will make an acquisition relative to other firms, is dependent only on its own characteristics. If this is an industry typified by a highly attractive acquisition set, then this will

Table 7.4. Summary statistics of variables used in acquirer duration model.

Variable	Min.	1st Q.	Median	Mean	3rd Q.	Max.
dur	0	3	7	8	14	16
history	0	0	0	0.6159	1	7
countpat	1	15	134	1219	1467	9547
share.pat	0.000	0.000	0.002	0.018	0.023	0.149
pct.ag	0.000	0.064	0.194	0.353	0.684	1.000
pct.yng	0.000	0.206	0.371	0.463	0.735	1.000
prvi.ag	0.003	0.008	0.015	0.020	0.026	0.116
hhi.ag	0.050	0.062	0.071	0.067	0.071	0.086

show up in the intercept term. Accordingly, we model the probability of a firm making an acquisition at time t as a function only of the firm's characteristics and the characteristics of the market (the HHI):

$$\lambda(t) = e^{x\beta+\varepsilon}$$

where X is a matrix of firm and market characteristics (given in Table 7.1),

$$\lambda(t) = \frac{f(t)}{1-F(t)}$$

is the hazard function, and $f(t)$ and $F(t)$ are the usual density and cumulative probability functions.

Our specification assumes a constant hazard: $\lambda(t) = \lambda$, so that the hazard function does not vary with time. That is, there is no duration dependence; the length of time a firm has gone without a merger does not, *ceteris paribus*, affect the likelihood of merger in the next interval of time. The hazard rate is constant in t if $1 - F(t)$ is distributed according to the exponential distribution.

Estimation involves maximum likelihood estimation where the censored observations are incorporated much like the Tobit model (see Greene, 1993):[2]

$$\ln L = \sum_{\text{uncensored}} \ln \lambda(t \mid \theta) + \sum_{\text{all}} \ln\big(1 - F(t \mid \theta)\big).$$

The results of the estimation are given in Table 7.5.

The signs of the coefficients can be interpreted in the usual way; since $\lambda = e^{X\beta}$, $\ln\lambda = X\beta$. So, a positive coefficient indicates a positive relationship with $\ln\lambda$, meaning a positive relationship with λ. The hazard rate is 4.4%, evaluated at the means of the independent variables. This rate means that – on average – the probability that a firm will make an acquisition in the following year is 4.4%.

From Table 7.5, we see that there are several factors that appear to be important in determining the rate at which firms acquire. The number of previous mergers is related to a higher rate of acquisition. That is, it appears that firms that merge more also merge increasingly frequently. This can be viewed as a firm-specific 'taste' for mergers. To see the quantitative effects of changes in *history*, it is instructive to graph the hazard rate as a function of *history*, with the other variables assumed to be equal to their means. We do this for each independent variable in Fig. 7.4. Note that variables that are not significant, like *pct.ag*, will not have a large effect on the hazard rate. From Fig. 7.4, we can see that increases in *history* quickly increase the likelihood of the next acquisition.

The inclusion of the HHI of agricultural patents is intended to control for the acquisition behaviour of other firms in the industry. If acquisitions are reactions to competitors' mergers, than we should see a

Table 7.5. Duration before making an acquisition.

Variable	Value	SE	*z*	*p*
(Intercept)	−1.845	1.2334	−1.50	1.35e–001
history	0.580	0.0953	6.08	1.19e–009
hhi.ag	−47.266	18.3613	−2.57	1.00e–002
share.pat	7.992	4.8423	1.65	9.88e–002
pct.ag	−0.738	0.6516	−1.13	2.57e–001
pct.yng	2.676	0.6636	4.03	5.51e–005
prvi.ag	20.620	7.1382	2.89	3.87e–003

Degrees of freedom: 133 total; 126 residual.
−2 × log-likelihood: 241.
λ = 0.0440 at the means of the independent variables.
SE, standard error.

[2] Estimation was performed with *S*-plus, using a Newton–Raphson method.

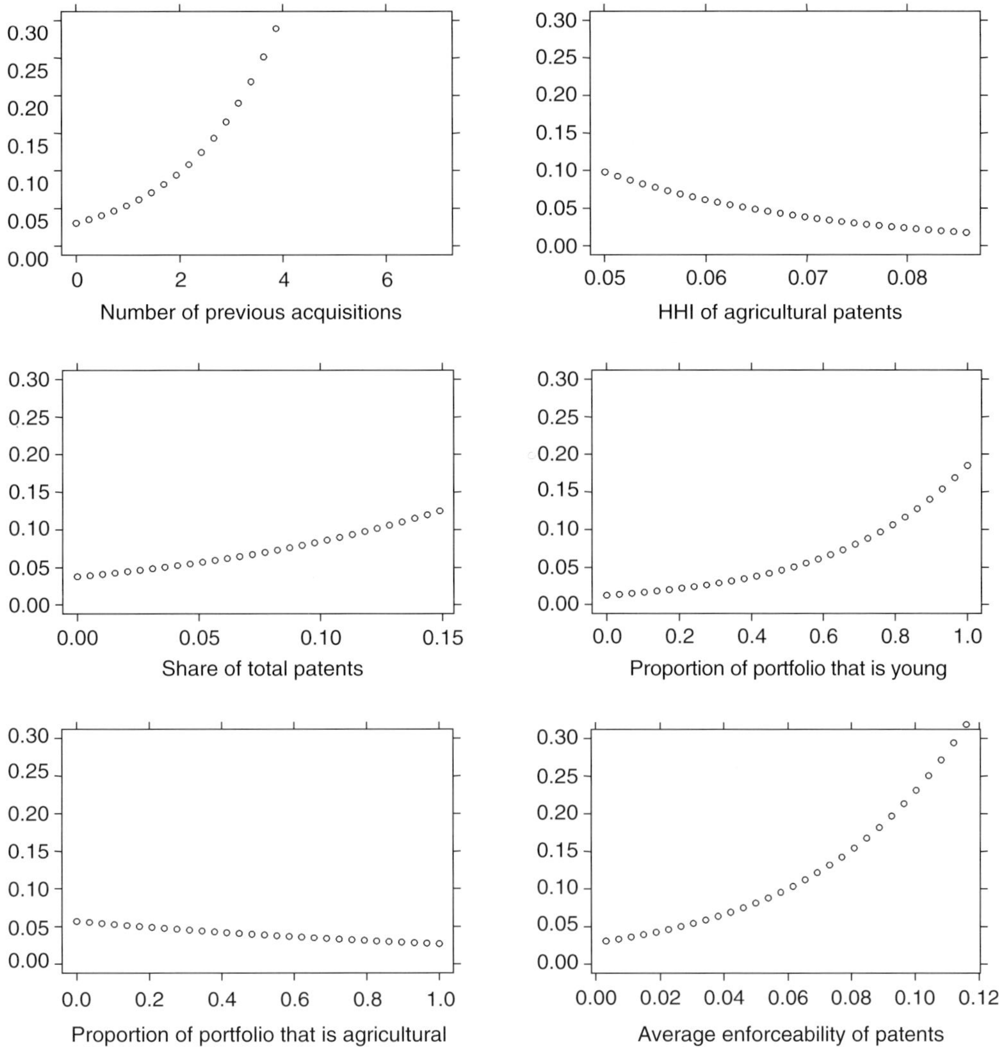

Fig. 7.4. Hazard rate of making an acquisition as a function of the independent variables. In each panel the hazard rate is evaluated at the means of the other independent variables.

positive influence between *hhi.ag* and the hazard rate. However, the opposite appears to be true. This can partly be explained by a reduction in the set of available targets as concentration grows.

Other factors that appear to be important are size (which increases the likelihood, as measured by the share of industry patents owned by the firm), recent patenting behaviour (which also increases the likelihood), as well as the derived explanatory variable *prvi.ag*. Agricultural intensity itself does not affect the likelihood of acquisition.

A large value for *prvi.ag* indicates a 'high-quality' patent portfolio, or one that is characterized by less uncertainty about both validity and scope. Note that a low value for *prvi.ag* does not indicate that the patented technologies are not valuable – only that they are difficult to enforce. As the results suggest, stronger intellectual property rights at the firm level are associated with higher rates of acquisition. We will return to this hypothesis when we examine the matching model.[3]

Likelihood of being acquired

The parallel analysis for the acquirer duration is the target duration. This model estimates the probability that a firm will be acquired, conditional upon its characteristics and industry characteristics. We use only the target's characteristics – as opposed to the acquirer's – for the same reason elaborated in the acquirer duration analysis. The data for the target duration analysis (the rate of being acquired) is very similar to the acquirer duration analysis above, with one additional restriction. We assume that independent firms are the only candidates for acquisition. That is, a firm can only be acquired once. Once it is a subsidiary, it is 'off the market'.

Clearly there is a market for the acquisition of assets, including wholly owned subsidiaries. We do not include these assets in the sample. Had we, it would have involved tracking the patenting behaviour of all subsidiaries. Unfortunately, different firms handle post-merger patenting differently. While some maintain independent patenting by the subsidiary, some absorb the R&D activities of the new subsidiary into those of the parent, making the entities inseparable. Since we cannot observe the difference from available information, we exclude sales of subsidiaries from the analysis.

Our data include 90 observations, of which 31 involve firms that are acquired, and 59 are truncated (or censored), because they are not acquired by the end of the sample period. Summaries of the variables used are in Table 7.6.

All independent variables are calculated as described in the acquirer duration analysis. We again estimate an exponential model for the target analysis. The results are summarized in Table 7.7.

The target duration results are similar to those of the acquirer duration model, with one exception: the size of the firm's patent portfolio was marginally important for making an acquisition. Here we find the converse. This result is to be expected since acquirers in this industry tend to be larger, more diversified firms. Somewhat surprisingly, intensity of agricultural patenting continues to be unimportant, and, since we are examining the agricultural biotechnology sector, we would expect that target firms will be ag-patent intensive. However, the patent classes are only rough guidelines for the actual technological uses of patents. Figure 7.5 shows the response of the hazard rate to changes in the independent variables. Again, insignificant independent variables show up as very flat curves.

Not surprisingly, recent patenters are also attractive targets as measured by the positive coefficient for *pct.yng*. Also, the industry concentration (*hhi.ag*) has a similar coefficient as in the acquirer analysis, which is to be expected because it is an

Table 7.6. Summary statistics for variables used in target duration analysis.

Variable	Min.	1st Q.	Median	Mean	2nd Q.	Max.
dur	0	7	13	11	16	16
countpat	1	14	87	927	848	9547
share.pat	0.000	0.000	0.001	0.014	0.014	0.149
pct.ag	0.000	0.068	0.225	0.359	0.672	1.000
pct.yng	0.000	0.176	0.318	0.432	0.703	1.000
prvmean	0.139	0.292	0.371	0.359	0.408	0.706
primean	0.022	0.121	0.232	0.214	0.271	0.537
prvi.ag	0.003	0.008	0.013	0.018	0.022	0.116
hhi.ag	0.050	0.063	0.071	0.068	0.071	0.085

[3] The coefficients are robust to a number of different distributional assumptions, including the Weibull. Those results are available upon request.

Table 7.7. Duration before being acquired.

Variable	Value	SE	*z*	*p*
(Intercept)	−2.9165	1.635	−1.784	0.074410
history	0.2650	0.151	1.759	0.078624
hhi.ag	−37.9462	23.278	−1.630	0.103079
share.pat	−5.5495	12.246	−0.453	0.650413
pct.ag	0.0936	0.730	0.128	0.897869
pct.yng	2.9045	0.828	3.510	0.000448
prvi.ag	18.1124	8.133	2.227	0.025944

Degrees of freedom: 90 total; 83 residual.
−2 × log-likelihood: 117.
$\lambda = 0.0321$ at the means of the independent variables.
SE, standard error.

environmental variable, and not firm specific. That is, if concentration leads to fewer acquisitions, then it will lead to both fewer acquirers and fewer targets, on average.

Patent enforceability appears to enter in the same direction (a positive effect on the hazard rate), and magnitude (a coefficient of 18.1 in the target analysis vs. 20.6 in the acquirer analysis). This result is interesting. We found that high expectations of enforceability led to a higher likelihood of making an acquisition. In the target analysis, we find that high values for enforceability also increase the likelihood of being acquired. The interpretation is that firms with more enforceable patents are more attractive targets, *and* more aggressive suitors.

Whether firms with enforceable patents align themselves one with another is a question that cannot be answered by examining acquirers and targets independently. Thus, we turn to a model of acquirer and target matching.

Matching

We are interested in addressing the question of who acquires whom. To do this we examine the acquirers who made acquisitions, and obtain information on their contemporaneous choice set. In this fashion, we can ascertain which characteristics of the targets made the realized target the best choice for the acquirer. Thus, we are conditioning on the acquirer having made a decision to acquire at date *t*.

Our methodology is to use a conditional – or fixed effects – logit. To do so requires the development of more explanatory variables, and to arrange the data in a particular way. The data consist of acquirer–target pairs for each acquirer at the date of an actual acquisition. One acquirer–target pair will be the actual consummated deal. The other acquirer–target pairs will consist of the actual acquirer matched with all possible targets at date *t*. A possible target is any independent firm as of the date of acquisition.

Our sample contains 31 acquisitions of independent firms. These events can be described by acquirer at date *t*: A_t. An acquirer may enter more than once, so that $A_t = A_{t+1}$, but acquirer–date combinations are unique. Each acquirer–date combination contains one observation for each potential target. In data, there are – on average – 62 available targets, yielding 1921 observations. At any given date there may be more or fewer available targets, due to entry and exit. An event equals 1 if the acquirer actually purchased the target, 0 otherwise.

The explanatory variables for the matching data are similar to those of the duration models. We use the target's values for *share.pat*, *pct.ag*, *pct.yng*. Furthermore, we create two new variables following Podolny *et al.* (1994). Specifically, let B_A be the set of patents that are cited by the acquirer's patent portfolio (where the time subscript is omitted). Similarly, let B_T be the set of patents that are cited by a poten-

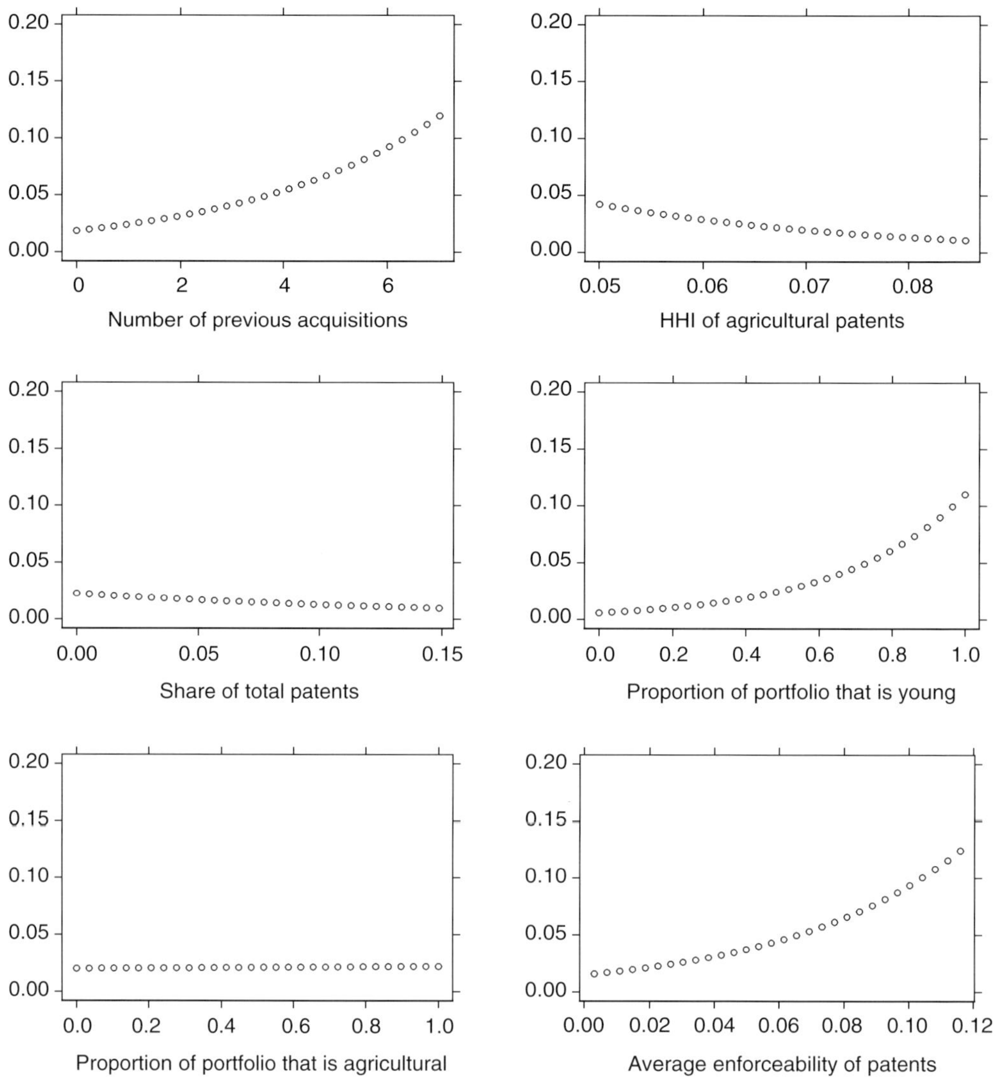

Fig. 7.5. Hazard rate of being acquired as a function of the independent variables. In each panel the hazard rate is evaluated at the means of the other independent variables.

tial target's portfolio. Then the overlap between the acquirer and target is O_{AT} = (number of patents in $B_A \cap B_T$)/(number of patents in B_A). Similarly the overlap between the target and acquirer is O_{TA} = (number of patents in $B_A \cap B_T$)/(number of patents in B_T). Table 7.8 summarizes the overlap variables.

The overlap variables are intended to measure the similarity in the research programmes of a pair of firms. To the extent that backward citations define a technology space, then the overlap variables measure whether the firms' research programmes lie in the same space. Essentially, this measure allows us to observe whether the firms lie in the same space or not. Note that the measures are not symmetric; if $B_T \subset B_A$, then $O_{TA} = 1$, and $O_{AT} < 1$. The overlap variables provide a basis for inferring whether overlapping property rights, complementarities or uncertain property rights

Table 7.8. Summary statistics for overlap variables.

Variable	Min.	1st Q.	Median	Mean	2nd Q.	Max.
over_at	0.000	0.000	0.000	0.009	0.007	0.268
over_ta	0.000	0.000	0.002	0.018	0.021	0.667

are at the heart of the matches between firms.

The matching data contain information for 23 acquirers making 31 acquisitions. For each acquisition there were on average 62 potential targets, so that each acquisition accounts for approximately 62 observations, for a total of 1921 observations. Our sample is narrow enough that we can feasibly include all possible targets in the industry.

For the available data, we estimate the matching model using a conditional, or fixed-effects, logistic regression following Greene (1993). Let A_t be the acquirer at time t, and let $T_t = \{T_{t1} \dots T_{tNt}\}$ be the set of potential targets at time t. y_{ti} is an event variable describing whether T_{ti} was acquired by A_t ($y_{ti} = 1$) or not ($y_{ti} = 0$). Importantly, the acquirer is restricted to making one and only one acquisition per period. So, we want to measure the probability that $y_{ti} = 1$, conditional on $\Sigma_{Nt} y_{ti} = 1$.

For A_t to find it worthwhile to acquire T_{ti}, it must be that the value (V) of making the acquisition is greater than that for the other potential targets.[4] Let the value of acquiring T_i at date t be V_{ti}. If A_t chooses target T_{tj}, then it must be that $V_t \geq V_{ti}$ for all $T_{ti} \in T_t$. If ε_{ti} is distributed with the Weibull distribution, we can write the relevant probability as

$$\Pr(y_{ti} = 1 \mid \sum y_{ti} = 1) = \frac{e^{\alpha_t + \beta' x_{ti}}}{\sum_{N_t} e^{\alpha_t + \beta' x_{ti}}}.$$

Since α_t enters all the terms, it drops out of the probability. That is, acquirer specific effects do not alter the probability that a particular target is chosen, *conditional on the fact that the acquirer has already chosen to make an acquisition.* Also, note that joint characteristics (which involve characteristics of A_t, but which vary with the target) remain. Of course, these characteristics include the overlap variables.

Estimation is done via maximum likelihood. The results are reported in Table 7.9, where '*_t*' represents values of a variable for the target.[5]

The independent variables, with the exception of *over_at* and *over_ta*, are the same as those of the target in the duration analysis. We see that the only variables with any independent explanatory power are *pctag_t* and *over_at*. The signs of the coefficients can be interpreted in the usual way: a positive coefficient increases the probability of a match.

The agricultural intensity appears to have positive effect on the probability of a match. However, the enforceability variable does not have any explanatory power. So, while enforceability helps to explain the likelihood of being acquired, it does not help to explain the likelihood of being acquired at a specific point in time. The same is true of the percentage of young patents. These variables tend to make the targets more attractive, but more attractive for all acquirers, not just one in particular.

The overlap variables are more revealing. The positive coefficient on *over_ta* shows that the greater the overlap in the backward citations (the technology space), the greater

[4] We assume that the acquirer makes a unilateral choice. It is possible that for the target there may be a more suitable acquirer, where the complementarities are greater. However, had this been the case, the target should already have merged with the more suitable acquirer. Hence, at time *t*, all potential targets should be willing to merge with the acquirer, assuming that they are able to bargain for a share of the surplus. Another caveat is our implicit assumption that the most profitable acquisition actually makes the acquirer better off. This assumption is subsumed in our condition that the acquirer has already decided to make an acquisition, implying that it must be made better off by doing so.

[5] The estimation was performed using Stata.

Table 7.9. Conditional (fixed-effects) logit.

Var.	Coef.	SE	*z*	$P > \|z\|$
shrpat_t	−8.490233	11.86213	−0.716	0.474
pctag_t	1.251405	0.6980167	1.793	0.073
pctyng_t	0.4393101	0.8009511	0.548	0.583
prviag_t	7.480979	14.75118	0.507	0.612
over_at	6.717609	5.554239	1.209	0.226
over_ta	7.796754	2.422257	3.219	0.001

Number of obs. = 1921.
LR $chi^2(6) = 22.47$.
Prob. > $chi^2 = 0.0010$.
Log likelihood = −116.57998.
Pseudo $R^2 = 0.0879$.
SE, standard error.

likelihood that the firms will match. The value of *over_ta* will generally be larger than *over_at*, since acquirers tend to have larger portfolios; hence the effect *over_ta* is larger than the effect of *over_at*. The result is not unexpected. However, it can be interpreted in two ways. First, a high value for *over_ta* may indicate complementarities in intellectual property. Second, a high value may also indicate overlapping property rights, which lead to mutually blocking technologies. Without additional information, it is impossible to determine which story explains matching. However, bringing together results from the three studies, inferences about the determinants of consolidation in agricultural biotechnology can be drawn.

Conclusion

Our results show, among other things, that the average enforceability of a firm's patent portfolio is positively related to the probability of acquiring and being acquired. Put another way, the estimated ability of a firm to enforce its property is important in its consolidation decision. At the same time, the amount of overlap in technology space helps to explain who will merge with whom. So, firms with enforceable intellectual property boundaries tend to be involved in mergers, and those with overlapping property rights tend to match up with one another.

While the causal connection with the overlapping variable suggests complementarities, the additional impact of our enforceability measure is consistent with industry anecdotal evidence which suggests that many of the mergers were rooted in conflicts over mutually blocking patents. In fact, a handful of mergers, e.g. Calgene/Monsanto and DeKalb/Monsanto, were completed in the midst of patent infringement suits.

References

Bacon, F.W., Shin, T.S. and Murphy, N.B. (1994) Factors motivating mergers: the case of rural electric cooperatives. *Journal of Economics and Business* 46, 129–134.

Graff, G., Rausser, G.C. and Small, A.A. (1999) Agricultural Biotechnology's Complementary Intellectual Assets. Working Paper, Department of Agricultural and Resource Economics, University of California at Berkeley.

Greene, W.H. (1993) *Econometric Analysis*, 2nd edn. MacMillan, New York.

Hall, B.H. (1988) Estimation of the Probability of Acquisition in an Equilibrium Setting. IBER Working Paper No. 8887, Berkeley, California (revised January 1991).

Hall, B.H. (1990) The Impact of Corporate Restructuring on Industrial Research and Development. *Bookings Papers on Economic Activity: Microeconomics* (1990), 85–124.

Marco, A.C. (2000) The economic consequences of uncertainty in intellectual property rights. PhD thesis, University of California at Berkeley.

Podolny, J.M., Stuart, T.E. and Hannan, M.T. (1996) Networks, knowledge, and niches: competition in the worldwide semiconductor industry, 1984–1991. *American Journal of Sociology* 102, 659–689.

Ravenscraft, D.J. and Scherer, F.M. (1991) Divisional sell-off: a hazard function analysis. *Managerial and Decision Economics* 12, 429–438.

Sinay, U.T. (1998) Pre- and post-merger investigation of hospital mergers. *Eastern Economic Journal* 24, 83–97.

Tremblay, V.J. and Tremblay, C.H. (1988) The determinants of horizontal acquisitions: evidence from the US brewing industry. *Journal of Industrial Economics* 37, 21–45.

Van de Gucht, L.M. and Moore, W.T. (1998) Predicting the duration and reversal probability of leveraged buyouts. *Journal of Empirical Finance* 5, 299–315.

Appendix 7.1. List of Companies Used in Merger Analysis

Advanced Genetic Sciences
Advanced Polymer Systems
AgriBio Technology
Agridyne/Native Plants Inc.
Agrigenetics
Agritope
Allelix Biopharmaceuticals
American Cyanamid
American Maize
Amoco
Astra
AstraZeneca
Aventis
Bayer
Biosource
Biosys
Biotechnica
Boswell
Calgene
Cargill
Celanese
Chevron
Ciba Geigy
Continental Grain
Copley Pharmaceuticals
Corn States International
Crop Genetics
Dekalb
Delta and Pine Lands
DNA Plant Technology
Dow Chemical
DuPont Chemical
Ecogen
EcoScience
Empressa La Moderna (ELM)/Savia
Epitope
Escagenetics
Espro
FMC
Genencor
Helena Chemical
Hoechst AG
Imperial Chemical Industries (ICI)
International Paper
Japan Tobacco
Jinro
Kirin
Limagrain
Lubrizol
Mallinckrodt
Merck
MGI Pharmaceuticals/Molecular Genetics Inc.
Mitsubishi Chemical
Mogen
Monsanto
Morganseeds
Mycogen
Nordisk
Northrup King
Novartis
Novo
Novo Nordisk
NPS Pharmaceuticals
Nunhems
Pfizer
Pharmacia
Pharmacia & Upjohn
Pioneer Hi-Bred
Plant Genetics Inc. (PGI)
Plant Genetic Systems (PGS)
Prodigene
Rhône-Poulenc
Rorer
Royal Dutch Shell
Sandoz
Sapporo
Scotts
Sepracor
Sumitomo Chemical
Sungene
Syntro
Systemix
Takara Shuzo
Thermo Ecotek/Thermo Trilogy
Tosco
Transgene
Unilever/Lever Bros.
Union Camp
Union Carbide
Upjohn
Westvaco
Weyerhaeuser
Wilbur-Ellis
W.R. Grace
Yissum Research Development Co
Zeneca

Chapter 8

Cost of Conserving Genetic Resources at *ex Situ* Genebanks: an Example of the ICARDA Genebank

Bonwoo Koo[1], Philip G. Pardey[1], Jan Valkoun[2] and Brian D. Wright[3]

[1] *International Food Policy Research Institute (IFPRI), 2033 K Street NW, Washington, DC 20006-1002, USA;* [2] *International Center for Agricultural Research in Dry Areas (ICARDA), Aleppo, Syria;* [3] *University of California, Berkeley, CA, USA*

Introduction

Belatedly following the lead of Vavilov's remarkable collection in St Petersburg, institutions for conservation of agricultural genetic resources (seeds and other germplasm such as roots and trees) have proliferated in the last few decades. Seed collections held by breeders and researchers have been expanded and organized into more comprehensive 'genebanks' focused on particular classes of crops. These genebanks are often located at crop breeding centres such as the Centers of the Consultative Group on International Agricultural Research (CGIAR). The decline in the resources available for such centers in recent years has engendered concern that the associated genebanks be funded in some more independent and permanent fashion. A question of interest to potential funders is whether such funding is economically justified.

A full cost–benefit analysis of a genebank is a truly daunting task. Even if pure existence value is ignored, obtaining a realistic economic value of conserved crop varieties for current and future breeding would be expensive in itself, if it is possible at all. An alternative is to begin by assessing the costs of conservation as indicated by recent experience. If this cost is below the consensus minimal estimate of benefits, then the cost calculation can be sufficient to justify continued conservation and set the level of necessary long-term financial commitments. This was the logic behind the initial study (Pardey *et al*., 2001) of the maize and wheat genebanks at the Center for the Improvement of Maize and Wheat (CIMMYT) in Mexico.

That study laid out a conceptual framework and developed practical estimation methods for collecting and analysing genebank costing data. It also estimated the costs involved in various genebank functions (e.g. storage, viability testing, regeneration and dissemination), provided total and average cost estimates, and documented the short- and long-term costs of conserving wheat and maize accessions, each of which is relevant for different management and policy purposes. This study is therefore the second in a series of studies on the cost of conserving germplasm in *ex situ* genebanks within the CGIAR system.

The structure of conservation costs depends heavily on: (i) the type of crops; (ii) the storage capacity of the genebank facility and the size of the holdings; (iii) the condition of the initial seed samples and the regeneration protocols; and (iv) the climate and infrastructure of the genebank location. Genebanks operate within different institutional and local market contexts, and thus vary in the size and structure of their crop holdings as well in the conservation protocols they follow. All these differences have different cost implications so that the cost of conserving a crop in one genebank may not be the same as the cost incurred for the same crop at a different site. Here, we provide estimates of the costs of conserving cereals (wheat and barley), forage legumes and food legumes (chickpea, lentil and faba bean) in the genebank maintained by the International Center for Agricultural Research in the Dry Areas (ICARDA) and compare the results with those derived for the CIMMYT genebanks in the previous study.

The ICARDA Genebank

ICARDA, established in 1977, is one of the 16 CGIAR centres (called CG centres) located throughout the world. From its main research station and offices based in Aleppo, Syria, ICARDA targets the entire developing world in its crop improvement research on barley, lentils and faba beans, and dry-land areas in developing countries in the research aimed at improving the management of rangelands and water resources, and the nutrition and productivity of small ruminants (sheep and goats). Among the CG centres, ICARDA also takes responsibility for breeding several crops (durum and bread wheat, chickpeas, pasture and forage legumes), improving farming systems, and for protecting and enhancing the natural resource base of water, land and biodiversity in the Central and West Asia and North Africa (CWANA) region.

Thus, ICARDA's research interests include the CWANA region, as well as the subtropical and temperate dry-land areas within developing countries, which collectively account for an estimated one-third of the world's agricultural land. The CWANA region is particularly rich in genetic resources, including three of the eight centres-of-origin identified by Vavilov. For this reason, since its inception, ICARDA has paid particular attention to conserving landraces and other primitive materials of crops originating in the region (Valkoun *et al.*, 1995), and these constitute a higher portion of genebank accessions than are found at the CIMMYT wheat genebank.

In 1983, ICARDA's various genetic resources activities were consolidated in a new Genetic Resources Unit (GRU), which was established to oversee the Centre's germplasm collection, conservation, documentation, characterization and distribution activities. The unit provides a centre-wide, seed-health service through its seed health laboratory. In 1989, with financial support from the Italian Government, the GRU completed construction of its new facilities, which include cold stores for medium- and long-term storage of seed samples. The International Board of Plant Genetic Resources (IBPGR) designated ICARDA to hold global base collections of barley, wild wheat relatives, durum wheat, lentils and faba beans, and a regional base collection of bread wheat and chickpeas in a coordinated network of base collections. Since 1994, the ICARDA collections have been held under the auspices of an in-trust agreement with the Food and Agriculture Organization (FAO) of the United Nations for the benefit of the global community.

Initially, the GRU mainly supported ICARDA's crop improvement programmes, but gradually, interactions with the national agricultural research systems (NARS) in the CWANA region became equally important. This process culminated in 1992, when ICARDA, jointly with IBPGR – today's International Plant Genetic Resources Institute (IPGRI) – and the FAO Commission on Plant Genetic Resources, established a network for collaboration in plant genetic resources in the CWANA region. In 1996, ICARDA and IPGRI, along with five Central Asian countries of the former USSR, set up another sub-regional network to collaborate on the conservation of genetic resources (CAN/PGR). Three trans-Caucasian countries joined this network in 1999 and its name was changed to CATN/PGR.

Germplasm collection and acquisition

ICARDA is located in the Near Eastern centres of origin and diversity, where wild ancestors and relatives of the ICARDA mandate crops may still be found in their natural habitats. ICARDA gives high priority to collecting and conserving indigenous germplasm, not only from the Near East, but also from other countries in the CWANA region. Local germplasm is well adapted to the harsh, stressful and highly variable environments, and may provide useful genes for breeding stress-tolerant varieties adapted to various target areas and farming systems. Another reason for building up the collection's holding of wild relatives and landraces is the rapid pace of genetic erosion throughout the region caused by the loss of natural habitats for wild species.

Approximately 70% of ICARDA's accessions came from major genebanks outside the CWANA region or directly from national programmes throughout the region. The major sources of ICARDA's holdings are the US Department of Agriculture's National Small Grains Collection in Aberdeen, Idaho (23,800 accession), the Germplasm Institute in Bari, Italy (16,500), and the N.I. Vavilov All-Russian Research Scientific Institute of Plant Genetic Resources.

Around 24,500 accessions stem from the 150 germplasm collection missions, in cooperation with national programmes in 28 countries, mostly in the CWANA region. This collection activity continues in areas of substantial genetic diversity as well as in 'under-sampled' areas throughout the CWANA region. Areas generally favoured for collection missions include those with a harsh climate, ecology, crop history and cultural practices that may endow the indigenous germplasm with certain traits that could be useful to future crop improvement research, e.g. tolerance to drought and extremes of temperature, disease and insect resistance, and good food or feed quality. In 1999, the GRU holdings numbered around 120,000 accessions, representing germplasm from 90 countries, but mostly from the CWANA region. About 50% (57,900 accessions) of which were cereals and the rest divided almost equally between food legumes (29,900 accessions) and forage legumes (31,700 accessions).

Germplasm use

The growing interaction between national programmes in CWANA, scientists at ICARDA, and partners in advanced institutions and other genebanks has resulted in an increasing number of requests for germplasm. From 1989 to 1998, ICARDA distributed more than 118,000 germplasm samples of cereals, forage and food legumes to external users, with 62% of these samples going to national programmes in developing countries.

ICARDA's crop breeding programme began by assembling a large collection of germplasm as a basis for genetic enhancement. The initial lines distributed to national programmes for testing consisted simply of selection from the germplasm collection, based on evaluation trials at ICARDA's primary evaluation sites in Syria and Lebanon. The most important use of the ICARDA germplasm collections has been as a source for resistance to and tolerance of biotic and abiotic stresses. Crops growing in dry sub-tropical and temperate areas, like crops grown elsewhere, are susceptible to yield losses due to diseases, insects, and adverse (often low-moisture) environmental conditions; consequently, screening procedures were developed to detect tolerance or resistance to the biotic and abiotic stresses prevalent in these areas.

Research activities

In recent years, the GRU has conducted research to study the genetic diversity within its collections of barley, durum wheat, lentil, chickpea and medicagos using agronomic, biochemical and molecular characterization techniques (e.g. RAPD, AFLP and SSR markers). This research provides a better understanding of the geographical distribution of crop diversity and is also being used to identify a 'core' collection that involves a reduced number of accessions to represent the global diversity in each respective crop.

Research on genetic diversity in natural populations of wild progenitors and relatives of mandate crops also receives high priority because of its role in helping to identify the most representative and appropriate populations to conserve in their natural habitats (*in situ* conservation). Such populations have already been identified for wild progenitors of cultivated wheat, *Triticum urartu* and *T. dicoccoides,* and for wild lentils.

To learn more about the factors important for *in situ* conservation, experiments have also begun at ICARDA in which self-regenerating populations of the wild ancestors of wheat, barley, lentil, chickpea and medicagos were established in two different environments: (i) long-term fallow; and (ii) rangeland, based on seed from ICARDA's *ex situ* genebank collections. Similarly, larger-scale experiments were established in collaboration with the Syrian Agricultural Research Centre at its three research stations.

To promote *in situ* and on-farm conservation of wild crop relatives and landraces in the Near Centre of crop origin and primary location of diversity, ICARDA, jointly with IPGRI and the national programmes of Jordan, Lebanon, the Palestinian Authority and Syria, developed a project proposal for funding by the Global Environmental Facility (GEF). The project was approved and its implementation started in 1999. ICARDA is the executing and coordinating agency for this project, which is based heavily on community and other national stakeholder participation.

Costing the ICARDA Genebank Operation[1]

The services provided by a genebank can be broadly grouped into three categories, generally involving an increasing amount of effort and cost. The first set of services involves conserving the genetic diversity in a base collection (long-term storage) for use in the distant future. Examples include maintaining the viability of a collection in long-term storage at low temperature with long regeneration cycles to minimize genetic drift, and duplicating the collection at other locations for security purposes. The second set of services comprises distribution activities involved in making accessions available upon request for more immediate use. This involves maintaining an active collection in medium-term storage, with more frequent regeneration requirements. Documentation of basic information such as passport and agronomic characterization data (related to the origin of the accession) is an essential requirement for fulfilling this function. The third group of activities relates to the provision of phenotypic, molecular and other data that characterize each accession in ways that expedite and enhance the use of the material for crop improvement purposes. Publishing catalogues and providing a standardized database of molecular characterization and evaluation details of each accession form part of this service.

[1] Detailed documentation on the cost elements is available from the authors on request.

In this costing exercise, we focus mainly on activities related to the first and second set of services, specifically storage (medium-term and long-term), viability testing, regeneration, dissemination, acquisition and safety duplication. We do not attempt to cost other functions such as germplasm collection, molecular characterization, and pre-breeding evaluation. Figure 8.1 gives a stylized representation of the activities undertaken by the ICARDA genebank. It illustrates a series of activities to conserve and distribute accessions from the time they are introduced to the collection to when they are disseminated to third parties. Activities in the dotted boxes in the figure are outside the scope of this study.

Capital input costs

Table 8.1 provides a breakdown of the capital input costs incurred by the GRU. We

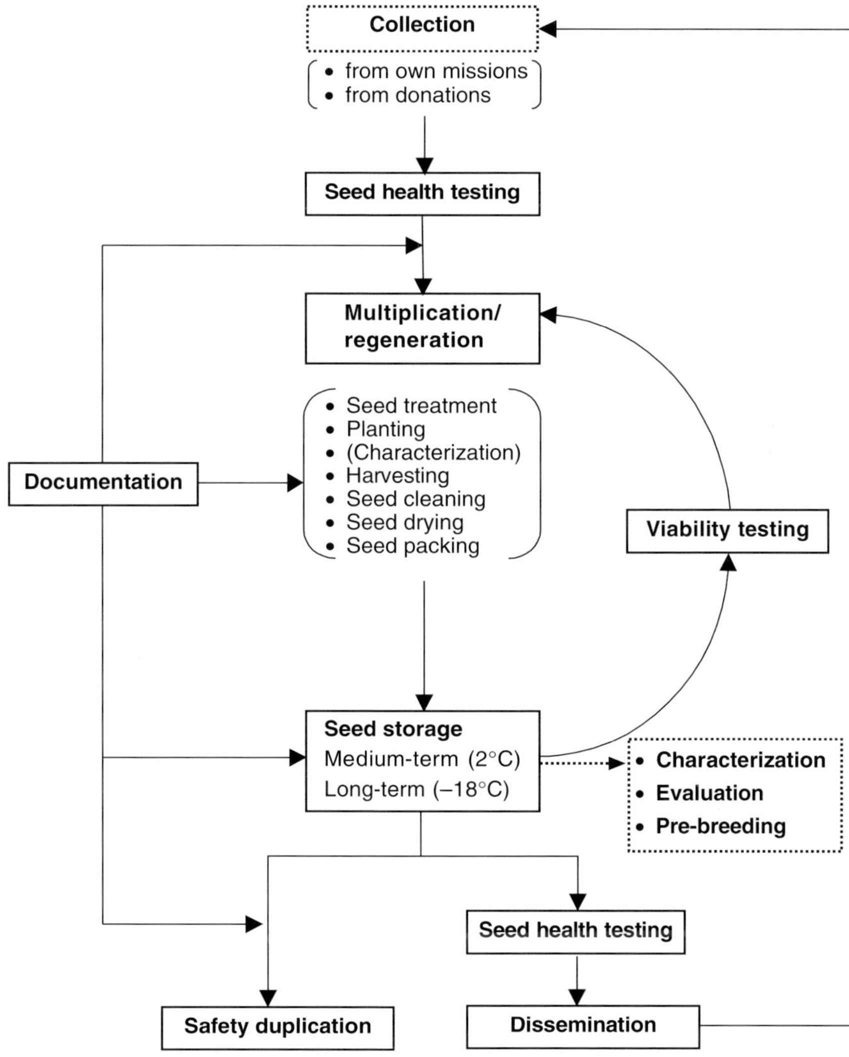

Fig. 8.1. Activities related to ICARDA's genebank operation.

Table 8.1. Capital costs (US$) of maintaining germplasm in the ICARDA genebank.

Items	Service life (years)	Replacement cost	Annualized cost[a]
Medium-term storage		**279,186**	**24,351**
Storage facility	40	104,500	5,077
Storage equipment	10	155,250	18,405
Seed container	50	19,436	870
Long-term storage		**317,479**	**22,796**
Storage facility	40	68,600	3,333
Storage equipment	10	107,800	12,780
Vacuum sealing device	10	5,000	593
Seed container	50	136,079	6,091
Viability testing		**40,000**	**3,133**
Viability testing facility	40	23,000	1,117
Incubator	10	12,000	1,423
Other equipment	10	5,000	593
Regeneration		**239,904**	**28,042**
Farming equipment	10	1,000	119
Greenhouse/screenhouse	10	112,554	13,343
Vehicle	7	63,000	10,093
Seed cleaning equipment	10	2,150	255
Seed drying facility	40	5,700	277
Seed drying equipment	10	16,000	1,897
Seed processing facility	40	37,500	1,822
Seed processing equipment	10	2,000	237
Seed health testing		**57,000**	**5,314**
Seed health facility	40	25,500	1,239
Greenhouse	10	13,600	1,612
Lab/office equipment	10	12,000	1,423
Vehicle	7	4,200	673
Computer	5	1,700	367
General capital		**192,500**	**19,737**
General facility	40	113,000	5,490
Office equipment	10	30,000	3,556
Computer	5	49,500	10,691
Total capital cost		**1,126,069**	**103,373**

[a] Calculated using a 4% rate of interest.

converted the historical purchase prices of various capital inputs into current replacement costs to reflect the contemporary cost structure of the genebank operation. The last column in Table 8.1 provides the annualized user costs of these capital items in constant dollars, using a four percent real rate of interest and the corresponding replacement cost and service life estimates.

The GRU building is divided into several offices and various rooms for seed testing, packing, drying and retrieval, as well as medium- and long-term seed storage rooms. The medium-term storage room (122 m^2) is equipped with a refrigeration system (dehumidifiers, cooling devices and control unit) and various shelves and trays. The long-term storage room (80 m^2) is also equipped with shelves and trays, and has a similar refrigeration system, but without dehumidifiers. Seeds in long-term storage are packed in vacuum-sealed containers (70 cents each), eliminating the need to control for humidity. Plastic containers (8–26 cents each) are used for medium-term storage.[2]

The seed health laboratory tests all incoming and outgoing accessions for seed-borne pathogens and pests to ensure a disease-free collection. Housed in a separate building (255 m^2), the seed health laboratory is equipped with essential laboratory equipment for laboratory testing, and two greenhouses for field testing and quarantine purposes. Part of its operation supports the Centre's breeding programme; we excluded the cost of those activities from the costs reported here.

Stored accessions are periodically tested for viability in an incubator. If they fail the testing, they are regenerated in the field or in a greenhouse or a screenhouse. The GRU has two greenhouses for regeneration of seeds that require special attention, and 20 screenhouses (called 'cages') for regenerating faba beans to control cross-pollination. The general capital items in Table 8.1 include inputs (such as office space, computers and other office equipment) that are commonly shared between crops or those that are difficult to allocate to a specific crop.

Seed storage

The seeds in the medium-term storage facility are dried to 6–7% moisture content and maintained at 0–4°C with relative humidity of 15–20%. Stored under these conditions, seeds can maintain their viability for 20–30 years. Seeds in the long-term storage are also dried to the same moisture content and hermetically vacuum-sealed in aluminium foil packets. They can maintain an acceptable level of viability for up to 50 years (or more) at a temperature of −20°C. Extending the viability of seeds by using appropriate temperature and humidity controls reduces the frequency of regeneration, and thus reduces the genetic drift that occurs in each seed sample during the regeneration process.

The non-labour storage costs in Table 8.2 include the costs of the electricity required to run the refrigeration system. The storage facility is checked daily by technicians from the engineering service unit. Part of the relevant labour cost includes the time required by the technicians to maintain and operate the storage facility and its equipment. However, a major part of the labour cost involves the labour cost of the genebank manager.

Seed health testing

Bringing new germplasm into the collection runs the risk of introducing exotic or otherwise undesirable pathogens. Similarly, disseminating germplasm can inadvertently spread pathogens to areas where they were not previously a problem. To avoid the exchange of infected accessions, the seed health laboratory tests all genetic materials entering ICARDA or disseminated elsewhere for seed-borne pathogens and pests (El-Ahmed, 1998). After fumigation or cold treatment, all incoming seeds are visually inspected and tested in the laboratory. All seeds deemed to be free of pathogens are planted in post-quarantine areas for one generation.

For accessions that are to be disseminated, random samples of seeds are tested, with particular attention being paid to samples from fields where potential seed-borne diseases were found during field inspection. If a recipient country has specified certain pathogens or pests as quarantined organisms, appropriate tests are conducted to insure freedom from these organisms. After all these procedures, the seed health laboratory applies for a phytosanitary certificate (issued by the Syrian government) and prepares a certificate of origin that accompanies the shipped seeds.

The operations of the seed health laboratory are supervised by a (part-time) plant

[2] A total of 83,197 accessions (53,982 cereals, 16,727 food legumes and 12,488 forage legumes) were held in long-term storage in 1998, compared with 119,522 accessions in the medium-term storage facility.

Table 8.2. Labour and non-labour costs (US$) of maintaining germplasm at the ICARDA genebank.

	Labour	Non-labour	Subtotal	Capital
Acquisition	**3,977**	**456**	**4,433**	**402**
Seed health testing	1,198	367	1,565	
Seed handling	2,007	–	2,007	
Overhead	772	89	861	
(No. of accessions)			(2,270)	
Medium-term storage	**10,716**	**2,549**	**13,265**	**24,351**
Storage management	5,933	–	5,933	
Temperature control	2,702	2,054	4,757	
Overhead	2,081	495	2,576	
(No. of accessions)			(119,520)	
Long-term storage	**4,690**	**1,700**	**6,389**	**22,796**
Storage management	1,978	–	1,978	
Temperature control	1,802	1,369	3,171	
Overhead	911	330	1,241	
(No. of accessions)			(83,200)	
Viability testing	**5,523**	**248**	**5,772**	**3,133**
Viability testing	4,451	200	4,651	
Overhead	1,073	48	1,121	
(No. of accessions)			(6,700)	
Dissemination	**33,291**	**9,692**	**42,983**	**4,911**
Diss. management	11,865	–	11,865	
Seed health testing	14,617	4,483	19,099	
Packing/shipping	344	3,327	3,671	
Overhead	6,465	1,882	8,347	
(No. of accessions)			(27,700)	
Duplication	**2,583**	**9,234**	**11,816**	
Packing/shipping	2,081	7,441	9,522	
Overhead	502	1,793	2,295	
(No. of accessions)			(8,410)	
Information management	**54,554**	**6,950**	**61,504**	
Maintaining database	43,960	2,400	46,360	
Publication	–	3,200	3,200	
Overhead	10,594	1,350	11,944	
General management	**77,029**	**11,838**	**88,868**	**19,737**
Managerial staff	62,070	–	62,070	
Electricity	–	2,739	2,739	
Other expenses	–	6,800	6,800	
Overhead	14,959	2,299	17,258	
Total	**192,364**	**42,666**	**235,030**	**75,331**

pathologist who is assisted by a research associate, a technician and five temporary labourers hired on a daily basis. About 20% of the seed health laboratory's activities are geared towards genebank operations. Since the costs for testing the health of incoming and outgoing accessions are about the same, we allocated these costs to the 'acquisition' and 'dissemination' categories in Table 8.2, based on the respective shares of acquired and disseminated accessions processed for the genebank.

Viability testing

The viability of seeds stored in the genebank needs to be checked periodically. About 70% of ICARDA's holding of those crops for which it has a global mandate (barley, lentils, and faba beans) and a regional mandate (chickpeas and wheat) have had at least one round of viability testing. For crops of lower priority (e.g. some forage legumes), the proportion of accessions that have been tested for viability is quite low. Over the past 10 years, a total of 67,405 accessions have gone through viability testing, averaging 6700 accessions per year.

To test for viability, 40 seeds from each accession are placed in four separate petri dishes, held in an incubator for a week, and then visually checked. A large share of the testing costs consists of the labour required to supervise and carry out the tests (Table 8.2). If the viability rate of an accession falls below 85%, it is regenerated in the next season.

Regeneration

One of the most labour-intensive aspects of conserving germplasm is to multiply and regenerate stored accessions when it becomes necessary to maintain an adequate stock of viable seeds. Multiplication to scale up accessions is done when the holding falls below a minimum sample size (500–700 seeds), and regeneration is performed when seed viability drops below 85%. In addition, some accessions are planted out for evaluation and characterization purposes.[3] Other accessions such as wild wheat were multiplied for prebreeding usage in 1998, the base year for a cost data. Newly acquired accessions are also planted in the field or greenhouse to increase the number of seeds per sample before placing them in storage.[4]

Regeneration activities are supervised by two germplasm specialists for each group of crops (i.e. cereals, food legumes, and forage legumes), and together they process around 4000–6000 accessions per year in each group. The number of accessions regenerated or multiplied in 1998 is summarized in Table 8.3 for each crop along with its purpose and the size of planting area. Most cultivated cereals and food legumes (lentils and chickpeas) meet the recommended storage quantity after the first cycle of regeneration. However, some crops such as wild cereals, faba beans and forage legumes often need more than one cycle. We assumed two rounds of regeneration were required for these crops to guarantee an appropriate sample size, and adjusted the number of regenerated accessions accordingly in Table 8.3.

Tables 8.4A and 8.4B detail the labour and non-labour costs incurred in the field operations and the seed processing activities during the regeneration process for each crop. The field management requires 80% of two crop specialists' labour, and the remaining 20% is allocated to characterization activities. The steps undertaken prior to planting include land preparation, dressing seeds with fungicides, and scarifying the hard cover from seeds of some vicia and other pasture species to ease the germination. The station operation unit maintains the farmlands and provides land preparation services on a $400 per hectare charge-back basis. A scarification machine is used to scarify some species with large-sized seeds, and a more labour-intensive hand-scarification technique is required for small-sized seeds of wild food legumes and pastures.

Most accessions are planted in field plots of 3–4 m^2 in size, which consist of two rows of planted seeds and two or three empty rows to prevent the mixing of seeds between adjacent plots.[5] Some crops require more

[3] For example, the main focus of the cereal germplasm group for the past 3 years has been the preparation of a catalogue of its bread wheat holdings, so substantial effort has been directed to characterizing and evaluating this crop.

[4] In most cases, newly acquired materials come in small quantities that require multiplication. Less than 1% of the incoming materials does not require bulking up.

Table 8.3. Regeneration activities at the ICARDA genebank in 1998 season.

Crops	No. of accessions	Adjusted no. of accessions	Purpose	Planting area
Cereals	**4800**	**4750**		**4 ha**
Bread wheat	2900	2900	Characterization	
Wheat crosses	1200	1200	Pre-breeding	
Barley	550	550	Characterization	
Aegilops	150	100	Characterization	
Food legumes	**3700**	**3250**		**4.5 ha**
Chickpea	1000	1000	Multiplication	
Lentil	900	900	Multiplication	
Faba bean	1800	1350	Characterization	
Forage legumes	**7200**	**5400**		**3.5 ha**
Medicago	1000	750	Multiplication	
Trifolium	1200	900	Multiplication	
Pasture	200	150	Multiplication	
Vicia	1000	750	Characterization	
Lathyrus, Pisum	3800	2850	Multipli./charact.	

specialized care and are planted in either plastic-covered greenhouses or meshed screenhouses. For example, *medicago* and other pasture species with a small number of seed samples (less than 50 seeds) are planted out in greenhouses, while faba beans are planted in screenhouses to control cross-pollination. Hand-planting is preferred in most cases to avoid inadvertently mixing seeds and some small-sized seeds are not suitable for machine-planting.

The amount of chemicals (fertilizers, herbicides and pesticides) applied each year depends on the incidence of disease infestations, which in turn depends on temperature and other environmental factors. Additionally, hand-weeding is also used when necessary because herbicides are efficacious only for some specific types of weeds (e.g. broadleaf or narrow-leaf weeds). Plants are inspected throughout the growing period to characterize their morphological traits. This activity requires a total of 2 months of a specialist's and a daily labourer's time for each group.

Harvesting is done with the help of the station operation unit which provides a driver for harvester machines. Cultivated cereals and chickpeas are harvested by machine, but lentils and faba beans are hand-harvested, as are some wild forage legumes, due to seed shattering problems. Harvested materials are packed in paper bags (for cereal) or cotton bags (for legumes) and transported to the seed cleaning area. Seeds are then threshed and cleaned either manually (some wild forage species and pastures) or mechanically (cereals and some food legumes).

Upon reception from the seed cleaning area, seed samples are first weighed, separated, and then stored in a dehumidifying room for 4–5 weeks. The room is held at a relative humidity level of 10–13% and a temperature of 21°C, designed to reduce the seed moisture content to a 6–7% range. After the seeds have been cleaned and dried, they are stored in plastic containers for medium-term storage and in aluminium bags for long-term storage.[6]

[5] One hectare can be divided into 1000 plots, so approximately 1000 accessions are planted per hectare.
[6] In most years, not all harvested accessions are cleaned, dried and stored: for example, some are used for testing and pre-breeding purposes after harvesting. However, we assumed that all of the multiplied accessions were processed and packed for storage when estimating the average costs of regeneration reported in Table 8.5.

Table 8.4A. Labour and non-labour costs (US$) for regenerating cereal and forage legumes.

	Cereal				Forage legumes			
	Labour	Non-labour	Subtotal	Capital	Labour	Non-labour	Subtotal	Capital
Regeneration	**23,254**	**10,746**	**34,000**	**4,955**	**27,014**	**13,383**	**40,397**	**16,255**
Field operation	**17,289**	**953**	**18,242**	**3,364**	**19,763**	**6,235**	**25,998**	**14,446**
Field management	12,320		12,320	3,364	12,320		12,320	3,364
Seed preparation	74	50	124		241	70	311	119
Land preparation	1,120	480	1,600		980	420	1,400	
Planting	37		37		333		333	
Screenhouse					1,090	578	1,668	10,963
Chemicals/weeding	296	142	438		370	2,372	2,742	
Harvesting	84	96	180		592	1,584	2,176	
Overhead	3,357	185	3,543		3,838	1,211	5,049	
Seed processing	**5,965**	**9,792**	**15,758**	**1,591**	**7,251**	**7,148**	**14,399**	**1,808**
Process management	4,206		4,206	730	4,781		4,781	830
Seed cleaning	389		389	90	858		858	103
Seed drying	53		53	771	81		81	876
Medium-term packing	70	803	874		56	360	416	
Long-term packing	89	7,088	7,176		67	5,400	5,467	
Overhead	1,158	1,902	3,060		1,408	1,388	2,796	
(No. of accessions)			(4,750)				(5,400)	
Characterization	**4,281**		**4,281**		**4,741**		**4,741**	
Recording traits	3,450		3,450		3,820		3,820	
Overhead	831		831		921		921	
(No. of accessions)			(4,750)				(5,400)	

Table 8.4B. Labour and non-labour costs (US$) for regenerating chickpea, lentil and faba bean.

	Chickpea				Lentil				Faba bean			
	Labour	Non-labour	Subtotal	Capital	Labour	Non-labour	Subtotal	Capital	Labour	Non-labour	Subtotal	Capital
Regeneration	**5,109**	**5,108**	**10,218**	**1,456**	**5,217**	**3,135**	**8,352**	**3,358**	**12,437**	**5,609**	**18,046**	**2,019**
Field operation	**3,910**	**1,137**	**5,047**	**1,121**	**4,127**	**1,236**	**5,364**	**3,056**	**10,502**	**2,761**	**13,262**	**1,566**
Field management	2,710	167	2,877	1,121	2,710	167	2,877	1,121	6,899	167	7,066	1,121
Seed preparation	19	10	29		33	10	43		37	25	62	
Land preparation	280	120	400		280	120	400		700	300	1,000	
Planting	17		17		17		17		74		74	
Screenhouse					192	102	294	1,935	641	320	961	445
Chemicals/weeding	37	400	437		37	400	437			1,017	1,017	
Harvesting	87	220	307		56	198	254		111	396	507	
Overhead	759	221	980		802	240	1,042		2,039	536	2,576	
Seed processing	**1,200**	**3,971**	**5,171**	**335**	**1,090**	**1,899**	**2,989**	**301**	**1,935**	**2,848**	**4,783**	**452**
Process management	885		885	154	797		797	138	1,195		1,195	207
Seed cleaning	37		37	19	37		37	17	296		296	26
Seed drying	11		11	162	11		11	146	20		20	219
Medium-term packing	15	200	215		15	180	195		19	270	289	
Long-term packing	19	3,000	3,019		19	1,350	1,369		30	2,025	2,055	
Overhead	233	771	1,004		212	369	580		376	553	929	
(No. of accessions)			(1,000)				(900)				(1,350)	
Characterization	**979**		**979**		**979**		**979**		**2,508**		**2,508**	
Recording traits	789		789		789		789		2,021		2,021	
Overhead	190		190		190		190		487		487	
(No. of accessions)			(1,000)				(900)				(1,350)	

Dissemination and safety duplication

Around 30,000 accessions have been disseminated to cooperators each year for the past 5 years, and about 40–50% of these accessions are distributed to scientists outside ICARDA. In 1998, a total of 132 shipments involving 27,700 accessions were made, including 54 shipments destined for overseas. Shipping costs depend on the locations of recipients and the amount of seeds shipped. The genebank normally incurs between $3000 and 4000 per year in shipping costs.

The genebank periodically sends a duplicate set of samples of its accessions for storage in various off-site locations, to insure against accidental loss of germplasm through natural disaster or other hazards. The labour cost of packing and handling duplicate seed samples is similar to that of disseminating seeds. Duplicate samples of barley and wheat are sent to the International Maize and Wheat Improvement Center (CIMMYT) genebank in Mexico. The chickpea collection is duplicated at the International Crop Research Institute for the Semi-Arid Tropics (ICRISAT) in India, the faba bean collection at the Austrian national genebank, and the lentil collection at the National Bureau of Plant Genetic Resources (NBPGR) in India. *Medicago* and *Vicia* are conserved at the Federal Institute of Agrobiology (FIA) in Austria, and *Lathyrus* is duplicated at the Station Federal de Recherches Agronomiques de Changing (RAC) in Switzerland.

Duplication activities are overseen by the genebank manager. Not all accessions are duplicated every year. For example, 2097 food legumes and 6313 forage legumes were sent as duplicates to these various locations in 1998. When the number of unique accessions requiring duplication reaches a certain level, seed samples are packed in aluminium bags, vacuum-sealed, and shipped to their respective locations under 'black-box' arrangements.

Information and general management

The documentation officer coordinates data gathered from the field and other sources, manages the information, and oversees publication of occasional catalogues describing the holdings of each crop. When new accessions are introduced into the genebank, passport data for each collection is entered in the GRU information system as well as ICARDA's data management and retrieval system. Passport, characterization and evaluation data are published regularly in germplasm catalogues.[7] The information is also available on the Internet as part of the System-wide Information Network for Genetic Resources supported by the CGIAR System-wide Genetic Resources Programme (SGRP). An important part of GRU's data management system is the seed stock control system, which assists germplasm specialists in managing the storage of seeds and distribution of samples to users.

We determined that about 20% of time was spent by the head of the genebank and his secretaries on activities other than the conservation and distribution costed here. The other 80% of their time was allocated to the general management category in Table 8.2. Various non-labour costs, including the costs of electricity and other miscellaneous operational expenses, were similarly allocated.

Economic Analyses of Genebank Costs

Representative annual costs of conservation

In 1998, the ICARDA genebank operated with 'core' support of $587,000 – its total budget for that year including all project-based support was $1,042,000. This budget, however, is not a representative figure

[7] The GRU also supports two international databases, one on wild wheat including *Aegilops*, and the other on forages from the Mediterranean Basin, which was initially created through collaboration with IBPGR.

Table 8.5. Annual average costs (US$) of conserving and distributing germplasm in the ICARDA genebank*.

	No. of accessions	TCC	TQFC[a]	TVC	**ACC**	**AQFC**	**AVC**
Medium-term storage	119,520	27,312	24,674	11,147	**0.23**	**0.21**	**0.09**
Long-term storage	83,200	23,783	8,225	5,683	**0.29**	**0.10**	**0.07**
Acquisition	2,270	1,389	9,459	2,493	**0.61**	**4.17**	**1.10**
Viability testing	6,700	5,106	16,449	4,359	**0.76**	**2.46**	**0.65**
Dissemination	27,700	9,846	58,638	21,938	**0.36**	**2.12**	**0.79**
Safety duplication	8,410	987	8,225	11,110	**0.12**	**0.98**	**1.32**
Regeneration							
Cereal	4,750	6,797	31 280	16,754	**1.43**	**6.59**	**3.53**
Forage legumes	5,400	18,097	31 995	22,437	**3.35**	**5.92**	**4.15**
Chickpea	1,000	2,070	8,053	6,843	**2.07**	**8.05**	**6.84**
Lentil	900	3,972	7,943	5,087	**4.41**	**8.83**	**5.65**
Faba bean	1,350	2,633	13,636	9,088	**1.95**	**10.10**	**6.73**
Characterization							
Cereal	4,750	461	6,515	1,275	**0.10**	**1.37**	**0.27**
Forage legumes	5,400	461	6,515	1,734	**0.09**	**1.21**	**0.32**
Chickpea	1,000	154	1,739	410	**0.15**	**1.74**	**0.41**
Lentil	900	154	1,739	410	**0.17**	**1.93**	**0.46**
Faba bean	1,350	154	3,038	639	**0.11**	**2.25**	**0.47**
Total costs		**103,373**	**238,123**	**121,406**	**16.20**	**58.01**	**32.86**

* Costs of information and general management in Table 8.2 are allocated to each activity according to the following share: Medium-term storage (15%), Long-term storage (5%), Acquisition (5%), Viability testing (10%), Regeneration (30%), Characterization (5%), Duplication (5%) and Dissemination (25%).
ACC, average capital cost; AQFC, average quasi-fixed cost; AVC, average variable cost; TCC, total capital cost; TQFC, total quasi-fixed costs; TVC, total variable cost.

of the annual cost of 'conserving' and 'distributing' germplasm. On the one hand, it omits some important cost elements such as overhead expenses (that cover the cost of general institutional and administrative support), various services provided by other units, and many of the relevant capital costs. On the other hand, it includes prebreeding, molecular characterization and evaluation activities, which are not explicitly costed as part of this study. Moreover, parts of the operating costs incurred by the seed health laboratory that falls within the administrative purview of the GRU should be allocated to the activities of breeding programmes.

Table 8.5 provides an overview of the total capital, quasi-fixed and variable costs grouped according to various activities for each crop, along with corresponding average costs per accession. By our reckoning, the total annual cost of conserving and distributing germplasm at ICARDA is about $462,902. Figure 8.2 gives a breakdown of this total cost into various cost components. The capital cost includes an estimate of the annualized costs of building facilities, vehicles, and various laboratory and office equipment. The quasi-fixed costs are the human capital (i.e. labour) costs of the senior scientific staff and genebank specialists, which are insensitive to changes in the scale of the genebank operation.

Figure 8.2 shows that far more than half (61%) of the annual expense of the genebank operation involves labour input while less than a quarter (22%) of the total cost is related to capital. Although the investment required to build a genebank is substantial, Fig. 8.2 illustrates that the overall genebank operation, when viewed

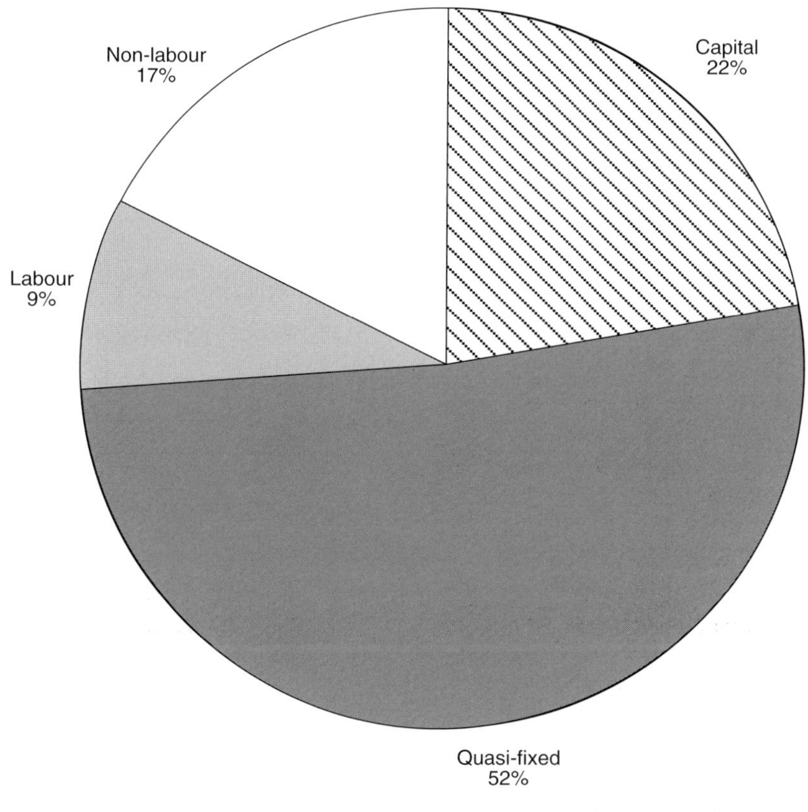

Fig. 8.2. A representative annual operation cost of the ICARDA genebank (US$462,902).

on a representative annualized basis, is not especially capital-intensive. Figure 8.2 also indicates that a substantial portion of the labour cost is lumpy in nature, i.e. the costs of senior scientific staff that are treated as quasi-fixed input are not easily divisible. (Scientists come in discrete units that cannot be shared by geographically separate locations.) This suggests that within a certain range of activity, the overall costs of conservation do not increase dramatically as the number of accessions increases, and thus the average cost per accession critically depends on the number of accessions.

Three columns from the right of Table 8.5 provide a breakdown of the per-accession costs for each crop by types. As one might expect, the share of the capital cost component of both the medium- and long-term storage operations is high (40–60% of the total cost), compared with other genebank functions whose share of capital cost is in the 10–20% range. The table also shows that the regeneration of seeds incurs a much higher cost than any other genebank operation, and that typically food legumes (chickpea, lentil and faba bean) are more expensive to regenerate than cereal or forage legumes. The comparatively larger-sized seeds for food legumes require considerably larger plot sizes for regenerating these seeds, which increases the per-accession costs accordingly.

Economic analyses

Costs of conserving for 1 year

Using the average cost estimates presented in Table 8.5, we can address the following question: what is the cost of storing an existing accession for one more year, or the benefit from eliminating a duplicate accession from the genebank? The answer depends, among other things, on the state of the sample which, in turn, depends on its time in storage (specifically the time from last regeneration or viability test). If a sample does not require regeneration, the marginal variable cost of holding over an accession of any crop for one more year is just 7 cents (Table 8.6). If we consider the average non-capital costs (which include both variable and quasi-fixed costs), the corresponding figure is still small, just 17 cents per accession. However, if the sample needs to be regenerated due to low viability, then the marginal variable cost of keeping it for another year jumps to between $4.25 (cereals) and $7.56 (chickpeas), while the average non-capital costs increase to between $13.35 (forage legumes) and $20.11 (faba beans).

What is the cost of conserving a newly acquired accession for 1 year, given the decision to store it is revisited after 1 year? The protocol followed by the ICARDA genebank is to check the health of every new accession before placing it into storage, and so the cost of seed health testing is included as part of the 'acquisition' cost. In addition, since most new acquisitions arrive with small quantities, they require bulking up. Including both the costs of acquisition and regeneration, the non-capital cost of storing a newly introduced accession for its first year in the genebank is between $22.32 and $31.17 per accession.

The cost of distributing an accession reported in Table 8.6 includes the cost of maintaining seeds in medium-term storage as well as the cost incurred in packing and shipping the seeds. If there is sufficient stock in the medium-term storage, the non-capital cost of distributing an accession is $3.21 per accession. However, if the stock is insufficient so that regeneration is required, the distribution cost jumps to between $16.39 and $23.15 per accession, depending on the crop. For a new accession, a one-off cost of characterizing the accession needs to be included, so the cost of distributing a new accession is in the range of $17.92 to $25.87. Notably, the cost of distributing accessions is greater than the cost of conserving them.

Costs in the long run

In the previous section we calculated the costs of conserving an accession for 1 year,

Table 8.6. Annual per-accession costs (US$) of conserving and distributing an accession.

	Variable cost			Non-capital cost		
	Existing acc.		New acc.	Existing acc.		New acc.
	w/o Reg.	w/ Reg.		w/o Reg.	w/ Reg.	
Conservation						
Long-term (LT) storage	0.07	0.07	0.07	0.17	0.17	0.17
New introduction						
Acquisition			1.10			5.27
Container (chickpea, faba bean)			3.5			3.5
Container (other crops)			1.4			1.4
Safety duplication			1.32			2.30
Viability testing		0.65	0.65		3.11	3.11
Regeneration						
Cereal		3.53	3.53		10.11	10.11
Forage legumes		4.15	4.15		10.08	10.08
Chickpea		6.84	6.84		14.90	14.90
Lentil		5.65	5.65		14.48	14.48
Faba bean		6.73	6.73		16.83	16.83
Cons. cost (cereal)	**0.07**	**4.25**	**8.07**	**0.17**	**13.39**	**22.35**
Cons. cost (forage legumes)	**0.07**	**4.87**	**8.69**	**0.17**	**13.35**	**22.32**
Cons. cost (chickpea)	**0.07**	**7.56**	**13.48**	**0.17**	**18.17**	**29.23**
Cons. cost (lentil)	**0.07**	**6.37**	**10.19**	**0.17**	**17.75**	**26.72**
Cons. cost (faba bean)	**0.07**	**7.45**	**13.37**	**0.17**	**20.11**	**31.17**
Distribution						
Medium-term storage	0.09	0.09	0.09	0.30	0.30	0.30
Dissemination	0.79	0.79	0.79	2.91	2.91	2.91
Viability testing		0.65	0.65		3.11	3.11
Regeneration						
Cereal		3.53	3.53		10.11	10.11
Forage legumes		4.15	4.15		10.08	10.08
Chickpea		6.84	6.84		14.90	14.90
Lentil		5.65	5.65		14.48	14.48
Faba bean		6.73	6.73		16.83	16.83
Characterization						
Cereal			0.27			1.64
Forage legumes			0.32			1.53
Chickpea			0.41			2.15
Lentil			0.46			2.39
Faba bean			0.47			2.72
Dist. cost (cereal)	**0.89**	**5.06**	**5.33**	**3.21**	**16.43**	**18.07**
Dist. cost (forage legumes)	**0.89**	**5.69**	**6.01**	**3.21**	**16.39**	**17.92**
Dist. cost (chickpea)	**0.89**	**8.38**	**8.79**	**3.21**	**21.21**	**23.36**
Dist. cost (lentil)	**0.89**	**7.19**	**7.64**	**3.21**	**20.79**	**23.18**
Dist. cost (faba bean)	**0.89**	**8.27**	**8.74**	**3.21**	**23.15**	**25.87**

assuming that the decision to conserve the material is revisited the following year. However, genebanks may well be required to guarantee safekeeping of accessions in perpetuity; for example those accessions held in trust by the CGIAR centres by way of their commitments to the FAO. The costs of such a guarantee depend on the state of future technology, input costs (including the rate of interest), storage capacity, and regeneration intervals.

Table 8.7 provides the present values of the costs of conserving and distributing an accession in perpetuity, assuming costs are constant over time in real (inflation-adjusted) terms. We considered different levels of interest rates to check the sensitivity of costs, and calculated the costs for both existing and new accessions. A new accession is assumed to go through an initial round of regeneration, based on the current protocol of the ICARDA genebank. The appropriate interval for viability testing has not yet been established for the current conservation technology, so we assumed that testing begins in the 10th year after acquisition, with re-testing every 5 years thereafter (consistent with the protocol used by the CIMMYT genebank). An average of around 25,000 accessions are disseminated each year, and so we assumed that an accession is disseminated once every 5 years.

The average variable costs of each activity critically depend on the real rate of interest; the present value of costs calculated using a 2% rate of interest are several times larger than the costs estimated using a 6% rate in Table 8.7. Using 4% as a baseline rate of interest, the average cost of conserving an existing accession of cereals in perpetuity is $21.22 per accession when there is no initial round of regeneration. The comparable figures for other crops are in the range of $21.22 and $22.32. For a newly acquired accession, the conservation costs range from $38.80 to $44.83 per accession.

Accessions maintained in the medium-term storage facility for distribution require more frequent regeneration than those held in long-term storage due to the rapid loss of seed viability. Following the standards for maintaining seed viability described by Hamilton *et al.* (1997), we assumed that the viable shelf-life of an accession in the active collection averages 25 years, while one held in long-term storage remains viable for an average of 50 years. The present value of the distribution cost is in the range of $32.06 and $36.10 if the size of seed samples is sufficient to negate the need for regeneration. Distributing a newly acquired accession incurs additional costs related to characterizing the sample, thereby increasing the present value to a range of $39.08–$47.16 per accession.

Total costs in the long run

In the previous section, we reported that the representative annual total cost of the ICARDA genebank operation (including both conservation and distribution activities) is $462,902 ($103,373 for capital and $359,529 for non-capital). A natural question to ask is how much will it cost to maintain the current genebank operations forever (i.e. in perpetuity)? The simple way to answer this question is to derive the in-perpetuity costs by multiplying the annual total expense by $(1 + r)/r$, where r is the real interest rate.[8] With a 4% rate of interest, the multiplier (i.e. $(1 + r)/r$) is 26, resulting in an in-perpetuity cost of $12,035,920 ($2,687,698 for capital and $9,347,754 for non-capital). In other words, it requires about $12.04 million invested at a real rate of interest of 4% to maintain the current level of ICARDA's genebank operation in perpetuity.

An alternative, and in many ways, a more accurate method for estimating the in-perpetuity cost is to calculate the costs of conserving and distributing an accession from acquisition to forever, and then multiplying these respective costs by the total number of accessions. Table 8.8 provides the *total costs* of conserving and distributing ICARDA's present holdings in perpetuity estimated using this method. It costs $6,422,321 to conserve and $6,052,751 to distribute the current level of holdings in the ICARDA genebank – a total of $12,475,073.

[8] The present value of outlays X spent every year from time zero is given by $PV = X + X/(1+r) + X/(1+r)^2 + \ldots = X(1/1\text{-}(1/(1+r)))$, where r is the interest rate.

Table 8.7. Present values of the costs (US$) of conserving and distributing an accession in perpetuity.

	Existing accession[a]			New accession		
	2%	4%	6%	2%	4%	6%
Conservation						
Long-term storage	8.53	4.35	2.95	8.53	4.35	2.95
New introduction						
Acquisition				5.27	5.27	5.27
Container (chickpea, faba bean)				3.5	3.5	3.5
Container (other crops)				1.4	1.4	1.4
Initial viability testing				3.11	3.11	3.11
Initial duplication				2.30	2.30	2.30
Viability testing	29.84	14.34	9.18	29.84	14.34	9.18
Safety duplication	1.36	0.38	0.13	1.36	0.38	0.13
Regeneration (50 years)[b]						
Cereal	7.81	2.16	0.76	10.52	7.69	6.99
Forage legumes	7.79	2.16	0.76	10.49	7.67	6.97
Chickpea	10.64	2.95	1.03	14.32	10.47	9.52
Lentil	10.40	2.88	1.01	13.99	10.23	9.30
Faba bean	11.79	3.27	1.14	15.86	11.60	10.54
Cons. cost (cereal)	**47.54**	**21.22**	**13.03**	**62.31**	**38.82**	**31.33**
Cons. cost (forage legumes)	**47.52**	**21.22**	**13.02**	**62.28**	**38.80**	**31.31**
Cons. cost (chickpea)	**50.37**	**22.01**	**13.30**	**68.22**	**43.70**	**35.96**
Cons. cost (lentil)	**50.12**	**21.94**	**13.28**	**65.78**	**41.36**	**33.63**
Cons. cost (faba bean)	**51.51**	**22.32**	**13.41**	**69.76**	**44.83**	**36.98**
Distribution						
Medium-term storage	15.29	7.79	5.29	15.29	7.79	5.29
Dissemination[c]	30.86	16.34	11.51	30.86	16.34	11.51
Regeneration (25 years)/Charact.[d]						
Cereal	20.63	7.93	4.02	24.98	15.10	11.89
Forage legumes	20.58	7.92	4.01	24.81	14.96	11.75
Chickpea	28.10	10.81	5.47	33.93	20.48	16.10
Lentil	27.45	10.56	5.34	33.43	20.30	16.02
Faba bean	31.12	11.97	6.06	37.92	23.03	18.18
Dist. cost (cereal)	**66.78**	**32.06**	**20.82**	**71.12**	**39.23**	**28.69**
Dist. cost (forage legumes)	**66.73**	**32.04**	**20.81**	**70.95**	**39.08**	**28.55**
Dist. cost (chickpea)	**74.24**	**34.93**	**22.27**	**80.07**	**44.61**	**32.90**
Dist. cost (lentil)	**73.59**	**34.68**	**22.15**	**79.57**	**44.42**	**32.82**
Dist. cost (faba bean)	**77.27**	**36.10**	**22.86**	**84.07**	**47.16**	**34.98**

[a] Existing accession is assumed to be freshly regenerated, but new accession needs initial regeneration.
[b] Regeneration cost includes the cost of viability testing which is performed after each regeneration.
[c] Dissemination of an accession is assumed to occur every 5 years.
[d] This cost includes one-time cost of characterization.

The present-value equivalent of $1,646,202 is needed to underwrite the capital costs of conserving ICARDA's current collection in perpetuity. Excluding the cost of capital, it takes a total of $4,776,119 in labour and operating costs to conserve the entire holdings in perpetuity. This figure includes much more than the labour and operational costs required to simply store the seeds in the genebank. It factors in the costs of checking the viability of the seeds, periodically regenerating the

Table 8.8. Total costs (US$) of conserving and distributing all ICARDA accessions in perpetuity.

		Per-accession cost		Total cost	
Crops	No. of acc.	Conservation	Distribution	Conservation	Distribution
Cereal	**57,900**	**52.11**	**48.62**	**3,017,330**	**2,815,339**
Non-capital		38.82	39.23	2,247,646	2,271,387
Capital		13.29	9.39	769,685	543,952
Forage legumes	**31,700**	**53.21**	**49.70**	**1,686,794**	**1,575,576**
Non-capital		38.80	39.08	1,229,974	1,238,958
Capital		14.41	10.62	456,820	336,618
Chickpea	**10,400**	**57.37**	**54.20**	**596,628**	**563,711**
Non-capital		43.70	44.61	454,508	463,934
Capital		13.67	9.59	142,120	99,777
Lentil	**8,700**	**56.39**	**55.89**	**490,579**	**486,252**
Non-capital		41.36	44.42	359,831	386,478
Capital		15.03	11.47	130,747	99,774
Faba bean	**10,800**	**58.43**	**56.65**	**630,991**	**611,873**
Non-capital		44.83	47.16	484,160	509,298
Capital		13.60	9.50	146,830	102,575
All crops	**119,500**	**277.51**	**265.08**	**6,422,321**	**6,052,751**

samples, plus the data management costs required to manage the collection.

Separate from these conservation costs are the costs of distributing the seeds, usually on request to breeders and others outside of ICARDA, although a sizeable share is taken from the genebank by the managers themselves for prebreeding and evaluation purposes. If the genebank continued to distribute seeds at the rate typical of the past few years, this distribution function alone would incur an in-perpetuity cost of about $6,052,751 ($1,182,697 for capital and $4,870,054 for non-capital) expressed in present-value terms.

Bundling all these costs together (i.e. including the seed storage, regeneration, duplication, information management and dissemination activities), we estimate that the capital, labour, and operational costs combined would total $12,475,073 in perpetuity. This represents the amount of money that would need to be set aside (at a 4% real rate of interest) to underwrite ICARDA's genebank activities at their current levels over the longer run, a sizeable but not an especially large sum of money.

Comparison with CIMMYT costs

Pardey *et al.* (2001) derived a series of cost estimates using comparable methods for the CIMMYT genebank. Comparison of the cost structures between CIMMYT and ICARDA provides more insights into the costs of conserving different or similar crops at various locations under different institutional arrangements. In terms of the number of stored accessions and overall annual costs, the two genebank operations are similar; ICARDA stores 119,500 accessions at an annual cost of $462,902 versus 140,000 accessions at an annual cost of $531,914 at CIMMYT. The composition of costs is also very similar between two genebanks. Capital cost account for about 24% of CIMMYT's annual costs and about 22% for ICARDA's; labour constitutes 62% of CIMMYT's costs and 61% of ICARDA's. The difference in the share of quasi-fixed cost (36% for CIMMYT and 52% for ICARDA) stems largely from the differences in the compositions and roles of the genebank personnel.

However, there are some substantial differences in the organizational structure and operational details of the two genebanks that give rise to differences in their costs. While the CIMMYT wheat and maize holdings are now stored in one facility, the genebank storage for wheat is supervised largely independently from those for maize under the direction of two genebank managers. This contrasts with ICARDA where one genebank head oversees the whole operation, with many of the day-to-day genebank activities being devolved to genebank curators whose labour costs are comparatively low.

A study of the CIMMYT and the ICARDA genebanks makes it possible to compare the costs of conserving an accession of wheat in different locations. The marginal variable cost of holding a wheat accession in long-term storage for one additional year is 11 cents at CIMMYT and 7 cents at ICARDA. The average non-capital cost is 19 cents per accession per year at CIMMYT and 17 cents at ICARDA. At first glance, the substantial differences in the cost of daily labour – $80 per month at ICARDA compared with $260 per month at CIMMYT – would lead one to suspect that the costs of storage at CIMMYT are substantially larger than those at ICARDA. However, the quasi-fixed component of non-capital costs is comparatively large at both genebanks, and the cost differentials for senior staff across these genebanks are comparatively small, thus making the cross-genebank difference negligible.

The present values of conserving an accession in perpetuity at ICARDA for crops other than wheat are bounded by the per accession costs of conserving wheat and maize in CIMMYT. For example, the costs of conserving an existing accession (without initial regeneration) at the ICARDA genebank range from $21.22 to $22.32 (Table 8.8), while the costs for wheat and maize at CIMMYT are $10.26 and $58.75, respectively. The costs of distribution are about $33–$38 per accession from ICARDA, compared with about $15 per accession for wheat and $185 for maize from CIMMYT.

The total in-perpetuity costs of conserving and distributing ICARDA's existing holdings is $6.42 million (compared with $8.9 million at CIMMYT) and $6.05 million ($4.2 million at CIMMYT), respectively. A large proportion (90%) of ICARDA holdings consist of landraces and wild species, which typically are acquired in small quantities that require bulking up before placing them in storage. CIMMYT's holdings consist mainly of improved breeding materials (more than 50%) that typically arrive in sufficiently large quantities that do not require bulking up. Thus, a comparison of costs per 'unit of genetic diversity' (assuming an appropriate measure) might yield results more favourable to ICARDA relative to CIMMYT.

Conclusions

This study extends the scope of the previous CIMMYT cost study (wheat and maize in Latin America) by examining different crops (barley, chickpeas, lentils, faba beans and forage legumes) in a different geographical area (West Asia and North Africa). The cost structure of the ICARDA genebank is somewhat different from those of the conservation of maize or wheat at CIMMYT. The per-accession costs of conserving ICARDA crops are, in general, higher than those of wheat and lower than those for maize at CIMMYT. The costs of endowing the ICARDA genebank in perpetuity is around $12.5 million, at a 4% real rate of interest, a modest sum for conservation of large numbers of landraces of food and forage crops important to worldwide agriculture.

Acknowledgements

This chapter was in part funded by and prepared for the System-wide Genetic Resources Program of the CGIAR. Additional support was provided by the Swedish International Development Cooperation Agency (SIDA).

This chapter would simply not have been possible without the generous help of a good number of colleagues. For assistance in collecting and interpreting data we are grateful to Bilal Humeid, Suresh Sitaraman, Jurgen Diekmann, Jan Konopka, Siham Asaad, Ali Ismail, Ali Shehaden, F. Sweid, and the staff at the Genetic Resource Unit at the ICARDA.

References

El-Ahmed, A. (1998) International Center for Agricultural Research in the Dry Areas. In: Kahn, R.P. and Mathur, S.B. (eds) *Containment Facilities and Safeguards for Exotic Plant Pathogens and Pests*. American Phytopathological Society, St Paul, Minnesota.

Fuccillo, D., Sears, L. and Stapleton, P. (1997) *Biodiversity in Trust: Conservation and Use of Plant Genetic Resources in CGIAR Centres.* Cambridge University Press, Cambridge.

Hamilton, N., Sackville, R. and Chorlton, K.H. (1997) *Regeneration of Accessions in Seed Collections: a Decision Guide.* Handbook for genebanks No.5, International Plant Genetic Resources Institute, Rome, Italy.

Pardey, P.G., Koo, B., Wright, B.D., van Dusen, M.E., Skovmand, B. and Taba, S. (2001) Costing the *ex situ* conservation of genetic resources: maize and wheat at CIMMYT. *Crop Science* 41, 1286–1299.

Valkoun, J., Robertson, L.D. and Konopka, J. (1995) Genetic resources at the heart of ICARDA mission throughout the Mediterranean region. *Diversity* 11, 11–12.

Chapter 9

Impact of Terminator Technologies in Developing Countries: a Framework for Economic Analysis

C.S. Srinivasan[1] and Colin Thirtle[2]

[1]*Department of Agricultural and Food Economics, University of Reading, Earley Gate, Whiteknights Road, Reading RG6 6AR, UK;* [2]*Environmental Policy and Management Group, T.H. Huxley School of Environmental, Earth Sciences and Engineering, Imperial College of Science, Technology and Medicine, RSM Building, Prince Consort Road, London SW7 2BP, UK*

Introduction

The emerging technology for inducing sterility in seeds – or 'terminator technology' as it has popularly come to be known – has the potential to bring far-reaching changes in the seed industry and the organization of agriculture. Terminator technology alters a fundamental characteristic of seed – its self-reproducing nature – and threatens to change agricultural practices that have been in vogue for centuries. Seldom does an innovation have the potential to alter a product market in such a fundamental way. Given its potential impact, it is understandable that terminator technology has attracted a virulent and polemical response from a range of quarters. It has been seen as an unethical or immoral technology that threatens the livelihood of millions of farmers, especially resource-poor farmers in developing countries. It is viewed as a technology that needs to be stopped in its tracks, or discarded completely, if its adverse consequences are to be avoided. Such a response has discouraged a more dispassionate economic analysis of the technology. Such an analysis must attempt to identify the economic forces that have induced the development of this technology. There may be insights to be gained in viewing terminator technology as an induced technological response to the inadequacies and weaknesses of existing intellectual property rights (IPR) institutions. The technology has important consequences for the implementation, enforcement and duration of IPRs for plant varieties. Consequently, the technology could have a significant impact on the appropriability of returns to investment in plant breeding and the level of investment in the development of new plant varieties. The potential impact on innovation in the plant-breeding sector must be taken into account when a balance sheet for the technology is drawn up. The current intense debate about the technology has obscured the fact that the choice for developing countries may not simply lie in accepting or rejecting the technology altogether. Other strategic responses, involving the regulation of the application of the technology may be available, that need to be explored.

The Technology

An understanding of the complex biotechnological processes involved in terminator technology is not readily accessible for non-technical people. The details of how these processes work are contained in lengthy technical patent documents, quite difficult to comprehend even for someone skilled in the art. Ironically, while seed multinationals have been severely criticized in their attempts to promote the technology, the origin of the controversy can be traced to a patent granted jointly to the United States Department of Agriculture and Delta and Pineland Company in 1996. The technology involves the following components:

1. Techniques for development of seeds that grow normally when first planted by farmers, but which produce sterile seeds at the end of the cycle, i.e. on harvest.
2. Techniques which allow companies to produce seed for sale to farmers that grows normally.
3. Techniques that ensure that other product characteristics remain unaffected. This implies that terminator technology should come into play at a late stage in seed development.

From the economic point of view, the most important implication of this technology is that farmers cannot save seeds from their crops – they have to buy fresh seeds from the seed companies every year. It must be noted the that this technology may be relevant only for self/open-pollinated varieties; in the case of hybrids farmers generally buy seed every year because of the loss in yield (owing to the loss of hybrid vigour) when second-generation seeds (F2) are used. There is no incentive for seed companies to put this technology into hybrid varieties, as there is already a mechanism to ensure repeat purchases of seed by farmers. However, a quick survey of the terminator literature shows that the development of the technology is not oriented only toward seed sterility. The same technology can be used to switch on or off specific traits in the seeds of a variety. Certain traits can be rendered dormant. These dormant traits may be expressed only when the seeds are used in conjunction with certain proprietary chemicals. The technology opens up new possibilities for companies to bundle together seeds and other inputs.

It is not clear in which crop species the implementation of terminator technology will ultimately prove feasible. However, if terminator emerges as a truly generic technology, capable of being implemented in a whole range of crops species and varieties, then it may become the biotechnologist's vehicle of choice for the delivery of innovations. All biotechnological innovations (e.g. transgenic varieties resistant to insects) would be delivered to the market bundled with terminator. Even the availability of existing varieties could be affected if companies choose to sell them only in the terminator version.

It appears that, unlike genetically modified varieties that offer agronomic benefits to farmers, terminator technology offers no such benefits to them. The technology is aimed at the elimination of loss of revenue to seed companies owing to the use of farm-saved seed. It is also aimed at facilitating the joint sale of different types of inputs and price discrimination in the seed market. If the technology has emerged not for agronomic reasons, but rather as a marketing device, it becomes necessary to explore the economic rationale for its emergence.

Plant Breeding and IPRs

It is generally recognized that the growth of agricultural production has become critically dependent on yield increases primarily based on the development of new high-yielding varieties. While traditionally, the development of new varieties was the result of informal innovation by farmers over several generations, in the last

125 years this process has been greatly accelerated by scientific plant breeding, which has emerged as one of the most important areas of agricultural research. Its distinguishing feature is the *planned incorporation* of specific desirable traits in new varieties using the range of *techniques* and *information* available to plant breeders (OECD, 1993). Modern plant breeding not only takes place in a different institutional context, but this process of evolving new varieties is fundamentally different from the farmer's process that relies on careful selection from randomly occurring mutations in nature.[1]

The transformation of plant breeding into an organized scientific activity raises questions of incentives for research. The two important questions are: (i) whether any form of IPRs can be made applicable to plant breeding, and (ii) whether the market mechanism will produce an optimal level of investment in plant breeding.

In the case of industrial products, the concept of IPRs as a device for promoting innovation has been around for more than 500 years (Machlup, 1958). The Paris Convention for the protection of industrial property dates back to 1860, but the idea of applying IPRs to plant varieties took another 100 years to emerge. An important reason for this was that until recently the prevailing paradigms of IPRs precluded their application to living materials. However, the most important difficulty in applying IPRs to plant varieties arose from the difficulties in segregating[2] and appropriating benefits from their use (Swanson, 1997). This difficulty arose directly from the self-reproducing nature of seed. Once a plant breeder released a new variety to farmers through the sale of seed, he had no further control over the use of the variety, as farmers could multiply it themselves. The benefits that the breeder could appropriate by the direct use of his new variety would be only a very small fraction of the total social benefit that could be derived if his variety were to be widely diffused to farmers. The vast discrepancy between the benefits that could be appropriated privately by the breeder and the total social benefits implied that the market mechanism would fail to produce a socially desirable level of investment or effort in plant breeding. The market mechanism would invariably lead to under-investment in the development of new varieties.

The conventional response to the problem of market failure, in developed and developing countries, has been to provide investment through public sector research – as evidenced by the establishment of land grant universities in the USA, public agricultural research institutions in Europe and the large National Agricultural Research Systems (NARS) in several developing countries. The contribution of public research systems to the development of improved varieties has been spectacular, but financial support for public sector agricultural research has suffered a serious setback since the 1980s in both developed and developing countries (Alston *et al.*, 1998). Alston *et al.* (1998) report that, in real terms, between 1945 and the mid-1970s in most developed countries, public expenditures on agricultural R&D grew more rapidly than in the rest of the post World War II period. Then in the mid-1970s, rates of growth in public R&D outlays slowed quite markedly and, in the 1980s, public R&D expenditures generally stagnated or declined. In the 1990s, however, public R&D expenditures recovered or began to increase again but at more modest rates of growth than in the 1960s or 1970s. This slowdown in the growth of R&D

[1] As farmers have adopted new varieties developed through agricultural research for economic reasons, the genetic uniformity of agricultural crops has considerably increased. This in turn has increased the vulnerability of agricultural crops to diseases and pests, which are constantly adapting and evolving. The development of new varieties has, therefore, become necessary not only to secure yield increases but also to keep ahead in the 'varietal relay race' against evolving pests (Plucknett *et al.*, 1987).

[2] The difficulties in segregating the benefits flowing from a new plant variety arise because productivity also depends on other inputs (e.g. fertilizer, agronomic practices, etc.) and it is difficult to separate out the contribution of a new variety to productivity.

agricultural spending has been a feature of developing countries as well. Maintaining previous levels of growth in public research expenditure may prove infeasible in both developed and developing countries. The fiscal constraints of the public sector, the emerging role of biotechnology in plant breeding (that calls for a magnitude of investment only very large players can afford), changing ideas about the applicability of IPRs to living materials and scientific advances that facilitate the enforcement of IPRs for seeds have brought into sharp focus the question of IPRs for new plant varieties to provide incentives for private sector research.

If private investment in plant breeding has to be encouraged, then the problem of appropriability has to be addressed. This implies what some authors (e.g. Kloppenburg, 1988) have called the 'commoditization' of seed. This thrust towards commoditization has probably been the most important factor in the emergence of the concept of IPRs over plant varieties. Successful commoditization requires that institutional barriers be placed on the self-reproducing characteristics of seeds-and this is precisely what plant variety protection (PVP) legislations attempted to do. Scientific progress itself facilitated commoditization – for instance, the emergence of hybrids provided a technological solution to the problem of appropriability. The emergence of hybrids encouraged the participation of the private sector in plant breeding for some crops, e.g. maize in the USA. PVP legislation can be seen as an attempt to address the problem of appropriability for self/open-pollinated varieties. The operationalization of PVP was facilitated by the following developments:

1. The development of systematic botany, which permitted the unique description of varieties, based on morphological characteristics. This enabled a particular case of plant variety rights to be distinguished from any other case. Better breeding techniques made it possible to breed genetically uniform varieties. This facilitated the identifiability of individual varieties, which was a key element for the application of IPRs.
2. The increase in farm size and a decline in the number of farmers in developed countries, which drastically reduced the transaction costs of enforcing IPRs in relation to farmers.
3. The development of molecular techniques for identification of varieties and their parentage that facilitated the enforcement of IPRs in relation to imitators and competitors (Godden, 1998).

It was in the 1960s that several European countries enacted PVP legislation in the context of growing private sector participation in plant-breeding activities. Special laws for PVP were enacted partly because of the technical difficulties in applying the patent system (designed for industrial products) to plant varieties, which were thought not to precisely reproduce themselves and whose appearance could vary depending upon the environment in which they were grown. Prior to this, the USA had enacted a Plant Patent Act in 1930 to provide protection to varieties of plants that reproduced themselves asexually. In the Netherlands, the Breeders Ordinance of 1941 granted a very limited exclusive right for breeders of agriculturally important species to market the first generation of certified seed. In Germany in 1953, the Law on the Protection of Varieties and the Seeds of Cultivated Plants gave breeders the exclusive right to produce seed of their varieties for the purposes of the seed trade and to offer for sale and market such seed. In the period prior to 1961, while a number of states provided limited rights to plant breeders, the criteria for grant of rights differed from state to state and even the concept of variety was not seen in the same light in all states. There was no guarantee that the rights that the state was prepared to grant to its own nationals would be extended to the nationals of other states. Where varieties were protected in one state but not in another, several distortions could result. It was the adoption of the International Convention for the Protection of New Varieties of Plants (UPOV) in 1961 that provided, for the first time, recognition of the rights of plant breeders on an international basis.

The UPOV Convention attempted to harmonize the PVP legislation of member countries. It specified uniform criteria for the protection of new varieties as *distinctness, uniformity* and *stability*. These criteria reflected the need for identifiability of a variety as a prerequisite for the application of IPRs. The Convention required member-states to accord the same treatment to nationals of other states as they accorded to their own nationals. It also provided for certain elements of reciprocity. Importantly, it defined the scope of the breeders' rights, which extended to *production for purposes of commercial marketing of the propagating material of the new plant variety*. The UPOV Convention of 1978 and the PVP legislation of most member countries had two important features, which distinguished the protection of plant varieties from patents. These were:

1. Farmers' privilege, which acknowledged the right of farmers to use farm-saved seed. The breeders' right extended only to the production of seed for commercial marketing and consequently the use of farm-saved seed was outside the purview of the breeders' right.
2. Research exemption, which provided that the use of the new variety as the initial source of variation for creating other new varieties and marketing them was free; that is, it did not require the breeder's authorization. Accordingly, the protection under the 1978 Act did not give the plant breeder any rights in the genes, the underlying genetic resource, contained in his variety. The research exemption meant that a protected variety could be used in the development of other new varieties, but it also facilitated 'cosmetic breeding'. A variety, which was only marginally different from a protected variety, could qualify for protection as a new variety. Such 'imitation' could deprive the original breeder of a substantial part of royalties from his protected variety.

Since the 1960s, almost all developed countries have enacted PVP legislation. Several developed countries have had the legislation for more than 20 years. Empirical studies on the impact of PVP have been scarce, but there have been a few studies that have assessed the impact of PVP in the USA and Latin American countries (Kalton and Richardson, 1983; Perrin *et al.*, 1983; Butler and Marion, 1985; Kalton *et al.*, 1989; Babcock and Foster, 1991; Jaffe and Van Wijk, 1995; Butler, 1996; Frey, 1996; Alston and Venner, 1998). While PVP does appear to have facilitated private sector participation in the breeding of non-hybrid crops, the available empirical evidence suggests that the incentive effects of PVP may be fairly weak, as farmers' and researchers' exemptions appear to constrain the appropriation of returns. PVP has not brought the appropriability of returns from investment in self-pollinated crops anywhere near the levels reached with hybrid crops. PVP does appear to play an important role in facilitating important changes in the institutional framework for agricultural research. It forces a reappraisal of the rationale for public sector intervention in plant-breeding research. This can lead to a redefinition of the role of the public sector and its relationship with the private sector.

Tackling the Problem of Appropriability

There have been several attempts to improve the appropriability of returns permitted by PVP laws. A comparison of the 1991 UPOV Convention with the 1978 UPOV Convention clearly brings out the nature of these efforts. The 1991 UPOV Convention extends the right of the breeder to all reproduction of the seed of the protected variety; but individual member-states may provide for farmers' privilege in their laws as an exception. The farmers' privilege, therefore, becomes available only as an exception to the breeder's right. In countries where legislation has continued to provide for farmers' exemption (as an exception), it may not have made much of a difference in practice, but an important change of principle was involved. To facilitate better enforcement of the rights of the breeder, the UPOV Convention extended the rights of the breeder to the harvested material (including whole plants) in cases where the breeder had not been able to exercise his right in relation to the propagating material. Member-states can optionally extend the right of the breeder even to products made from the harvested material in certain circumstances. The loss to breeders through 'cosmetic breeding' was sought to be tackled by introducing the concept of 'essential derivation'. The right of a breeder was extended to all varieties that could be defined as being 'essentially derived' from the protected variety. While the definition of what constitutes an essentially derived variety is controversial and is still unsettled, the concept is clearly intended to make PVP as close as possible to patents.[3] The 1991 Convention also seeks to improve appropriability for breeders by extending protection to all species and by increasing the duration of protection.

The changes in the UPOV Convention have been reflected in the changes in national laws. In almost the whole of the EU, farmers' privilege no longer exists except in the case of small farmers. Other farmers have to pay a royalty to the breeder (or the PVP title-holder) even when they use farm-saved seed of a protected variety. In the USA, the PVP Act originally provided that farmers could sell farm-saved seed to other farmers. The sale of such farms' saved seed became fairly extensive and came to be known has 'brown-bag' sales. While such sales may seem innocuous, their aggregate effects cut severely into seed companies' profits. In 1990, Pioneer Hi-Bred, one of the world's largest seed companies, decided to cease production of a variety of winter wheat in Kansas when it discovered that only 8% of the variety grown in Kansas had been raised from seed actually purchased from Pioneer. The secondary brown-bag market had swallowed 92% of Pioneer's market share. The right to sell farm-saved seed was circumscribed in the USA by a series of judicial decisions (Goss, 1996). In the Asgrow vs. Winterboer case, the US Supreme Court held that a farmer could sell only the amount of seed that he would need to replant his own acreage. The 1994 amendments to the PVP Act removed the provision for sale from the farmers' exemption altogether. Farmers could now save seed only for replanting their own land. This measure was recognition of the impact of brown-bag sales on the seed industry.

The seed industry has also been using various other techniques to improve appropriability, such as the use of purchase contracts and label notices. A purchase contract specifically prohibits a farmer from using the harvested seed of a protected variety for replanting or for selling for breeding or variety improvement purposes. Sometimes, restrictions on the use of seed are sought to be imposed by affixing notices on the label of a product. The purchaser is deemed to agree to the restrictions when he opens the packet and uses the seed. The enforceability of such purchase contracts and label notices, especially when the product is covered by PVP and not by patents, is not very clear and will probably have to be decided by judicial decisions. However, it is clear that in the quest for better appropriability, seed companies have been exploring a range of measures. Such measures have become extremely important in the context of the sale of transgenic or genetically modified varieties. These varieties are covered by utility patents rather than by PVP. For making available the seed of transgenic varieties to farmers, seed companies like Monsanto require each farmer to enter into an agreement with the company, which prohibits the farmer from using farm-saved seed. The company can carry out of physical checks or what are known as 'audits' for a period of 3 years after the sale to ensure that the farmer has not replanted farm-saved seed. Seed companies have also

[3] In the case of patents the exploitation of an invention which builds upon an earlier invention requires the consent of the patent-holder of the earlier invention (during the life of the earlier invention).

been using different pricing systems. For instance, the price of transgenic seeds comprises a basic seed price and a technology fee. In certain cases the company offers to refund a portion of the technology fee depending on the actual benefit derived by the farmer by way of lower use of pesticides, etc. (Lindner, 1999). This implies that the company can effectively charge a different price to every farmer, which approaches perfect price discrimination.

The quest for stronger protection has led to plant varieties being protected through patents in a number of countries. In the USA, plant varieties became patentable as a result of a series of judicial decisions reinterpreting the existing patent laws (*Diamond vs. Chakrabarty* and *ExParte Hibberd* are two landmark cases; Goss, 1996). Patents provide stronger protection because they are not subject to farmers' and researchers' exemptions. The scope of utility patents is wide because it allows a breeder to exclude others from *making, using* or *selling* the seeds of the patented variety. The 'doctrine of equivalents ' protects a breeder from imitation of his variety. Furthermore, a breeder is also protected against inventors who independently come up with the same invention. Patent protection is expensive to obtain. A breeder must prove several elements: (i) that the variety is novel and useful; (ii) that it is 'enabled' (currently this requirement can be satisfied by a deposit of the variety in the USA); and (iii) that it is not an obvious improvement upon an earlier protected variety. On account of the difficulty of proving that these requirements have been met, patents are frequently more difficult to obtain and take longer to obtain than PVP certificates. In the USA during the last 2 years, there has been a decline in the number of PVP certificates, while utility patents for plants have maintained their rising trend.

We have seen above that PVP does not significantly improve the appropriability of returns from investment in self-pollinating varieties, even with changes in laws that provide for more stringent enforcement of IPRs. Measures to restrict the use of farm-saved seed through the use of contractual arrangements, such as those described above, may be feasible only when varieties are protected by patents rather than by PVP. Utility patents are inherently more difficult to obtain because of the criteria of 'non-obviousness' and 'usefulness' that they involve (which are not applicable to PVP). Besides, contractual arrangements can be difficult to monitor to detect infringements and costly to enforce. Under contractual arrangements, seed companies are restricted to contract remedies when a breach occurs and these remedies may cover only a fraction of the losses suffered by a vendor seed company due to infringement. The costs of IPR-related litigation to seed companies runs into millions of dollars each year. It is against this background that terminator technology can be seen as a technological solution to the problem of appropriability. In the context of changes in PVP laws that circumscribe the use of farm-saved seed and the contractual arrangements which have come into vogue for transgenic varieties, the terminator technology represents only a better technique of enforcement of IPRs, *rather than a change of principle governing what is legitimately appropriable*. The technology has emerged because the existing institutional mechanisms for enforcement were inadequate, expensive to use and imperfect. It promises drastically to reduce the transaction costs for enforcement of IPRs *vis-à-vis* farmers. It is a technological response to an institutional problem.

Economic Implications

The impact of terminator technology on appropriation of returns from investment in a self-pollinated variety is shown in Appendix 9.1, using a simple model adapted from Scherer (1986). In the simplest variant of the model, the development of a new variety by an innovator opens up a completely new market for seeds. This happens only when farmers using farm-saved seeds of traditional varieties switch to bought-in seeds of the new

Table 9.1. Summary of innovator's discounted quasi-rents for different scenarios.

	No imitation	With imitation
Full replacement by farmers	396 (100%)	240 (60%)
No replacement by farmers	39 (9%)	33 (8%)
Replacement by 20% farmers	118 (30%)	75 (19%)
Replacement every 5 years by farmers	58 (15%)	41 (10%)

Figures in brackets give the percentage to the (maximum) quasi-rents in the situation with no imitators and full replacement by farmers.

variety. In the second variant, the development of a new variety opens up a new market for seeds, but the innovator's market share is affected by imitators developing very similar varieties with a certain time lag. In the third and most likely scenario, the introduction of a new variety somewhat alters the market shares between innovators and competitors, and does not lead to an overall expansion of market. In all three cases, the innovator/competitor does not realize the whole of the annual quasi-rent potential from the new variety instantaneously because it takes time for a variety to diffuse to farmers and build its market share. For all three cases, the appropriation of quasi-rents from the introduction of a new variety is assessed on the basis of different assumptions about farmer seed-replacement behaviour, namely: (i) farmers do not replace seed after adopting a new variety; (ii) farmers replace seed of new variety every year; (iii) a proportion of the adopters of a new variety replace seed every year; (iv) adopters of a new variety replace seed at periodic intervals (say every 5 years). Farmer seed-replacement behaviour makes a dramatic difference to the appropriation of returns from a new variety. Terminator technology is equivalent to the case in which all farmers who adopt a new variety replace seed every year. For certain plausible values of the parameters, the discounted value of quasi-rents to the innovator are four times higher with terminator technology compared to the case in which farmers do not replace seed at all after initial purchase of a variety. Assuming that the maximum annual potential quasi-rent from the introduction of a new variety is 100 (i.e. when there are no imitators or competitors), the appropriation of returns from a new variety by an innovator under different scenarios is summarized in Table 9.1.

The important question is whether this dramatic difference in appropriation of returns from a new variety when farmers replace seed every year gets reflected in the resources allocated to research. The evidence from the USA suggests that there is in fact a strong link between appropriability and private plant-breeding efforts as may be seen from Table 9.2.

It may be seen from the above table that private plant-breeding expenditures for maize, a predominantly hybrid crop, are almost four times the expenditures on a self-pollinated crop like wheat. Share of purchased seed for maize is 95%, while for wheat it is only 40%. The growth in yields of hybrids is also significantly higher than that of non-hybrids. In another study, Thirtle (1985) estimated the contribution of biological inputs to the growth in farmers' yields in the USA after accounting for changes in fertilizer, labour, machinery and land use, and allowing for substitution among inputs. It was estimated that biological inputs increased maize yields by 1.7% per year, wheat yields by 1.5%, soybean yields by 1.1% and cotton yields by 0.5%. Contribution of biological inputs to yield gain appears to increase with the appropriability of returns. Fuglie *et al.* (1996) estimate that for crops grown with hybrid seed, seed companies appeared to capture 35–48% of the value of improved seed with the remainder going to farmers. For non-hybrid crops (wheat, soybeans and cotton) seed companies obtained even lower shares of yield gains, from 12–24%

Table 9.2. Seed sales, private plant breeding and trends in seed prices and yields of major field crops in the USA (1975–1992).

Crop	Seed sales (million 1989 US$)	Private plant breeding (million 1989 US$)	Seed cost (US$ acre^{-1})	Share of seed purchased (%)	Growth in seed price (%)	Annual growth in crop yields (%)
Hybrid seed						
Corn	1031	112.9	21.09	95	4.75	1.33
Sorghum	90	12.6	5.13	95	5.08	1.54
Non-hybrid seed						
Wheat	256	13.5	8.92	40	0.97	1.13
Soybean	610	24.9	12.03	73	1.92	1.23
Cotton	256	4.6	14.93	74	4.46	2.23

Source: Fuglie *et al.* (1996).

as shown in Fig. 9.1. This figure shows the share of genetic yield growth for these crops captured by seed companies in the form of higher seed prices and the share kept by farmers, on the assumption that half the growth in farmer yields can be attributed to genetic improvements and the other half to other factors. For the hybrid seed crops, seed companies invested over 10% of seed sales in research. For the non-hybrid seed crops, only 4–5% of seed sales was re-invested in research. 'The inability to capture a larger share of the gains from breeding non-hybrid crops served as a disincentive for seed companies to invest more in research' (Fuglie *et al.*, 1996, p. 41).

If terminator technology were to be applied to self-pollinated crops, appropriability of returns would increase dramatically and the level of research expenditures could potentially go up to the level of hybrid crops, i.e. increase to four times the current level. The resultant increase in the rate of yield gain and the cumulative impact on production could be very substantial.

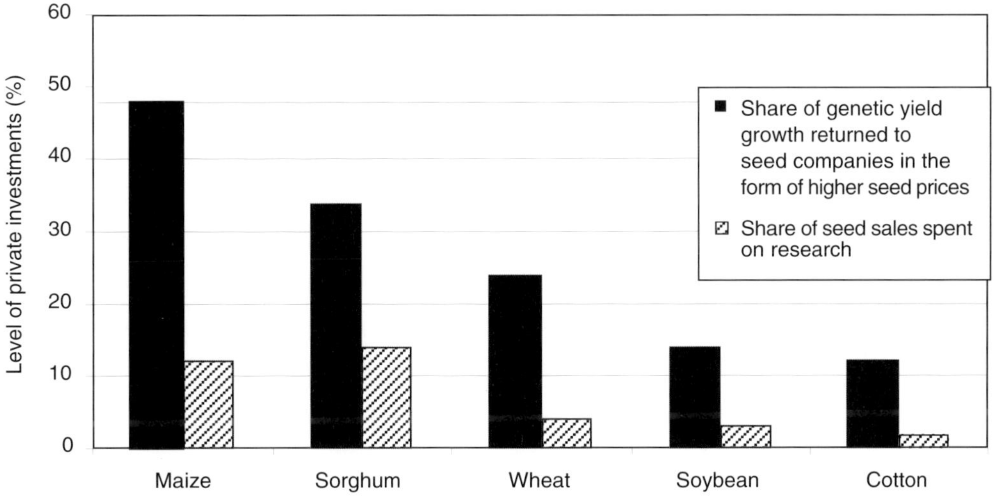

Fig. 9.1. Appropriability and private investment in plants. Source: Fuglie *et al.* (1996).

In combination with certain types of contractual arrangements discussed above, terminator technology could facilitate perfect price discrimination by seed companies. Each farmer would pay a price for seed that would be related to the benefits he derives from its use. Perfect price discrimination would allow a seed company to cream off the entire consumer surplus. However, it is also a well-known result in economic theory that under perfect price discrimination the monopolist would produce the competitive level of output and the allocation of resources would be 'efficient'. Price discrimination could result in an expansion of output.

Terminator technology could also open up some interesting possibilities for IPR policy. The duration of protection under PVP currently ranges from 18 to 25 years, which is longer than the duration for patents. With improved appropriability, it may be possible to reduce the period of protection to, say, 10 years. This would reduce the 'dead-weight' loss associated with IPRs and enable faster diffusion of varieties. However, this would be possible only if companies use PVP/patents to protect their varieties. Terminator renders PVP irrelevant for protection against unauthorized reproduction by farmers. Companies still need protection against imitation by rivals. However, companies could choose to protect terminator processes through trade secrets, in which case protection would be theoretically available in perpetuity – at least until competing varieties are developed. If this happens, then two key elements of IPR policy – the duration of protection and the requirement of 'disclosure' – would no longer be available to governments to influence the development and diffusion of the technology. It must be noted that trade secrets do not protect against independent discovery of a process or product, while patents do.

Theoretically, it can be argued that farmers will buy terminator seeds only if they provide higher returns even after taking into account the higher costs of seeds and other inputs that they would be compelled to buy. This assumes that farmers would have a choice and that seed markets would at least be 'contestable' if not competitive. However, the seed industry has seen consolidation through takeovers and mergers at a frightening pace. In the first wave of acquisitions and mergers in the 1960s and 1970s, many large chemical, oil and food corporations acquired many medium- and small-sized seed companies. Another round of mergers in the 1980s saw many of these food, oil and chemical companies sell their interest in the seed business to agricultural chemical firms. The 1990s have seen the takeover of many small biotechnology firms by the large seed firms and mega-mergers involving the seed giants (Butler, 1996). In developed countries, the new life science companies that combine crop biotechnology with agrochemicals and seed production dominate the seed industry. The big six multinationals – AgrEvo, Dow, Dupont, Monsanto, Novartis and Astra-Zeneca – have a very large market share for certain crops. For instance, Monsanto could have controlled 85% of the cotton seed market if its merger with Delta and Pine had gone through. If these companies all introduce terminator technology, there is a real danger that farmers may not be able to have a choice. They may not be able to choose improved agronomic characteristics without also choosing terminator technology. Local adaptation of varieties and community-based seed supply systems would cease to exist, with adverse consequences for biodiversity on farmers' fields.

The maintenance of some kind of competitive pressure, so necessary for innovation would depend on the 'freedom to operate' that other players (e.g. public research systems, universities, competing companies, etc.) have. 'Freedom to operate' will depend greatly on how IPRs are applied to biotechnological processes. Terminator technology by itself confers no agronomic benefits. To be commercially successful, it must be bundled with other useful agronomic traits and these traits must not be available otherwise. Increasingly, these useful agronomic traits will be ones that are engineered through biotechnology with the associated processes being patented. It is only then that companies will be able to retain exclusive use of the desirable agronomic traits. The development of new varieties in the future may require access to a number of biotechnological processes, all patented by different companies. Unless an efficient system is developed for licensing the component biotechnological processes, there could be gridlock in the development of new varieties. The scope of protection accorded to biotechnological processes needs to be carefully modulated if such a gridlock is to be avoided. There are indications that many International Agricultural Research Centres (IARCs) are finding that they cannot work on biotechnology-based varietal improvement without licences for component processes from seed companies (Cohen *et al.*, 1998). The ability of universities and public sector institutions to license processes often depends on their having a portfolio of patents for cross-licensing. Without the ability to cross-license, their 'freedom to operate' would be severely restricted. The important issue is whether the public sector and other players will be able to provide improved varieties without terminator genes and offer a measure of competition.

Choices Before Developing Countries

In acceding to the TRIPs Agreement,[4] all developing countries have committed themselves to provide some form of 'effective' IPRs for plant varieties. Several developing countries are in the process of drafting their legislation. Some are attempting to formulate a *sui generis* system that balances breeders' rights with farmers' rights and recognition of their contribution to the maintenance and enhancement of biodiversity. The debate about the terminator, therefore, is not about the fundamental desirability of IPRs for plant varieties but about a particular technique of enforcement. The fact that terminator technology has emerged in response to inadequacies IPR institutions must make developing countries ponder about the kind of IPRs they need to put in place, if they really intend to encourage private investment. Stronger IPR systems may even obviate the need for development of such technologies.

An important factor, which will determine the response of developing countries to the new technology, will be the approach followed by the CGIAR[5] institutions. This will be especially true of those countries that rely on the IARCs for new varieties. The CGIAR system has apparently taken a decision that none of its institutions would use the terminator technology for the development of new varieties for developing countries in view of the potential deleterious effects of the technology. This could have important implications for the commercialization of the new technology in developing countries.

Quite independently of the CGIAR system, developing countries will have to decide on their response to the emerging technology. Very little is known about the ecological impacts of terminator varieties. The nature of interaction of terminator varieties with other crops and species is not known. It remains to be assessed whether terminator varieties are, accidentally or otherwise, capable of rendering the seeds of (adjacent) non-terminator varieties also sterile. This is a critical issue for developing countries where millions of subsistence or resource-poor farmers depend on farm-saved seed. If terminator varieties can render the farm-saved seed of non-terminator varieties sterile even in limited number of situations, there will a potential for a great deal of social unrest. Similarly, if terminator varieties have artificially lowered resistance to certain diseases or infestations, then the unintended build-up of pathogens in the fields of both users and non-users would be a crucial

[4] Agreement on Trade-related Aspects of Intellectual Property Rights which is part of the agreement establishing the WTO.

[5] Consultative Group on International Agricultural Research.

issue. The large-scale use of antibiotics that currently appears to be a feature of the technology is also an important matter of concern. What developing countries require at this stage is capacity building for making detailed ecological and economic impact assessments. If improved varieties of the private sector in developed countries are going to be invariably bundled with the terminator in the future, then developing countries will have to assess the trade-off between increased productivity through the use of these varieties and the retention of farmers' privileges. The existence and severity of such a trade-off will depend on the sustained capability of NARS and IARCs to produce improved varieties without the terminator technology. Their 'freedom to operate' in the brave new world of IPRs will be a critical issue. There may be a case for selective adoption of the new technology for certain crop species. If the capacity for ecological and economic impact assessment is not built up, then policy will inevitably be determined by emotive, populist rhetoric.

The introduction of terminator technology in developing countries would involve interface with several elements of the regulatory framework:

1. Quality control legislation.
2. Varietal release and notification procedures.
3. IPR legislation.
4. Policies on export and import of seed.
5. Quarantine regulations.
6. Approval procedures and marketing regulations for agrochemicals.
7. Policy governing foreign direct investment (FDI) in the seed industry.

The existing regulatory framework in developing countries may be simply not adequate to deal with terminator technology. Several elements of the regulatory framework may have to be modified substantially in a co-ordinated and coherent fashion to respond to the challenges posed by terminator technology. Wherever the regulatory framework is weak, developing countries will have to contend with the possibility of surreptitious introduction (or introduction through informal channels) of terminator varieties. This would be particularly important in countries where varietal registration is not mandatory for marketing of seeds – and would be more difficult to handle than introduction through the organized sector.

The policy in most developing countries has been to push for increasing the seed replacement rate (SRR) and quicker adoption of new varieties. However, even in many Green Revolution areas like the Indian Punjab, SRRs have been very slow to increase. This happens either because farmers cannot afford to replace seed every year, or because farmers do not perceive the advantages of replacing seed or because farm-saved seed is almost equal in quality to bought-in seed. Terminator technology by its very nature can swiftly push up SRRs and also increase varietal turnover, as farmers will be more likely to change to new varieties if they replace seed every year. Developing countries have a very difficult decision to make in deciding whether terminator technology should be used as an instrument to improve SRRs and encourage varietal replacement.

Developing countries will also have to consider the impact of terminator technologies on FDI in the seed industry. There may be a tendency for the investment to flow to those countries that accept the use of this technology. Even if better appropriability generates more research investment, the location of that investment would still be an important issue for developing countries. Approval of terminator technology can greatly increase volumes for the domestic seed industry provided it has access to the technology. The important question will be whether the domestic industry will be able to compete with terminator varieties introduced from outside. The competitive position of the domestic seed industry may depend on whether and on what terms the technology is made available to them.

The public sector in developing countries could see a major revival of its role in developing countries as the developer and producer of non-terminator varieties that may be very important for resource-poor farmers. The important question for the public sector will be whether lack of access to private sector material owing to IPRs will hamper it. Or will access to germplasm in international collections such as those of the CGIAR system be adequate for the development of competing varieties?

In assessing terminator technology, developing countries will have to address the Green Revolution questions as well. Which farmers and which regions will be the early adopters of the new technology in developing countries? Will the adoption of the new technology be different between rich and poor farmers? Will the adoption of terminator technology follow the Green Revolution pattern of adoption of high-yielding varieties? What will be the implications for credit requirements? Is the technology a coercive one in the sense of compelling unwilling farmers to adopt it?

Concluding Observations

Terminator technology has emerged as a response to the inadequacies and imperfections of existing IPR institutions. By providing a technological response to the problem of inadequate appropriability, it can be a corrective to market failure. At a time when public research systems everywhere are faced with declining budgets and yield gains are becoming increasingly dependent on the application of expensive biotechnology, it has the potential to increase private research spending on self-pollinated crops to the level of hybrids. The resultant productivity gains may well be substantial. While better appropriability can provide additional resources for investment, there are legitimate concerns about the use of the technology. The environmental consequences of the technology can be far-reaching and complex and need to be carefully assessed. The use of the technology in an industry that is oligopolistic and highly concentrated may imply that farmers lose their right to choose and competitive pressures that are so necessary for innovation may cease. Besides, the gridlock creating potential of IPRs can stifle, rather than promote, the development of new varieties. The scope of protection for biotechnological processes in the IPR system needs to be carefully modulated if such adverse impacts are to be avoided, but the most important lesson to be drawn from the emergence of this technology is this: in a process of market-led development, technological change will respond not only to relative factor scarcities but also to institutional bottlenecks. The development of a technology with no agronomic benefit to farmers may appear to be a huge waste of resources, but the market *will* devote resources to the reduction of transaction/appropriation costs.

Developing countries face many difficult choices in responding to this technology. If a large part of varietal improvement in the future is going to be bundled together with terminator technology, then it may not be possible for them to ignore the technology altogether. Productivity is of even greater concern to them than it is to developed countries. Potential productivity gains will have to be weighed against the social costs arising from the coercive element of the technology. What they need at this stage is the capacity to evaluate its ecological, economic and social impacts, selectively apply it where appropriate and upgrade their regulatory system to deal with the technology. They will need to seriously think about appropriability and IPRs.

References

Alston, J.M. and Venner, R.J. (1998) The effects of U.S. Plant Variety Protection Act on wheat genetic improvement. Paper presented at the symposium on Intellectual Property Rights and Agricultural Research Impact, sponsored by NC 208 and CIMMYT Economics Program, El Batan, Mexico, 5–7 March.

Alston, J.M., Pardey, P.G. and Smith, V.H. (1998) Financing agricultural R&D in rich countries: what's happening and why? *Australian Journal of Agricultural and Resource Economics* 42, 51–82.

Babcock, B.A. and Foster, W.E. (1991) Measuring the potential contribution of plant breeding to crop yields: flue-cured tobacco, 1954–87. *American Journal of Agricultural Economics* 73, 850–859.

Butler, L.J. (1996) Plant Breeders' Rights in the U.S.: update of a 1983 study. In: Van Wijk, J. and Jaffe, W. (eds) *Proceedings of a Seminar on The Impact of Plant Breeders' Rights in Developing Countries*, Santa Fe Bogota, Colombia, 7–8 March 1995. University of Amsterdam, Amsterdam, pp. 17–33.

Butler, L.J. and Marion, B.W. (1985) *The Impact of Patent Protection on the U.S. Seed Industry and Public Plant Breeding.* Food Systems Research Group Monograph 16. University of Wisconsin-Madison, Madison, Wisconsin.

Cohen, J.I., Falconi, C., Komen, J. and Blakeney, M. (1998) *Proprietary Biotechnology Inputs and International Agricultural Research.* ISNAR Briefing Paper 39, International Service for National Agricultural Research, The Hague.

Foster, W.E. and Perrin, R.K. (1991) *Economic Incentives and Plant Breeding Research.* Faculty Discussion Paper 91–05, Department of Agricultural and Resource Economics, North Carolina State University, Raleigh.

Frey, K.J. (1996) *National Plant Breeding Study -I: Human and Financial Resource Devoted to Plant Breeding Research and Development in the United States in 1994.* Special Report 98, Iowa Agriculture and Home Economics Experiment Station, Iowa State University, Iowa.

Fuglie, K., Ballenger, N., Day, K., *et al.* (1996) *Agricultural Research and Development: Public and Private Investments Under Alternative Markets and Institutions.* AER-735, Economic Research Service, United States Department of Agriculture.

Godden, D. (1998) Growing plants, evolving rights: plant variety rights in Australia. *Australian Agribusiness Review* 6(2).

Goss, P.J. (1996) Guiding the hand that feeds: toward socially optimal appropriability in agricultural biotechnology innovation. *California Law Review* 84, 1395–1435.

Jaffe, W. and Van Wijk, J. (1995) *The Impact of Plant Breeders' Rights in Developing Countries: Debate and Experience in Argentina, Chile, Colombia, Mexico and Uruguay.* Inter-American Institute for Co-operation in Agriculture and University of Amsterdam, Amsterdam.

Kalton, R.R. and Richardson, P.A. (1983) Private sector plant breeding programmes: a major thrust in U.S. agriculture. *Diversity* (Nov–Dec), 16–18.

Kalton, R.R., Richardson, P.A. and Frey, N.M. (1989) Inputs in private sector plant breeding and biotechnology research programs in the United States. *Diversity* 5, pp. 22–25.

Kloppenburg Jr, J.R. (1988) *First the Seed: the Political Economy of Plant Biotechnology 1492–2000.* Cambridge University Press, Cambridge.

Lesser, W. (1994) Valuation of plant variety protection certificates. *Review of Agricultural Economics* 16, 231–238.

Lindner, B. (1999) Prospects for public plant breeding in a small country. Paper presented at a seminar on The Shape of the Coming Agricultural Transformation: Strategic Investment and Policy Approaches from an Economic Perspective, Rome, 17–18 June.

Machlup, F. (1958) *An Economic Review of the Patent System.* Sub-Committee on Patents, Trademarks and Copyrights, Committee on the Judiciary, U.S. Senate, Study no. 15, U.S. Government Printing Office, Washington, DC.

Nordhaus, W. (1969) *Invention, Growth and Welfare: a Theoretical Treatment of Technological Change.* MIT Press, Cambridge, Massachusetts.

OECD (1993) *Traditional Crop Breeding Practices: an Historical Review to Serve as a Basis for Assessing the Role of Biotechnology.* OECD, Geneva.

Perrin, R.K., Kunnings, K.A. and Ihnen, L.A. (1983) *Some Effects of the U.S. Plant Variety Protection Act of 1970.* Economics Research Report No. 46, Department of Economics and Business, North Carolina State University.

Plucknett, D.L. *et al.* (1987) *Gene Banks and the World's Food.* Princeton University Press, Princeton, New Jersey.

Scherer, F.M. (1986) *Innovation and Growth: Schumpeterian Perspectives.* MIT Press, Cambridge, Massachusetts.

Swanson, T. (1997) *Global Action for Biodiversity: an International Framework for Implementing the Convention on Biodiversity.* Earthscan Publications, London.

Thirtle, C.G. (1985) Technological change and productivity slowdown in field crops: United States, 1939–1978. *Southern Journal of Agricultural Economics* 17 , 33–42.

Appendix 9.1. Analysis of the Economic Rationale of Terminator Technology

It has been argued in the text that terminator technology has emerged in response to the inadequacies of IPR institutions for plant varieties. In particular, IPR institutions, as they operate in practice, permit only very low levels of appropriability of returns from investment in plant-breeding research. The incentive effects of existing IPR institutions (e.g. plant variety rights) in generating research investment tend to be rather weak.

The problem of low appropriability of returns from research investment can be analysed using the framework developed by Nordhaus (1969) and Scherer (1986) for the economic analysis of patents and the allocation of R&D resources under rivalry. The framework can be adapted to plant breeding with minor modifications. The key feature of the analysis of plant breeding arises from the self-reproducing nature of seed. Most PVP laws provide for 'farmers exemption' which acknowledges the right of farmer to use farm-saved seed. This implies that once a variety has been developed and released in the market, farmers can use seed produced from the harvest to plant their subsequent crops. They need not return to the seed company every year for fresh seed. They need to do so only after 5–7 years when farm-saved seed starts losing its vigour (giving lower yields) or suffers from 'genetic drift'. This has a dramatic impact on the appropriability of returns from a new variety. This argument does not apply to F1 hybrids (as farmers generally have to buy seed every year) and, therefore, the framework can also be used to illustrate the difference in appropriability between hybrids and open/self-pollinated varieties.

The other feature of plant variety rights, which affects the appropriability of returns, is the 'research exemption'. This provides that any protected variety can be used as a basis for developing other varieties without infringing the right of the protection holder. This affects appropriability by drastically reducing the time required by imitators and competitors to develop similar competing varieties. The speed of imitation is an important factor affecting the appropriation of benefits from a new variety.

Framework for the seed market

In assessing the appropriability of returns for a seed company (innovator) developing a new variety, we can conceive of three types of situations.

1. *New market:* This situation arises when a new variety opens up a completely new market for seeds. This is unlikely to arise in practice, as for any crop there will already be varieties existing in the market. Only when farmers who are completely dependent on farm-saved seeds of traditional varieties switch to bought-in seeds of modern varieties is a new market situation likely to arise.
2. *New market with imitation:* In this situation, the development of a new variety opens up a completely new market for seeds, but the innovator's market share is affected by imitators developing very similar varieties with a certain time lag.
3. *Market sharing:* This is the most likely scenario in seed markets as the area under a crop, seed use rates and SRRs change rather slowly over time. Therefore, the introduction of a new variety is more likely to alter market shares between the innovator and competitors rather than lead to an overall expansion of the market. During its period of technological leadership, an innovator captures more and more of its rivals' share (and hence a larger and larger share of the quasi-rent potential of its variety), but imitators can recover market share once they have imitated.

Each of the above situations can be combined with different assumptions about farmer seed-replacement behaviour – e.g. 100% replacement of seed every year, replacement every 6 years, replacement by

20% adopting farmers each year, no replacement etc. – to ascertain the returns appropriated by the innovator from a new variety. Though the new market situation is not a likely one for the seed market (even in developing countries), it can be used to illustrate the impact of farmer seed-replacement behaviour and imitation lags on the innovator's discounted stream of returns. The results from the new market situation will apply with greater force in the market-sharing situation.

Variables

Following Scherer we define the following variables for different market situations. Subscript L denotes the innovator and subscript F denotes the imitation competitors.

General

V = Potential quasi-rents (total revenues less costs) attainable from the sale of a new variety. It is initially assumed that costs and prices are constant and hence V does not depend upon the number of firms. V is the annual potential quasi-rent when *all* farmers who will eventually use the variety buy fresh seeds from the innovator in each year.

TL = Innovator's new variety introduction date.

TF = Imitator's product innovation date.

ρ = Discount rate.

New market case

γ = Penetration coefficient: the innovator will not realize the whole of the annual potential quasi-rent immediately because it takes time for a variety to diffuse to farmers and to build a market. All farmers who are eventually going to use a variety will not adopt it in the first year after introduction. γ is a coefficient that indicates the proportion of unexploited market potential captured each year. The value of γ depends upon the yield or agronomic superiority of the variety, marketing efforts etc. The innovator moves toward realization of full market potential at the rate of 100γ per cent per year following introduction.

S_F = Imitators target market share. It is assumed that this depends on the share of the imitators' in related markets and other strengths.

μ = Rate at which the imitator moves toward its market share. It is assumed that the imitator moves toward full realization of its share at the rate of 100μ per cent per year following imitation.

ε = The imitator's permanent share erosion rate. This is assumed to vary with the lag between the introduction of a new variety by the innovator and introduction of a competing variety by the imitator. The imitator's permanent market share gets eroded at the rate of 100ε per cent per year of lag.

α = Percentage of farmers who buy replacement seed in the current having bought fresh seeds the previous year.

Market sharing case

δ = The innovator's (temporary) market takeover rate in market sharing rivalry. This is assumed to be 100δ per cent per year following new variety introduction per year of lag between innovator and imitator.

β = The imitator's share recovery rate in a market sharing rivalry (after it has introduced the competing variety).

New market without imitation

Full replacement by farmers

The innovator's quasi-rent in the absence of imitation and assuming farmers replace seed every year ($\alpha = 1$) is:

$$V_L = \int_{TL}^{\infty} \left(1 - e^{-\gamma(t-TL)}\right) V e^{-\rho t} dt.$$

This can be shown to be equal to

$$V_L = V\left[\frac{1}{\rho}e^{-\rho TL} - \frac{1}{\gamma+\rho}e^{-\rho TL}\right].$$

This is also the innovator's quasi-rent in the absence of imitation when the new variety is a hybrid.

No replacement by farmers

If farmers do not replace seed at all (after having bought the variety once) then the market for the innovator will comprise only the new adopters every year. In this case the innovator's quasi-rent becomes:

$$V_L = \int_{TL}^{\infty}\left[\left(1-e^{-\gamma(t-TL+1)}\right)-\left(1-e^{-\gamma(t-TL)}\right)\right]Ve^{-\rho t}dt$$

$$= \int_{TL}^{\infty}\left(e^{-\gamma(t-TL)}-e^{-\gamma(t-TL+1)}\right)Ve^{-\rho t}dt.$$

On simplification this yields:

$$V_L = V\frac{1}{\gamma+\rho}\left[e^{-\rho TL}-e^{-\gamma-\rho TL}\right].$$

Replacement by proportion of farmers

If we assume that a proportion α of the farmers who have adopted the seed replace their seed every year then the total market for seed for the innovator in each year becomes:

Demand from new adopters + Replacement demand from previous adopters.

The quasi-rent for the innovator can then be expressed as:

$$V_L = \int_{TL}^{\infty}\left[\left(1-e^{-\gamma(t-TL+1)}\right)-\left(1-e^{-\gamma(t-TL)}\right)\right]Ve^{-\rho t}dt \quad \longleftarrow \quad \text{New adopters}$$

$$+ \int_{TL}^{\infty}\left(1-e^{-\gamma(t-TL)}\right)\alpha Ve^{-\rho t}dt \quad \longleftarrow \quad \text{Replacement demand from old adopters}$$

On simplification this yields:

$$V_L = V\frac{1}{\gamma+\rho}\left[e^{-\rho TL}-e^{-\gamma-\rho TL}\right]$$

$$+\alpha Ve^{-\rho TL}\left[\frac{1}{\rho}-\frac{1}{\gamma+\rho}\right].$$

Replacement of seed every n *years*

If farmers replace seed say every 5 years (i.e. farmers who adopt the variety in the first year replace it in the sixth year) then again the total market for the innovator in each year becomes:

Demand from new adopters + Replacement demand from previous adopters.

In this case the quasi-rent for the innovator can be expressed as:

$$V_L = \int_{TL}^{\infty}\left[\left(1-e^{-\gamma(t-TL+1)}\right)-\left(1-e^{-\gamma(t-TL)}\right)\right]Ve^{-\rho t}dt$$

$$+ \int_{TL+6}^{\infty}\left(e^{-\gamma(t-TL-6)}-e^{-\gamma(t-TL-5)}\right)Ve^{-\rho t}dt$$

$$= V\frac{1}{\gamma+\rho}\left[e^{-\rho TL}-e^{-\gamma-\rho TL}\right]$$

$$+V\frac{1}{\gamma+\rho}\left[e^{-\rho(TL+6)}-e^{-\gamma-\rho(TL+6)}\right].$$

New market with imitation

Full replacement by farmers

In this case, we assume that the innovator introduces a new variety at *TL* but that after a time lag imitators are able to introduce a competing variety at TF, i.e. with a time lag of $(TF - TL)$. We assume that imitators have a target market share of S_F and proceed towards the realization of this market share at the rate of 100μ per cent per year. With all farmers replacing seed every year, the innovator's quasi-rent becomes:

$$V_L = \int_{TL}^{\infty} \left(1 - e^{-\gamma(t-TL)}\right) V e^{-\rho t} dt$$

$$- \int_{TF}^{\infty} \left(S_F - S_F e^{-\mu(t-TF)}\right) V e^{-\rho t} dt$$

$$= \frac{Ve^{-\rho TL}}{\rho}\left(\frac{1}{\rho} - \frac{1}{\gamma+\rho}\right)$$

$$- S_F V e^{-\rho TF}\left(\frac{1}{\rho} - \frac{1}{\gamma+\rho}\right).$$

No replacement by farmers

When farmers do not replace seed after the initial purchase of a new variety, the imitator can expect a share of only the new adopters' market every year after imitation. If the time lag for imitation is large, the imitator will have very narrow market because almost all the adopters of the new variety would have made the initial purchase before the imitator enters the market. In such a situation the innovator's discounted quasi-rents will be:

$$V_L = \int_{TL}^{\infty} \left[\left(1 - e^{-\gamma(t-TL+1)}\right) - \left(1 - e^{-\gamma(t-TL)}\right)\right]$$

$$V e^{-\rho t} dt - \int_{TF}^{\infty} S_F \left(1 - e^{-\varepsilon(TF-TL)-\mu(t-TF)}\right)$$

$$\left[\left(1 - e^{-\gamma(t-TL+1)}\right) - \left(1 - e^{-\gamma(t-TL)}\right)\right] V e^{-\rho t} dt.$$

The second expression shows the loss of quasi-rents to the innovator because the imitator takes up a share of the new adopters in each year following imitation. For simplicity in derivation, we can assume that the imitator reaches its target market share of new adopters instantaneously after imitation. (This amounts to assuming that $\varepsilon = 0$ and that the value of μ is sufficiently high such that $(1 - e^{-\varepsilon(TF - TL) - \mu(t - TF)})$ is approximately equal to 1.) The expression for V_L reduces to:

$$V_L = \int_{TL}^{\infty} \left(e^{-\gamma(t-TL)} - e^{-\gamma(t-TL+1)}\right) V e^{-\rho t} dt$$

$$- \int_{TF}^{\infty} S_F \left(e^{-\gamma(t-TL)} - e^{-\gamma(t-TL+1)}\right)$$

$$V e^{-\rho t} dt = \frac{Ve^{-\rho TL}}{\gamma+\rho}\left(1 - e^{-\gamma}\right)$$

$$- \frac{VS_F}{\gamma+\rho}\left(e^{-\gamma(TF-TL)-\rho TF}\right)\left(1 - e^{-\gamma}\right).$$

Similar expressions can also be derived when the imitator takes time to build up market share.

Replacement by a proportion of farmers every year

Let us assume that 20% of the adopters of a variety replace seed every year. In this case the imitator can expect to sell not only to some of the new adopters but also to some of the old adopters who come to the market for replacement seed. Though the imitator will take time to reach its target market share, again for simplicity we assume that it reaches its target instantaneously. The innovator's discounted quasi-rent becomes:

$$
\begin{aligned}
V_L = & \int_{TL}^{\infty} \left[\left(1 - e^{-\gamma(t-TL+1)}\right) - \left(1 - e^{-\gamma(t-TL)}\right) \right] V e^{-\rho t} dt + \int_{TL}^{\infty} \left(1 - e^{-\gamma(t-TL)}\right) \alpha V e^{-\rho t} dt \\
& - \int_{TF}^{\infty} S_F \left[\left(1 - e^{-\gamma(t-TL+1)}\right) - \left(1 - e^{-\gamma(t-TL)}\right) + \alpha\left(1 - e^{-\gamma(t-TL)}\right) \right] V e^{-\rho t} dt \\
= & \frac{V}{\gamma+\rho}\left(e^{-\rho TL} - e^{-\gamma-\rho TL}\right) + \alpha V e^{-\rho TL}\left(\frac{1}{\rho} - \frac{1}{\gamma+\rho}\right) \\
& - V S_F e^{-\rho TL}\left[\frac{(1-\alpha)e^{-\gamma(TF-TL)}}{\gamma+\rho} - \frac{e^{-\gamma(TF-TL+1)}}{\gamma+\rho} + \frac{\alpha}{\rho} \right].
\end{aligned}
$$

Replacement of seeds every n *years*

If farmers replace seed every 5 years, then again the imitator can expect to sell to some of the new adopters every year in the years following imitation and also to some farmers who come to the market periodically for replacement seed. The innovator's discounted stream of quasi-rents can be derived as:

$$
\begin{aligned}
V_L = & \int_{TL}^{\infty} \left[\left(1 - e^{-\gamma(t-TL+1)}\right) - \left(1 - e^{-\gamma(t-TL)}\right) \right] V e^{-\rho t} dt - && \leftarrow \text{New adopters} \\
& + \int_{TL+5}^{\infty} \left(e^{-\gamma(t-TL-5)} - e^{-\gamma(t-TL-4)}\right) V e^{-\rho t} dt && \leftarrow \text{Replacement demand from old adopters} \\
& - \int_{TF}^{\infty} S_F \left(1 - e^{-\varepsilon(TF-TL)-\mu(T-TF)}\right) \left[\left(1 - e^{-\gamma(t-TL+1)}\right) - \left(1 - e^{-\gamma(t-TL)}\right) \right] V e^{-\rho t} dt && \leftarrow \text{Quasi-rents from new adopters lost to imitator} \\
& - \int_{TL+5}^{\infty} S_F \left(1 - e^{-\varepsilon(TF-TL)-\mu(T-TF)}\right) \left(e^{-\gamma(t-TL-5)} - e^{-\gamma(t-TL-4)}\right) V e^{-\rho t} dt. && \leftarrow \text{Quasi-rents from replacement farmers lost to imitator}
\end{aligned}
$$

Once again if we assume that the imitator reaches its target market share instantaneously after imitation, the expression for V_L simplifies to:

$$V_L = V\frac{1}{\gamma+\rho}\left[e^{-\rho TL} - e^{-\gamma-\rho TL}\right] + V\frac{1}{\gamma+\rho}\left[e^{-\rho(TL+5)} - e^{-\gamma-\rho(TL+5)}\right] - \frac{VS_F e^{-\rho TL}}{\gamma+\rho}\left[e^{-\gamma(TF-TL)} - e^{-\gamma(TF-TL+1)}\right] - \frac{VS_F e^{-\rho(TL+5)}}{\gamma+\rho}\left[1-e^{-\gamma}\right].$$

Impact of seed replacement patterns and imitation

The impact of farmer seed-replacement behaviour on the innovator's discounted stream of quasi-rents can be seen with the help of an example in which the following values of different parameters are assumed:

$TL = 4$
$\gamma = 0.4$
$\alpha = 0.2$ (or in the alternative farmers replace seed every 5 years)
$V = 100$.

As shown in Table 9.1, with an annual potential quasi-rent of 100, the discounted value of the innovator's quasi-rent, with no imitators and full replacement by farmers, is 396. However, with no replacement by farmers the discounted value of quasi-rents falls drastically to just 9% of the value with full replacement. When farmers replace seed every 5 years the discounted value is 30% of the value with full replacement, while it is 15% when 20% of farmers replace seed every year. The undiscounted stream of quasi-rents for different cases is shown in Fig. 9A.1, while the discounted stream is shown in Fig. 9A.2. The appropriation of returns in a situation with competitors is still lower for each of these cases.

If the discounted value with full replacement is taken to be the appropriation of returns from hybrids, then it is clear that self-pollinated varieties yield only one-quarter of the returns from hybrids (assuming for the moment that development costs are similar). The low appropriability of returns from self-pollinated varieties acts as a disincentive to research. It is not surprising that according to data for the US, hybrids attract almost four times the investment attracted by self-pollinated crops.

The analysis can be similarly carried out for the market rivalry situation, where the innovator's incremental rents will be even less. The analysis can be further modified for the case when there is monopoly in the seed market to begin with, but the key point is that patterns of seed-replacement behaviour have a drastic impact on appropriation of rents by the innovator.

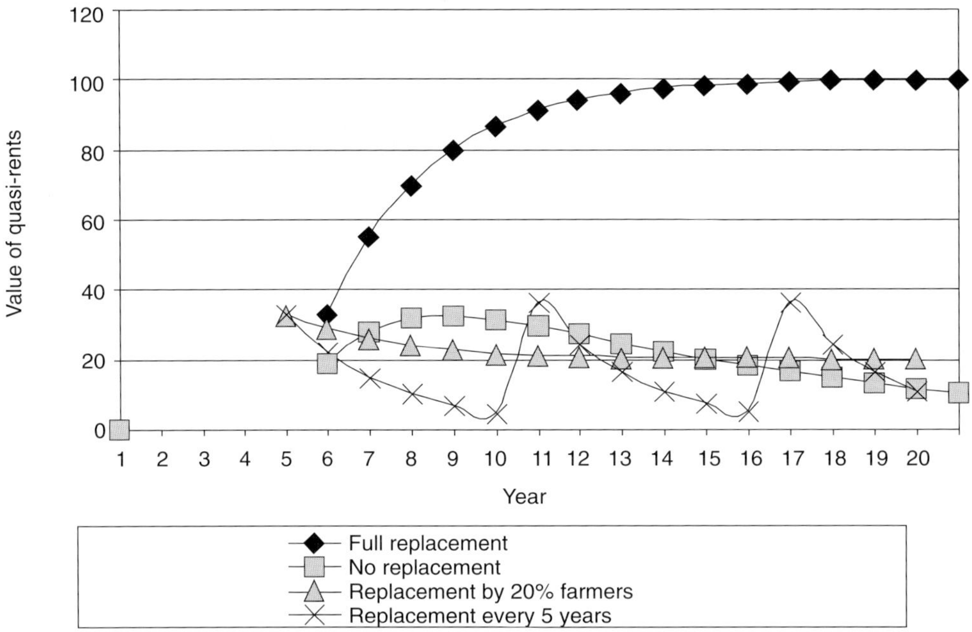

Fig. 9A.1. Undiscounted quasi-rents for different seed-replacement practices.

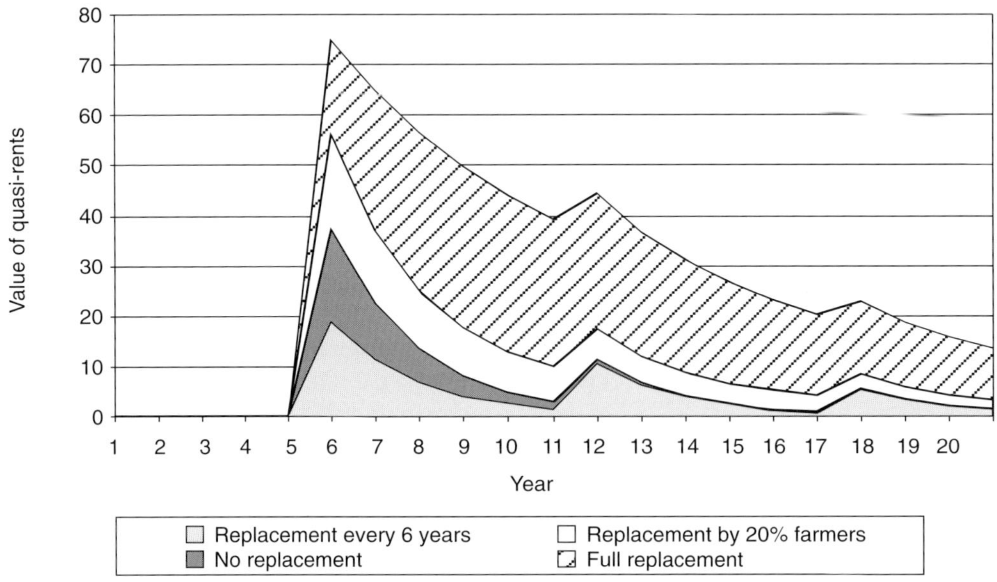

Fig. 9A.2. Discounted value of quasi-rents under different seed-replacement patterns.

Chapter 10

The Impact of Genetic Use Restriction Technologies on Developing Countries: a Forecast

Timo Goeschl[1] and Tim Swanson[2]

[1]*Department of Land Economy, University of Cambridge, 19 Silver Street, Cambridge CB3 9EP, UK;* [2]*Department of Economics, Faculty of Laws and CSERGE, University College, Gower Street, London WC1E 6BT, UK*

Abstract

Advances in biotechnology have made available gene-manipulation techniques that enable the protection of genetic material from unauthorized use and the prevention of self-supply of commercial seeds by farmers – in order to allow enhanced appropriation of the values of innovation in agricultural R&D. These techniques have become known as genetic use restriction technologies (GURTs).

This chapter estimates the potential impact of widespread adoption of GURTs by the providers of HYV seeds on the yield development in developing countries. To do so, it assesses: (i) the effects of enhanced appropriation through GURTs on the technological expansion at the yield frontier; and (ii) the effects of technological protection of value-adding traits through GURTS on the diffusion of yield gains from the frontier to developing countries. These assessments are based on a particular hypothesis, which is that GURTs will replicate across most staple crops the experiences that were made with a previous use restriction technology (hybridization) in only a few crops. The estimation of impacts is carried out as a simulation and is based on expansion and diffusion parameters estimated for hybrid seeds over a 39-year period. It shows that the impact of GURTs on developing countries' yields will vary considerably. Specifically, GURTs are likely to affect unfavourably the countries that currently have the lowest yields.

Introduction

Proponents of biotechnologically modified crops generally argue that the arrival of advanced techniques of genetic manipulation in plants can be expected to deliver significant productivity increases in agriculture. At the same time, the same techniques have delivered the means of protecting the genetic information responsible for the productivity improvements against unauthorized reproduction and extraction. These techniques have been termed 'genetic use restriction technologies' (GURTs) in reference to the control of unauthorized use of these genes.[1]

[1] In the general public, they have become better known by the name 'terminator genes'. This epithet was coined by Rural Advancement Foundation International (RAFI), a non-governmental organization (NGO); 'traitor technology' has been another suggestion.

 Economic and Social Issues in Agricultural Biotechnology (eds R.E. Evenson, V. Santaniello and D. Zilberman)

Although still at the patent stage, the concerns about the potential implications of GURTs are manifold. Some observers worry about the environmental effects of geneflow from crops thus sterilized to other plants, causing the potential sterilization of seeds beyond the confines of the individual field (Crouch, 1998; Jefferson, 1999). The distribution of economic rents between farmers, seed companies and consumers is another area of possibly undesirable consequences (Srinivasan and Thirtle, 2000). Others are concerned about the impacts of these technologies on the livelihoods of subsistence farmers that predominantly rely on saved seed for replanting their fields (RAFI, 1999).

The focus of this chapter is the analysis of GURTs from the perspective of agricultural productivity growth through crop-improving innovations and the diffusion of these innovations to developing countries. This focus situates the problem in the context of economic development and agricultural R&D. The fundamental idea is that GURTs represent a novel mechanism for appropriating the rents from innovation in the plant-breeding industry. This mechanism radically enhances the plant breeder's scope for rent capture. This is likely to result in increased private investment into agricultural R&D and hence in a higher rate of innovation in the plant-breeding industry. On the other hand, a consequence of this novel mechanism is that it significantly complicates the dissemination of crop improvements through adaptive breeding and informal seed trade

Technologies that inhibit the dissemination of innovations are likely to have adverse impacts on countries for which innovations created abroad are the major source of crop improvement and that are therefore dependent on access to this flow for productivity increases.[2] Among these countries are the least developed of the world (Coe *et al.*, 1997). One of the possible consequences of GURTs is hence that they may lead to a distinct downward shift in the growth trajectories of agricultural productivity in developing countries. This downward shift would occur despite the potential increase in private R&D for crop improvement stimulated by GURTs. It would be caused by restrictions in the flow of innovations to which developing countries have had access over the period of pronounced yield growth in the last 50 years. If this restriction arises, then developing countries are likely to face cumulative losses in agricultural productivity growth as a result of widespread adoption of GURTs by crop innovators.

How can the possibility of this link between GURTs and a reduction in the rate of diffusion be substantiated? In this chapter, we draw on the experiences with a previous use restriction technology in agriculture in order to establish that the mechanism for rent appropriation in agricultural R&D has measurable effects in terms of both the rate of innovation at the technological frontier and the rate of diffusion from the frontier to developing countries. This empirical evidence allows us to make simulation-based inferences regarding the probable impact of GURTs on developing countries and on the diversity of experiences that this technology will bring about among these countries based on structural differences in their capacity to capture the flow of innovations.

Although GURTs are of recent origin, technologies with similar characteristics have existed for many decades, specifically the hybridization of cultivated varieties. This technique has been available for commercial seeds since the 1920s. Hybridization of cultivars has two implications: one is that the replanting of seeds from a hybrid results in a rapid deterioration of yield potential.[3] The other is that it protects against unauthorized reproduction by farmers and that the composition of the

[2] The National Agricultural Research Centres (NARCs) and the Consultative Group on International Agricultural Research (CGIAR) system currently provide mechanisms for facilitating this flow of innovations through adaptation of advanced material to local conditions (Pardey *et al.*, 1991).

[3] The first-generation loss is normally in the order of 25–30%.

hybrid can be withheld from other breeders if the innovator does not disclose the inbred lines that make up the hybrid crop. GURT crops share these two characteristics with hybrid crops, albeit in more extreme forms: replanting of GURT seeds results in an expected yield loss of close to 100%, and the reproduction of the crop's underlying genetic structure by a third party is currently not feasible since reproduction of the seed itself is impossible (DfID, 1999). The application of hybridization in the commercial seed sector also suggests that the availability of use restriction through this technique has been widely used by private companies when investing in R&D in those crops in which hybrids are feasible (Butler and Marion, 1985; Butler, 1996). This means that hybrid crops share fundamental features of use restriction with GURTs (although these features operate to different degrees of perfection in these two applications) and that industry has made significant use of these features as a form of rent protection.

In a previous paper (Goeschl and Swanson, 2000), we have estimated the rate of diffusion of innovations in hybrid and non-hybrid crops over the last 40 years. These estimates can be regarded as indicative of the likely impacts of the adoption of GURTs by crop innovators. Here we use these estimates as the basis for forecasting what probably constitutes the lower bound on the impact of GURTs in developing countries. These impacts are expected to arise out of the application of GURTs to those crops for which hybridization is currently not carried out on a significant scale, such as wheat and rice.

The remainder of the paper is organized as follows: the following section reviews the results of a 39-year panel study of yield developments in the most important hybrid and non-hybrid crops. We then apply these results in order to forecast the yield development in currently non-hybridized crops in selected countries. The likely development in the absence of GURTs, i.e. a perpetuation of the current regime, is then compared with the expected growth of yields when GURTs are widespread. Subsequently, we discuss the nature and robustness of the results, and in the concluding section, we relate these results to the question of a global system of intellectual property protection for agricultural R&D.

Review of Panel Study on Diffusion

In this section, we survey the results of a panel study on yield development in the eight most widely cultivated crops,[4] barley, cotton, maize, millet, rice, sorghum, soybeans and wheat covering 39 years, from 1961 to 1999. For a full description of the data, econometric methodology and modelling, we refer to (Goeschl and Swanson, 2000, and forthcoming). In the past, use restriction has been crop-specific: in two of the eight crops, namely maize and sorghum, the vast majority of improved varieties are hybridized. In the remaining six, hybridization is rare. This crop specificity enables us to compare the performance of hybrid and non-hybrid crops with respect to diffusion.

The method used is a fixed-effect panel estimation model that allows for heterogeneity among the countries through variable intercepts (Hsiao, 1986). The specific model that is being estimated is common in the empirical estimation of productivity convergence in other sectors and widely used in the literature on economic growth (Barro and Sala-i-Martin, 1995).[5] It presumes that all developing countries are subject to the same exogenous stochastic shock, in this case the event of an innovation that sets countries back in their relative yields. It then estimates for each crop the rate at which this shock is compensated for, allowing for heterogeneity in the intrinsic 'rate of recovery' between countries. The model has the form

[4] The criterion applied here is the global acreage of a crop.

[5] For a full development of the model in the context of fixed effects such as agroecological factors, see Goeschl and Swanson (2000).

$$\Delta G_{it} = a_i + \beta \cdot G_{i,t-1} + \varepsilon \qquad (1)$$

where G is the gap (difference) in logarithm between the yields in a specific country and the lead country and Δ signifies the change in the gap. The intercept term a_i denotes the long-term difference in productivity growth in equilibrium. One way of interpreting a_i is to regard it as a country-specific intercept that captures the agro-ecological and institutional factors that influence the overall productivity development of the country. In this it captures the content of the hypotheses that claim country-specific factors are responsible for the disproportionate yield gap that exists in the case of maize and sorghum. The coefficient β that is to be estimated then reports the diffusion coefficient of the particular crop. Empirically, we perform Fisher's test as proposed by Maddala and Wu (1999) as a panel data unit root test. We then estimate the diffusion coefficient β according to Equation (2) as a GLS (generalized least squares)-regression correcting for the residuals being cross-section heteroskedastic by down-weighting each pool equation by an estimate of the cross-section residual standard deviation.[6]

Econometric results

Each of the estimations delivers a coefficient β that is statistically highly significant. We also report the *average* intercept for all countries in the estimation denoted by $\hat{a}$. Before interpreting the results, it is convenient to perform some algebra in order to bring the model into a simpler form. Re-arranging (1), we arrive at the following equation for the growth rate of yield, $\Delta\hat{y}_t$ in the average developing country:

$$\Delta\hat{y}_t = \Delta y_t^* - (1+\beta) \cdot G_{t,t-1} + \hat{a} + \varepsilon. \qquad (2)$$

This formulation highlights the separate components that drive the growth rate of yields in the average developing country: The first component is the yield gain at the frontier Δy^*. This reflects the expansion of the set of technological possibilities. The second component captures the extent to which an innovation can diffuse in the country. We define the gap G to take on positive values. Therefore, we would expect that the coefficient β is negative (indicating that innovations do not have a negative effect on growth) and that the closer the coefficient is to -1, the more rapid the gains dissipate from the frontier to the average developing country. The third parameter, $\hat{a}$, summarizes the country-specific growth lags as an average. A positive value would indicate that on average, developing countries have a higher 'intrinsic' rate of yield growth in this crop and vice versa.

Interpreting the results: diffusion

Table 10.1 shows the results of the econometric estimation of Equation (1). The results indicate that hybrids have had a lower rate of diffusion of innovations from the frontier to developing countries than non-hybrids. While non-hybrids carried over roughly 69% of the gap opened by an innovation into the next year, hybrids carried over about 76%. This means that developing countries retained 7% more of the yield gap each year in hybrids than in non-hybrids. This explains an important part of the cumulative yield gap that has developed in hybrids. The results also indicate that there is merit to the idea that structural effects, such as agroecological conditions, have contributed to inhibiting yield growth of hybrids in developing countries. The parameter $\hat{a}$ is the mean of the individually estimated parameters a_i. The means computed for hybrids and non-hybrids indicate that in hybrids, the average developing country has had a greater negative long-term deviation from the

[6] The presence of heteroskedasticity tends to lead to higher diffusion coefficients. This weighting procedure corrects for that. The White test for cross-section heteroskedasticity is performed for all estimations and reports consistent parameters for all crops.

Table 10.1. Regressions for diffusion of innovations in different crops.

Crop	Hybrid crops	Non-hybrid crops
β^a	−0.242	−0.312
	(0.008158)***	(0.007711)***
α	−0.33611	−0.28171
R^2	0.136	0.167
(Number of obs.)	(6004)	(8664)
DW-statistic	2.438933	2.374825

[a] The figure in parentheses is the standard error.
* Indicates significance at the 5% level, ** at the 1% level.

growth rate of the frontier than in non-hybrids. The combination of structural and diffusion effects is therefore responsible for the significant gap in yields that persists between developed and developing countries in hybrid crops.

The results on the rate of diffusion have an intuitive economic interpretation: For a farmer cultivating different crops, an important criterion for evaluating crops is the loss of yield suffered as a result of slow diffusion. This loss can be assessed as the present value of the cumulative process of an innovation arriving at developing countries in a delayed fashion rather than arriving immediately.[7] Figure 10.1 reports the multiplier to the initial shock in order to estimate the present value of the loss at a 10% discount rate.[8] The interpretation of figure is the following: consider an innovation at the frontier in period *1* that results in an increase in profits by, say, US$100 for the average farmer. Within this period, the farmer in the developing country does not receive any of the innovation, thus incurring a loss of US$100. In the next period, the first benefits start to trickle down from the frontier, thus decreasing the loss relative to the frontier, with more of the benefits becoming available in subsequent periods. The curve depicts the present value of the total accumulated losses as a factor that can be applied to the initial 'loss'. Figure 10.1 shows the economic loss of the frontier shifting away from the developing country by virtue of an innovation at the frontier as a function of the rate of diffusion. Based on the parameters estimated for hybrids and non-hybrids, the discounted cumulative loss in the case of hybrids is about four times the initial shock while for non-hybrids it is around three times. This implies cumulative losses being about a third higher in hybrids as opposed to non-hybrids.

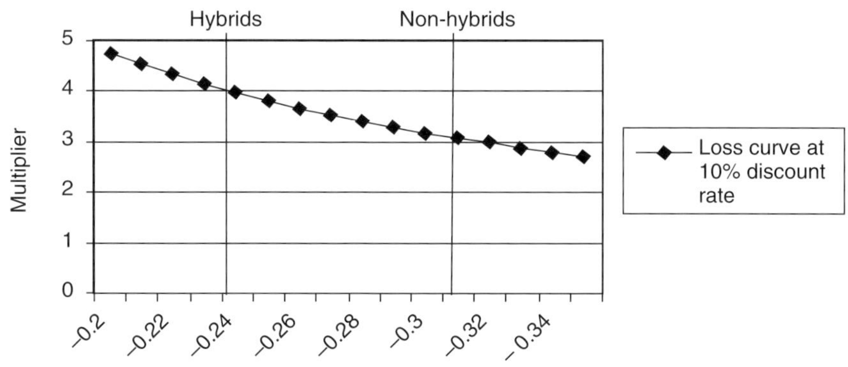

Fig. 10.1. Loss multiplier as a function of the diffusion rate.

[7] This curve is constructed on the assumption that the demand curve for agricultural output has constant and equal demand elasticity in the developed and the developing countries.
[8] The curve is fairly robust against changes in the discount rate. A higher rate pushes the curve down slightly as future losses have a lower value, and vice versa.

Interpreting the results: country-specific lags

A second set of important differences arises from the country-specific data on 'individual growth capacity'. The first result is that, on average, developing countries would experience slower growth in all crop yields as the coefficient $\hat{a}$ is below zero for all crops. However, these impediments to growth are quite different between crops, ranging from rice, a crop with good intrinsic growth potential in developing countries at $\hat{a} = -0.230$, to wheat, with high average barriers to growth at $\hat{a} = -0.384$. This captures whether innovations have been diffusing to countries where the local conditions are beneficial or adverse to the successful cultivation of the plant.[9] Interestingly, there is no correlation between parameter estimates of $\hat{a}$ and β, which indicates that the processes of diffusion are disjoint from the effects of local conditions.[10]

Forecasting the Impact of GURTs

With respect to its impact on the unauthorized reproduction of advanced cultivars, hybridization is a technological precursor of GURTs. It shares fundamental features with GURTs, although these features are exhibited in GURTs to a higher degree of 'perfection' than in hybrids.

Crops that have so far not been marketed as hybrids are the most likely target of GURTs (DfID, 1999). It will be in these crops, therefore, where the impact of GURTs will be the most significant as the regime of intellectual property protection will shift from an essentially public-domain provision to a use restricted regime. On the basis of the empirical estimates on hybrids, we carry out some forecasts regarding the likely impact of adopting use restriction technologies in these crops.

Baseline scenario and parameters

The panel study reported is based on a leader–follower framework of innovation and diffusion (Barro and Sala-i-Martin, 1995). The simulation extends this framework into a forecasting situation where the three components determining the growth of yields in the developing country are the expansion at the frontier, the rate of diffusion from the frontier to the developing country and the long-term differences in growth rates in crop yields. The baseline scenario for this forecast is the continued absence of GURTs from seed production. This baseline is established by assuming a perpetuation of the estimated growth rate at the frontier and rate of diffusion from the frontier. We refer to the baseline as the scenario 'in the absence of use restriction'.

The experience from hybrid crops suggests that expansion at the frontier will benefit from increased scope for rent capture by stimulating private R&D (Srinivasan and Thirtle, 2000). We therefore assume that the rate of expansion at the frontier will be higher in the presence of GURTS. This is important because it implies that *ultimately* every developing country's yields will be higher under use restriction than under the continuation of the present public domain regime. This underlines the importance of GURTs providing private industry incentives for the long-term development of yields. Due to the presence of a discount rate, however, the welfare effects of GURTs are not time-invariant. The crucial question is *how long* it takes for yields under GURTs to overtake the baseline scenario without use restriction. This is determined by the rate of diffusion, the country-specific long-term

[9] There are for each crop countries in which the intrinsic growth rate of the yield is basically equal or above that prevalent in the frontier countries. In the case of barley, this holds for Zimbabwe; in the case of cotton, for Israel and Syria; in the case of maize, for Chile; in the case of millet, for China; in the case of rice, for Egypt and Korea; in the case of sorghum, for Egypt and Israel; in the case of soybeans, for Ethiopia; and in the case of wheat, for Egypt and Zimbabwe.

[10] The correlation coefficient between $\hat{a}$ and β is 0.02 across all crops.

deviation from the growth rate at the frontier and the current gap in yields between the individual developing country and the frontier.

Our forecasting simulation looks at a 20-year time horizon for the development of yields in non-hybrids crops under the two scenarios. In order to simplify the comparison, the average values of the six non-hybrid crops are used. To illustrate the various possible impacts of GURTs, we present the simulation results of three countries, namely China, Ethiopia and Tanzania, and contrast the expected impacts there with the development of yields at the productivity frontier.

For the productivity frontier, we assume a starting yield in non-hybrids of 4 t ha^{-1} for the year 2000 in both the baseline and the use restriction scenario.[11] For the baseline scenario, we assume a continued growth at 1.58% per annum in the yields of non-hybrid crops in developed countries. This is the mean of past growth rates in the six different crops at the frontier (Goeschl and Swanson, unpublished). For the use restriction scenario, we initially assume that yield growth will be take place at the rate experienced in hybrids in developed countries over the last 40 years, which has been 2.175% on average (Goeschl and Swanson, unpublished). The parameters estimated through Equation (1) and reported in Table 10.1 are used as parameters for the rates of diffusion under the baseline (−0.312) and the use restriction (−0.242) scenarios.

Table 10.2 reports the country specific parameters that enter into the simulations. Yields in the initial period are below developed countries' yields by the average yield gap in non-hybrids. Across these six crops, China has the lowest average shortfall in yields relative to developed countries with a 15% gap while Ethiopia has a little over half the yields of developed countries. Tanzania does particularly poorly with a gap of about 75%. What is also important for the simulation is the country's long-term deviation from the yield growth rate in developing countries across the six crops. This data is generated by the estimation of Equation (2) on a crop and country-specific basis.

Simulation results

Figures 10.2–10.5 report the simulation output graphically. The forecasts show that the individual country experiences vary quite considerably. In developed countries (Fig. 10.2), the adoption of user restriction results in higher growth rates in yield and a more favourable yield development over the 20-year time horizon. There are developing countries where the experience is similar to developed countries, but arises in a more delayed fashion: in China (Fig. 10.3) for instance, yields in the first 10 years are expected to be very similar under both scenarios before the impact of use restriction on the yield frontier begins to push yields in China above the baseline. The case of Ethiopia illustrates a country that in the short run would be better off under the current regime, as the flow of innovations would be more easily appropriable. However, towards the end of the 20-year horizon, the faster expansion at the

Table 10.2. Country-specific simulation parameters.

	China	Ethiopia	Tanzania
Average yield in non-hybrids in percentage of developed country yield	85.1%	56%	25%
Average country-specific long-term deviation from developed countries' growth rate in non-hybrids	−0.094	−0.2128	−0.338

[11] The simulation is not sensitive to particular numerical yield values. The initial yield is only chosen for illustrative purposes.

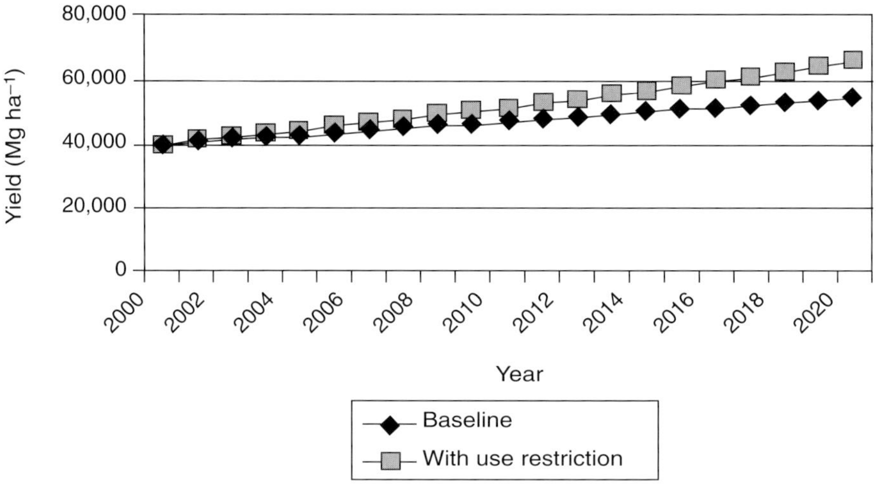

Fig. 10.2. Developed countries: comparison of yields under the use restriction and baseline scenarios, 2000–2020.

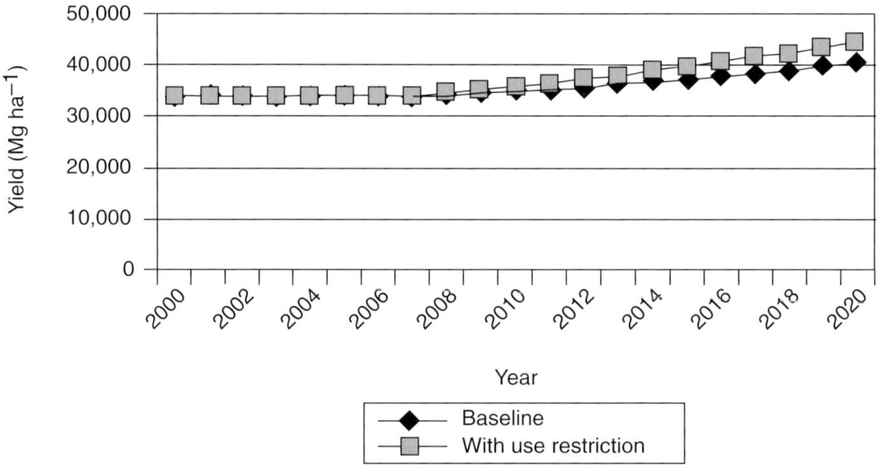

Fig. 10.3. China: comparison of yields under the use restriction and baseline scenarios, 2000–2020.

technological frontier under user restriction has compensated for the slower diffusion inherent in this regime. Lastly, the case of Tanzania illustrates a case where for the foreseeable future, the country would be worse off under a use restriction scenario than under a perpetuation of the current regime.

These four cases illustrate the diversity of outcomes that can be expected as a result of a potential adoption of GURTs. This diversity implies that over a policy-relevant time horizon of 20 years, countries will not be indifferent as to the regime adopted, depending on the current state of a country's agriculture. The simulations suggest that the most advanced countries stand to benefit most from use restriction while the least advanced stand to lose. As stressed before, when projected sufficiently far into the future, the productivity gains that the stimulation of private R&D through use restriction delivers result in the baseline scenario being overtaken. However, the net present value of these gains may currently be insufficient for developing countries to outweigh the short-term losses. It is interesting to note that even if

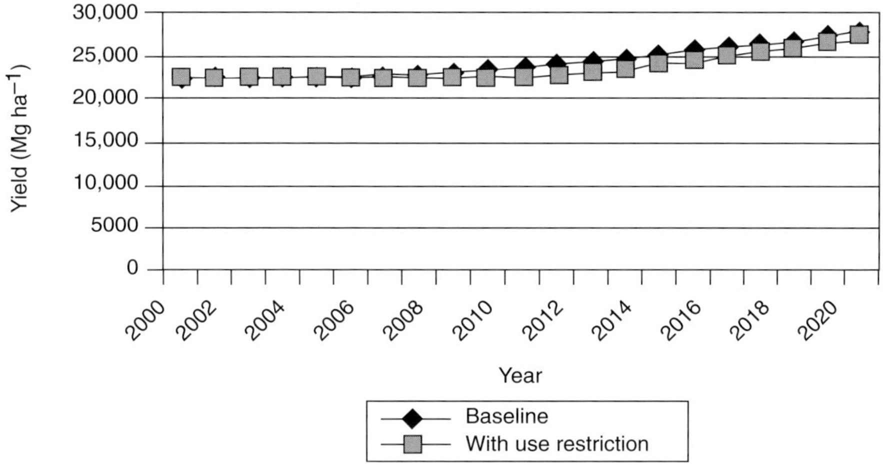

Fig. 10.4. Ethiopia: comparison of yields under the use restriction and baseline scenarios, 2000–2020.

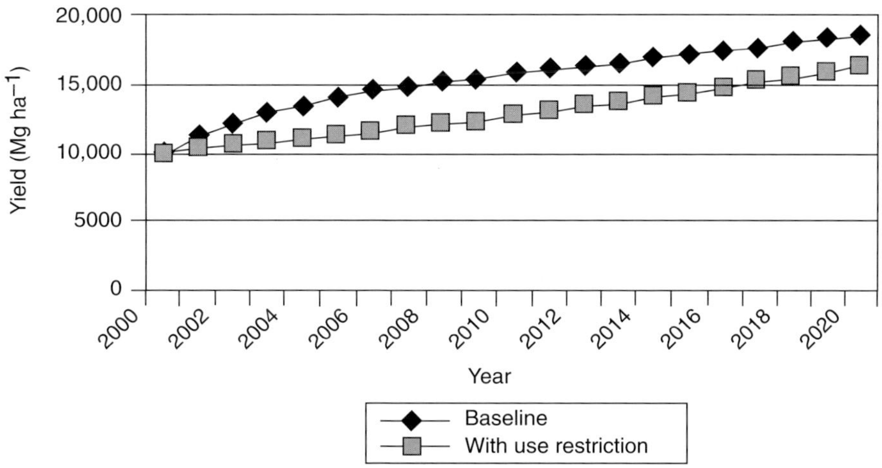

Fig. 10.5. Tanzania: comparison of yields under the use restriction and baseline scenarios, 2000–2020.

GURTs led to a doubling of the rate of innovation seen in hybrids at the same rate of diffusion, it would take more than 10 years in the case of Tanzania for yields under use restriction to outperform the baseline yields.

Some implications

The shift in the growth trajectory that developing countries are likely to experience as a result of a widespread adoption GURTs will lead in the long run to higher yields everywhere. However, most countries, and particularly the least developed ones, will have to pass through a phase of loss relative to the present regime for the diffusion of agricultural innovations. One of the implications is that this will lead – at least in the short to medium term – to the emergence of new 'growth clubs' in agricultural development, as countries will be put on different yield growth trajectories. The distributional consequences of this development over time may be considered undesirable. Another implication is that developing countries will need to develop new approaches to capturing the results of international agricultural R&D in order to maintain the flow of innovations into the country and thus to mitigate the adverse transitional period to higher yield growth. This leads to the wider question about the role of the public sector at a national and international level when use restriction technology is widespread. For some countries, such in the case of Tanzania, new technological, contractual and regulatory measures would be required in order to avert significant welfare losses.

Qualifications

There are a number of qualifications to this forecast, some of which are generic to any forecasting exercise, some of which are case-specific. The generic qualifications are that any *ex ante* appraisal of a technology has a wide error margin as there is no observable evidence from the technology *per se* and that data available on antecedent technologies has its limitations. Specific qualifications are that the country impacts estimated are unweighted averages across all non-hybrids. This tends to downplay the impact of crops where a country has a comparative advantage in production. The simulation also does not take into account any endogenous adjustments that might take place as a result of use restriction technology becoming available. Obvious examples would be a change in the portfolio of crops produced. It is not clear, however, whether including such adjustments would necessarily decrease the error margin. The final qualification is that the simulation is based on empirical estimates derived from a comparative study of innovation and diffusion in hybrids and non-hybrids. Although the *direction* of the impact of GURTs is likely to be the same as the impact of hybrids, the *volume* of the impact is harder to assess. Since GURTs present a more advanced form of use restriction, the impact of hybrids is likely to represent the lower bound on the impact of GURTs. Interestingly, though, the timing of the impacts illustrated by Figs 10.2–10.5 is not significantly altered if there is a parallel decrease in the rate of diffusion when the rate of innovation increases.

Conclusions

If – as is reasonable to expect – GURTs will replicate the experiences in hybrid-based agriculture across all other staple crops, this will lead in the first instance to a higher rate of investment by private industry in crop improvement motivated by enhanced scope for rent capture. This could result in a net increase or decrease in the amount of crop improvement produced globally, depending on the degree to which this private spending will crowd out public expenditures on agricultural R&D. We have assumed for the purpose of this simulation that the rate of innovation will rise to the level experienced in hybrid crops over the last 40 years. The impact of GURTs on the rate of diffusion of these new innovations is less ambiguous. On the basis of the experiences with hybrid crops, we can predict that GURTs will negatively impact on the rate of diffusion of innovations in those crops for which developing countries could previously rely on an inflow of innovations from abroad. This is because the advanced germplasm incorporated within advanced commercial cultivars will be much more costly to extract, reproduce and disseminate, impeding the adaptation of the latest generation of value-adding traits to local farming systems. This means that developing countries that predominantly use varieties that are currently not hybridized will experience a discontinuous shift on to a different trajectory of yield growth from the one they are currently moving along. For most countries, this implies a period of deteriorating net growth relative to the current regime. It is distributionally problematic that this period is the longest for the least-developed countries. It is for this reason that GURTs present a challenge to the global regulation of biotechnologies and to the role of the public sector in generating and diffusing productivity gains.

Acknowledgements

This research has its origins in a research grant by the Department for International Development (CNTR 99 8215) to investigate the impacts of genetic use restriction technologies on developing countries. We are grateful to Robert Carlisle from DfID for encouraging us to explore this area. We are particularly grateful to James Symons for helpful discussions on the econometrics and comments, and grateful to Mark Rogers for helpful discussions and comments without implicating them in any way in all the remaining errors. This research has benefited from discussions with Ed Barbier, William Fisher, Jonathan Jones, Michael Lipton, C.S. Srinivasan and Colin Thirtle that took place as part of the research.

References

Alston, J.M., Pardey, P.G. and Smith, V.H. (1998) Financing agricultural R&D in rich countries: what's happening and why? *Australian Journal of Agricultural and Resource Economics* 42, 51–82.

Alston, J.M. and Venner, R.J. (1998) The effects of U.S. Plant Variety Protection Act on wheat genetic improvement. Paper presented at the symposium on Intellectual Property Rights and Agricultural Research Impact, sponsored by NC 208 and CIMMYT Economics Program, El Batan, Mexico, 5–7 March.

Barro, R.J. and Sala-i-Martin, X. (1995) *Economic Growth*. McGraw-Hill, New York.

Butler, L.J. (1996) Plant breeders' rights in the U.S.: update of a 1983 study. In: van Wijk, J. and Jaffe, W. (eds) *Proceedings of a seminar on The Impact of Plant Breeders' Rights in Developing Countries,* held at Santa Fe Bogota, Colombia, 7–8 March, 1995, University of Amsterdam, Amsterdam, pp. 17–33.

Butler, L.J. and Marion, B.W. (1985) *The Impact of Patent Protection on the U.S. Seed Industry and Public Plant Breeding.* Food Systems Research Group Monograph 16. University of Wisconsin, Madison, Wisconsin.

Capalbo, S.M. and Antle, J.M. (1989) Incorportating social costs in the returns to agricultural research. *American Journal of Agricultural Economics* 71, 458–463.

CIMMYT (1999) *A Sampling of Impacts 1999. New Global and Regional Studies.* CIMMYT, Mexico.

Coe, D.T., Helpman, E. and Hoffmaister, A. (1997) North–South R&D spillovers. *Economic Journal* 107, 134–149.

Crouch, M. (1998) *How Terminator Terminates*, revised edition. Edmonds Institute Occasional Papers, Edmonds Institute, Washington, DC.

Dalton, T.J. and Guei, R.J. (1999) *Ecological Diversity and Rice Varietal Improvement in West Africa.* A report submitted to the CGIAR impact assessment and evaluation group.

Department for International Development (DfID) (1999) *Costs and Benefits to the Livelihoods of the Rural and Urban Poor arising from the Application of so-called 'Terminator Genes' and Similar Technologies in Developing Countries.* Report submitted by GS Consulting. DfID, London.

Evenson, R.E. (1989) Spillover benefits of agricultural research: evidence from US experience. *American Journal of Agricultural Economics* 71, 447–452.

Evenson, R.E. and Kislev, Y. (1973) Research and productivity in wheat and maize. *Journal of Political Economy* 81, 1309–1329.

Falck-Zepeda, J.B. and Traxler, G. (1998) *Rent creation and distribution from transgenic cotton in the U.S.* Paper prepared for the symposium, Intellectual Property Rights and Agricultural Research Impacts. Sponsored by NC-208 and CIMMYT Economics Program, CIMMYT Headquarters, El Batan, Mexico, 5–7 March.

Frey, K.J. (1996) *National Plant Breeding Study – I: Human and Financial Resource Devoted to Plant Breeding Research and Development in the United States in 1994.* Special Report 98, Iowa Agriculture and Home Economics Experiment Station, Iowa State University, Iowa.

Fuglie, K., Ballenger, N. and Day, K. (1996) *Agricultural Research and Development: Public and Private Investments Under Alternative Markets and Institutions.* AER-735. Economic Research Service, United States Department of Agriculture, Washington, DC.

Goeschl, T. and Swanson, T. (2000a) Of terminator genes and developing countries: what are the impacts of appropriation technologies on technological diffusion. In: Qaim, M., Krattiger, A. and von Braun, J. (eds) *Agricultural Biotechnology in Developing Countries. Towards Optimizing the Benefits for the Poor.* Kluwer Academic Publishers, Dordrecht, pp. 237–254.

Goeschl, T. and Swanson, T. (2000b) *The Diffusion of Innovations to Developing Countries: the Case of Crop Varieties.* Mimeo, Department of Land Economy, University of Cambridge, Cambridge.

Hsaio, C. (1986) *The Analysis of Panel Data.* Econometric Society Monographs No. 11, Cambridge University Press, Cambridge, UK.

Huffman, W.E. and Evenson, R.E. (1993) *Science for Agriculture: a Long-Term Perspective.* Iowa State University Press, Ames, Iowa.

Jaffe, W. and van Wijk, J. (1995) *The Impact of Plant Breeders' Rights in Developing Countries: Debate and Experience in Argentina, Chile, Clombia, Mexico and Uruguay.* Inter-American Institute for Co-operation in Agriculture and University of Amsterdam. Amsterdam.

Jefferson, R.A., Byth, D., Correa, C., Otero, G. and Qualset, C. (1999) Genetic use restriction technologies. Technical assessment of the set of new technologies which sterilize or reduce the agronomic value of second generation seed as exemplified by U.S. Patent No. 5,723,765 and WO 94/03619. Expert Paper, prepared for the Secretariat of the Convention for Biological Diversity, Subsidiary Body on Scientific, Technical and Technological Advice.

Kalton, R.R. and Richardson, P.A. (1983) Private sector plant breeding programmes: a major thrust in U.S. agriculture. *Diversity* 1, 16–18.

Kalton, R.R., Richardson, P.A. and Frey, N.M. (1989) Inputs in private sector plant breeding and biotechnology research programs in the United States. *Diversity* 5, 22–25.

Komen, J. and Persley, G.J. (1993) *Agricultural Biotechnology in Developing Countries: a Cross-Country Review.* ISNAR Research Report No.2. International Service for National Agricultural Research, The Hague.

Lesser, W. (1990) Sector issue II: seeds and plants'. In: Siebeck, W.E. (ed.) *Strengthening Intellectual Property Rights in Developing Countries: a Survey of Literature.* World Bank, Washington, DC, pp. 59–68.

Maddala, G.S. and Wu, S. (1999) A comparative study of unit root tests with panel data and a new simple test. *Oxford Bulletin of Economics and Statistics,* Special Issue 61, 631–652.

Pardey, P.G., Roseboom, J. and Anderson, J.R. (eds) (1991) *Agricultural Research Policy: International Quantitative Perspectives.* Cambridge University Press, Cambridge, UK.

Perrin, R.K, Kunnings, K.A. and Ihnen, L.A. (1983) *Some Effects of the U.S. Plant Variety Protection Act of 1970.* Economics Research Report No. 46. Department of Economics and Business, North Carolina State University, Raleigh.

Pray, C.E., Ribeiro, S. and Mueller, R.A.E. (1991) Private research and public benefit – the private seed industry for sorghum and pearl-millet in India'. *Research Policy* 20, 315–324.

Pray, C.E., Knudson, M.K. and Masse, L. (1993) *Impact of changing intellectual property rights on US Plant Breeding R&D.* Mimeo, Department of Agricultural Economics, Rutgers University, New Brunswick.

RAFI (1999) Traitor Technology. The Terminator's wider implications. RAFI Communiqué, 30 Jan. 1999. Rural Advancement Foundation International, Winnipeg, Manitoba.

Srinivasan, C.S. and Thirtle, C. (2000) Understanding the emergence of terminator technologies. *Journal of International Development* 12, 1147–1158.

Thirtle, C.G. (1985) Technological change and productivity slowdown in field crops: United States, 1939–1978. *Southern Journal of Agricultural Economics* 17, 33–42.

Thirtle, C.G., Ball, V.E., Bureau, J.C. and Townsend, R. (1994) *Accounting for Efficiency Differences in European Agriculture: Cointegration, Multilateral Productivity Indices and R&D Spillovers.* Mimeo, University of Reading.

Chapter 11

Managing Proprietary Technology in Agricultural Research

John Komen,[1] Joel I. Cohen,[1] Cesar Falconi[2] and Silvia Salazar[3]
[1] *International Service for National Agricultural Research (ISNAR), PO Box 93375, 2509 AJ The Hague, The Netherlands;* [2] *Inter American Development Bank, 1300 New York Avenue NW, Washington, DC 20577, USA;*
[3] *PO Box 91-3100, Santo Domingo de Heredia, Costa Rica*

Abstract

Significant events have occurred in the past decade that affect the use and exchange of genetic resources and the management of intellectual property, including the establishment of the Convention on Biological Diversity (CBD) in 1993 and the conclusion of the Agreement on Trade Related Intellectual Property Rights (TRIPs) in 1995. As a result, a growing number of developing countries are revising, or setting up their systems to protect intellectual property rights (IPRs). As members of the World Trade Organization (WTO), they are bound to introduce international standards for the protection of IPRs. At the same time, new inputs derived from biotechnology, especially those coming from the private sector, are widely used in agriculture. IPRs protect most of these inputs. However, it is not only the commercial sector that is using and developing materials for which intellectual property protection is being sought. The development and use of protected materials is also occurring among public, national, and international agricultural research organizations working for and with developing countries. International agricultural research, including the centres of the Consultative Group on International Agricultural Research (CGIAR) and the national agricultural research organizations (NAROs) of developing countries, is affected by these trends.

This chapter presents insights from studies on proprietary biotechnology inputs that are used by national and international agricultural research organizations. It reviews the most important findings derived from these inventories. It documents the difficult and often confusing situation that public research institutes face regarding the use and dissemination of products resulting from proprietary science where others hold rights.

Introduction

Developing-country national agricultural research organizations (NAROs), research centres of the Consultative Group on International Agricultural Research (CGIAR), and advanced research institutes are all affected by changes in proprietary rights related to agricultural research. Two developments in particular have affected them: the increasing importance of biotechnology in agricultural research and the growing position of the private sector in international agricultural research. These

 Economic and Social Issues in Agricultural Biotechnology (eds R.E. Evenson, V. Santaniello and D. Zilberman)

developments prompted the International Service for National Agricultural Research (ISNAR) to study the magnitude of, and procedures for using inputs that are protected by intellectual property rights (IPRs)[1] at selected CGIAR centres and NAROs.

The studies aimed to: (i) assess the extent to which proprietary applications of biotechnology (technologies and materials) are being used in NAROs and CGIAR centres; (ii) present potential legal implications regarding use of the identified proprietary technologies and materials; and (iii) synthesize findings and recommendations to stimulate further discussion.

Proprietary technologies and materials are those that are privately owned, managed or protected through some sort of IPR. Such materials and technologies may have restrictions placed on their use during the research stage or in a later stage, when products derived from the protected materials are ready for wide dissemination. A growing number of research inputs are protected as intellectual property. This chapter focuses on the use of such protected or proprietary materials and technologies in seven CGIAR centres, and in 13 NAROs from five countries in Latin America.

Results from the CGIAR Study

Since the early 1980s, the centres of the CGIAR have invested in strengthening their infrastructure and human resources for biotechnology research. Most of them now have specialized units or divisions to do research on molecular biology and other techniques covered by the term biotechnology. In 1996, the CGIAR adopted 'guiding principles' on genetic resources and intellectual property. These principles reaffirm that the resources maintained in the centers' genebanks should be freely available and that the centres should not seek legal protection for their innovations unless it is absolutely necessary to ensure that developing countries have access to new technologies ('defensive patenting'). The centres should not seek intellectual property protection for income-generating purposes and will not view potential returns from intellectual property protection as a source of operating funds. The 1996 document also states that any IPR acquired by a centre should be exercised without compromising in any manner the fundamental position of the CGIAR regarding free access by developing countries to knowledge, technology, materials and genetic resources.

At the annual CGIAR Mid-term Meeting in May 1997, an expert panel was established to focus on issues of proprietary science and technology. The panel explored legal issues and their ramifications regarding proprietary science and the complex partnerships arising in agricultural research (TAC, 1998). The panel felt that gaining an understanding of the current technologies and practices employed by the various centres would be an important first step in this process. The study of the seven CGIAR centres was done by ISNAR on behalf of the panel.

All the CGIAR centres that responded to the survey currently use proprietary inputs for biotechnology research. ISNAR recorded 166 applications of proprietary research inputs. In total, 46 different technologies and materials were reported over the eight technology categories introduced above. Most centres apply these technologies and materials in research on several mandated commodities. Of the technology categories surveyed, three had the broadest utility across centres: selectable marker genes, promoters and transformation systems (Table 11.1). This clearly demonstrates the important role that proprietary technologies and materials have assumed in research in the CGIAR, as is true for advanced research centres globally.

[1] 'Intellectual property' is a broad term for the various rights which the law gives for the protection of economic investment in creative effort. The principal categories of intellectual property relevant to agricultural research are patents, plant variety rights and trademarks.

Table 11.1. Applications of proprietary technologies and materials by research category.

Technology category	Number of applications per research category			
	Cereals	Non-cereals	Other	Total
Selectable markers	17	25	2	44
Promoters	18	14	3	35
Transformation systems	12	14	3	29
Insect-resistance genes	8	11	0	19
Disease-resistance genes	6	5	0	11
Genetic markers	4	4	2	10
Diagnostic probes	0	0	3	3
Others	7	6	2	15
Total	72	79	15	166

Means of protection and permission for use

Figure 11.1 describes the means by which the technologies and materials are protected. Not all proprietary inputs pose difficulties related to intellectual property or the dissemination and use of resulting products. Results from the survey help centres explore applications where potential difficulties could be foreseen. For many applications of proprietary technology, centres were not able to obtain clear knowledge or information regarding the type of IPRs provided for a particular proprietary tool. The rapid consolidation among biotechnology providers seeking ownership of technologies complicates obtaining such information. The extensive litigation occurring among commercial, public and farmers regarding details of licence and use agreements is another complicating factor (Barton, 1998).

Depending on the means of protection involved, institutes can therefore inadvertently infringe upon legal conditions

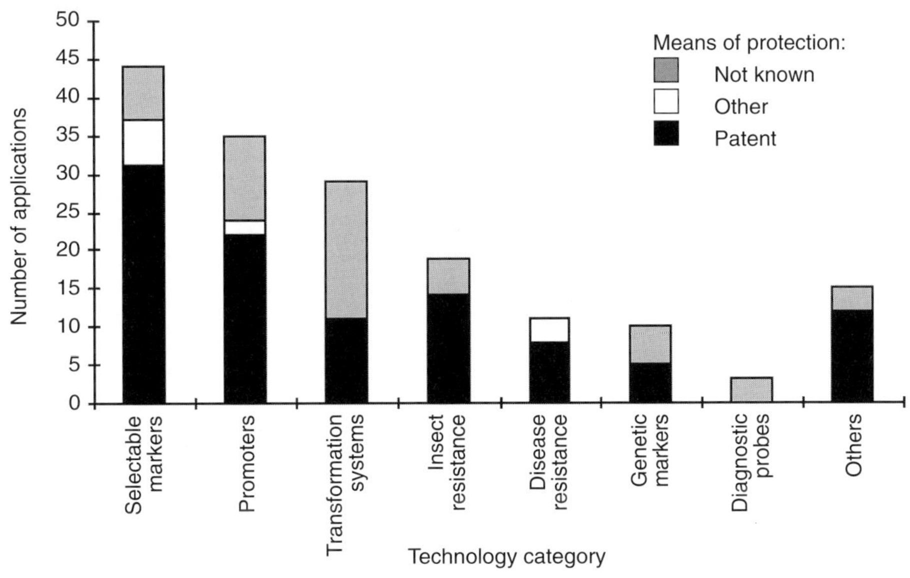

Fig. 11.1. Applications of proprietary technologies in CGIAR centres and their means of protection (by reported number of applications).

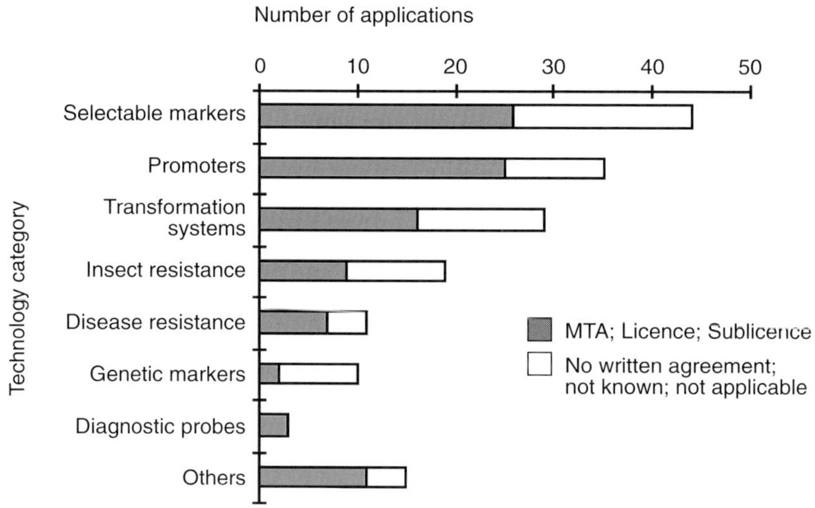

Fig. 11.2. Agreements and potential difficulties in permission to use proprietary technologies and materials (by reported number of applications).

regarding the dissemination of future products derived from these inputs. In addition, written agreements on use are often lacking when collaborating scientists exchange technologies or materials that are needed to address a particular research objective. This leaves centres in an unclear position regarding legal obligations, both to the owner of the technology and to other research partners. CGIAR centres are now studying situations where a lack of information exists about conditions for using these inputs.

Figure 11.2 shows that when permission for use was obtained, it is generally achieved through material transfer agreements (MTAs), licences or sublicences. However, for each technology category, except diagnostic probes, the centres are using a number of technologies and materials for which written agreements are lacking, or for which permission status is unknown. Obtaining permission is the general rule, but exceptions exist. For example, in some transformation systems, the permission to use specific systems is unclear. Similar problems exist for the use of selectable markers, genetic markers, and various insect- and disease-resistance genes. However, it should also be noted that while material transfer agreements may secure permission for research, they do not generally confer permission for subsequent dissemination.

Expected products from research and ability for dissemination

The survey also sought information on research products that were expected to encounter difficulties in their dissemination. As inputs are proprietary, restrictions may exist for their use, dissemination or further production. Survey results showed that the centres expected a total of 58 outputs or products (Table 11.2).

More than 40% of the responses indicated that centres often lack critical information or knowledge needed to anticipate difficulties in post-research use and the dissemination of outputs that are generated from proprietary technologies. Consequently, some outputs developed with proprietary technologies are likely to encounter problems in their use and dissemination, especially when they are exported to countries where the technologies are protected. Many CGIAR centres have yet to take into account extensions of

Table 11.2. Products expected from the application of proprietary tools.

Product category	Number of products expected	Examples
Improved crops	36	Improved cereal and non-cereal varieties with enhanced insect, fungal and virus resistance
Diagnostics	11	Diagnostic tests for tropical livestock diseases
Vaccines	1	Vaccine for East Coast fever
Others	10	Transformation protocols, genetic markers

protection and permission requirements for such technologies.

However, centres have considered the implications of legal agreements for disseminating research outputs, with 14% of the responses foreseeing some limitations. For example, a contractual arrangement between a CGIAR centre and a private multinational as owner of the input technology specifies that outputs can be distributed only in certain countries. In this case, there is *a priori* understanding of the restrictions and limitations for dissemination that should be studied further by the CGIAR. Such studies have already begun on a centre-by-centre basis.

CGIAR-centre patents

The study assessed the degree to which centres are planning to patent or otherwise protect inventions. Only three outputs were identified that may be patented. Centres anticipated the use of other protective measures, e.g. copyright, trademarks or plant variety rights, for another 11 outputs. This limited amount of intellectual property protection being sought by centres can be attributed to many factors, including lack of familiarity with IPR issues, the fact that suitable IPR options are not yet developed and approved, and the tradition that goods and services are developed as international public goods. Further, many bilateral donor and civil society organizations are opposed to applying IPR protection to products of CGIAR research.

Results from the Latin American Study

Between July and September 1998, ISNAR conducted a similar survey among NAROs in Brazil, Chile, Colombia, Costa Rica and Mexico. At the time of the survey, none of the institutions that were surveyed had suitable institutional or legal frameworks for dealing with IPR matters. With the exception of two research centres, none of the institutions had an office or person responsible for assisting the researchers in issues of intellectual property, access to adapted technologies, technology transfer or ways to protect their own inventions. There was little coordination between institutions in the same country and even between researchers in the same institution. For example, within the same institution, there were cases where the same vector was requested from different sources, all under different conditions. There were also cases where one scientist has an agreement for the transfer of biological material for a vector, while a neighbouring scientist used the vector without permission. The researchers were functioning without the institutional support needed to address these issues for their research.

Figure 11.3 summarizes the range of proprietary technologies and materials used by the research organizations and the number of applications reported in each category. In total, 34 different technologies and materials as well as 386 specific applications of proprietary research inputs were reported. Of the eight technology categories surveyed, selectable marker genes (GUS,

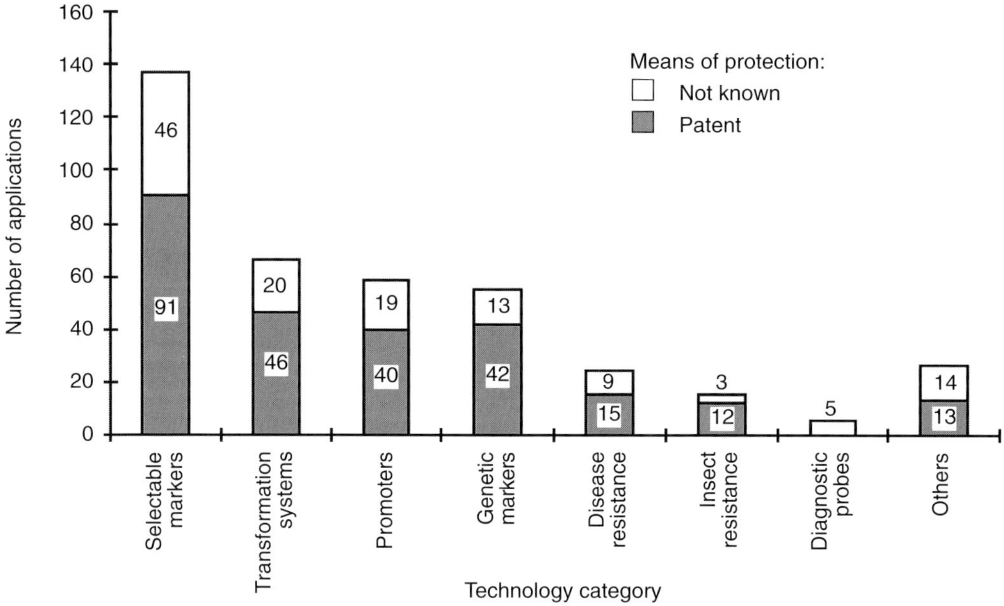

Fig. 11.3. Applications of proprietary technologies and their means of protection (by reported number of applications).

kanamycin), promoters (CaMV/35S) and transformation systems (*Agrobacterium*) show the broadest utility across the institutes involved in the survey.

Means of protection and permission for use

Figures 11.3 and 11.4 show what means of protection have been given to the technologies and materials analysed in the survey and how permission for use was obtained. Again, while not all proprietary inputs pose difficulties regarding intellectual property or the dissemination and use of resulting products, this study has helped research organizations explore areas where potential difficulties may occur.

As indicated in Fig. 11.3, in more than 30% of the applications of proprietary technology, NARO respondents indicated that they lack clear knowledge or information on the type of means of protection provided for a particular proprietary tool. These cases need to be further examined to avoid potential infringement of legal conditions on the use of these proprietary technologies and materials. Figure 11.4 indicates that MTAs are the most common means for acquiring technologies, as they are in CGIAR centres. However, it also highlights the importance of international collaboration as an important mechanism for acquisition, and it indicates that some of the NAROs purchase proprietary technologies for their own use. Many applications that are being used lack written agreement, leaving the institute wondering about its legal responsibilities and how the use of the proprietary tool has been achieved. In addition, the use of licence, as a legal alternative for technology transfer, is very limited.

NARO patents

The results show that the Latin American NAROs have high expectations to obtain intellectual property protection for their new products. They expect that 74% of the 50 products that are anticipated from the application of proprietary technologies will

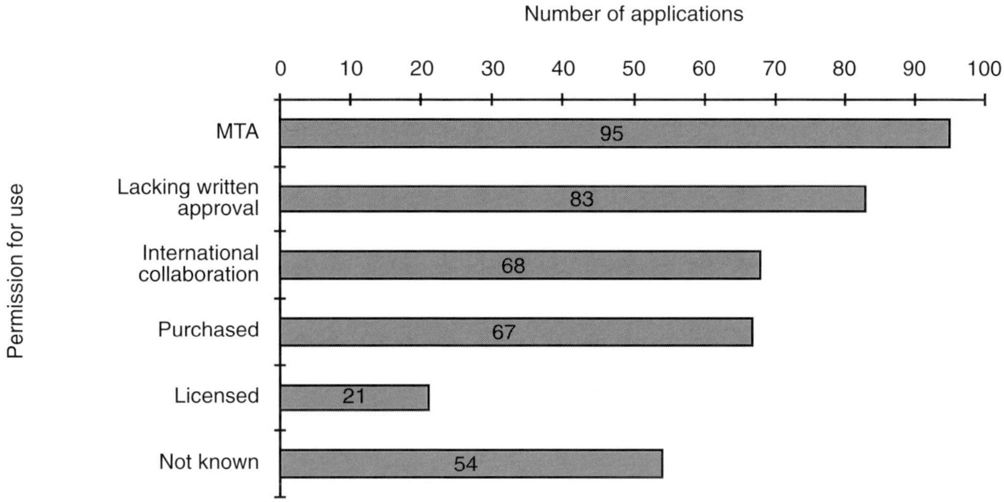

Fig. 11.4. Permission for use of proprietary technologies (by reported number of applications).

be protected either by patents or by plant variety protection (see Table 11.3).

Conclusions and Management Lessons

The ISNAR studies indicate a pressing need for establishing competent legal expertise in issues of proprietary rights for the CGIAR centres and their partners, including NAROs and other agricultural research institutes. Based on our surveys, a number of specific managerial lessons and recommendations can be made.

First approximation of proprietary inputs

The inventories provide a first approximation of the use of third-party innovations that are protected through patents or other forms of intellectual property. They show the extent to which international and national research institutes are using and integrating biotechnology into their ongoing research programmes, with a wide use of selectable markers, promoters and transformation systems.

These first approximations provide a source of information regarding the use of proprietary technologies and materials in agricultural research programmes. When several international centres or research organizations in one country are found to use the same proprietary tools, it may be advantageous to collaborate in the acquisition of such technologies and to seek common understanding for the IPRs involved. The surveys provide one means for CGIAR centres and NAROs to analyse the use of proprietary technologies and materials in their research programmes and follow up with more formal IPR audits.

Table 11.3. Expectations to protect products.

	Brazil	Chile	Colombia	Costa Rica	Mexico	Total
Protection anticipated	11	9	0	0	17	37
No protection anticipated	0	3	6	0	0	9
Not yet decided	0	0	2	2	0	4
Total	11	12	8	2	17	50

Extensive use of material transfer agreements

The most common legal arrangement by which CGIAR centres obtain permission for using proprietary inputs is through MTAs, followed by licensing. The extensive use of agreements and licences is a new fact of life for publicly funded agricultural research organizations. Proprietary technologies and materials that are developed by or that are under the protection of private-sector research organizations have made important contributions to CGIAR-centre and NARO research. The use of such materials means that, as centres transform mandate crops, develop vaccines, do diagnostic probes and provide for marker-assisted breeding, their dependence on licences, MTAs and other agreements with the private sector increases. The experiences with and conditions of MTAs should therefore be analysed and exchanged among CGIAR centres, NAROs and other agricultural research institutes to explore possibilities for standard formats and to determine legal implications regarding possible restrictions on use and dissemination. Where CGIAR centres or NAROs in one country are using the same category of proprietary tools, it may be advantageous to cooperate in acquiring such technologies.

Enforcement of obligations

Where MTAs impose legal obligations, the receiving institute must be in a position to honour its obligations under the agreement. For example, where the proprietary technology involves the supply of a trade secret and imposes confidentiality obligations, the institute must be able to police the handling of the material supplied. This may require the establishment of a secure system of operation, and placing researchers and visitors under confidentiality obligations. The increasing legal complexity involved in the supply of proprietary technologies may raise matters of contract law, intellectual property law, biodiversity and biosafety law, technology transfer, and competition law (where restrictive provisions are imposed). A second point is whether research institutes anticipate any dissemination constraints due to their current legal arrangements. Some centres were not clear if such constraints exist. This information confirms the need for CGIAR centres and NAROs to: (i) develop IPR expertise to analyse potential limitations; and (ii) designate an officer to administer legal obligations.

These increasing legalities indicate the need for research institutes to have a primary legal administrator. The administrator would ensure the compliance of staff with obligations generated by MTAs and with the terms of intellectual property licences. Because MTAs and licences are among the most common ways to acquire proprietary technologies, it should be reiterated that MTAs, licences and sublicences impose direct legal obligations upon the research institutes.

Protection sought by CGIAR centres and NAROs

All of the centres surveyed indicated that they use applications of biotechnology that are protected by others. While the centres receive numerous technologies, the number of products that they expect to be patented or otherwise protected are very few. However, the NAROs that were surveyed expected a far greater number of products for patent filing, which requires an interface such as a research liaison office or intellectual property management unit.

Developing IPR competence

International and national research organizations using agricultural biotechnology are caught in a complex environment, reflecting a transition from earlier periods where products and processes for research resided in the public domain. The increasing acquisition of proprietary technologies, their use in research serving the public good, and the vast array of developing countries where such use occurs raises questions regarding appropriate IPR arrangements. However, for many scientists and institutions, such concerns are overwhelming. Yet their work continues, trusting that as final products are developed, no legal instruments will block the dissemination of improved materials to their clients.

To adopt a more proactive strategy requires significant time and investment in taking steps to find institutional mechanisms for addressing these complex challenges. Sampaio and Brito da Cunha (1999) discuss an example from a large national agricultural research organization, Embrapa in Brazil. In the CGIAR, several developments have occurred to enhance centre management of intellectual property and related issues. These have included implementing the CGIAR Central Advisory Service (CAS) for Proprietary Technology, individual centres undertaking formal intellectual property audits, and the first steps toward 'Intellectual Property Management Units' at selected CGIAR centres.

References

Barton, J.H. (1998) The impact of contemporary patent law on plant biotechnology research. In: *Global Genetic Resources: Access and Property Rights.* CSSA Special Publication. Crop Science Society of America, Madison, Wisconsin.

Cohen, J.I., Falconi, C., Komen, J. and Blakeney, M. (1998) *Proprietary Biotechnology Inputs and International Agricultural Research.* ISNAR Briefing Paper 39. International Service for National Agricultural Research, The Hague.

Salazar, S., Falconi, C., Komen, J. and Cohen, J.I. (2000) *The Use of Proprietary Biotechnology Research Inputs at Selected Latin American NAROs.* ISNAR Briefing Paper 44. International Service for National Agricultural Research, The Hague.

Sampaio, M.J.A. and Brito da Cunha, E.A.B. (1999) Managing intellectual property in Embrapa: a question of policy and change of heart. In: Cohen, J.I. (ed.) *Managing Agricultural Biotechnology – Addressing Research Program Needs and Policy Implications,* chapter 20. Biotechnology in Agriculture Series, No. 23. CAB International, Wallingford, UK.

TAC (1998) *Report of the CGIAR Panel on Proprietary Science and Technology.* SDR/TAC:IAR/98/7.1. Technical Advisory Committee of the CGIAR, Rome.

Chapter 12
Is Marker-assisted Selection Cost-effective Compared with Conventional Plant Breeding Methods? The Case of Quality Protein Maize

Kate Dreher,[1] Michael Morris,[1] Mireille Khairallah,[2] Jean Marcel Ribaut,[2] Shivaji Pandey[3] and Ganesan Srinivasan[3]
[1]*Economics Program;* [2]*Applied Biotechnology Center;* [3]*Maize Program, International Maize and Wheat Improvement Center (CIMMYT), Apdo. Postal 6-641, Mexico City, Mexico*

Abstract

This chapter presents the findings of a case study designed to compare the cost of using conventional plant breeding methods with the cost of using a new DNA-based crop selection technique known as marker-assisted selection (MAS). The case study, which was carried out in Mexico at the International Maize and Wheat Improvement Center (CIMMYT), focused on a narrowly defined breeding objective – transferring the quality-protein maize (QPM) phenotype, controlled primarily by a mutant allele of a gene called *opaque2*, from one elite maize inbred line to another elite inbred line.

Costs associated with use of conventional breeding methods and MAS for QPM line conversion were estimated using a spreadsheet-based budgeting approach. First, field and laboratory operations involved in conventional and MAS breeding were identified and costed out. Second, representative conventional and MAS breeding schemes were identified. Third, the unique laboratory and field parameters set forth in each breeding scheme were used to calculate the total cost of implementing that particular scheme.

Results of the budgeting exercise suggest that currently at CIMMYT, the relative cost-effectiveness of conventional breeding methods as compared to MAS for QPM line conversion differs depending on the circumstances. In cases where it is possible to identify segregating materials by visually inspecting ears in the field, conventional breeding methods can be very cost-effective, but in cases where visual selection is not possible, use of molecular markers can lead to significant cost savings.

CIMMYT's experience with MAS parallels the experience of many other breeding programmes. Even though MAS has come to play a prominent role in the field of plant breeding, for many practical applications the economics of MAS are still being worked out on a case-by-case basis. The continuing uncertainty concerning the utility of MAS in specific applications should not give rise to undue pessimism, however. Everything that made MAS attractive in the first place still holds true; the key to successfully integrating the technology into applied breeding programmes will lie in identifying applications in which molecular markers offer real advantages over conventional breeding

 Economic and Social Issues in Agricultural Biotechnology (eds R.E. Evenson, V. Santaniello and D. Zilberman)

methods. MAS should be able to offer significant advantages in cases where phenotypic screening is particularly expensive or difficult, including breeding projects involving multiple genes, recessive genes, late expression of the trait of interest, seasonal considerations or geographical considerations. In addition to reducing the cost of breeding, MAS also has the potential to generate time savings. Depending on the benefits that a breeding programme realizes from earlier release of its breeding products (which typically differ between the private and public sectors), the value of these time savings can be enormous – often justifying the additional cost involved in using MAS.

Continuing refinement of molecular marker technologies will make MAS cheaper and more effective in coming years, but at the same time it would be a mistake to assume that marker technologies represent a 'silver bullet' solution to every breeding problem. As this case study has revealed, conventional breeding methods still provide a cost-effective option for many types of breeding project, and they will continue to be attractive for many years to come.

Executive Summary

This chapter presents the findings of a case study designed to compare the cost of using conventional plant breeding methods with the cost of using a new DNA-based crop selection technique known as marker-assisted selection (MAS). The case study, which was carried out in Mexico at the International Maize and Wheat Improvement Center (CIMMYT), focused on a narrowly defined breeding objective – transferring the quality-protein maize (QPM) phenotype, controlled in part by a mutant allele of a gene called *opaque2*, from one elite maize inbred line to another elite inbred line.

The case study had three objectives:

1. To generate detailed information about the cost of carrying out conventional breeding operations as well as the cost of implementing MAS procedures at CIMMYT's research facilities in Mexico,
2. To determine the cost-effectiveness of using MAS for a particular breeding application, namely, QPM line conversion, and
3. To provide insights into the likely cost-effectiveness of potential future applications of MAS whose costs and benefits are not yet known.

Costs associated with use of conventional breeding methods and MAS for QPM line conversion were estimated using a spreadsheet-based budgeting approach. The analysis proceeded in three stages. First, field and laboratory operations involved in conventional and MAS breeding were identified and costed out. Second, representative conventional and MAS breeding schemes were identified. Third, the unique laboratory and field parameters set forth in each breeding scheme were used to calculate the total cost of implementing that particular scheme.

CIMMYT maize breeders and molecular geneticists were asked to design representative breeding schemes for QPM line conversion. Four hypothetical stylized breeding schemes were selected. Two of the schemes rely solely on conventional breeding methods and phenotypic evaluation, and two incorporate MAS.

The results of the budgeting exercise suggest that, currently at CIMMYT, the relative cost-effectiveness of conventional breeding methods as compared to MAS for QPM line conversion differs depending on the circumstances. In cases where it is possible to identify segregating materials by visually inspecting ears in the field, conventional breeding methods can be very cost-effective, but in cases where visual selection is not possible, use of molecular markers can lead to significant cost savings.

In attempting to extrapolate from this finding to derive conclusions about the cost-effectiveness of MAS in general, it is important to remember two things. First, the case study focused on one particular application of MAS; since QPM line conversion in maize has a number of distinctive characteristics that affect the breeding process, the cost-effectiveness of MAS will differ for other applications. Second, the budgets developed for the case study were based on the CIMMYT maize-breeding programme in Mexico; where conventional breeding practices and/or laboratory procedures differ significantly from those found at CIMMYT, and where input costs differ significantly from those in Mexico, the cost-effectiveness of MAS will differ.

CIMMYT's experience with MAS parallels the experience of many other breeding programmes. Even though MAS has come to play a prominent role in the field of plant breeding, for many practical applications the economics of MAS are still being worked out on a case-by-case basis. The continuing uncertainty concerning the cost-effectiveness of MAS in specific applications should not give rise to undue pessimism, however. Everything that made MAS attractive in the first place still holds true; the key to successfully integrating the technology into applied breeding programmes will lie in identifying applications in which markers offer real advantages over conventional breeding methods or complement them in novel ways. MAS should be able to offer significant advantages in cases where phenotypic screening is particularly expensive or difficult, including breeding projects involving multiple genes, recessive genes, late expression of the trait of interest, seasonal considerations or geographical considerations.

In addition to reducing the cost of breeding, MAS also has the potential to generate time savings. For example, using background markers to perform full MAS, breeders can accelerate a back-cross scheme by definitively identifying plants that contain higher percentages of the recurrent recipient parent genome than would be possible using conventional phenotypic selection methods based on visual observation of plants in the field or laboratory analysis of tissue samples. Given this greater degree of selection precision, the rate of progress can be significantly increased, potentially allowing entire cycles of back-crossing to be eliminated and thereby reducing the time required to reach a particular breeding objective. Depending on the context, the value of these time savings can be enormous – often far larger than the additional cost involved in using MAS.

Further advances in biotechnology will make MAS cheaper and more effective in the coming years, but as most breeders and biotechnologists clearly recognize, it would be a mistake to assume that marker technologies represent a 'silver bullet' solution to every breeding problem. As this case study has revealed, conventional breeding methods still provide a cost-effective option for many types of breeding project, and they will continue to be attractive in the future. Additional case studies are needed to build up the body of empirical evidence that research managers will require to identify cost-effective opportunities for combining conventional and marker-assisted methods.

Introduction

Considerable disagreement exists within the research community as to the eventual impacts of biotechnology on crop genetic improvement efforts. Proponents claim that biotechnology has the potential to revolutionize the plant-breeding industry by allowing breeders quickly and cost-effectively to produce a wide range of novel plants with vastly enhanced production and consumption traits. Opponents continue to express doubts, pointing out that the potential benefits of biotechnology for agriculture have consistently been overestimated and arguing that genetic gains in major food and fibre crops will for the foreseeable future continue to come from conventional breeding methods.

In the absence of empirical evidence relating to the benefits and costs of alternative breeding methods, little can be done to evaluate the validity of these opposing viewpoints. Computer simulations have been conducted to estimate the genetic gains that plant breeders theoretically could realize through the use of new biotechnological tools, but few efforts have been made to determine the cost of putting these new tools to work in an applied breeding programme. This is an important omission, because cost considerations play a major role in shaping the strategies of plant breeders. Information about the relative costs of comparable conventional and biotechnology-based breeding strategies will help policy makers determine the circumstances under which emerging biotechnology tools can be deployed efficiently.

This paper presents the findings of a case study designed to compare the cost of using conventional plant-breeding methods with the cost of using a new DNA-based crop selection technique known as marker-assisted selection (MAS). The case study, which was carried out in Mexico at the International Maize and Wheat Improvement Center (CIMMYT), focused on a narrowly defined breeding objective – transferring the Quality Protein Maize (QPM) phenotype, controlled in part by a mutant recessive allele of a gene called *opaque2*, from one elite maize inbred line to another elite inbred line.[1]

The case study had three specific objectives:

1. To generate detailed information about the cost of carrying out conventional breeding operations as well as the cost of implementing MAS procedures at CIMMYT's research facilities in Mexico,
2. To determine the cost-effectiveness of using MAS for a particular breeding application, namely QPM line conversion, and
3. To provide insights into the likely cost-effectiveness of potential future applications of MAS whose costs and benefits are not yet known.

Since the costs associated with any particular breeding project will depend on the project objectives, on the breeding programme that performs the work, and on the sites at which the work is performed, among other factors, the results of a single case study cannot be used to draw conclusions about the cost-effectiveness of MAS in general. None the less, our results provide the type of empirical evidence that managers of applied plant-breeding programmes need to plan effectively. As additional case studies are conducted, it should be possible to build the empirical database needed to formulate general principles concerning the cost-effective application of MAS to plant breeding.

Maize Genetic Improvement

Conventional breeding

Since ancient times, farmers have influenced the genetic characteristics of their crops by selectively saving seeds collected from high yielding, healthy, well-adapted plants for use in future planting cycles. In the absence of rigorous selection methods, however, farmer breeding tended to be extremely slow and highly unpredictable. Not until the advent of science-based selection strategies based on Mendelian principles of inheritance did genetic improvements begin to be realized at a rapid rate. What later came to be known as 'conventional plant-breeding methods' have evolved considerably since their emergence around the beginning of the 20th century, but they continue to revolve around three basic steps: generation of a population of plants having desirable traits, evaluation and selection of superior individuals, and recombination of the superior individuals to generate a new population for subsequent cycles of selection and improvement (Allard, 1960; Simmonds, 1979; Poehlman, 1987).

Conventional plant-breeding methods have produced important genetic gains in maize. The contribution of conventional breeding to improved yield potential has been documented in studies that have separated pure genetic gains from environmental and management impacts in analysing long-term changes in the yield potential and stress tolerance of commercial maize hybrids in the USA. Using an experimental design that controlled for changes in crop management practices, Duvick (1984) estimated that from 1930 to 1980, the genetic potential of US hybrids rose by 92 kg ha^{-1} $year^{-1}$, which accounted

[1] An allele is one of several forms of a gene, each of which provides alternatives of inheritance.

for 89% of the total yield gain observed in farmers' fields. Other studies examining slightly different time periods estimated genetic gains in US hybrids ranging from 33% to 79%, with an average value of 60.5% (see Duvick, 1984).

Conventional breeding methods have produced impressive results, but it is not clear that historical rates of genetic gains can be sustained using these methods alone. Evidence is beginning to emerge that improvements in yields of major cereals, including maize, have begun to slow in some countries (Pingali and Heisey, 1999). Should these trends continue, genetic gains generated through conventional breeding would not be able to keep pace with future growth in demand for cereals. Certainly this would disrupt grain markets in industrialized countries. Even more worrisome, the inhabitants of developing countries with high population growth rates and high human maize consumption levels will suffer acutely if grain yields stagnate. To ensure that productivity growth keeps pace with rising demand for food and feed, new tools are therefore required.

Biotechnology

Many experts predict that the new tools needed to reinvigorate the science of plant breeding will come from the rapidly evolving field of biotechnology. Some of these new tools allow breeders to do things that cannot be done using conventional breeding methods. For example, genetic engineering techniques allow genetic material to be transferred directly from one plant to another and even from one species to another. Other new biotechnological tools play a different role. Instead of helping breeders to bypass conventional breeding methods, these tools can be used to complement them. For example, MAS and DNA fingerprinting techniques can potentially increase the efficiency of traditional breeding programmes by speeding up the time of varietal release, lowering plant population requirements and eliminating costly field evaluation.

Biotechnology-based breeding tools have many potential applications, but as with any new technology, research managers must determine the circumstances under which the benefits are likely to justify the considerable costs. In some cases, questions of cost are less relevant, for example when a new biotechnology-based tool allows breeders to accomplish something that would be impossible to accomplish using conventional breeding methods (e.g. transfer a gene from a bacterium into maize). In other cases, questions of cost are highly relevant, because biotechnology may simply offer an alternative way of doing something that can also be done using conventional breeding methods. Research managers often have little information to guide the choice of plant-breeding technology, because little is known about the cost of using the new biotechnology-based tools. Empirical case studies can help to fill this information gap. By detailing the costs and benefits involved in using a particular biotechnology-based tool to address a range of applied breeding problems, case studies can generate the information that research managers need to decide whether that particular tool is worth integrating into a breeding programme.

Case Study: QPM

Brief history of QPM breeding

In 1963, after years of searching, researchers at Purdue University discovered that a recessive mutation in the gene known as *opaque2* resulted in maize kernels with elevated levels of two essential amino acids, lysine and tryptophan. This finding was of more than academic interest. Despite its widespread use as human food and animal feed, maize does not provide an adequate amount of lysine and tryptophan, which humans and monogastric animals cannot synthesize for themselves. Breeders subsequently tried to incorporate the mutant *opaque2* allele and its associated phenotype into their breeding materials, initially with limited success. Early varieties of maize with the *opaque2* mutation suffered from numerous problems, including soft endosperm (which is unappealing to many consumers), increased susceptibility to certain grain pests and ear rots, slower drying time in the field, and, perhaps most importantly, a 10–15% drop in yield (Villegas, 1994). Over time, breeders found modifier genes that restored the desirable hard endosperm phenotype in materials containing the recessive *opaque2* mutation. These agronomically acceptable, nutritionally enhanced materials later came to be known as QPM.

Recognizing the potential benefits of nutritionally enhanced maize for developing countries, in the late 1960s CIMMYT launched a QPM breeding programme. Over 30 years, numerous elite QPM materials were developed using conventional selection methods, including both improved open-pollinating varieties (OPVs) and elite inbred lines (CIMMYT, 1999).

While the QPM materials continue to undergo further improvement, CIMMYT breeders also are interested in transferring the mutant *opaque2* allele and the associated modifiers into other elite lines that lack the QPM phenotype. Typically this is done using some type of back-cross breeding scheme (Fig. 12.1). In back-cross breeding, the donor parent containing the allele of interest is crossed with the recurrent recipient parent, usually an elite material that contains many desirable characteristics but that lacks the desirable allele. Of the resulting progeny, those containing the desirable allele that most resemble the recurrent recipient parent are then 'back-crossed' to the recurrent recipient parent. This process is normally repeated several times. After repeated cycles of back-crossing and self-pollination, the breeder ends up with plants that are nearly identical to the recurrent recipient parent and that also contain the desirable allele introduced from the donor parent.

QPM line conversion using conventional breeding methods

Breeders who use conventional breeding methods to accomplish QPM line conversion face two main constraints.

First, breeders looking at young plants in the field are unable to determine if they will produce ears with elevated levels of lysine and tryptophan. Therefore, they face a disagreeable choice. Either they wait until the end of the season to identify desirable plants (which may force them to lose a season), or they pollinate an unnecessarily large number of plants because they do not know which ones carry the mutant *opaque2* allele (which means they have to incur the field costs of growing out a lot of plants that will later turn out to be useless). Even at the end of the season, simple inspection of mature ears will not always be sufficient to identify QPM materials. To identify QPM plants definitively, maize grain samples must be sent to the laboratory for evaluation. At CIMMYT, this evaluation can be done in the Soils and Plant Laboratory (SPAL) using two alternative procedures. In the first procedure, the levels of nitrogen (a proxy for protein) and tryptophan present in the grain endosperm are directly measured. In the second procedure, enzyme-linked immunosorbent assay (ELISA) analysis is used to quantify the concentration of the EF1A protein, which is highly correlated with the level of lysine present in the grain endosperm. Since lysine and tryptophan almost always occur in a 4:1 ratio in QPM materials, only one of the two laboratory procedures is needed to determine the phenotype of the plant.

Second, the mutant allele that confers the desirable QPM phenotype is recessive, so heterozygous plants containing one copy of the mutant *opaque2* allele cannot be identified phenotypically, because they do not express the QPM trait. To identify heterozygous plants, breeders often self-fertilize all of the plants in a population and plant the seed from the self-fertilized plants in the next season to allow the QPM phenotype to be expressed; then they can go back and identify which plants in the previous generation carried the mutant *opaque2* allele.[2]

[2] Under certain circumstances, it may be possible to avoid having to take an extra season to grow out the seed from the self-fertilized plants (see section on Representative breeding schemes).

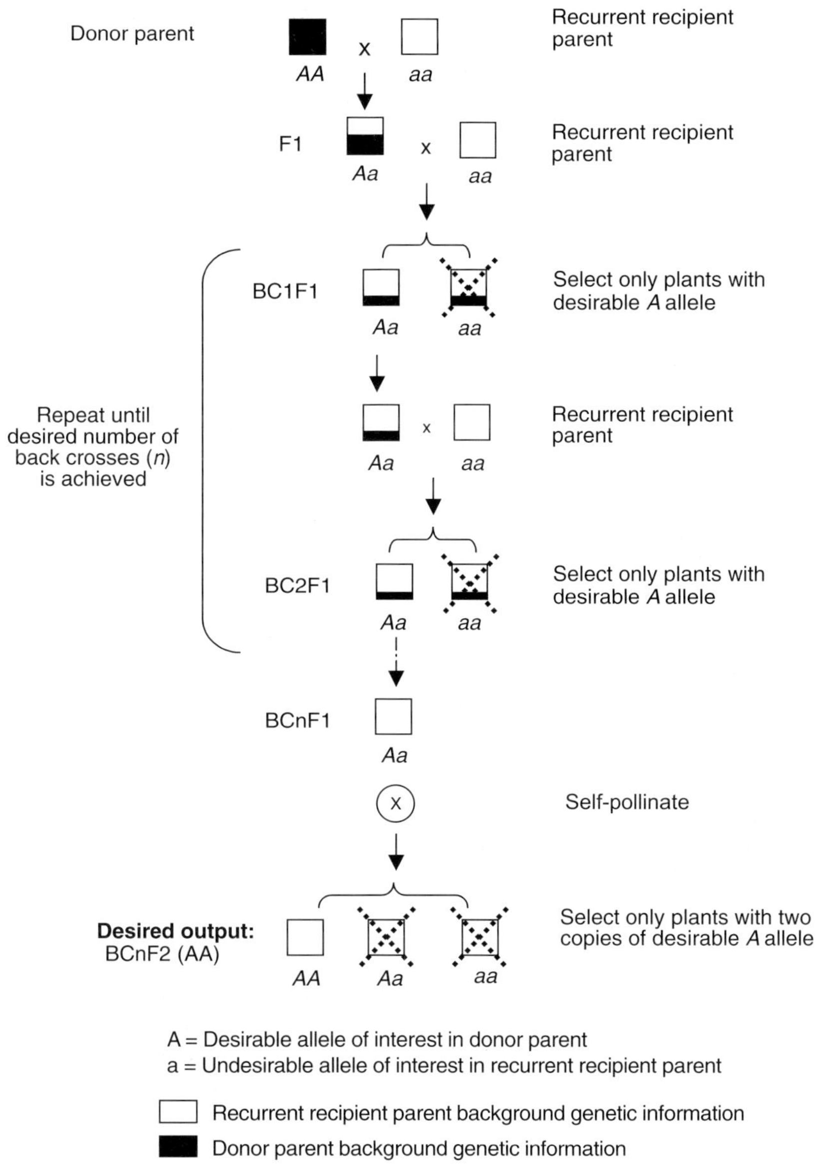

Fig. 12.1. Standard back-crossing scheme for a dominant trait.

QPM line conversion using marker-assisted selection (MAS)

A basic objective of any breeding scheme is to identify within a differentiated population of plants elite individuals that can pass on their desirable characteristics to their progeny. Since observable differences in the physical characteristics of plants are usually caused by non-observable differences in the plants' genetic make-up, what breeders are really trying to do when they make selections in the field is to choose individual plants that contain favourable alleles in desirable combinations. Conventional breeders are forced to rely on phenotypic evaluation, which does not always accurately indicate the underlying information present in a plant's genome. Environmental influences and genetic interactions can obscure the presence or absence of specific alleles, making it difficult for breeders to identify the plants that they really seek.

Even in cases in which a plant's phenotype provides reliable evidence about the plant's underlying genetic characteristics, phenotypic evaluation can be costly and/or time-consuming. For example, a breeder who wants to select plants with resistance to a particular grain storage pest must grow a large population of plants in the field, make controlled crosses among all of the plants, wait until the end of the season and harvest ears from all of the plants, rear a large number of insects, apply the insects to the ears in a controlled manner, evaluate the resulting damage, and then select the most resistant plants. If the insects have to be left on the ears for an extended period before their feeding activity can be observed, the next cycle of planting may have to be delayed for a season. In such cases, breeders obviously would prefer to avoid the cost (measured in terms of money and time) associated with phenotypic evaluation.

Molecular markers provide a potential solution to some of these problems. Markers, short pieces of DNA that correspond to particular sequences of DNA in the plant genome, can inform breeders about the presence of desirable alleles in individual plants. If the presence of the desirable alleles can be confirmed through the use of markers, breeders do not have to resort to costly and time-consuming phenotypic evaluation to determine whether or not the alleles are present.

Among the many different types of molecular markers that can be used in plant breeding, for reasons of cost and simplicity, researchers at CIMMYT often use a particular type of marker known as simple sequence repeat (SSR) markers. SSRs are short, repeated DNA sequences found in most plant genomes (Weber and May, 1989). Since the number of SSRs present at a particular location on the genome tends to differ among plants, SSRs can be analysed to determine the absence or presence of specific alleles.

SSR-based molecular marker analysis for QPM breeding begins with the extraction of a DNA sample from the plant to be studied (Fig. 12.2). Primers that flank a region containing SSRs within the *opaque2* gene are then used to amplify the sample using the well-known polymerase chain reaction (PCR) method (PCR amplification involves making millions of copies of the DNA segment flanked by the primers). When the number of repeated sequences between the flanking primers differs in the normal *opaque2* and mutant *opaque2* alleles, the amplified DNA fragments vary in length. By electrophoretically separating the amplified DNA fragments on gels and analysing them, researchers can determine if the plant possesses two copies of the normal allele, two copies of the desirable mutant *opaque2* allele, or one copy of each allele.

In the context of QPM line conversion, MAS offers the possibility of overcoming the two main constraints faced by conventional breeders.

First, since marker analysis can be done using DNA samples extracted from leaf tissue of very young plants, it allows QPM plants to be identified early in the breeding cycle. This allows the breeder to discard plants that do not contain mutant *opaque2* alleles prior to pollination, reducing the size of the breeding population and saving both time and money.

Second, with molecular markers, breeders can distinguish between homozygous recessive plants that carry two copies of the mutant *opaque2* allele (these will express the QPM trait) and heterozygous plants that carry only one copy (these will not express the trait, because the mutant *opaque2* allele is recessive). Armed with this knowledge, breeders do not have to go through the laborious and costly process described at the end of the previous section.

QPM line conversion makes an appropriate choice for this case study because conventional breeding methods and MAS methods represent alternative approaches

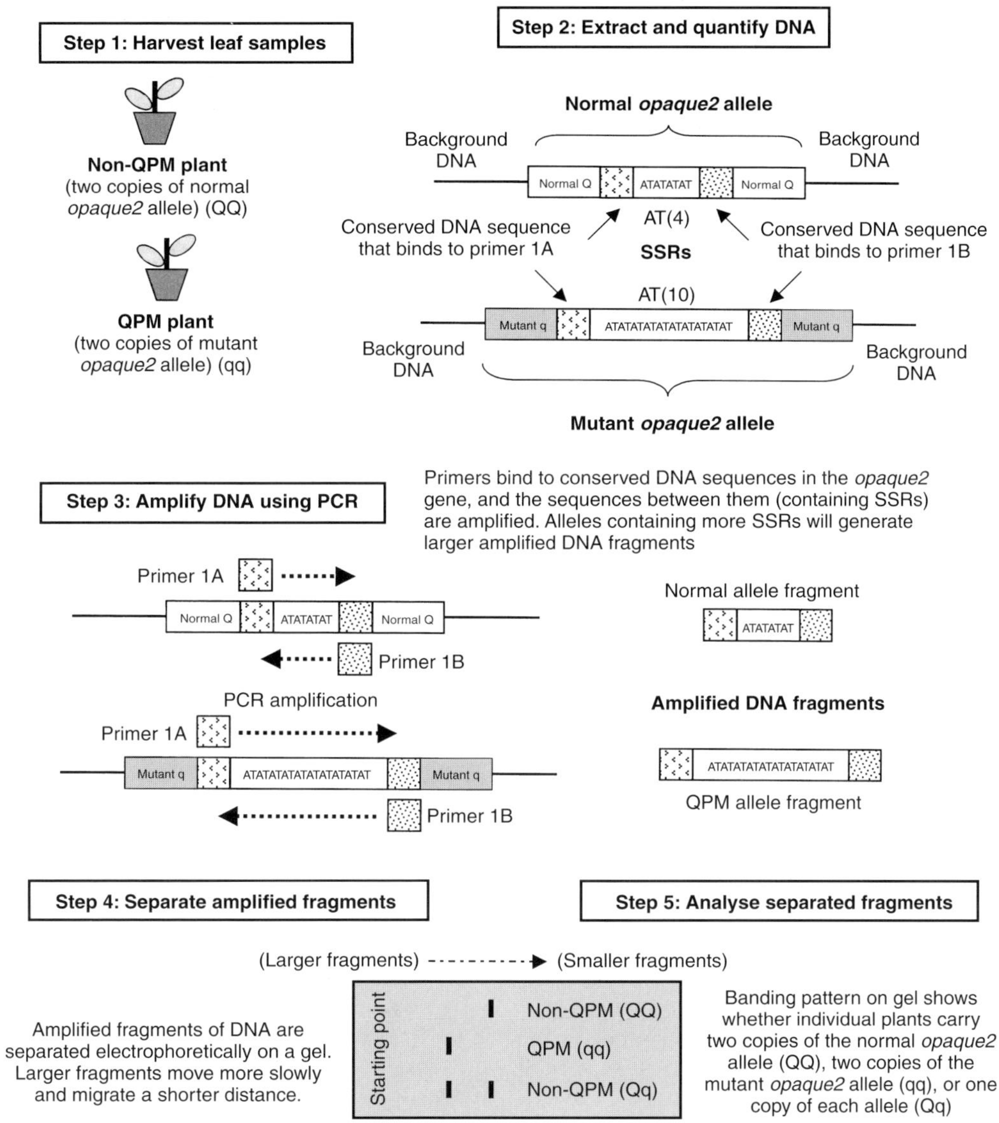

Fig. 12.2. Molecular marker analysis using SSR markers.

for achieving the same objective. It should be noted, however, that MAS offers additional potential advantages over conventional breeding methods that are also worth considering. The discussion thus far has focused on use of one molecular marker (or possibly a small number of markers) to identify the presence or absence of a specific allele that otherwise would have to be identified phenotypically. This procedure is known as *partial MAS*. Another procedure, known as *full MAS*, involves the use of a large number of molecular markers to characterize a plant's entire genome. With full MAS, a large number of 'background' molecular markers are used to develop unique genetic fingerprints of both the donor line and the recurrent recipient line. After the lines are crossed, the progeny are screened using the allele-specific and background markers to identify the individual plants that contain the allele of interest *and* whose overall genetic fingerprint most resembles that of the recurrent recipient parent.

Because the full MAS procedure allows selection based on direct knowledge of the genome of individual plants, it can significantly increase the rate of breeding gains. Ribaut and Hoisington (1998) estimate that a complete line conversion for a single, monogenetic trait theoretically can be achieved in three cycles of back-crossing using MAS, whereas at least six cycles of back-crossing would be required to accomplish the same goal using traditional breeding methods.

Study Methods and Data Sources

Costs associated with use of conventional breeding methods and MAS for QPM line conversion were estimated using a spreadsheet-based budgeting approach. The analysis proceeded in three stages. First, all field and laboratory operations involved in conventional and MAS breeding were identified and costed out. To facilitate the subsequent analysis, field costs were expressed in terms of cost per 5-m row of plants, and laboratory costs were expressed in terms of cost per individual sample. Second, representative conventional and MAS breeding schemes were identified. Third, the unique laboratory and field parameters set forth in each breeding scheme were used to calculate the total cost of implementing that particular scheme.

Stage 1: Costing field and laboratory operations

Identifying field operations

Research station managers outlined the field operations involved in growing a normal crop of maize at two research stations operated by CIMMYT in Mexico: El Batán and Tlaltizapán. In addition to providing general planting parameters (e.g. number of 5-m rows per hectare, row spacing, plant spacing), they also detailed the specific management operations required to grow a crop of maize (e.g. soil preparation, planting, irrigation, fertilization, insecticide application, pollination and harvesting).

Identifying laboratory operations

Information about the molecular marker analysis procedures used at CIMMYT is available in ABC laboratory manuals. Supplementary information regarding laboratory protocols, supplies and labour requirements for each stage of analysis was provided by laboratory technicians or obtained through direct observation. Two other laboratory procedures employed in phenotypic analysis of QPM materials – nitrogen and tryptophan analysis, and ELISA – are performed at CIMMYT in a separate laboratory, the SPAL. Information about these two procedures was collected in essentially the same manner.

Estimating costs

Once the supplies and labour requirements had been estimated, the cost of each operation was calculated. Cost data were collected from various sources.

MATERIALS. Experiment station managers and purchasing department staff furnished the prices of fuel, disposable supplies and chemical inputs used in the field operations. The ABC inventory manager provided price data for most of the supplies used in molecular marker analysis. Prices for most disposable supplies and chemical inputs used in the SPAL were provided by the SPAL manager; in a few cases where actual purchase prices were unavailable, commercial quotes for purchases of bulk quantities were substituted.

LABOUR. Information about salaries and benefits of senior scientists, laboratory technicians and administrative staff came from CIMMYT's Human Resources Office. Station managers and the CIMMYT Finance Office assembled equivalent figures for field workers. Hourly wages were calculated on the assumption of a 2000-h work year.

All costs were denominated in current US dollars. To account for the fact that products may be purchased either domestically or in the USA, a flexible exchange rate parameter was introduced into the spreadsheet.

A number of simplifying assumptions were made to facilitate the costing exercise.

CAPITAL EQUIPMENT. To avoid an exhaustive accounting exercise to estimate hourly use charges for every piece of field machinery (e.g. tractors, implements, irrigation equipment) and laboratory equipment (e.g. autoclaves, PCR thermocyclers), capital equipment coefficients were estimated for the Maize Program and for the ABC based on the total value of the capital equipment maintained by each programme. The total estimated variable costs for each operation were then inflated using the relevant capital equipment coefficient to factor in the cost of fixed capital equipment. This method of incorporating capital equipment costs implicitly assumes that the use of capital equipment is roughly equal across all operations within each programme.

LAND. Since CIMMYT owns its field experiment stations and as an international organization does not have to pay taxes on them, land costs were not explicitly included in the case study analysis. Nor was an opportunity cost imputed to land, which might be appropriate where field space is limited and increasing the number of rows devoted to one breeding project necessarily means decreasing the number of rows devoted to others. Land costs (either explicitly based on purchase or rental prices, or implicitly estimated as some form of opportunity cost) could easily be included in future studies.

ECONOMIES OF SCALE AND DIVISIBILITY OF INPUTS. Since field operations at CIMMYT are routinely carried out on a large scale (e.g. entire fields are cultivated at once; individual trial plots within the field are not cultivated separately), cost estimates for most field operations were made at the large scale and then adjusted as necessary. Implicit in this approach is the assumption that most field operations are divisible. On the other hand, laboratory analyses at CIMMYT are sometimes performed using machinery that is operating at less than full capacity (e.g. an ELISA analysis might be performed on 20 samples, even though the plate contains 96 wells). Therefore, in a number of operations associated with molecular marker analysis and QPM phenotypic analysis, some inputs were considered indivisible and treated as lumpy.

WASTAGE. A small additional cost was added to each budget to account for wastage and to allow a margin of safety to offset possible errors in data collection.

Stage 2: Developing representative breeding schemes

The spreadsheets can be used in isolation as tools for budgetary analysis, since they generate estimates of the cost of performing any one of a large number of well-defined activities (e.g. grow a row of maize plants at Tlaltizapán station, analyse one molecular marker in 50 leaf samples in the ABC laboratory, estimate the lysine content of 200 maize grain samples using ELISA). For an effective comparison of the costs of alternate breeding schemes, however, the information generated by the spreadsheets must be combined with additional information about the actual steps involved in a particular breeding scheme.

One characteristic of plant breeding is that many different strategies can be adopted in pursuit of a given breeding objective, no matter how specific and well defined the objective. In practice, the breeding scheme actually implemented to pursue a particular goal will depend upon each breeder's specific knowledge, experience, subjective opinions and practical limitations with regard to land, labour, machinery and other inputs. For the purposes of this study, CIMMYT breeders and biotechnologists worked separately and together to develop several hypothetical stylized breeding schemes. To the extent possible, an effort was made to identify schemes that can be considered comparable (e.g. all the schemes were designed to produce similar end products). In practice, designing comparable schemes proved more difficult than anticipated, because conventional breeding methods and MAS differ in certain respects that have implications for overall breeding strategy. It is important to note also that the representative schemes do not exactly reflect breeding schemes currently being used at CIMMYT.

Stage 3: Estimating costs of breeding schemes

Once the stylized breeding schemes had been identified, key parameters were quantified for each scheme, including the number of plants grown in each season, the number and types of molecular marker analyses undertaken, and the number and types of phenotypic tests performed. Constant baseline values were selected for the remainder of the variable parameters.[3] This information was entered into the spreadsheets, which were used to calculate the costs for field and laboratory work performed during each season of a given breeding scheme. The costs incurred during each season were then summed to obtain an estimate of the total cost of each breeding scheme.

Empirical Results

Stage 1: Field and laboratory unit costs

Field unit costs

Field unit costs were estimated for both the El Batán and Tlaltizapán stations. At El Batán only one maize crop can be grown per year, while at Tlaltizapán two maize crops can be grown per year, an irrigated winter crop (Cycle A) and a rain-fed summer crop (Cycle B). Representative budgets were prepared for 1-ha plots at each of the two stations. In the baseline scenario, it was assumed that all plants were covered with glassine bags, that one-half of the plants were fertilized, and that one-quarter of the ears were harvested.

Field costs per row are lower at El Batán ($3.64) than at Tlaltizapán (Table 12.1). At Tlaltizapán, per row costs are higher during the dry season (Cycle A, $4.63) than during the rainy season (Cycle B, $4.49), a difference that is caused mainly by additional costs associated with irrigation. The marked difference of almost 25% between the two research stations can be attributed mainly to differences in fuel costs; fuel requirements for a number of operations are higher at Tlaltizapán, and fuel prices are also slightly higher. Fertilizer and herbicide regimens are quite different between the two stations, but the associated costs tend to offset one another, so the cost of chemicals is not a decisive factor. Labour costs are slightly lower at El Batán, mainly because hourly wages are slightly lower, particularly for temporary field labourers. Finally, since the El Batán experimental fields are located at the main CIMMYT campus, no travel costs are incurred for work done there. Scientists and their assistants who travel to Tlaltizapán incur fuel, food and accommodation expenses.

The breakdown of unit costs by input category is relatively stable across all three estimates. Labour consistently accounts for approximately 60% of the total cost of field operations. The next largest cost component is supplies, with glassine bags and pollinating bags comprising close to 15% of the total. Surprisingly, chemical inputs (3%) and fuel (>3%) do not contribute significantly to the overall costs of field operations.

[3] In the baseline scenario, it was assumed that no operations had to be repeated in the laboratory or field, that no cycles had to be duplicated, and that the alternative breeding schemes were equally likely to produce the intended result. Sensitivity analysis was subsequently used to explore the consequences of changing various technical parameters and of relaxing some of the assumptions used in the baseline scenario.

ABC laboratory unit costs

Laboratory unit costs were calculated for two MAS procedures, partial MAS and full MAS, used by staff of the ABC to examine QPM populations.

In the case of partial MAS, after leaf samples are harvested, DNA is extracted using a sap extractor, quantified using a spectrophotometer and separated following amplification on agarose gels. In the case of full MAS, an attempt is made to limit the number of samples analysed with background markers by first analysing each DNA sample and selecting for further screening only those that contain the target allele of interest. Consequently, the costs for leaf sample harvest, DNA extraction and DNA quantification were left out of the background marker budget. The procedure was costed assuming that all amplified DNA fragments are separated on agarose gels, even though some of the background markers used at CIMMYT require the use of more expensive acrylamide gels.

Given that the two procedures differ substantially in complexity, it is not surprising that they also differ substantially in cost (Table 12.2). The cost of analysing one plant using the partial MAS procedure (one marker) comes to \$2.13, while the cost of analysing one plant using the full MAS procedure (55 markers) comes to \$68.73.

Since the two procedures are used for different purposes, there is little point in comparing them directly. None the less, several aspects of the unit cost estimates are noteworthy. In both procedures, reagents comprise the largest proportion of the total cost, accounting for more than one-half of the total in the case of the full MAS procedure. Among reagents, by far the most expensive is one required for PCR amplification, *Taq* polymerase, which is sold under patent by the manufacturer at a very high price. *Taq* polymerase accounts for 26% of the cost of the partial MAS procedure and 44% of the full MAS procedure. Labour costs are also significant, accounting for 34% of the cost of the partial MAS procedure and 26% of the cost of the full MAS procedure. Laboratory supplies accounted for 10% and 3%, respectively, of the cost of the partial and full MAS procedures.

SPAL laboratory unit costs

The costs of the two laboratory analyses performed in the SPAL, nitrogen and tryptophan determination and ELISA, which can be used to identify maize samples exhibiting the QPM phenotype, were also calculated. In costing out the nitrogen and tryptophan analysis, it was assumed that 15 whole kernels are analysed per sample and that the kernels need not be washed to clean off chemical treatments. In costing out the ELISA analysis, it was assumed that nine samples can be analysed per plate.[4]

The unit cost (per sample cost) for the nitrogen and tryptophan determination (\$3.64) significantly exceeds that of the ELISA analysis (\$2.62; Table 12.3). The cost difference is attributable mainly to differences in labour costs, which are considerably higher in the nitrogen and tryptophan determination. Both processes involve a laborious initial step of grinding and weighing the maize kernel samples, but the nitrogen and tryptophan determination involves additional labour-intensive steps. Labour costs for laboratory supervision are also higher for the nitrogen and tryptophan determination, because twice as many data points must be analysed for each sample. The absolute cost of the reagents used in the nitrogen and tryptophan procedure is more than double the

[4] The protocol used for the ELISA analysis was based partly on the protocol used by the SPAL during their first experience with ELISA and partly on the published protocol recommended at Colorado State University. Wherever possible, labour estimates obtained through observation were used, but some parameters concerning the preparation of the reaction plates were taken from the published protocol because they provide a more realistic idea of the costs that will be incurred in the future when the procedure becomes more standardized.

Table 12.1. Field costs (US$) at El Batán and Tlaltizapán research stations.

	El Batán[1,8]			Tlaltizapán (cycle A)[1,2,7,8]			Tlaltizapán (cycle B)[1,3,7,8]		
	Cost per ha ($)	Cost per row ($)	Per cent of total	Cost per ha ($)	Cost per row ($)	Per cent of total	Cost per ha ($)	Cost per row ($)	Per cent of total
Total cost	8708.61	3.64	100	9857.66	4.63	100	9560.27	4.49	100
Capital	305.57	0.13	4	345.88	0.16	4	335.45	0.16	4
Overheads	763.91	0.32	9	864.71	0.41	9	838.62	0.39	9
Variable inputs									
Chemicals	244.87	0.10	3	258.73	0.12	3	292.58	0.14	3
Fertilizer	86.86	0.04	1	179.53	0.08	2	179.53	0.00	0
Nitrogen	63.81	0.03	1	152.70	0.07	2	152.70	0.07	2
Phosphate	23.05	0.01	0	26.83	0.01	0	26.83	0.01	0
Insecticide	33.85	0.01	0	58.45	0.03	1	58.45	0.03	1
Herbicide	121.53	0.05	1	20.00	0.01	0	20.00	0.01	0
Fungicide	0.31	0.00	0	0.27	0.00	0	0.27	0.00	0
Other chemicals	2.32	0.00	0	0.48	0.00	0	20.40	0.01	0
Fuel	150.21	0.06	2	366.77	0.17	4	395.76	0.19	4
Diesel	86.38	0.04	1	319.61	0.15	3	278.88	0.13	3
Butane	63.83	0.03	1	47.15	0.02	0	47.15	0.02	0
Electricity for irrigation[4]	94.79	0.04	1	0.00	0.00	0	0.00	0.00	0
Supplies	1,738.01	0.73	20	1541.64	0.72	16	1618.72	0.76	17
Envelopes	77.23	0.03	1	29.60	0.01	0	29.60	0.01	0
Bags	1,486.85	0.62	17	1321.65	0.62	13	1321.65	0.62	14
Other supplies	173.93	0.07	2	190.39	0.09	2	190.39	0.09	2

Labour	5364.08	2.24	62	5883.23	2.76	60	5652.37	2.66	59
Scientist	2141.29	0.89	25	1982.67	0.93	20	1982.67	0.93	21
Ayudante	1298.00	0.54	15	1025.72	0.48	10	1025.72	0.48	11
Ingienero	927.41	0.39	11	1030.46	0.48	10	1030.46	0.48	11
Tractor driver	58.22	0.02	1	49.27	0.02	0	60.22	0.03	1
Permanent labourer	417.22	0.17	5	570.78	0.27	6	558.18	0.26	6
Temporary field labourer	521.95	0.22	6	1217.20	0.57	12	987.99	0.46	10
Other field labourer	0.00	0.00	0	7.13	0.00	0	7.13	0.00	0
Travel costs	0.00	0.00	0	418.27	0.20	4	418.27	0.20	4
Extra variable inputs[6]	47.17	0.02	1	169.92	0.08	2	160.74	0.08	2
Irrigation permit[5]	0.00	0.00	0	8.51	0.00	0	8.51	0.00	0

[1] Assumptions: 1 ha of maize is planted. One-half of the plants are fertilized. One-quarter of the ears are harvested. Flowering date is measured but no other special scoring procedures are performed. One package of bulk seed is prepared per row for the following season. The separation of QPM seed using the light box is not performed.

[2] Cycle A is the dry-season cycle at Tlaltizapán and requires more irrigation.

[3] Cycle B is the rainy-season cycle at Tlaltizapán and requires fewer irrigations but additional chemical treatments.

[4] At El Batán, irrigation water is pumped using an electric pump.

[5] At Tlaltizapán, irrigation water is free, but a permit for use is required; diesel fuel is used to pump the water around the station.

[6] Experts gave opinions on the amount of excess variable materials consumed during a cycle. The coefficient for the extra fuel costs at Tlaltizapán is high (25%) because some fields are physically separated from the main station, so workers, materials and equipment must be transported to and from the 'Annex'.

[7] Although Tlaltizapán does have the Annex, which has slightly different chemical treatment requirements, these budgets assume that all of the plants were sown at the main station.

[8] El Batán has 2394 rows and 50,274 plants per hectare. Tlaltizapán has 2128 rows and 44,688 plants per hectare.

Table 12.2. Costs (US$) of partial MAS and full MAS procedures, ABC laboratory.

	Molecular markers used for target allele analysis[1]			Molecular markers used for background genome analysis[2]		
Number of samples	210			53		
Markers per sample	1			55		
Total cost	$448.16			$3642.58		
Total cost per sample	$2.13			$68.73		
Total cost per data point	$2.13			$1.25		
	Cost per analysis ($)	Cost per sample ($)	Per cent of total	Cost per analysis ($)	Cost per sample ($)	Per cent of total
Total cost	448.16	2.13	100	3642.58	68.73	100
Capital	23.18	0.11	5	188.41	3.55	5
Overheads	38.63	0.18	9	314.02	5.92	9
Variable inputs						
Reagents	158.02	0.75	35	1936.02	36.53	53
Taq polymerase	114.66	0.55	26	1591.59	30.03	44
Other reagents	43.36	0.21	10	344.43	6.50	9
Supplies	44.34	0.21	10	101.01	1.91	3
Pipette tips	8.62	0.04	2	33.21	0.63	1
Eppendorf tubes	4.39	0.02	1	0.15	0.00	0
PCR plates	17.19	0.08	4	28.75	0.54	1
Field supplies	6.94	0.03	2	0.00	0.00	0
Other supplies	7.20	0.03	2	38.90	0.73	1
Labour	151.25	0.72	34	938.15	17.70	26
Scientist	58.70	0.28	13	814.78	15.37	22
Laboratory technician	79.19	0.38	18	123.37	2.33	3
Field labourer	9.97	0.05	2	0.00	0.00	0
Field supervisor	3.39	0.02	1	0.00	0.00	0
Travel costs	16.36	0.08	4	0.00	0.00	0
Extra variable inputs	16.38	0.08	4	164.98	3.11	5

[1] The molecular marker analysis assumes that the DNA is quantified using a spectrophotometer, that all of the samples are separated using 110 ml agarose gels (2% Metaphor, 1% Seakem), and that one trip is made to Tlaltizapán to harvest leaf samples.
[2] The background molecular marker analysis assumes that the leaf samples have been harvested and the DNA has already been extracted for target allele analysis. Consequently, the cost estimation begins with the PCR amplification. It is presumed that 280 ml agarose gels are used for the DNA separation.

costs of the chemicals used in the ELISA.[5] On the other hand, while the ELISA procedure relies primarily on disposable plastic supplies, the nitrogen and tryptophan determination is performed almost entirely using reusable glassware, which explains the discrepancy in the absolute cost of supplies between the two analyses.

The major cost component in both analyses is labour, which accounts for 60% of the total cost of the nitrogen and tryptophan determination, and 49% of the total cost of the ELISA procedure. Supplies are the second most important cost component for the ELISA procedure, but they are a negligible component in the case of nitrogen and tryptophan determination. Reagents are the second most important component of the nitrogen and tryptophan determination.

[5] The reagent cost is low in part because CIMMYT is able to produce a primary antibody at very low cost. Commercial preparations of the antibody can be quite expensive, however, so reagent costs for the ELISA procedure would rise if the antibody were to be outsourced.

Table 12.3. Costs (US$) of kernel analysis to determine protein quality, SPAL laboratory.

	Nitrogen and tryptophan determination			ELISA[1] protein content analysis		
Number of samples	105			105		
	Cost per analysis ($)	Cost per sample ($)	Per cent of total	Cost per analysis ($)	Cost per sample ($)	Per cent of total
Total cost	382.29	3.64	100	275.39	2.62	100
Capital	19.77	0.19	5	14.24	0.14	5
Overhead	32.96	0.31	9	23.74	0.23	9
Variable inputs						
Reagents	95.20	0.91	25	38.52	0.37	14
Supplies	1.83	0.02	0	61.59	0.59	22
Pipette tips				17.77	0.17	6
Eppendorf tubes				15.19	0.14	6
PCR plates				24.15	0.23	9
Other supplies	1.83	0.02	0	4.48	0.04	2
Labour	228.66	2.18	60	134.42	1.28	49
Technician	214.05	2.04	56	127.32	1.21	46
SPAL supervisor	14.61	0.14	4	7.09	0.07	3
Extra variable inputs	3.88	0.04	1	2.89	0.03	1

[1] The ELISA analysis assumes that nine samples are analysed per plate.

Economies of scale

Consistent with actual practice at CIMMYT, all of the unit cost estimates assume that field operations are performed in an efficient manner; consequently, costs per row tend not to vary as the number of rows devoted to a particular breeding project changes. In contrast, the unit cost estimates sometimes assume that specialized laboratory procedures required for QPM line conversion are performed on an 'as-needed' basis, so labour and supply inputs are occasionally considered to be lumpy. Consequently, unit costs for laboratory analyses increase when fewer samples are analysed (Fig. 12.3).

Because the unit cost of some laboratory procedures varies with sample size, in comparing the relative costs of phenotypic and molecular marker analyses, it is necessary to take into account the sample size. The cost of the nitrogen and tryptophan analysis is always higher than that of the ELISA analysis, regardless of the number of samples being analysed. When fewer than 40 samples are analysed, the cost of the ABC partial molecular marker procedure is higher than the costs of both SPAL analyses. When more than 40 samples are analysed, the cost of the ABC partial molecular marker procedure falls below the cost of the nitrogen and tryptophan analysis, and when more than 60 samples are analysed, it falls below the cost of the ELISA analysis as well. Economies of scale considerations thus make molecular marker analysis relatively more attractive compared to phenotypic evaluation (analysis of grain amino acid content) when large numbers of samples are analysed.

Stage 2: Representative breeding schemes

For some breeding projects that could be carried out using either conventional breeding methods or MAS, it would be possible to use the same basic breeding scheme; exactly the same procedures would be performed in the field, but in one case desirable plants would be identified

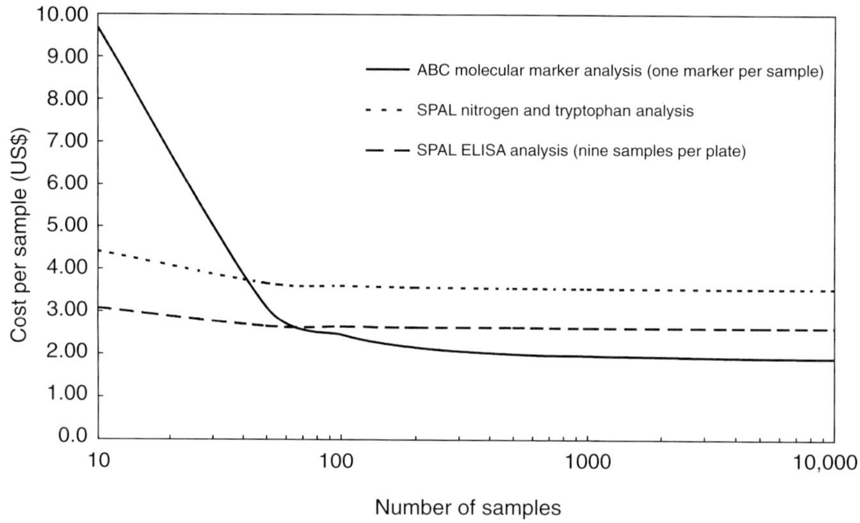

Fig. 12.3. Unit cost of laboratory analyses as a function of sample size.

using phenotypic evaluation, and in the other case they would be identified using molecular markers. In such cases, since the only difference between the two schemes would be the evaluation method, the cost-effectiveness of conventional breeding relative to MAS could be determined simply by comparing the unit cost of phenotypic evaluation to the unit cost of molecular marker analysis (taking into account possible economies of scale effects).

For other breeding projects, however, including QPM line conversion, it is unlikely that a breeder using conventional methods would choose exactly the same breeding scheme as a breeder using MAS. Instead, each breeder would design the breeding scheme taking into account the particular characteristics of the tools available (e.g. cost of laboratory analysis procedures, timing in the crop cycle when the procedure can be performed, nature of the information provided by the procedure). In such cases, a conventional breeding scheme is likely to differ from a MAS scheme not only in terms of the evaluation methods, but also in terms of the procedures performed in the field (numbers of cycles, sizes of plant populations, numbers and types of crosses made, etc.). Consequently, in order to evaluate the cost-effectiveness of conventional breeding relative to MAS, it would be necessary not only to compare the unit cost of phenotypic evaluation to the unit cost of molecular marker analysis, but also to cost out the entire breeding scheme, including all field operations and laboratory analyses.

CIMMYT maize breeders and molecular geneticists were asked to design representative breeding schemes for QPM line conversion. Four hypothetical stylized breeding schemes were selected for this case study. Two of the schemes rely solely on conventional breeding methods and phenotypic evaluation, and two incorporate MAS. As noted earlier, while the schemes were designed to be as comparable as possible, they differ in certain respects because conventional breeding methods and MAS techniques are not always exact substitutes. It is important to remember also that these schemes do not exactly reflect breeding schemes currently being used at CIMMYT.

The first breeding scheme, Conventional 1 (Accelerated), depicts a fast and inexpensive way to transfer the mutant recessive *opaque2* allele using a back-cross line conversion approach (Fig. 12.4). After the initial cross has been made between the donor and the recurrent recipient parent,

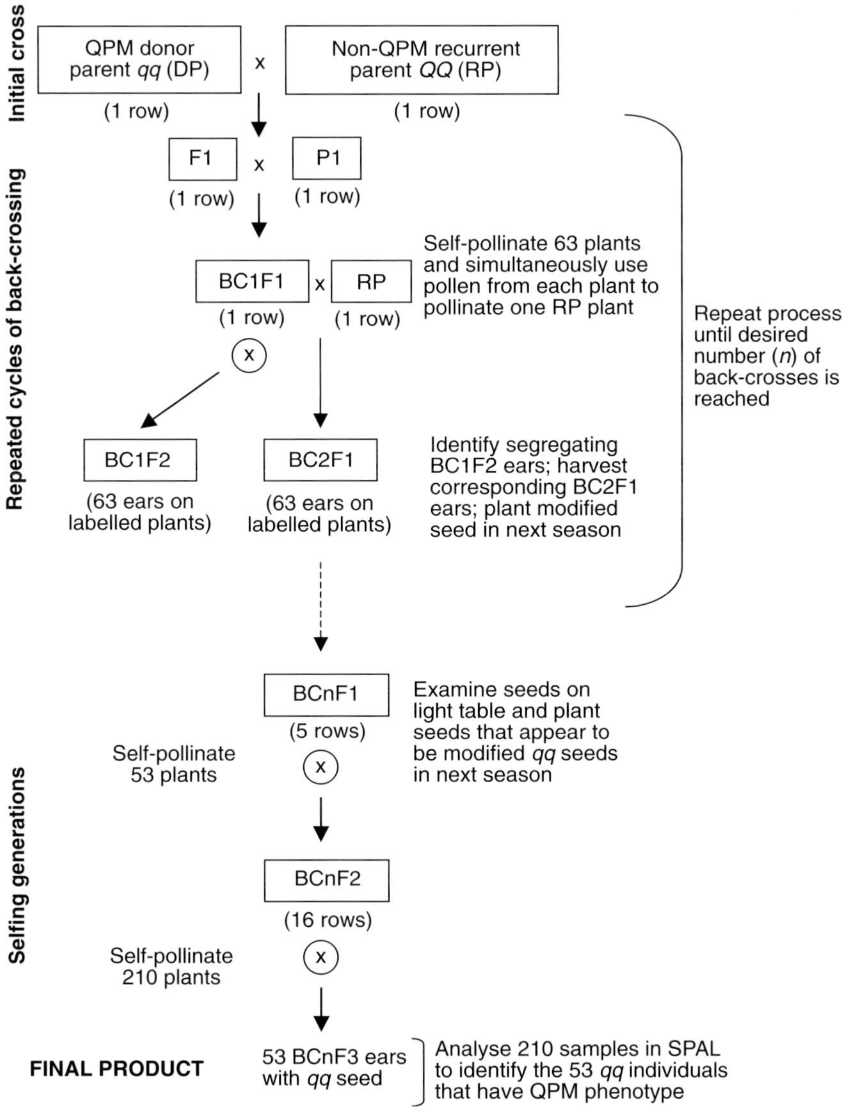

Fig. 12.4. Conventional 1 (Accelerated) back-cross breeding scheme.

back-crossing is initiated. During each cycle of back-crossing, every BCnF1 plant is self-pollinated, and pollen from the same plant is also used to pollinate one ear on a uniquely numbered recurrent parent plant. At harvest, the breeder examines the BCnF2 ears on the self-fertilized plants and visually identifies ears that have segregating seeds (some of the seeds will have soft endosperm, and others will have hard endosperm). If the seed on the BCnF2 ear is segregating, the plant must be a carrier of the recessive *opaque2* allele. This information is then used to guide the selection of desirable recurrent parent plants that were pollinated by BCnF1 plants carrying the mutant *opaque2* allele. Using this approach, a back-cross cycle can be completed in one season. In the last season, samples are sent to the SPAL for final confirmation that the selected plants exhibit elevated levels of nitrogen and tryptophan.

It is important to note that Conventional 1 will work only if the breeder can visually identify ears with segregating seeds among the BCnF2 plants. At CIMMYT, breeders have found that in some cases it is possible to identify ears with segregating seeds, but in other cases, it is not. When ears with segregating seeds cannot be identified visually in the field, some other scheme will be required. A good example of an alternative scheme is Conventional 2 (Classic), which resembles the type of scheme that would be recommended in most plant breeding textbooks for transferring a single recessive allele via a back-cross line conversion approach (Fig. 12.5). Conventional 2 is relatively slow, because in each back-cross generation the plants are self-fertilized, and seed is then grown out to observe the QPM phenotype in the next generation. Kernel

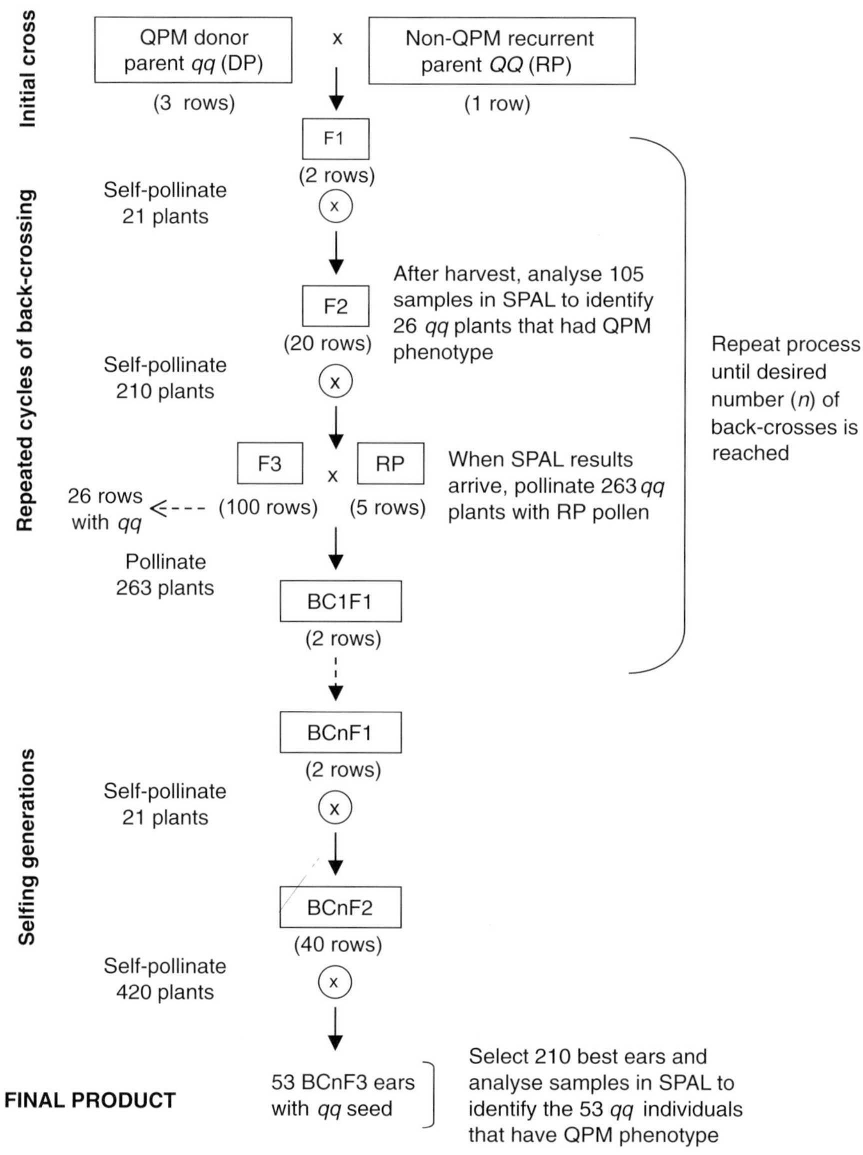

Fig. 12.5. Conventional 2 (Classic) back-cross breeding scheme.

samples are analysed in the SPAL during every cycle of back-crossing, and SPAL phenotypic evaluation is performed in the final season.

The third breeding scheme, MAS 1 (Partial) involves molecular marker analysis only of the target allele (Fig. 12.6). In the first season, the two parental lines are analysed using the three *opaque2* markers available at CIMMYT to determine which of the three markers is most appropriate for the lines being used. Subsequently, in every cycle of back-crossing, each plant is analysed using the one marker; heterozygous individuals identified with the help of the marker are pollinated using the recurrent recipient parent and passed on to the next cycle of back-crossing. In the last season, all of the plants are analysed using the marker to identify plants that are homozy-

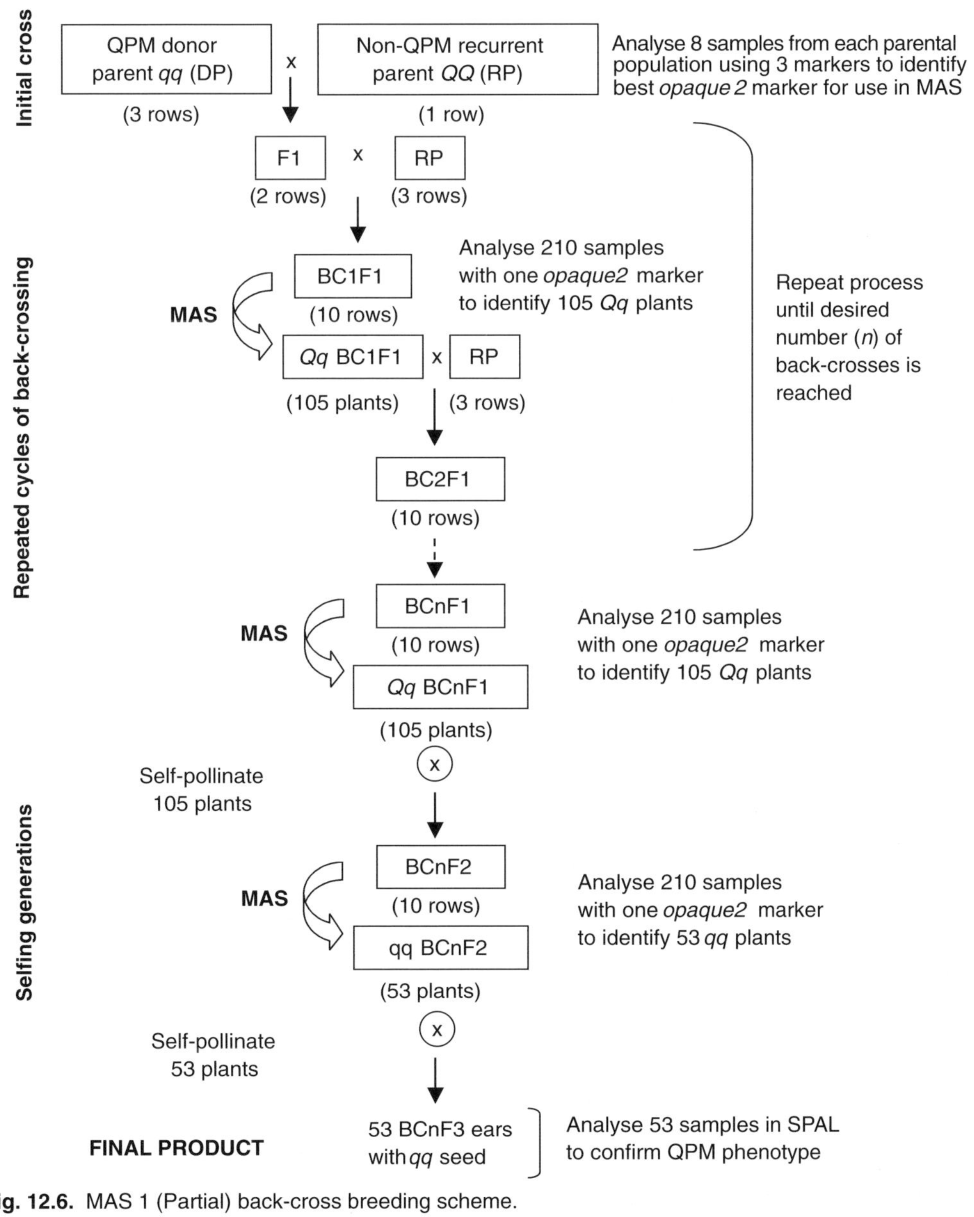

Fig. 12.6. MAS 1 (Partial) back-cross breeding scheme.

gous for the mutant *opaque2* allele. Kernel samples from the homozygous mutant plants are sent to the SPAL for phenotypic evaluation to ensure that the marker analysis has reliably identified plants that possess elevated levels of lysine and tryptophan.

The fourth breeding scheme, MAS 2 (Full), is similar to MAS 1, but it also includes full molecular marker analysis using background markers (Fig. 12.7). In the first season, the two parental lines are analysed using both the three *opaque2* markers (used for tracking the target allele) and a large number of additional markers that correspond to sequences of DNA scattered throughout the genome. Information

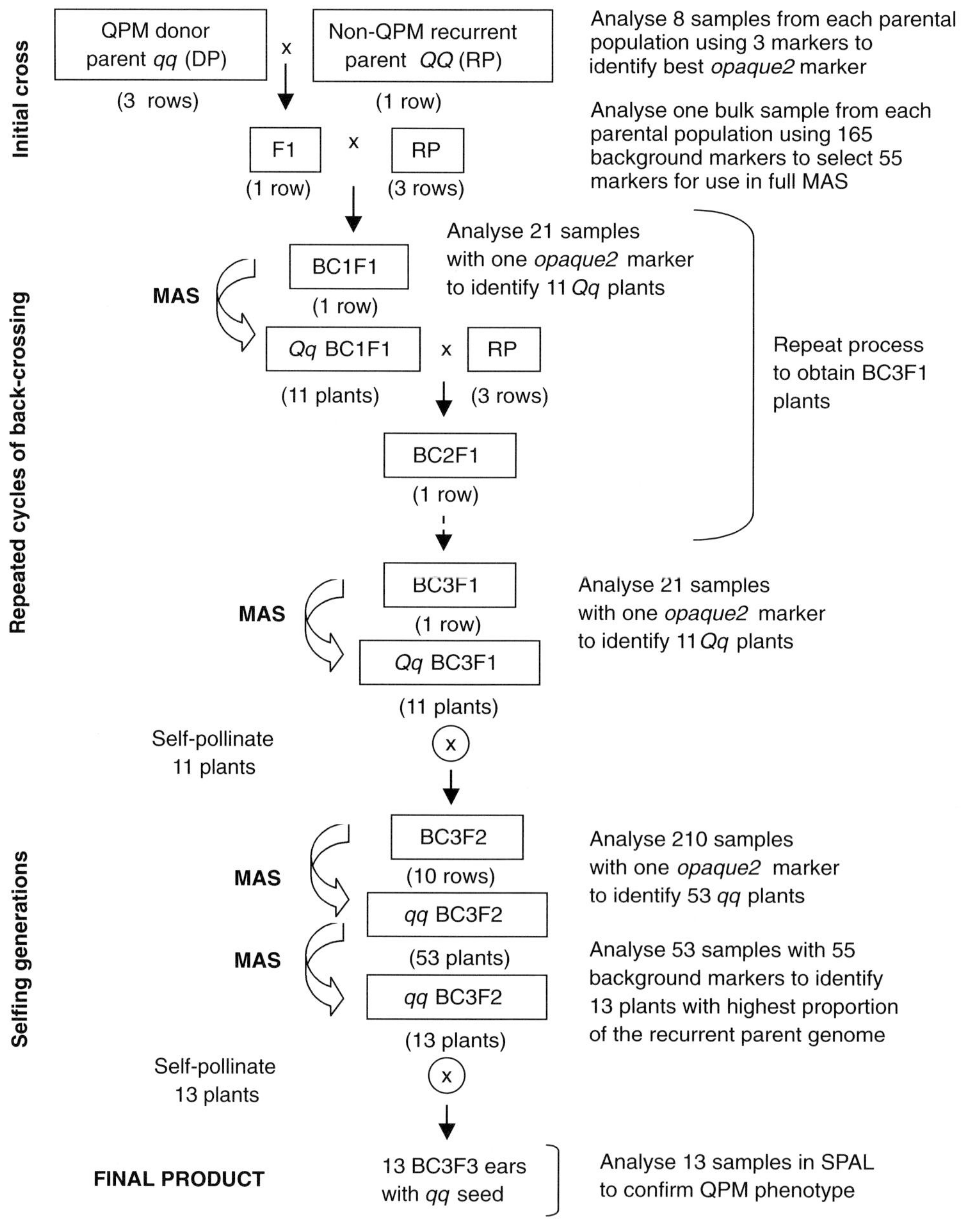

Fig. 12.7. MAS 2 (Full) back-cross breeding scheme.

from the background markers is used to develop a genetic 'fingerprint' of each parent. As in MAS 1, a series of back-crosses is then initiated; in each cycle of back-crossing, the *opaque2* marker is used to select heterozygous plants to cross to the recurrent parent. In the last season, the *opaque2* marker is again used to identify QPM plants. In addition, the background markers are used to fingerprint the plants that have the QPM phenotype; based on these fingerprints, it is possible to select individual QPM plants that at the DNA level most closely resemble the recurrent parent. Kernel samples from these plants are sent to the SPAL for a confirmatory phenotypic evaluation of lysine and tryptophan levels.

In the initial baseline scenario, it was assumed that the final product of all four schemes would be approximately 50 ears of maize fixed for the mutant recessive *opaque2* allele and containing elevated levels of nitrogen and tryptophan.[6] On average, these ears would be 93.75% genetically identical to the recurrent parent. Theoretically, this requires four cycles of back-crossing for the two conventional breeding schemes and for the partial MAS scheme. Under the full MAS scheme, plants that are approximately 94% identical to the recurrent parent can be identified after only three cycles of back-crossing. In addition, under the full MAS scheme it is possible to identify the individual plant(s) that contain the highest proportion of the recurrent recipient parent genome.

Stage 3: Costs of representative breeding schemes

Baseline scenario

The total cost of each of the four representative breeding schemes was calculated using unit cost data generated by the spreadsheet-based budgets.

Summary data for the four representative breeding schemes appear in Table 12.4. Conventional 1 is cheap (total cost: $975) and fast (total time requirement: seven planting seasons), but as was pointed out in the previous section, it may not always work for QPM line conversion. Conventional 2 is considerably more expensive (total cost: $4367) and much more time-consuming (total time requirement: 15 planting seasons), but it is extremely reliable. MAS 1 falls in between the two conventional schemes in terms of expense (total cost: $2761), but it is relatively fast (total time requirement: seven planting seasons). MAS 2 is the most expensive scheme of all (total cost: $5084), but it is also the fastest (total time requirement: six planting seasons).

The breeding schemes differ significantly in terms of the relative importance of field vs. laboratory costs. Conventional 1 is characterized by a modest level of field costs and very low SPAL costs, reflecting its overall efficiency and relative quickness. Conventional 2 is characterized by high field costs and high SPAL costs, reflecting its relative inefficiency (in terms of plant populations) and its long duration, as well as the large numbers of samples that must be subjected to phenotypic evaluation. MAS 1 is characterized by slightly higher field costs, but it has a high level of ABC laboratory costs. MAS 2 features extremely low field costs, but the ABC laboratory costs are extremely high, reflecting the considerable expense of the background marker analysis.

Labour and reagents are the most significant cost categories in all four schemes. However, while labour accounts for close to 60% of the total costs in both conventional schemes, it comprises only 39% of the total costs in MAS 1 and only 29% in MAS 2. Reagent costs are more important in the two marker-assisted schemes. Over one-half (52%) of the total cost of MAS 2 consists of reagent costs, mainly the cost of *Taq* polymerase.

6 Some of the parameters originally suggested by the scientists were adjusted to make the outputs similar in terms of quality and quantity.

Table 12.4. Summary of technical parameters and total costs, conventional and MAS breeding schemes.

	Conventional breeding schemes		Marker-assisted breeding schemes	
	Conventional 1 (Accelerated)	Conventional 2 (Classic)	MAS 1 (Partial)	MAS 2 (Full)
Number of back-crosses performed	4	4	4	3
Average percentage of recovered recurrent recipient parent genotype	93.75%	93.75%	93.75%	94%
opaque2 genotype	qq	qq	qq	qq
Total cost (US$)	975	4367	2761	5084
Total number of seasons required	7	15	7	6
Total number of years required (two seasons per year)	3.5	7.5	3.5	3
Number of BC3F3 *qq* ears harvested in final cycle	53	53	53	53[1]
Total field cost (US$)	224	2087	266	103
(Field cost as a percentage of total cost)	23%	48%	10%	2%
Total ABC cost (US$)	–	–	2295	4927
(ABC cost as a percentage of total cost)	–	–	83%	97%
Total SPAL cost (US$)	752	2281	199	54
(SPAL cost as a percentage of total cost)	77%	52%	7%	1%
Total cost (US$)	975	4367	2761	5084
of which:				
Capital	4.8%	4.4%	5.0%	5.1%
Overhead	8.7%	8.7%	8.6%	8.6%
Labour	59.7%	62.2%	39.4%	28.6%
Chemical supplies	0.5%	1.4%	0.3%	0.1%
Fuel	0.5%	1.2%	0.2%	0.1%
Supplies	5.8%	7.6%	3.0%	0.4%
Irrigation	0.2%	0.4%	0.1%	0.0%
Travel costs	0.1%	0.8%	0.7%	0.4%
Reagents	19.7%	13.5%	34.4%	51.7%
Laboratory supplies	0.0%	0.0%	8.3%	5.1%

[1] Fifty-three (53) ears are pollinated and available for selection, but due to the precision screening that can be performed using background markers, it is possible to identify the 13 plants that most closely resemble the recurrent recipient parent.

Sensitivity analysis

NUMBER OF CYCLES OF BACK-CROSSING. In the baseline scenario, all four representative breeding schemes generate QPM lines that contain approximately 94% of the same germplasm as the recurrent recipient parent. Except for the elevated levels of lysine and tryptophan found in the grain, these (fully) converted lines would be virtually indistinguishable from the original recurrent recipient parent. Not all breeding programmes would attempt to recover such a high proportion of the original recurrent recipient parent genome. For example, CIMMYT breeders normally perform fewer cycles of back-crossing, with the objective of producing (partially) converted QPM lines containing a lower percentage of the recipient parent genome but incorporating desirable traits from the donor parent. The cost-effectiveness of using MAS for partial line conversion was explored by calculating the costs of the four representative breeding schemes under two alternative scenarios assuming two and three cycles of back-crossing (Table 12.5). Although the cost of the MAS schemes falls relatively faster than the costs of the conventional schemes as the number of back-cross cycles is reduced, the ranking of the four schemes does not change.

NITROGEN AND TRYPTOPHAN DETERMINATION VS. ELISA. In the baseline scenario, it was assumed that phenotypic evaluation of grain protein quality is performed using nitrogen and tryptophan determination. The cost savings that could potentially be realized through the use of the recently introduced ELISA procedure were explored under an alternative scenario in which the ELISA procedure was substituted for nitrogen and tryptophan determination. At current levels of laboratory efficiency (analysis of nine samples per ELISA plate), the total cost of Conventional 1 drops by 21%, and that of Conventional 2 by 15% (Table 12.6). These cost reductions would increase significantly if the SPAL were able to achieve higher levels of efficiency (analysis of 21 samples per plate); total costs would fall by 41% for Conventional 1 and by 28% for Conventional 2 (Table 12.6). For the two MAS schemes, since SPAL analyses account for a small proportion of the total costs, the cost savings realized through introduction of the ELISA procedure would be minor. The ranking of the four breeding schemes in terms of total cost is not affected by introduction of the ELISA procedure.

VISUAL IDENTIFICATION OF QPM SEEDS. Breeders involved in QPM line conversion are inter-

Table 12.5. Cost (US$) of nitrogen and tryptophan determination for two-, three- and four-cycle back-crossing schemes (number of years needed to complete each scheme).

Average percentage of recurrent parent genome recovered	Conventional breeding schemes		Marker-assisted breeding schemes	
	Conventional 1	Conventional 2	MAS 1	MAS 2
75.00	$909 (2.5)	$2698 (4.5)	$1805 (2.5)	
87.50	$942 (3)	$3532 (6)	$2238 (3)	
93.75[1]	$975 (3.5)	$4367 (7.5)	$2761 (3.5)	$5084 (3)

[1] In MAS 2, scientists recover a population of plants that match the recurrent parent over approximately 94% of the genome, and they can identify individual plants with even higher percentages of similarity to the recurrent parent.

Table 12.6. Cost (US$) of ELISA analysis for two, three and four-cycle back-crossing schemes (years required to complete each scheme).

Average percentage of recurrent parent genome recovered	Conventional breeding schemes		Marker-assisted breeding schemes	
	Conventional 1	Conventional 2	MAS 1	MAS 2
Case 1: Assuming nine samples analysed per plate				
75.00	$708 (2.5)	$2282 (4.5)	$1745 (2.5)	
87.50	$741 (3)	$3009 (6)	$2223 (3)	
93.75[1]	$774 (3.5)	$3737 (7.5)	$2700 (3.5)	$5067 (3)
Case 2: Assuming 21 samples analysed per plate				
75.00	$525 (2.5)	$1916 (4.5)	$1702 (2.5)	
87.50	$558 (3)	$2553 (6)	$2179 (3)	
93.75[1]	$591 (3.5)	$3189 (7.5)	$2657 (3.5)	$5056 (3)

[1] In MAS 2, scientists recover a population of plants that match the recurrent parent over approximately 94% of the genome, and they can identify individual plants with even higher percentages of similarity to the recurrent parent.

ested in identifying materials that possess two copies of the desirable mutant *opaque2* allele, as well as the modifiers that increase the opaqueness and endosperm hardness of the kernel. With certain lines of maize, experienced breeders are able to identify modified seeds that contain two copies of the *opaque2* recessive allele by placing them on a light table and carefully observing their appearance. Although this method does not always work (because sometimes it is impossible to distinguish between homozygous *opaque2* seeds that have the appropriate modifiers, and heterozygous or homozygous normal seeds that have a typical, non-opaque phenotype), when it does work breeders can plant only those seeds that are known to be fixed for the mutant *opaque2* allele, thus reducing the size of populations in subsequent generations. Visual selection, when possible, can significantly increase the efficiency of conventional breeding schemes, because it can greatly reduce field costs and laboratory phenotypic evaluation costs (Table 12.7).

CHANGES IN COST STRUCTURE. The budgets for the four representative breeding schemes are based on laboratory and field protocols currently used at CIMMYT and on input prices currently prevailing in Mexico. It is unlikely that these will remain static. Protocols continue to evolve (particularly in the ABC laboratory, since many MAS procedures are still relatively new at CIMMYT, and laboratory personnel are continually discovering more efficient ways to work), and prices change constantly. Generally speaking, however, minor adjustments to research protocols and modest changes in input prices are unlikely to affect the relative cost-effectiveness of the four representative breeding schemes (Table 12.8). Break-even analysis shows that field costs would have to fall by 89% for Conventional 2 to become cheaper than MAS

Table 12.7. Cost savings (US$) achieved under the Conventional 1 breeding scheme when modified homozygous mutant *opaque2* seeds can be selected visually.

Average percentage of recurrent parent genome recovered	Laboratory analysis (baseline scenario)	Visual selection (alternative scenario)
75.00	$909	$295
87.50	$942	$328
93.75	$975	$361

Table 12.8. Changes in field and laboratory costs (US$) required to alter current cost rankings (ranking of each scheme)[1].

	Conventional breeding schemes		Marker-assisted breeding schemes	
	Conventional 1	Conventional 2	MAS 1	MAS 2
Baseline scenario	$975.28 (1)	$4367.28 (3)	$2760.79 (2)	$5083.70 (4)
Field costs fall 89%	$776.20 (1)	$2510.29 (2)	$2523.77 (3)	$4991.95 (4)
Field costs rise 37%	$1058.05 (1)	$5139.29 (4)	$2859.33 (2)	$5121.84 (3)
ABC costs fall 15%	$975.28 (1)	$4367.28 (4)	$2416.49 (2)	$4344.63 (3)
ABC costs fall 78%	$975.28 (2)	$4367.28 (4)	$970.45 (1)	$1240.55 (3)
ABC costs rise 70%	$975.28 (1)	$4367.28 (2)	$4367.51 (3)	$8532.68 (4)
SPAL costs fall 78%	$389.04 (1)	$2588.28 (2)	$2605.45 (3)	$5041.97 (4)
SPAL costs rise 33%	$1223.30 (1)	$5119.93 (4)	$2826.51 (2)	$5101.36 (3)

[1] Conventional 1, Conventional 2 and MAS 1 involve four cycles of back-crossing; MAS 2 involves three cycles of back-crossing.

1, and they would have to rise by 37% for MAS 2 to become cheaper than Conventional 2. The one area where a change in the current cost structure could make a difference concerns laboratory costs. Laboratory costs would have to fall by only 15% for MAS 2 to become less costly than Conventional 2. This is potentially important, because a single reagent, *Taq* polymerase, accounts for 44% of the cost of background marker selection in MAS 2. Thus a significant drop in the price of *Taq* polymerase could conceivably affect the ranking of the four representative breeding schemes.

Discussion

This chapter has presented summary results of a case study carried out at CIMMYT to determine the cost-effectiveness of using SSR-based MAS for a particular breeding application (QPM line conversion) and to provide insights into the likely cost-effectiveness of additional applications of MAS whose costs and benefits are not yet known. What conclusions can be drawn from these results?

QPM case study results

One basic finding of the study is clear: currently at CIMMYT, for QPM line conversion projects the relative cost- effectiveness of conventional breeding methods as compared to MAS differs depending on the circumstances. Of the four representative QPM line conversion schemes analysed, Conventional 1 generates the desired output at the lowest cost. By implication, whenever Conventional 1 is feasible, conventional breeding methods will be most cost-effective, and use of markers will not reduce research expenditures. Unfortunately, Conventional 1 does not always work. In some cases, breeders are unable visually to identify ears with segregating seeds, making Conventional 1 unfeasible. Whenever this happens, they must resort to alternative means of tracking the mutant recessive *opaque2* allele, such as those used in Conventional 2, which is considerably more expensive and much more time-consuming than MAS 1. In this case, use of molecular markers will yield significant cost savings. In summary, for QPM line conversion projects at CIMMYT, the relative attractiveness of conventional breeding methods compared to MAS must be evaluated on a case-by-case basis.

Before any attempt is made to extrapolate from this finding to more broad conclusions about the usefulness of MAS in general, it is important to remember that the case study results are limited for at least two reasons.

First, the QPM case study focused on one particular application of MAS among many potential applications. This is important, because QPM line conversion in maize has a number of distinctive characteristics that influence the relative cost-effectiveness of conventional breeding methods and MAS:

- maize is an annual, cross-pollinating crop with one diploid genome;
- elevated levels of lysine and tryptophan are controlled largely by a single gene;
- the mutant *opaque2* allele is recessive;
- the QPM trait is expressed in grain kernels;
- expression of the QPM trait is influenced by modifier alleles;
- experienced breeders can sometimes visually identify homozygous mutant *opaque2* seed even in the presence of masking modifier alleles;
- phenotypic evaluation can be done in the laboratory with relative ease;
- comprehensive molecular marker maps for maize are publicly available, as are corresponding primer sequences; and
- primer sequences for PCR amplification of SSR regions within the *opaque2* gene are publicly available.

To the extent that other breeding applications involve traits that differ in one or more of these characteristics (which they nearly always will), the cost-effectiveness of MAS is likely to vary.

Second, the budgets developed for this case study were based on the CIMMYT maize breeding programme in Mexico. The technical parameters used to cost out field operations and laboratory procedures reflect standard operating procedures at CIMMYT, and the cost data used to estimate the budgets reflect input prices currently prevailing in Mexico. This, too, is important, because the cost structure of CIMMYT's Mexican maize breeding programme is influenced by a number of factors:

- CIMMYT's conventional breeding programme is large and well established; efficient field operation protocols and consistent breeding methodologies have evolved over time;
- CIMMYT's biotechnology facilities are modest by industry standards, so it is not always possible to perform procedures at the most cost-efficient scale;

none the less, cost-savings are being realized as higher-throughput capacity is established;
- marker-assisted breeding is a relatively new technology; laboratory procedures continue to improve as new techniques emerge;
- land costs incurred by CIMMYT for its research stations are minimal;
- labour costs in Mexico are low by international standards (especially for field labourers and for field and laboratory technicians); and
- costs of laboratory supplies and reagents in Mexico are relatively high.

To the extent that other case studies are carried out at facilities where conventional breeding practices and/or laboratory procedures differ significantly from those found at CIMMYT, and to the extent that they are carried out in locations where costs of key inputs (especially land, labour and laboratory supplies) differ significantly from costs in Mexico, the cost-effectiveness of MAS is likely to vary.

Implications for MAS

The fact that neither conventional breeding methods nor MAS offer an unequivocal cost advantage under all circumstances for QPM line conversion projects at CIMMYT should come as no great surprise. CIMMYT's experience closely mirrors the experience of many other breeding programmes (Young, 1999). Even though MAS has come to play a prominent role in the field of plant breeding, for many practical applications the economics of MAS are still being worked out on a case-by-case basis.

The continuing uncertainty concerning the cost-effectiveness of MAS in specific applications should not give rise to undue pessimism, however. Everything that made MAS attractive in the first place still holds true; the key to successfully integrating the technology into applied breeding programmes will lie in identifying applications in which molecular markers offer real advantages over conventional breeding methods.

Based on the characteristics of the two technologies, and taking into account insights derived from the QPM case study, it is possible to identify a number of areas in which MAS should offer significant advantages over conventional breeding methods. These are described below.

Expensive phenotypic screening

Molecular markers offer potential savings compared to the cost of conventional breeding when phenotypic screening is expensive. As this study has shown, QPM phenotypic screening can be relatively cheap; in some materials, kernels containing two copies of the mutant *opaque2* allele along with the desirable modifiers can be identified visually on a light table or even on the ear. In other materials, however, visual screening for the QPM phenotypic is not possible, and it is necessary to send samples for laboratory analysis, which is more expensive than using markers. More generally, as the cost of phenotypic screening rises, markers are more likely to represent a cost-effective alternative.

Multiple genes, one trait

Molecular markers offer potential savings compared to the cost of conventional breeding when they allow breeders to identify the presence of multiple alleles related to a single trait when the alleles do not exert an individually detectable influence on the expression of the trait (Melchinger, 1990). Phenotypic screening may be inadequate in such cases, if the breeder seeks to ensure that all of the complementary desirable alleles are present in a plant. The ability to identify the incorporation of unobservable alleles is important, especially in breeding for resistance to diseases and pests, because multi-genic resistance achieved through so-called 'gene pyramiding' is much more desirable than mono-genic resistance (which leaves plants vulnerable when the disease or pest evolves resistance to the single gene). For example, three alleles may separately impart resistance to a foliar disease. A plant in the field will demonstrate the same resistance whether it contains one, two or three of the alleles, confounding phenotypic selection for multi-genic resistance, but a breeder armed with molecular markers can easily identify the individual plants possessing all three resistance alleles and thus advance materials that will have more durable multi-genic resistance.

Multiple gene, multiple traits

Molecular markers offer potential savings compared to the cost of conventional breeding when there is a need to select for multiple traits simultaneously. For example, if the goal of a breeding programme is to improve a population for drought tolerance, insect resistance and disease resistance, often it will be necessary to conduct separate trials to screen for each trait. (Even if the stresses can be applied at the same time, interactive effects on the plant may make it impossible to determine what is happening at the genetic level.) By using molecular markers, breeders can screen simultaneously for the presence of the alleles linked to all three traits.

Seasonal considerations

Molecular markers offer potential savings compared to the cost of conventional breeding when it is necessary to screen for traits whose expression depends on seasonal considerations. For example, if the goal of a breeding programme is to improve a population for heat tolerance, and if screening can take place only during the hottest of two or more annual growing seasons, then the rate of breeding progress will be slowed. By using molecular markers, breeders can screen for the presence or absence of the allele or alleles associated with heat tolerance during every growing season, regardless of the actual temperatures.

Geographical considerations

Molecular markers offer potential savings compared to the cost of conventional breeding when it is necessary to screen for traits whose expression depends on geographical considerations. For example, if the goal of the CIMMYT breeding programme is to improve a population for tolerance to *Striga* spp., a parasitic pest of maize in Africa, then it will not be possible to screen for resistance in Mexico. By using molecular markers, breeders in Mexico can screen for the presence or absence of the allele or alleles associated with *Striga* resistance.

Early detection

Molecular markers offer potential savings compared to the cost of conventional breeding when they allow alleles for desirable traits to be detected early, well before the trait is expressed and can be detected phenotypically. In the case of QPM, markers can be used to detect the presence of the mutant *opaque2* allele when plants are still in the seedling stage, long before the QPM trait is expressed in the form of elevated levels of lysine and tryptophan in the kernel. Early detection of the presence of an allele can reduce field costs by allowing plants that do not contain the allele to be discarded and/or by potentially reducing population size requirements in subsequent generations. This benefit can be particularly important in species that grow slowly. In tree crops, for example, if markers can be linked to fruit quality traits, breeders may be able to select desirable individuals without having to wait several years until the trees produce fruit.

Recessive genes

Molecular markers offer potential savings compared to the cost of conventional breeding when they allow breeders to identify heterozygous plants that carry an allele of interest whose presence cannot be detected phenotypically. In many cases involving recessive traits, it is not possible to identify heterozygous plants, so extra cycles of selfing may have to be included in breeding schemes to produce phenotypically identifiable homozygous recessive plants in order to monitor the inheritance of the recessive allele. As in the case of QPM Conventional 2, this complication can significantly retard the process of line conversion. By using molecular markers, breeders can identify heterozygous plants and use them to pass the recessive allele of interest into subsequent generations more quickly (as in MAS 1).

Indirect benefits of MAS: time savings

Throughout this chapter, the focus has been on the potential cost savings associated with the use of MAS relative to conventional breeding methods. What has not been mentioned explicitly are time savings. The question of time savings is extremely important, because over the long run the greatest benefit offered by MAS almost certainly will be its ability to reduce the time required for plant breeding.

As discussed earlier, by using background markers to perform full MAS, breeders can accelerate a back-cross scheme by definitively identifying plants that contain higher percentages of the recurrent recipient parent genome than would be possible using conventional phenotypic selection methods based on visual observation of plants in the field or laboratory analysis of tissue samples. Given this greater degree of selection precision, the rate of progress can be significantly increased, potentially allowing entire cycles of back-crossing to be eliminated and thereby reducing the time required to reach a particular breeding objective.

Depending on the context, the value of these time savings can be enormous – often far larger than the additional cost involved in using MAS. For most private-sector breeding programmes, time considerations are paramount. The ability to get a new crop variety into an increasingly competitive commercial market earlier than the competition will often translate into substantial additional profits. Although the amount will of course vary depending on the performance advantage offered by the new variety, on the size of the seed market, and on the price of seed, among other factors, for important commercial crops of global importance such as maize, the value of being the first into the market can be measured in millions of dollars (D.N. Duvick, personal communication).

For CIMMYT and other not-for-profit research organizations that distribute their germplasm products free of charge, time considerations often appear less urgent, mainly because accelerated release of improved germplasm does not bring additional financial benefits to the organizations themselves. This does not mean, however, that the benefits do not exist, nor that they are unimportant. If MAS can reduce the time needed by public breeding programmes to develop improved germplasm, and if farmers consequently end up adopting improved varieties developed with the help of that germplasm sooner than they otherwise would have, the benefits can be considerable. For example, Pandey and Rajatasereekul (1999) estimate that the economic benefit from shortening the average length of the breeding cycle for a rice variety in Thailand is on the order of US$18 million over the useful life of the variety. In an earlier study of the returns to wheat breeding in Nepal, Morris *et al.* (1992) concluded that an important factor contributing to the success of the national wheat breeding programme has been the accelerated rate of adoption of modern varieties.

When millions of dollars of potential benefits are at stake, the additional cost of using MAS will often seem insignificant. This explains why the large multinational corporations that currently dominate the commercial plant-breeding industry have invested so heavily in the development of molecular marker technologies: they expect to recoup their investment by bringing improved germplasm products to market more quickly. It also partly explains why public research organizations have been relatively slow to follow suit: since they are by nature less attentive to profit considerations, public organizations often have difficulty in justifying the large expenditures that will be needed if they are to become major players in this rapidly evolving field.

As part of the case study, no attempt was made to calculate the additional benefits that might be realized by the accelerated release of converted QPM lines. Such benefits would in any case be difficult to calculate, since converted QPM lines developed by CIMMYT are not used directly by farmers but instead go to national breeding programmes and private seed companies as intermediate inputs. Nevertheless, it is worth noting that despite its relatively high cost, MAS 2 requires one fewer cycle than the other schemes.[7] If this means that the benefits of QPM varieties will be realized substantially sooner, then the time savings might well justify the additional expense.

Final thoughts: the way forward

Without question, molecular markers represent a powerful tool with the potential to influence the field of plant breeding. MAS procedures already offer opportunities for reducing costs, saving time and accomplishing breeding tasks that cannot be accomplished using conventional methods. Yet despite the many demonstrated advantages of the technology, the potential of MAS has yet to be fully realized. The relatively slow integration of MAS into many plant-breeding programmes can be explained by a number of factors, including the considerable expense of establishing biotechnology research facilities, the difficulty and expense of identifying useful markers, the limited usefulness of markers when traits of interest are controlled by a myriad of minor genes and the continuing high cost of laboratory equipment and materials.

Notwithstanding these difficulties, the prospects for molecular markers remain bright. It is important to remember that markers have been around for only a short time. As with any new technology, applications are evolving at a rapid pace, and as they do, the cost structure is changing accordingly. In some cases, the consequences of technological innovations can be dramatic; for example, the shift from RFLP to PCR-based markers slashed the cost of MAS virtually overnight. In other cases, impacts are more gradual, as for example with the introduction of improved statistical software for mapping analysis, incorporation of labour- and material-saving innovations into laboratory procedures, and fine-tuning of breeding strategies. Additional cost savings are sure to emerge in coming years from technologies currently in the pipeline (Ribaut and Hoisington, 1998). ‘Molecular beacons’ could end the need for electrophoretic separation of samples, and further development of DNA chip technology should allow scientists to study simultaneously the expression of thousands of genes related to a trait of interest. As research related to functional genomics surges forward, much more information regarding intra- and inter-specific genes and their sequences will become available; increased access to genomic data will help scientists to use markers more effectively.

Over the longer term, marker-based technologies could fundamentally alter traditional approaches to plant breeding. Rather than being seen merely as a new tool that can be substituted into traditional breeding schemes, markers may allow the breeding process to be approached in entirely new ways. Already novel breeding strategies are being proposed that would take conventional breeding into new directions (e.g. combining molecular mapping techniques and non-traditional wide crossing procedures to introduce and track alle-

[7] In principle, additional time savings could be achieved by screening with more of the background markers during additional cycles of selection (of course, this would also increase the cost of the overall scheme).

les from wild relatives of domesticated species) and/or draw upon the benefits offered by MAS in the most efficient manner possible (e.g. performing large-scale MAS in one early generation to fix multiple alleles of interest while maintaining genetic variability for future cycles of selection).

Further advances in biotechnology will make MAS cheaper and more effective in coming years, but as most breeders and biotechnologists clearly recognize, it would be a mistake to assume that marker technologies represent a 'silver bullet' solution to every breeding problem. As this case study has revealed, conventional breeding methods still provide a cost-effective option for many types of breeding project, and they will continue to be attractive in the future. Additional case studies are needed to build up the body of empirical evidence that research managers will require to identify cost-effective opportunities for combining conventional and marker-assisted methods.

Acknowledgements

Any interdisciplinary study requires input from a large number of people, and this one was no exception. We are grateful to have received assistance from many CIMMYT colleagues, including Ognian Bohorov, Margarita Crosby, David Hoisington, Scott McLean and many laboratory technicians from the Applied Biotechnology Center; Alejandro López, Daniel Jeffers, David Beck, Narciso Vergara, Hugo Cordova, David Bergvinson, Francisco Magellanes, Martin Cardenas and Rodolfo Caballero from the Maize Program; Jaime López Cesati, Jorge González Uribe, Silvia Palacios and many technicians from the Soil and Plant Analysis Laboratory; Claudio Cafati from the Director General's office; Salvador Fragoso and Martha Duarte from the Finance Office; Krista Baldini from the Human Resources Office; Patricia Galicia from the Purchasing Department; and Kelly Cassaday of Information Services.

References

Allard, R.W. (1960) *Principles of Plant Breeding.* John Wiley & Sons, New York.

Bolaños, J. and Edmeades, G.O. (1993) Eight cycles of selection for drought tolerance in lowland tropical maize. III. Responses in drought-adaptive physiological and morphological traits. *Field Crops Research* 31, 269–286.

CIMMYT (1999) Quality protein maize: food of the poor becomes inexpensive, accessible protein source. In: *CIMMYT Annual Report 1998–99: Science to Sustain People and the Environment.* CIMMYT, Mexico, D.F. pp. 38–39.

Duvick, D.N. (1984) Genetic contributions to yield gains of U.S. hybrid maize, 1930–1980. In: Fehr, W.R. (ed.) *Genetic Contributions to Yield Gains of Five Major Crop Plants.* Proceedings of a symposium sponsored by Division C-1 of the Crop Science Society of America, 2 December, 1981, Crop Science Society of America, Atlanta, Georgia. CSSA Special Publication Number 7, American Society of Agronomy, Madison, Wisconsin, pp. 15–48.

Melchinger, A.E. (1990) Use of molecular markers in breeding for oligogenic disease resistance. *Plant Breeding* 104, 1–19.

Morris, M.L., Dubin, H.J. and Pokhrel, T. (1992) *Returns to Wheat Research in Nepal. CIMMYT Economics Program Working Paper 92/04*, CIMMYT, Mexico, D.F.

Pandey, S. and Rajatasereekul, S. (1999) Economics of plant breeding: the value of shorter breeding cycles for rice in Northeast Thailand. *Field Crops Research* 64, 187–197.

Pingali, P.L. and Heisey, P.W. (1999) *Cereal Crop Productivity in Developing Countries: Past Trends and Future Prospects. CIMMYT Economics Program Working Paper 99-03*, CIMMYT, Mexico, D.F.

Poehlman, J.M. (1987) *Breeding Field Crops*, 3rd edn. AVI, Westport, Connecticut.

Ribaut, J.M. and Hoisington, D. (1998) Marker-assisted selection: new tools and strategies. *Trends in Plant Science* 3, 236–239.

Simmonds, N.W. (1979) *Principles of Crop Improvement.* Longman, London.

Villegas, E. (1992) Quality protein maize—what it is and how it was developed? In: Mertz, E.T. (ed.) *Quality Protein Maize.* The American Association of Cereal Chemists, St Paul, Minnesota, pp. 27–48.

Villegas, E. (1994) Factors limiting Quality Protein Maize (QPM) development and utilization. In: Larkins, B.A. and Mertz, E.T. (eds) *Quality Protein Maize: 1964–1994. Proceedings of the International Symposium on Quality Protein Maize.* 1–3 December, EMBRAPA/CNPMS. Sete Lagoas, Minas Gerais, Brazil, pp. 79–88.

Weber, J.L. and May, P.E. (1989) Abundant class of human DNA polymorphisms which can be typed using the polymerase chain reaction. *American Journal of Human Genetics* 44, 388–396.

Young, N.D. (1999) A cautiously optimistic vision for marker-assisted breeding. *Molecular Breeding* 5, 505–510.

Chapter 13
Can Biotechnology Reach the Poor? The Adequacy of Information and Seed Delivery*

Robert Tripp
Overseas Development Institute, 111 Westminster Bridge Road, London SE1 7JD, UK

Abstract

This chapter examines the expectation that biotechnology can provide significant benefits for smallholder farmers. It uses evidence from current seed systems and variety use in developing countries. In particular, it examines the adequacy of information flow and the performance of seed markets. Many of the biotechnology innovations proposed for use by smallholders feature cryptic qualities that may not be immediately obvious to farmers; implications are drawn for the potential demand for these transgenic varieties. The adequacy of seed systems is also examined, including the characteristics of local seed diffusion and the experience of commercial seed enterprises. The chapter concludes that investments in public biotechnology must be accompanied by policies that encourage commercial seed system development and that empower farmers to be able to take full advantage of new technology.

Introduction

In debates about the future of biotechnology, supporters who emphasize the importance of matching population growth with increased agricultural production are often confronted by critics whose concern is the severe inequities in current access to food and productive resources. As with many such debates, both sides have a point. The challenge is certainly to increase food production and at the same time ensure that the increased production contributes to widespread economic benefits and poverty alleviation. To do so will require attention not only to new technology but also to the mechanisms that allow such technology to be delivered to the smallholder farmers who constitute the majority of many developing nations' agricultural sectors. This chapter is concerned with issues of delivery and in particular with the information requirements and seed sector characteristics that will determine equitable access to biotechnology.

Any discussion of biotechnology's prospects for smallholders leads to

*A version of this chapter has already appeared in *Food Policy* (2001) 26(3), 249–264. Reproduced courtesy of Elsevier Science.

inevitable comparisons with the Green Revolution. A universally recognized difference is the dominance of private ownership of the new technology. This has a significant influence on incentives for technology development and argues strongly for developing public sector research capacities in biotechnology and for brokering effective public–private linkages. Other changes that have taken place since the Green Revolution are less commonly acknowledged. Public agricultural research in general is in a perilous state and public extension is being disbanded in many countries. In some countries, public seed companies are being challenged and overtaken by a vibrant private seed sector, but in other countries, seed parastatals are in disrepair with no viable alternative in sight. In addition, many observers, including some of biotechnology's strongest supporters (e.g. Nuffield Council on Bioethics, 1999), acknowledge that the Green Revolution was most successful in more favoured environments. Finally, the basic Green Revolution technology of easily distinguishable, short-statured, fertilizer-responsive varieties may be contrasted to many of the proposed benefits of biotechnology that incorporate more cryptic qualities such as disease or pest resistance.

All of these contrasts with the Green Revolution scenario argue for increased attention to the mechanisms of information diffusion and seed delivery that must support technological change. The point of this chapter is simply to emphasize the importance of these delivery issues when weighing the potential for biotechnology in poverty alleviation. Because there are currently few opportunities to study the adoption of transgenic crops in developing countries, the arguments in this paper are based on observations on conventional crop varieties and studies on the performance of the seed sector in developing countries.

The next section provides a brief review of the types of biotechnology that will most likely be of relevance to smallholders in developing countries. This is followed by an examination of farmer knowledge, including the abilities to recognize production problems and to distinguish among varieties, and a review of the special challenges presented by nutritionally enhanced varieties. The following section examines evidence on seed provision, focusing on the seed industry's capacity to serve smallholders and its ability to deliver information. The final section presents conclusions.

The Technology

The rapid expansion of agricultural biotechnology, and its seemingly limitless possibilities, often makes it difficult to focus on concrete examples, but the delivery of biotechnology to smallholders requires some thought regarding the most likely innovations. What type of technology can be anticipated to be accessible and useful in the early stages of directing transgenic crops to smallholder environments, either directly through public research or indirectly through policies that affect the incentives of private biotechnology firms?

The most widely used products of genetic engineering involve herbicide tolerance, a trait accounting for about 80% of the current area of transgenic crops. Although herbicide use is growing in developing countries, it is unlikely that herbicide tolerance would be high on the initial list of priorities for directing biotechnology to smallholders. It would only be recommended in cases where the cost of labour for weeding is an important limitation and (more important) where farmers already have considerable experience with chemical weed control.

More likely candidates for smallholder-friendly biotechnology include the many examples of disease and pest resistance that have been developed or are the subjects of research. Pests and diseases cause significant losses to smallholders and even when chemicals for combating these are available they are often environmentally harmful or prohibitively expensive. The most widely known examples to date are the crop varieties with transgenes derived from strains of *Bacillus thuringiensis* (*Bt*), but *Bt* is just one of many innovations, and a wide range of plant genes controlling enzymes, enzyme inhibitors and other metabolites with pesticidal properties is under investigation (Hilder and Hamilton, 1994). Similarly, the control of many disease-causing bacteria, fungi and viruses is possible through genetic engineering (e.g. Khush and Brar, 1998).

One of the examples increasingly used by the proponents of biotechnology in debates about poverty impact is the development of nutritionally enhanced varieties. The recent announcement of a transgenic vitamin-A enhanced rice is the most prominent example. Vitamin A deficiency is an exceptionally serious problem, often associated with rice-eating populations, and the availability of a nutritionally superior rice variety seems a logical candidate for a pro-poor transgenic crop. There are a number of other projects under way that use genetic engineering to enhance nutritional content (DellaPenna, 1999).

There is also considerable research devoted to applying biotechnology to address abiotic stresses. Some of this research may be able to take advantage of relatively simple genetic modifications to increase tolerance to conditions such as soil acidity or salinity. Other advances (such as drought tolerance or nutrient use efficiency) are likely to involve more complex genetic modification. It is difficult to predict the poverty impact of such research. In some cases, such as addressing soil acidity, the innovations would likely be directed to marginal environments (often associated with higher levels of poverty). The development of drought tolerance would presumably benefit farmers in marginal areas but could transform agricultural production in more favoured areas as well.

Although this paper is concerned with transgenic crops in general, it assumes that the most likely innovations for smallholders in the near term are varieties with increased resistance to biotic stresses. The paper also considers the possibilities of nutritional enhancement. However, the arguments are valid for many other types of transgenic crop. It is often said that genetic engineering is simply a more efficient method of pursuing the goals of conventional plant breeding, and this is supported by the fact that most transgenic varieties feature qualities obtainable by traditional methods. This justifies our strategy of exploring the acceptability of transgenics by examining farmers' information regarding conventional varieties. What type of information is required for farmers to recognize and make decisions about the production problems addressed by new varieties?

However, genetic engineering is also different from conventional breeding in a number of respects. One difference is the precision with which functional genes can be identified and transferred. This would seem to offer the promise of an end to the trade-offs that affect much of conventional plant breeding. Instead of producing a new variety with increased disease resistance but somewhat less acceptable grain quality, for instance, the variety with superior grain quality can be targeted to receive the disease-resistance gene. This could lower the learning costs plaguing the adoption of many current varieties, where farmers must learn about (and balance) various advantages and disadvantages of a new variety. However, these potential gains must be balanced against the problem of distinguishing the new variety. What type of information is required to assure the farmer that the seed for sale contains the additional gene? What type of industry is required to deliver this seed and the necessary information?

Farmer Knowledge

The rapid and widespread diffusion of Green Revolution varieties of rice and wheat, as well as many other examples of variety adoption, would seem, at first glance, to eliminate any doubts about farmers' capacity to recognize superior varieties. However, many modern varieties (MVs) diffuse very slowly or are not adopted at all. Blame is usually apportioned among inappropriate breeding programmes, uninspired extension and unappreciative farmers. All of these factors play a role, but this section focuses on farmers' perspectives and incentives. It examines farmers' capacities for recognizing production problems and their abilities to distinguish among varieties. It also briefly examines the special case of nutritionally enhanced varieties.

Recognizing production problems

If a new variety (transgenic or conventional) incorporates resistance to a particular stress, its likelihood of adoption will be enhanced if farmers are aware of the nature of the production problem and can easily recognize the variety's advantage. There are many examples where this is the case, such as the widespread adoption of downy mildew-resistant maize varieties in Southeast Asia (Morris, 1998a), but there are many other instances where farmers are less certain. Bean farmers in Central Africa do not recognize losses caused by disease, even though they may account for up to half of total yield loss (Trutmann *et al.*, 1996). Most Pakistani wheat farmers recognize rust as a problem, but few of them appreciate differences in varietal susceptibility (Heisey, 1990).

Part of the problem is smallholders' complex production environment and their lack of access to adequate information. Conelly (1987) describes a typical maize field in western Kenya where the crop may suffer simultaneously from stem-borer damage, maize streak virus, *Striga* infestation, poor soil fertility and drought. Conelly questions both scientists' and farmers' abilities to assign proportional yield loss to the various stresses. Such uncertainty also makes us speculate about the introduction of a new maize variety that addresses only one of these serious problems. Its reception is likely to be conditioned by the severity and variability of the particular stress, as well as by farmers' abilities to observe cause–effect relationships in the field. The investment required to develop transgenic varieties demands that agricultural research be able to identify the highest priority production constraints for specific groups of farmers.[1]

Farmers' detailed knowledge of their environment has received justifiable emphasis in many studies on traditional agriculture, but there are limits to farmers' understanding. Bentley (1992) presents a useful way of classifying farmers' knowledge of production problems in two dimensions: ease of observation and perceived importance. Those problems that are easiest to observe and have high importance (such as weeds or termites) receive the most concerted search for solutions. Important problems that are difficult to observe receive less attention. Nematode damage in some crops is an example of the latter; even in situations where nematodes cause a high average yield loss, considerable variability between fields and between seasons, coupled with farmers' lack of recognition of the problem, could hamper the rapid diffusion of a nematode-resistant variety. The conclusion is not to abandon research on nematode resistance, but to accompany it with investments in farmers' education that allows them to recognize these problems and test new varieties.

Of course, useful solutions to production problems do not necessarily depend on farmer understanding of biological relationships. Many aspects of traditional farm management represent trial-and-error innovations. Central African farmers regulate bean planting distances ostensibly to control humidity, which they perceive as a problem, but the practice also serves to limit disease transmission (Trutmann *et al.*, 1996).

In many cases farmers adopt a new variety without appreciating the contribution of its specific disease resistance to the variety's superior performance, but diffusion of new disease-resistant varieties may be

[1] See Pingali and Gerpacio (1997) for an example of the contrast between researchers' perceptions of insect damage and the actual situation in farmers' fields.

delayed, as comparisons of varietal replacement rates can show (Brennan and Byerlee, 1991). Farmers are often slow to replace susceptible varieties with newer ones, even in some 'Green Revolution' environments (e.g. Heisey, 1990). Farmers get accustomed to certain varieties and are reluctant to change. Part of the problem is the learning costs associated with distinguishing among new varieties and seeking information on their availability.

Recognizing new varieties

In discussing the adoption of MVs, Byerlee (1994) makes the useful distinction between Type A varieties (the first generation of varieties that initiated the Green Revolution) and Type B varieties (subsequent generations of MVs that incorporate additional improvements, such as disease resistance). Farmers had no trouble distinguishing Type A varieties from their traditional varieties, but the more subtle changes incorporated in Type B varieties certainly contribute to greater confusion about their names and are often responsible for slower rates of diffusion.

It is difficult to make general assertions on this issue. There are areas where farmers are very articulate about the range of MVs on offer and are keen to test the latest releases. For instance, rice farmers in Andhra Pradesh are generally well informed about the latest varieties and are often ahead of the public and private seed companies in multiplying and diffusing recent releases (Tripp and Pal, 2001). On the other hand, there are many cases where farmers are relatively unfamiliar with the MVs that are available and may even be unable to identify the particular variety they are growing. Table 13.1 presents data that illustrate such situations. The table makes no pretence of being representative, but the examples are drawn from a wide enough range of environments to encourage the argument that this is not an isolated problem. MV names are often complicated letter–number codes that anyone would have trouble remembering, but the examples in the table count any generally recognized local name for a variety as correct identification.[2] In addition, the ability to provide a name does not necessarily mean that it is the correct one; Goodell (1984) describes a study in the Philippines showing that over 40% of the farmers claiming to be growing a particular disease-resistant variety were mistaken. The implications of this type of uncertainty

Table 13.1. Farmer knowledge of modern varieties (MVs).

Location	Crop	% of farmers who learned of MV from another farmer	% of farmers who cannot name the MV they are growing
Punjab, Pakistan[a]	Wheat	48–69	5–20
NWFP, Pakistan[a]	Wheat	57	33–50
Nepal[b]	Wheat	N/A	50
Rajasthan, India[c]	Pearl millet	39–48	59–89
Western Province, Zambia[d]	Sorghum	60	42
Ghana[e]	Maize	48	34

N/A, not available.
Sources: [a]Heisey (1990); [b]Morris *et al.* (1992); [c]Tripp and Pal (2000); [d]Lyoba and Tripp, unpublished; [e]Morris *et al.* (1999).

[2] In the Zambia sorghum example, two MVs are available. Both are shorter and earlier maturing than most local varieties. Farmers are often uncertain of the identity of the particular one they are growing and few can describe the differences (in adaptation to soil type and rainfall regime) that distinguish the two MVs.

for proposed transgenic varieties that feature a single additional resistance gene are obvious.

A single new variety, especially if its agronomic performance is outstanding, is usually recognized by farmers and assigned a name. However, subsequent varieties may find themselves lumped in the same category. A common example is the tendency to call any MV a 'government' variety; in Ghana all maize MVs are known as 'agric' (i.e. Agriculture Department). In Kenya, early maturing varieties of various crops tend to be called 'Katumani', after the popular early maturing maize variety of that name. In India, *sankar* signifies 'hybrid' to a Hindi-speaking plant breeder, but the term is used by many farmers to describe any MV, hybrid or open-pollinated.

This confusion about 'hybrids' has particular relevance to a consideration of biotechnology. The 'terminator gene', or genetic use restriction technology (GURT), is sometimes promoted as the equivalent of hybrid technology: a biological innovation that limits seed saving and hence increases incentives for investment in research. Leaving aside the obvious difference that hybrid technology, on its own, is responsible for significant yield enhancement, there are also different implications for seed management in smallholder environments. It is true that hybrid technology has been responsible for the widespread adoption of public and private varieties of crops such as maize, pearl millet and sorghum, and that private investment in these crops in developing countries is significant. However, the level of farmers' understanding and the rigour with which they manage hybrid seed are subject to considerable variation. In the first place, although the second generation of hybrid seed almost invariably yields less than the first, the extent of yield depression is quite variable. This helps to explain the widespread practice of recycling hybrid seed, either because of unfamiliarity with the nature of hybrids or concern about the price of fresh seed (Morris *et al.*, 1999).[3]

If farmers in developing countries often decide to recycle their hybrids, the parallels with GURT begin to break down. The problem is fundamentally related to information. For farmers in highly commercialized settings, where information is widely available and the entire crop harvest is sold, GURT would be no more difficult to accept than the increasingly common buyer agreements that prohibit seed saving, or than the clear understanding of the consequences of recycling hybrids in high-yielding environments. For farmers in less-favoured circumstances, however, the consequences of a non-germinating crop of a GURT-protected variety would be vastly different from the yield depression (perhaps not even noticeable) of a recycled hybrid or the conscious decision to compromise by saving money for seed and accepting a lower yield.

The potential problems of GURT extend to the issue of seed source. Even if a farmer understands the nature of GURT, this does not necessarily protect other farmers who obtain seed from neighbours or grain markets. These are common practices. Many bean farmers in Central Africa regularly get their seed from grain markets (Sperling *et al.*, 1996), and 30% of the smallest wheat farmers in Bangladesh get their seed from local markets (O'Donoghue, 1995). If a proportion of this grain was GURT varieties, and if these were not easily distinguished from other varieties, the results could be unfortunate. Farmers who are short of seed or who wish to change varieties often acquire seed from neighbours. Even in the relatively advanced rice system of Andhra Pradesh, between 30% and 60% of off-farm seed is acquired from other farmers.

Because GURT represents a significant investment it is unlikely that it would be used in smallholder areas where a significant seed industry was not already in place. Indeed, it would be irresponsible to

[3] The extent of yield depression in advanced generation hybrids depends on a number of factors, including the type of hybrid and the specific parental lines. There are also differences among crops; advanced generations of hybrid rice appear to show much higher yield depression than those of many maize hybrids.

introduce GURT into areas that did not have long experience with a robust, diverse and easily accessible commercial seed market. The potential problems in variety identification underline the importance of adequate information in seed provision. The next section examines the performance of the seed industry, but first we shall look at the special case of nutritionally enhanced varieties.

Nutritionally enhanced varieties

The recent attention given to the development of a transgenic vitamin A-enhanced rice variety justifies a brief discussion of the relevance of our previous arguments for nutritional goals. Genetic engineering can not only be used to provide agronomic characteristics of importance to smallholders: it can also be used to improve the nutrition of poor consumers (including farmers). However, many of the same concerns about information apply to this case.

In the first place, a variety's nutrient content is usually a cryptic quality. (The vitamin-A rice, which is yellow, is an exception.) Enhanced protein, vitamin or mineral content will not be recognized before or after consumption, no matter what contribution the improved variety might make to the consumer's nutritional status. Thus the farmer who plants the variety, or the consumer who purchases it, must (first) be aware of the advantages and (second) be able to identify the correct variety. Both of these requirements may be difficult to fulfil.

A relevant case is the experience in Ghana with quality protein maize (QPM). QPM varieties have an altered endosperm protein that is of higher quality than normal maize protein. (The degree to which this increased protein quality is of relevance to the diets of the poor is a matter of debate among nutritionists, but need not concern us here.) An open-pollinated QPM variety ('Obatanpa') was released in 1992 (one of a series of maize MVs released since 1984 in Ghana). The variety has excellent agronomic characteristics and received the strong support of Sasakawa Global 2000. Both its good performance and its special promotion made it the dominant variety in the formal seed market by the mid-1990s. A study of maize production practices in Ghana in 1998 included questions about the QPM variety (Morris *et al.*, 1999).

The results are summarized in Fig. 13.1. The majority of Ghana's maize farmers have experience with one or more MVs. Although most farmers can discuss the characteristics (positive and negative) of MVs as a class and compare them to local varieties, far fewer are able to distinguish among the MVs. Obatanpa was featured in a nation-wide campaign by extension agents, radio and print media that explained its nutritional advantages and emphasized that it should be substituted for normal maize in porridge used for weaning children. The survey indicated that a relatively modest 29% of farmers had heard about the existence of a nutritionally superior variety. Of those who had heard, 55% were able to correctly identify the variety.[4] Of the farmers who correctly identified the variety, 44% claimed to be using it for feeding children.

It can be argued that those farmers who could identify the nutritional opportunity, the correct variety and its appropriate use were taking advantage of a plant-breeding innovation to improve their children's nutrition. However, this amounts to only 7% of the population. In addition, because of problems in identifying the correct variety, it may be that up to 4% of the population have changed their child-feeding practices for the worse, i.e. they have had the idea reinforced that normal maize porridge on its own is an adequate weaning food.

There are several lessons for biotechnology directed towards nutritional improvement. First, although it is sometimes possible to slip a nutrient into the food supply unnoticed (e.g. iodized salt), in most cases useful dietary improvement

[4] The majority of the incorrect responses were the generic name for MVs, 'agric'.

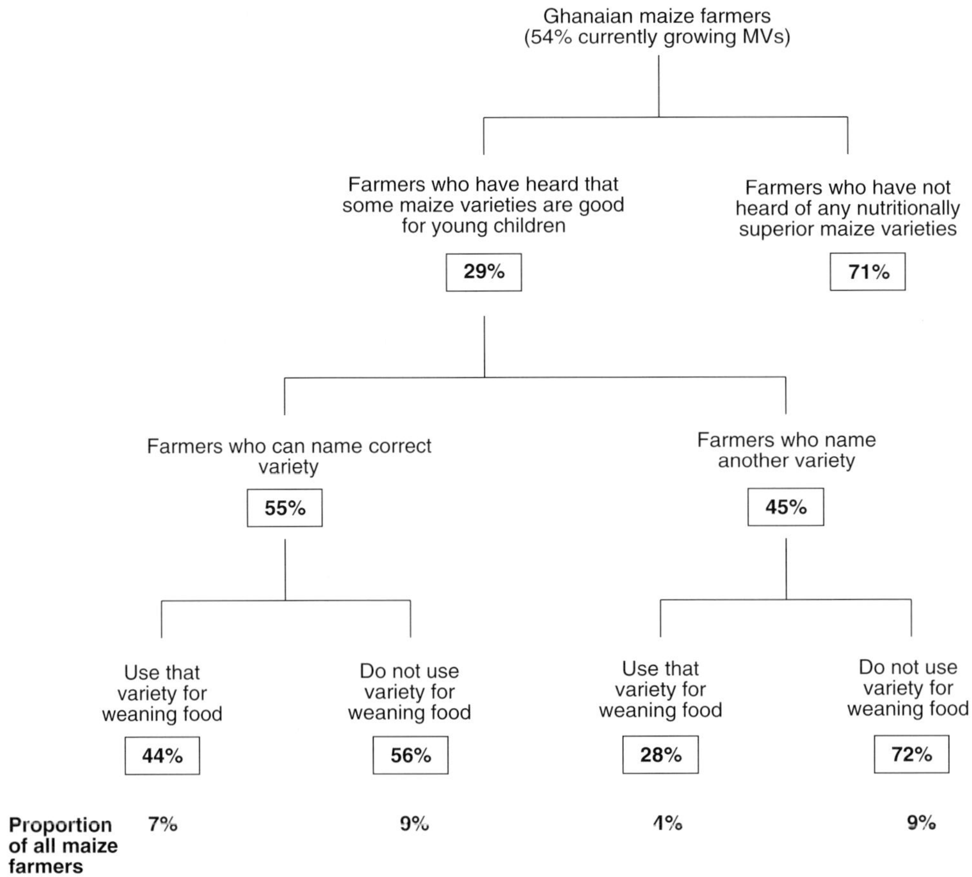

Fig. 13.1. Knowledge and use of quality protein maize (QPM) by Ghanaian farmers. Source: Morris *et al.* (1999) and unpublished data.

also requires consumer education. Second, even when people understand the nutritional problem, they must be able to identify the relevant varieties. Third, in those cases where the variety is easily distinguishable (e.g. vitamin-A rice), without adequate information there may be an opposite problem of resistance to varieties 'for poor people'.[5] Some nutritionists calculate that a significant amount of vitamin A deficiency in Central America could be eliminated if people switched from white to yellow maize, but no one seriously puts this forward as a solution.

There is an additional concern that must be addressed about genetic engineering (or conventional plant breeding) addressed at nutritional enhancement. Such projects may be able to make a contribution to tackling problems of malnutrition, but they must guard against becoming simply a technical fix. Malnutrition is a complex problem, linked to poverty and lack of information. Nutritionally enhanced varieties may bring publicity and support to public plant-breeding programmes but there is a danger that the emphasis will be on the release of the variety rather than on the validation of

[5] Early promotions of QPM in Latin America, more than 20 years ago, sometimes featured demonstrations of its truly remarkable results when fed to pigs. However, some people were insulted at being persuaded to eat 'maize for pigs'.

actual nutritional impact. In addition, the 'technical fix' attribute can tempt governments to believe they don't have to worry about nutrition because the plant breeders are handling it, thus ignoring the more deep-seated causes of malnutrition.

Seed Supply

In addition to the challenge of variety identity, the diffusion of transgenic varieties will also be affected by the efficiency of seed delivery. The growth of the private seed sector in many developing countries and the incentives provided by the establishment of intellectual property rights (IPRs) would seem to promise improved access to seed, but a number of problems remain. This section briefly examines the status of the seed industry in developing countries and the industry's ability to deliver information to farmers.

The seed industry

The past decade has seen the demise of many parastatal seed enterprises; perhaps the major exceptions are the public seed companies in some Asian countries that continue to deliver seed of rice, wheat and a few other crops. In some cases private companies have replaced the parastatal seed companies, but in other cases no viable alternative has yet appeared. The performance and prospects of the private seed industry are of utmost importance to the success of delivering transgenic varieties (even those developed by public research) to smallholders.

There are already a number of instances of the private seed industry serving smallholders. Hybrid maize is probably the outstanding example (Morris, 1998b), but when the conditions are appropriate, the private seed industry can even play an important role in marketing seed of self-pollinated varieties, as illustrated by the case of rice in Andhra Pradesh (Tripp and Pal, 2001). The expansion of private seed enterprise will be linked to effective IPRs and to the existence of efficient input and output markets that stimulate demand for seed. Where these conditions are less likely to be met, because of the nature of the crop or the economy, then private seed activity is less feasible and the delivery of new varieties is more problematic. Such adverse conditions are, of course, characteristic of many of the smallholder environments that biotechnology claims to target.

An *ex ante* economic analysis of the potential of transgenic virus-resistant potatoes in Mexico shows positive returns for smallholders (Qaim, 1998). However, the analysis assumes an efficient seed potato system to serve those farmers, something that currently does not exist. There are several reasons why Mexican smallholders do not use commercial seed potato, including the high cost of seed, the location of the seed industry close to the areas where large farms predominate, and smallholders' lack of familiarity with the advantages and management of certified seed potato.[6]

There are significant differences among crops and markets that help determine the course of seed industry development. When Peru liberalized its seed sector about a decade ago, two of the most affected crops were rice and potatoes, both of which had been the subject of state-controlled seed systems (J. Bentley *et al.*, unpublished paper). Despite a short-lived flourish of private seed potato production in the early years of liberalization, most seed potato in Peru is now provided by an indigenous trading system that manages both ware and seed potato. For rice, on the other hand, a number of viable private seed companies have emerged. The differences in experience between the crops can be explained by factors that include seed biology, the concentration of the growers and the demands of the market. Unfortunately, many of these factors are weighted against

[6] Potato is a particularly difficult case because seed costs are such a high proportion of total costs of production.

the dispersed smallholders in marginal environments whose relation to biotechnology we are considering in this chapter.[7]

The development of a private seed industry in sub-Saharan Africa deserves special attention, because many of the hopes for biotechnology to address 'the needs of the hungry' (Nuffield Council on Bioethics, 1999, p. 58) are focused here. With the exception of hybrid maize (and South Africa's seed industry), there is still little evidence of a viable private seed sector. There are numerous problems, not the least of which are national regulatory systems that discourage seed enterprise development and large numbers of uncoordinated and often ill-conceived donor seed projects that further discourage private initiative (Tripp, 2000).

An improved regulatory environment and more rational donor activities are only part of the answer for strengthening seed provision in Africa. A number of dilemmas remain. Public biotechnology research will likely concentrate on subsistence crops with few commercial markets. Public transgenic varieties should ideally be channelled through private seed companies, but will it be possible to ensure sufficient repeat sales (or to enforce IPRs) to provide sufficient incentives for the private sector? If not, is the answer to wait until the commercial seed system develops (based on more attractive varieties and markets), when sufficient skills and capacity are available for some entrepreneurs to take on the less profitable crops? The time required for such an evolution is surely unacceptably long.

The prospect of biotechnology only lends urgency to the need to develop adequate seed policies for sub-Saharan Africa. Alternative seed distribution schemes for new varieties must be designed in a way that does not discourage private initiative. If a strategy is based on farmer-to-farmer seed movement of a public transgenic variety (after an initial introduction of seed), we have already seen the importance of addressing information inadequacies at the farm level. Attractive new varieties do not necessarily spread rapidly among farmers. In one area of Kenya, an early maturing pigeonpea variety diffused very quickly, while an early maturing groundnut variety in Malawi experienced relatively little adoption, despite a decade of promotion (Tripp, 2000). One important difference in the two cases is the high market demand for the former variety, a fact that emphasizes how market development stimulates technology demand and variety recognition.

The seed industry as a source of information

Even when a private seed industry is well established, it may not provide sufficient information to help farmers distinguish among varieties. The expansion of the private pearl millet seed industry in India in the past decade is based mostly (but not entirely) on hybrids, initially from the public sector but increasingly the products of private plant breeding. Only 3 or 4 kg of pearl millet seed is sufficient to plant a hectare, so even though the hybrid seed price may be 15 or 20 times that of grain, most farmers can afford it.

A study in two areas of Rajasthan showed widespread utilization of purchased pearl millet seed (from private and public companies) but considerable uncertainty regarding varietal identity (Tripp and Pal, 2000). Table 13.2 summarizes farmers' abilities to identify the pearl millet hybrids they were planting. Although the company name is fairly well known (especially in Behror, where farmers have longer experience with using commercial inputs), the ability to name the precise variety is less evident. Most companies offer several pearl millet hybrids with different agronomic characteristics, and the

[7] This chapter does not consider the problems of distributing planting material for root and tuber crops. Such crops are often particularly important for the poor, and the diffusion of new varieties can be severely limited by inadequate mechanisms for multiplying and distributing planting material. See Gatsby Charitable Foundation (1997) for a discussion of the effectiveness of different modalities for diffusing planting material for virus-resistant cassava varieties in Uganda.

Table 13.2. Farmers' knowledge of pearl millet hybrid in two areas of Rajasthan.

Farmer knowledge	Shekhawati ($n = 100$)	Behror ($n = 80$)
Company and hybrid (%)	23	10
Company only (%)	35	76
Hybrid only (%)	18	1
No knowledge (%)	24	13

Source: Tripp and Pal (2000).

study indicates that farmers are often unaware of these differences.

Table 13.3 indicates that farmers often choose which variety to plant based on advice from an input merchant. The study found that most seed dealers are familiar with the basic characteristics of the varieties they are selling, but most advice to farmers consists of a general recommendation for a particular variety rather than an explanation of the advantages and disadvantages of alternative choices. The dealers are generally responsible in their recommendations, hoping to build a loyal patronage, but their advice is influenced by the size of mark-up they can charge and by the relatively undiscriminating demands of their clientele. How this situation might change with the introduction of transgenic varieties, especially those with cryptic qualities or particular management requirements, is a matter for speculation.

The capacity of agricultural input merchants is a cause for concern. For instance, a recent study cites the inadequate knowledge and skills of fertilizer dealers in Kenya (Yanggen *et al.*, 1998). In any vision of private seed supply input, dealers provide a vital link to the farmer. The advent of biotechnology has caused some observers to predict the demise of local farmer-dealers for seed in the USA because of the increased demand for specialist, technical advice on transgenics (Shimoda, 1996).

Another concern for any commercial seed system is ensuring consumer confidence. The seed of transgenic crop varieties is likely to be expensive and farmers will want to be sure that they are getting good quality seed of the correct variety. In theory, private companies will invest in building their reputations, but this is not done overnight. Experience in Rajasthan and Andhra Pradesh indicates that even those farmers who regularly buy commercial seed are often unable to identify the company that produced the seed and can rarely identify more than one or two seed companies in the market. (Not surprisingly, farmer knowledge of seed companies is highly correlated with education level.) Another possible solution is seed certification, but certification systems in many countries are ineffective (Tripp, 1997). In addition, the studies in India showed that very few farmers are familiar with the meaning of a certification tag.[8]

Finally, it is worth mentioning the concern of many critics that biotechnology will

Table 13.3. Source of information on pearl millet hybrids in Rajasthan.

Farmer knowledge	Shekhawati ($n = 100$)	Behror ($n = 80$)
Another farmer (%)	48	39
Shopkeeper (%)	36	46
Extension (%)	12	15
Other (%)	4	0

Source: Tripp and Pal (2000).

[8] A study in Zambia showed a similarly low level of understanding of certification tags among farmers who regularly purchased certified seed and even among the majority of seed stockists (Andren *et al.*, 1991).

result in domination of national seed markets by multinational corporations. The image of slick advertising overwhelming innocent farmers is a common debating point for biotechnology's critics, but it needs to be examined in the context of real seed markets. In the first place, there is little evidence that farmers are unduly impressed by the promotional campaigns currently addressed to them by seed companies. More important, aggressive promotion requires functioning markets. To the extent that local seed markets are diverse and competitive, and local companies understand their clients and have loyal customer bases, then foreign biotechnology companies will be motivated to establish partnerships or licences that combine external technology with local seed production and marketing.[9] The best way to take advantage of foreign technology is to ensure that a viable local seed industry is in place.

Conclusions

This chapter has identified a number of factors that can limit the relevance of biotechnology for smallholder farmers. The purpose is not to be unnecessarily pessimistic about the prospects of biotechnology for contributing to poverty alleviation but rather to identify specific challenges that must be overcome. It is difficult to avoid the conclusion that the successful utilization of biotechnology by resource-poor farmers depends crucially on the delivery of information and the performance of seed systems. All of the deficiencies discussed in this chapter are well known and most are potentially surmountable, but their combined impact is sufficient to thwart many current aspirations for delivering biotechnology to the poor.

It is therefore inaccurate to see biotechnology as some type of magic bullet that can break through barriers previously separating smallholder farmers from productive technology. Biotechnology has the potential of improving productivity and welfare for a wide range of farmers, but if smallholders are to be targeted then attention must be directed to a number of structural deficiencies.

Many of these deficiencies are related to information. Farmer observation and experimentation have been responsible for the widespread diffusion of many new varieties. However, farmers' understanding of new technology, and their ability to distinguish among options, particularly if they embody cryptic qualities, is often inadequate. Although transgenic varieties may be described in terms of minor genetic modifications, they often imply differences in management and performance that require increased knowledge. If smallholders cannot be served by increasingly demoralized and under-funded extension services, perhaps support should be focused on the basic education provided by rural schools. In the case of nutritionally enhanced varieties, attention needs to be given to adequate accompanying nutrition education.

Information is also provided by effective markets. The areas in which the Green Revolution was successful are not only distinguished by favourable growing conditions but also by adequate markets that are able to deliver the necessary inputs and purchase the increased production. Most smallholders are currently not well served by seed markets; there are inadequate incentives for seed enterprise development and local input marketing is often deficient. Yet most biotechnology will have to be delivered through private input systems. Without policies that support the development of strong and equitable local input markets, biotechnology is unlikely to reach many smallholders.

Information flows in two directions. If resource-poor farmers are to participate in commercial input markets, they need to know how to voice their opinions and to

[9] Responsive local seed firms also lower the chances of foreign companies 'dumping' inappropriate seed. See Gould and Cohen (2000) for a discussion of the dangers of using *Bt* varieties in environments different from the ones for which they have been developed.

pursue complaints. Better consumer protection and education programmes are required to support farmers who will become increasingly dependent on private input markets. In addition, if resource-poor farmers are a target for public biotechnology research, these farmers require increased representation in the decision-making of public research institutes.

Although this chapter does not address public agricultural research, the implications of the discussion of information deficiencies should be obvious. If most of the hopes for bringing biotechnology to resource-poor farmers rest on the performance of public research, then public research institutes bear a significant responsibility for addressing crops and production problems that are high priorities for smallholders. They also are responsible for ensuring that farmers have access to sufficient information to allow them to understand, test and evaluate biotechnology innovations. In addition, public research institutes must support the development of efficient market mechanisms for the delivery of seed and planting materials.

Agricultural biotechnology still faces an uphill battle for public acceptance. The potential to improve the livelihoods of resource-poor farmers is a strong point in its favour. This advantage will be lost if it is seen as simply a cynical strategy to gain support, or if it turns out to be a vain hope, unable to realize its good intentions. The balance will be determined by the degree to which new technology is accompanied by policies that make markets and political representation more effective for smallholders, in order for them to gain control of the process that develops technology in their name.

References

Andren, U., Nkomesha, A., Singogo, L.P. and Sutherland, A. (1991) *National Seed Availability Study: Seed Problems, Practices and Requirements Among Small-Scale Farmers in Zambia.* Ministry of Agriculture, Adaptive Research Planning Team, Lusaka.

Bentley, J. (1992) The epistemology of plant protection: Honduran *campesino* knowledge of pests and natural enemies. In: Gibson, R. and Sweetmore, A. (eds) *Proceedings of a Seminar on Crop Protection for Resource-Poor Farmers.* Natural Resources Institute, Chatham.

Brennan, J. and Byerlee, D. (1991) The rate of crop varietal replacement on farms: measures and empirical results for wheat. *Plant Varieties and Seeds* 4, 99–106.

Byerlee, D. (1994) *Modern Varieties, Productivity and Sustainability: Recent Experience and Emerging Challenges.* CIMMYT, Mexico, D.F.

Conelly, W.T. (1987) Perception and management of crop pests among subsistence farmers in South Nyanza, Kenya. In: Tait, J. and Napompeth, B. (eds) *Management of Pests and Pesticides: Farmers' Perceptions and Practices.* Westview Press, Boulder, Colorado.

DellaPenna, D. (1999) Nutritional genomics: manipulating plant micronutrients to improve human health. *Science* 285, 375–379.

Gatsby Charitable Foundation (1997) *Mastering Mosaic: the Fight for Cassava Production in Uganda.* The Gatsby Charitable Foundation, London.

Goodell, G. (1984) Challenges to international pest management research and extension in the Third World: do we really want IPM to work? *Bulletin of the Entomological Society of America* 30, 18–26.

Gould, F. and Cohen, M. (2000) Sustainable use of genetically modified crops in developing countries. In: Persley, G.J. and Lantin, M.M. (eds) *Agricultural Biotechnology and the Poor.* CGIAR, Washington, DC.

Heisey, P. (ed.) (1990) *Accelerating the Transfer of Wheat Breeding Gains to Farmers: a Study of the Dynamics of Varietal Replacement in Pakistan.* CIMMYT Research Report No. 1, CIMMYT, Mexico, D.F.

Hilder, V. and Hamilton, W. (1994) Biotechnology and prospects for improving crop resistance. In: Black, R. and Sweetmore, A. (eds) *Crop Protection in the Developing World.* BCPC Monograph No. 61, British Crop Protection Council, Farnham.

Khush, G.S. and Brar, D.S. (1998) The application of biotechnology to rice. In: Ives, C.L. and Bedford, B.M. (eds) *Agricultural Biotechnology in International Development.* CAB International, Wallingford, UK.

Morris, M. (1998a) Thailand. In: Morris, M. (ed.) *Maize Seed Industries in Developing Countries.* Lynne Rienner, Boulder, Colorado.

Morris, M. (ed.) (1998b) *Maize Seed Industries in Developing Countries.* Lynne Rienner, Boulder, Colorado.

Morris, M., Dubin, J. and Pokhrel, T. (1992) *Returns to Wheat Research in Nepal.* CIMMYT Economics Working Paper 92–04, CIMMYT, Mexico, D.F.

Morris, M., Risopoulos, J. and Beck, D. (1999a) *Genetic Change in Farmer-Recycled Maize Seed: a Review of the Evidence.* CIMMYT Economics Working Paper 99–07, CIMMYT, Mexico, D.F.

Morris, M., Tripp, R. and Dankyi, A.A. (1999b) *Adoption and Impacts of Improved Maize Production Technology: a Case Study of the Ghana Grains Development Project.* Economics Program Paper No. 99–01, CIMMYT, Mexico, D.F.

Nuffield Council on Bioethics (1999) *Genetically Modified Crops: the Ethical and Social Issues.* Nuffield Council on Bioethics, London.

O'Donoghue, M. (1995) *The Whole Family Training Program on Post-Harvest Technologies. A Program Review.* A report for the Bangladesh-Australia Wheat Improvement Project, Dhaka.

Pingali, P. and Gerpacio, R. (1997) Living with reduced insecticide use for tropical rice in Asia. *Food Policy* 22, 107–118.

Qaim, M. (1998) *Transgenic Virus Resistant Potatoes in Mexico: Potential Socioeconomic Implications of North–South Biotechnology Transfer.* ISAAA Briefs No. 7, ISAAA, Ithaca, New York.

Shimoda, S. (1996) The seeds of change. *Seed World* 6, 8–9.

Sperling, L., Scheidegger, U. and Buruchara, R. (1996) *Designing Seed Systems with Small Farmers: Principles Derived from Bean Research in the Great Lakes Region of Africa.* Agricultural Research and Extension Network Paper No. 60, ODI, London.

Tripp, R. (ed.) (1997) *New Seed and Old Laws. Regulatory Reform and the Diversification of National Seed Systems.* Intermediate Technology Publications, London.

Tripp, R. (2000) *Strategies for Seed System Development in sub-Saharan Africa. A Study of Kenya, Malawi, Zambia and Zimbabwe.* ICRISAT Working Paper, ICRISAT, Patancheru, India.

Tripp, R. and Pal, S. (2000) Information and agricultural input markets: pearl millet seed in Rajasthan. *Journal of International Development* 12, 133–144.

Tripp, R. and Pal, S. (2001) The private delivery of public crop varieties. Rice in Andhra Pradesh. *World Development* 29, 103–117.

Trutmann, P., Voss, J. and Fairhead, J. (1996) Local knowledge and farmer perceptions of bean diseases in the Central African Highlands. *Agriculture and Human Values* 13, 64–70.

Yanggen, D., Kelly, V., Reardon, T. and Naseem, A. (1998) *Incentives for Fertilizer Use in sub-Saharan Africa: A Review of Empirical Evidence on Fertilizer Response and Profitability.* International Development Paper No. 70, Michigan State University, East Lansing, Michigan.

Chapter 14

Value of Engineered Virus Resistance in Crop Plants and Technology Cooperation with Developing Countries

S. Flasinski,[1] V.M. Aquino,[2] R.A. Hautea,[3] W.K. Kaniewski,[1] N.D. Lam,[4] C.A. Ong,[5] V. Pillai[5] and K. Romyanon[6]

[1]*Monsanto, 700 Chesterfield Parkway North, St Louis, MO 63198, USA;* [2]*Institute of Plant Breeding, University of the Philippines at Los Baños, College, Laguna 4031, Philippines;* [3]*ISAAA-SE Asia Center, c/o IRRI, MCPO Box 3127, 1271 Makati City, Philippines;* [4]*National Center of Natural Science and Technology, Institute of Biotechnology, Hoang Quoc Viet, Caugiay, Hanoi, Vietnam;* [5]*Malaysian Agricultural Research and Development Institute, Strategic, Environment and Natural Resources Research Center, MARDI, GPO Box 12301, 50774 Kuala Lumpur, Malaysia;* [6]*Plant Genetic Engineering Unit, Kasetsart University, Kamphaengsaen, Nakhon Pathom 73140, Thailand*

Abstract

Modern biotechnology has significant potential to increase agricultural productivity to meet the demand for food from an increasing world population. Transformation of plants with viral genes has been proven in many cases to produce resistance to the virus from which the genes were derived. The technology has been successfully used to produce resistance in agriculturally important crops such as papaya, potato, tomato, squash, wheat and others. The benefits of transgenic virus resistance include increased yield, reduced pesticide use to control the vectors of viruses, and improved crop and food quality. The coat protein (CP) gene is most often used to confer resistance. In some cases, the expression of CP correlated with resistance, and strong evidence for prevention of uncoating was shown. For some viruses there can be both CP and RNA mechanisms that can confer resistance in transgenic plants. High levels of resistance can be produced in plants transformed with a viral replicase gene, which includes the full-length gene as well as various deletions or sequence modifications. The mechanism of resistance in replicase-expressing plants is complex and may involve expression of a protein that blocks virus replication and/or movement, as well as post-transcriptional gene silencing. In general, it has been demonstrated that plants resistant to mechanical inoculation are also resistant to vector transmission.

The development of transformation techniques broadens the possibility of use of engineered virus resistance in plant breeding. There are many destructive virus diseases of crop plants worldwide, and biotechnology may be the fastest and most efficient way to produce resistant cultivars. One example of this is production of transgenic papaya in Hawaii that showed field resistance to a potyvirus, specifically papaya ringspot virus (PRSV). However, Hawaiian resistant papaya did not confer resistance to Asian isolates of PRSV. To address the problem of PRSV impact on Asian papaya production, the Papaya Biotechnology Network was formed and is sponsored by five Southeast Asian

(SE Asian) countries (Malaysia, Thailand, the Philippines, Vietnam and Indonesia) and by ISAAA, the International Service for the Acquisition of Agri-biotech Application, with technical and financial support from Monsanto. A comprehensive plan was developed for genetic enhancement of papaya and subsequent technology transfer to SE Asia. The CP and replicase (NIb) genes from the most common PRSV isolates from four participating countries were cloned and sequenced. The sequences revealed a high degree of PRSV divergence in SE Asia that implicates the need for using genes from local strains in generation of resistant transgenic papaya. Several binary vectors for *Agrobacterium*-mediated transformation were constructed, using both CP and replicase genes and their modifications. A large effort is currently being undertaken in Network laboratories to adopt *Agrobacterium* transformation systems of papaya. Transgenic plants are now being produced and will soon be tested for PRSV resistance. Strategies for resistant line selection, agronomy tests, assays for regulatory approvals, and finally breeding and commercialization are under development. Prior to commercialization, extensive field testing and regulatory approvals are required to address agronomic performance, preservation of cultivar characteristics, and food/environmental safety. If successful, the technology would improve the yield and quality of papaya fruit in SE Asia.

Introduction

Powell-Abel *et al.* (1986) produced, in R. Beachy's laboratory, tobacco mosaic virus (TMV)-resistant tobacco plants through the expression of the gene encoding the TMV coat protein (CP). Since that initial success, many examples of coat protein-mediated resistance (CPMR) have been reported (reviewed by (Wilson, 1993; Baulcombe, 1996). More recently, the expression of other genes besides the CP gene has been demonstrated to give high levels of resistance in transgenic plants (Maiti *et al.*, 1993; Cooper *et al.*, 1995; Palukaitis and Zaitlin, 1997). With the commercial introduction of papaya (Fuchs and Gonsalves, 1995), squash (Tricoli *et al.*, 1995) and potato (Thomas *et al.*, 2000) varieties resistant to different virus infections, this technology has finally reached the market-place. Genetic engineering offers an alternative in crops where no effective resistance genes have been identified, or as a way of creating durable multiple resistances by combining natural and transgenic resistance (Hammond, 1998).

Together with the demonstration of the applicability of this technology, extensive research has been conducted to develop an accurate description of the mechanism(s) responsible for protection (Baulcombe, 1996). However, a wide number of viruses from different families and the numerous virus genes being expressed in different plants have resulted in quite distinct plant phenotypes and potential mechanisms being proposed for the resistance to virus infection (Lomonossoff, 1995). CPMR, as demonstrated for TMV in tobacco (Powell *et al.*, 1990) and cucumber mosaic virus (CMV) in tomato (Kaniewski *et al.*, 1999), correlated well with the level of protein expression in transgenic plants (Fig. 14.1a). CP may inhibit early infection processes, such as uncoating of viral RNA (in CP transgenic plants) resulting in a protection phenotype (Powell *et al.*, 1990).

In another scenario, defective mutant protein expressed in plants can compete with wild-type virus protein for the target of action but cannot support functions needed for the virus life cycle (replication or movement). This resistance phenotype is also dependent on the accumulation of protein. However, in some cases there is little correlation between the CP expression level and the degree of protection because the mRNA of the transgene is more critical for induction of resistance. Examples of this mechanism have been described for some viruses (Goodwin *et al.*, 1996).

RNA-mediated resistance to plant viruses was shown when a high level of transgene RNA was produced in the plant cell nucleus, as assayed by nuclear run-on, but the accumulation of the transgene in

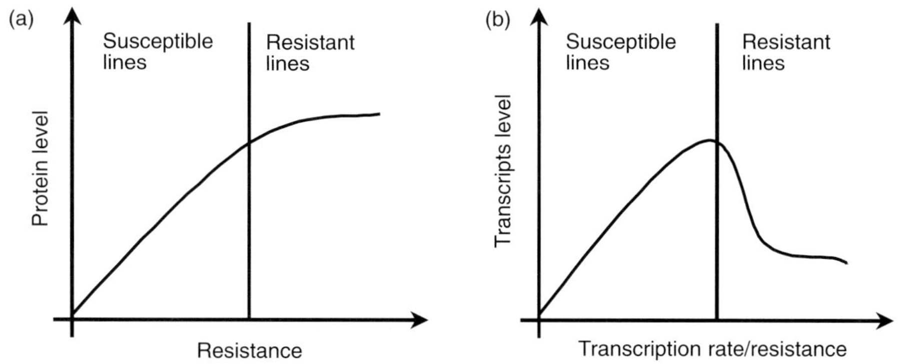

Fig. 14.1. Proposed mechanisms of resistance in transgenic plants expressing virus genes. (a) Protein-mediated resistance; (b) RNA-mediated resistance.

the cytoplasm was low (Dougherty *et al.*, 1994; English *et al.*, 1996; Fig. 14.1b). It was suggested that a high transcription rate might activate cytoplasmic procsses which target this RNA for degradation (Lindbo *et al.*, 1993) due to post-transcriptional gene silencing (Baulcombe, 1996). There could also be a combination of different mechanisms that act in different cells (tissues) of transgenic plants that involve inhibition of virus cell-to-cell and long-distance movement (Hellwald and Palukaitis, 1995).

Replicase-mediated resistance may be more suitable for the generation of commercial products in cases where a CP gene approach has not been as effective, such is the case for production of potato resistant to potato leafroll virus (PLRV; Thomas *et al.*, 1997, 2000). Protein expression may be necessary for expression of defective replicase genes; however, in some cases the replicase-mediated resistance may be RNA mediated (Baulcombe, 1996; Palukaitis and Zaitlin, 1997). The spectrum of resistance to virus strains and isolates in CP or replicase-transformed plants may differ, with replicase-mediated resistance being more narrow (Palukaitis and Zaitlin, 1997). However, differences in the spectrum of resistance can be seen even in transgenic lines generated with the same gene construct (Donson *et al.*, 1993; Gonsalves, 1998; Kaniewski and Lawson, 1998).

The data discussed above led to the conclusion that to obtain virus resistant plants, a strong constitutive promoter is needed to drive the gene expression, regardless of whether resistance is conferred by protein or by transgene RNA. In the case of protein-mediated resistance, transgene transcripts will be transported to the cytoplasm and translated to produce the gene product resulting in a resistance phenotype. In the case of RNA-mediated resistance, transgene transcripts will be translocated to the cytoplasm and possibly upon reaching a specific threshold level will trigger a plant mechanism leading to the degradation of the transgene RNA. The incoming virus from which the transgene was derived will be subjected to degradation because of sequence similarity (Dougherty *et al.*, 1994; Baulcombe, 1996).

Line (event) selection after transformation is critical for the development of resistant crop plants since a wide range of phenotypes can be observed, from very mild or no protection to extreme resistance to the virus from which the gene was derived. Lines with low resistance (in Fig. 14.1, left of the vertical lines) are discarded and lines with high resistance (right of the vertical lines) are maintained to perform extensive testing under field conditions (Kaniewski and Thomas, 1998; Thomas and Kaniewski, 1998). Low accumulation of virus as determined by enzyme-linked immunosorbent assay (ELISA), together with the absence of virus symptoms and the occurrence of a majority of non-infected individuals within the plant population, are good indications of a high resistance phenotype (Kaniewski and Lawson, 1998; Kaniewski *et al.*, 1999; Lawson *et al.*, 2001).

Transgenic virus resistance, like genetic resistance, is of great value for modern agriculture and can create even greater savings for subsistence farmers, who cannot afford other measures of crop protection (Qaim, 1998). It can eliminate or reduce the need for alternative measures of virus protection, such as seed certification, vector control by pesticides or cross protection, which may be less acceptable and more difficult to apply in developing countries.

Papaya suffers from several diseases and pests, the most widespread and destructive of which is papaya ringspot virus (PRSV; Gonsalves, 1998). The virus has significantly reduced papaya production in the Southeast Asia (SE Asia) region, causing substantial losses of income to farmers, a relative scarcity of the fruit in the market and higher costs to consumers (Hautea *et al.*, 1999). Transgenic papaya expressing the CP or replicase genes may be the optimal solution since no genetic resistance to PRSV has been identified (Gonsalves, 1998). The Papaya Biotechnology Network works closely with the International Service for the Acquisition of Agribiotechnological Application (ISAAA) to ensure effective transfer of this technology to different countries (Hautea *et al.*, 1999).

Monsanto has chosen to participate in the Network and in the development of papaya resistant to PRSV because the company is committed to 'sharing', that is, bringing the knowledge and advantages of all forms of modern agriculture to resource-poor farmers in the developing world thereby improving food security while protecting the environment. Monsanto scientists are involved in technology sharing and agricultural development collaborations with public institutions, non-profit groups and local industry around the world. Private companies have a role to play in projects such as this one, that can bring the value and benefits of biotechnology to farmers and consumers, particularly in developing countries.

Materials and Methods

PRSV isolates

PRSV, a member of the Potyviridae family (Berger *et al.*, 2000), has flexuous, filamentous particles of 780 × 12 nm and contains a positive-sense ssRNA (De La Rossa and Lastra, 1983). The virus has a single type of CP approximately 36 kDa in size (Gonsalves and Ishii, 1980). The genome of the virus contains only one open reading frame (ORF), which encodes a polyprotein of about 330 kDa (Yeh *et al.*, 1992) that is cleaved into at least seven functional proteins by three virus-encoded proteinases (Fig. 14.2a). Based on the target host for infection, PRSV is grouped into type P (infects cucurbits and papaya) and type W (infects cucurbits but not papaya; Purcifull *et al.*, 1986).

PRSV was isolated from infected papaya plants in four countries: Malaysia (PRSV-M), the Philippines (PRSV-Ph), Thailand (PRSV-TSb) and Vietnam (PRSV-Vn). In each case the most common and/or severe isolate for the country was chosen. Virus was inoculated into courgette and after symptoms appeared on plants, leaf samples were harvested and dried under calcium chloride. After USDA import permits were obtained, the scientist shipped dry plant tissue to the USA. Scientists from the respective Network laboratories used the samples for future experiments during their research internships at Monsanto.

For the sequence comparison studies, several PRSV sequences were obtained from GenBank.

Cloning and sequencing of PRSV-M from Malaysia

PRSV-M was inoculated into courgette and virus was isolated using a protocol previously described (Gonsalves and Ishii, 1980). Purified virus was used to isolate viral RNA (White and Kaper, 1989) and cDNA was synthesized using Life Technologies Gibco-BRL cDNA synthesis kit (Gaithersburg, Maryland) according to the

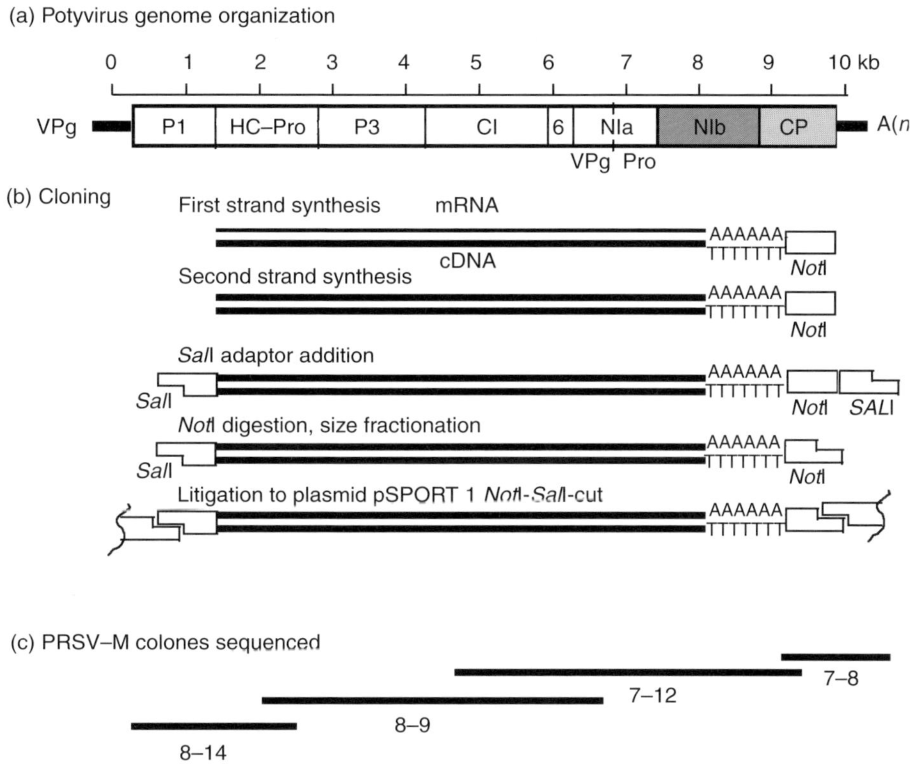

Fig. 14.2. (a) Schematic representation of potyvirus genome. Non-translated regions are shown as lines. Coding region of the polyprotein is shown as a long rectangle divided into smaller parts (representing single proteins). P1, proteinase; HC-Pro, helper component-proteinase; P3, protein; CI, cylindrical inclusion protein (helicase); NIa, nuclear inclusion protein a from which the *N*-terminus part is VPg (genome-linked protein) and the *C*-terminus is a proteinase; NIb, nuclear inclusion protein b (replicase); CP, coat protein. Genes used to produce plant expression vectors are highlighted. (b) Steps in cloning of PRSV-M genome. (c) Selected overlapping clones of PRSV-M were fully sequenced and used to assemble nearly full-length sequence. The oligo-dT primer used in first-strand synthesis primed not only at the 3′ end where the polyA track is present, but also internally in A-rich regions of the genome.

manufacturer's protocol. After size fractionation, cDNA was ligated into the pSPORT1 vector (Fig. 14.2b).

Recombinant clones were transformed into *E. coli* DH5α competent cells and plated on LB-Agar containing 100 mg l^{-1} ampicillin and X-gal. The next day, white colonies were inoculated into 2 ml of liquid LB medium containing 100 mg l^{-1} ampicillin. Plasmid DNA was isolated using the Qiagen miniprep kit (Valencia, California) and digested with different restriction enzymes to confirm the presence of inserts. Several clones were sequenced from both ends to verify the presence of PRSV-M sequences. Four overlapping clones that cover almost the entire genome of PRSV-M (except 16 nucleotides from the 5′ end) were sequenced completely (Fig. 14.2c). The sequence from these clones was used to assemble the full-length sequence of PRSV-M using University of Wisconsin GCG programme package (Devereux *et al.*, 1984). The full-length sequence of PRSV-M was compared with two published PRSV sequences (PRSV-HA isolate from Hawaii, acc. no. S46722, and PRSV-YK from Taiwan; acc. no. X97251).

Cloning PRSV CP and replicase (NIb) genes into plant expression vectors

The NIb and CP genes are located at the 3′ end of the genome with CP being at the *N*-terminus of the polyprotein (Fig. 14.2a). To clone the NIb and CP from PRSV-M, two sets of PCR primers were designed based on the complete genome sequence data obtained in this study (primer sets 7, 8 and 9, 10; Table 14.1B).

The sequence alignment of the three known full-length PRSV genome sequences (Hawaiian, Taiwanese and Malaysian isolates) was used to pick several conserved regions at the 3′ part of the PRSV genome. Three sets of polymerase chain reaction (PCR) primers (reverse and forward) for amplification of junctions between the 3′ end genes were produced (Table 14.1A). Total RNA was isolated from infected dry tissue containing different virus isolates (White and Kaper, 1989). The cDNA was synthesized using Superscript II Reverse Transcriptase (Life Technologies) and oligo-dT_{15}. The cDNA was used in PCR with three sets of primers that amplified regions between the proteinase (NIa) and replicase (primer 1 and 2), between replicase and CP (primer 3 and 4) and between CP and the 3′ UTR (primer 5 and 6). Similar PCR amplification products were produced for PRSV-Ph, PRSV-TSb and PRSV-Vn. The PCR products were gel isolated using the Qiagen Gel Isolation kit (Valencia, California) and sequenced to identify junctions between the 3′ end genes.

Binary vector construction

To clone the genes into binary vectors for each virus isolate, two sets of primers were made (Table 14.1C–E). The first set of primers amplified the full-length replicase gene and the second set was utilized for the amplification of the full length CP gene (primers 11–14, PRSV-TSb; primers 15–18, PRSV-Ph; primers 19–22, PRSV-Vn).

The ATG start codon was added to the forward primers at the 5′ end of both genes (CP and NIb) to initiate translation and the TGA stop codon was added to the reverse primer at the end of replicase gene to terminate translation. In addition, convenient restriction site sequences were added to the primers (*Nco*I or *Pci*I that is compatible with *Nco*I, in forward primer; *Bam*HI or *Xba*I in reverse primer) to facilitate cloning. Fragments were cloned into a plant expression binary vector (pMON54904B) containing the 35S promoter with a duplicated enhancer region, the hsp17.9 leader from soybean, and the 35S 3′ UTR (Guilley *et al.*, 1982; Kay *et al.*, 1987; Fig. 14.3).

The vector also contains the neomycin phosphotransferase (NPTII) plant selectable marker gene (Sanders *et al.*, 1987). For cloning, the vector was digested with *Nco*I and *Bam*HI or *Nco*I and *Xba*I. The PCR products were digested with appropriate restriction enzymes as indicated in Fig. 14.3. Recombinant clones were transformed into DH5α competent cells and plated on LB-Agar containing 100 mg l^{-1} spectinomycin. Bacterial colonies were screened for the presence of plasmids with PRSV sequences by digestion of plasmid DNA with restriction enzymes and gel electrophoresis. Appropriate clones were selected and sequenced to confirm the presence of the NIb or CP genes.

In addition, plant expression binary vectors were made containing CP genes with ~250 nt of the inverted repeat of the CP gene (ΔCP-IR) inserted just downstream of the stop codon (Fig. 14.3). The ΔCP-IR fragment of the CP gene was amplified using primers 23 and 24 and pMON65301 (CP, Thai isolate), pMON65304 (CP, Vietnamese isolate), pMON65306 (CP, Filipino isolate) or pMON54910 (CP, Malaysian isolate) as a template. Primers were designed in the conserved core of the CP gene for all four isolates of PRSV used in this study. PCR products were digested with *Xba*I and *Avr*II restriction enzymes and ligated into an appropriate vector containing the corresponding CP gene, pre-digested with *Xba*I, and dephosphorylated with shrimp alkaline phosphatase (Roche, Indianapolis, Indiana). Having *Xba*I and compatible *Avr*II restriction sites at the end of the ΔIR-

Table 14.1. Sequences of the primers used for polymerase chain reaction (PCR) and cloning of papaya ringspot virus (PRSV) replicase (NIb) and coat protein (CP) genes from four isolates of PRSV.

Primer name		Sequence
A	Primers in the 3′ conserved regions of RPSV genome	
1	PRSVNIaF	TACAATGGTATAGCTTTCGTGGTTTGC
2	PRSVNIbR	GCTCAGCAAATGAACTAGAGAATGAGTC
3	PRSVNIbF	TATCAATGGTGATGATCTCTGTATTGC
4	PRSVCPR	CGAGCCCTATCAGGTGTTTTCGAATT
5	PRSVCPF	ATGCCGCGGTATGGAATCAAGAGA
6	PRSVUTRR	TCAATTGCGCATACCCAGGAGAGAGT
B	Gene specific primers for PRSV-M	
7	PRSV-M-NIbF	GGATGGACCATGGCGAGTGGAAGATGGCTT
8	PRSV-M-NIbR	CACTTGGATCCTCATTCATGATACACCAGCAGTT
9	PRSV-M-CPF	GCTGGAGATCTCCATGGCGTCGATAAGCACTGGCGAT
10	PRSV-M-CPR	CACACTATCTAGAATTCAATTGCGCATA
C	Gene specific primers for PRSV-TSb	
11	PRSV-TSb-NIbF	*TGAGCAACC**ATG**GCG*AGTGGAAGTCGATGGCTATT
12	PRSV-TSb-NIbR	*TACGGATCC**TCA***TTCGTGATACACTAATAACTCAG
13	PRSV-TSb-CPF	*ACACGAACC**ATG**GCG*TCGAGAAGCATCGACGATTA
14	PRSV-TSb-CPR	*TACGGATCC***TCA**ATTGCGCATACCCAGGAGAGA
D	Gene specific primers for PRSV-Ph	
15	PRSV-Ph-NIbF	*TGAGCAAAC**ATG**TC*GAGTGGAAGTCGGTG
16	PRSV-Ph-NIbR	*CGGGATCC**TCA***TTCATGATACACCAGCAA
17	PRSV-Ph-CPF	*TCATGAACC**ATG**GCG*TTAAGGAGCACCGAT
18	PRSV-Ph-CPR	*CGGGATCC***TCA**ATTGCGCATACCCAGGA
E	Gene specific primers for PRSV-Vn	
19	PRSV-Vn-NIbF	*GGATGGACC**ATG**GCG*AGTGGAAGTCGATGGCTT
20	PRSV-Vn-NIbR	*GCTCTAGA**TCA***TTCATGATACACCAGTA
21	PRSV-Vn-CPF	*TCATGAACC**ATG**GCG*TCAAGAAGCATAGAC
22	PRSV-Vn-CPR	*AGGGATCC***TCA**ATTGCGCATACCCAGGA
F	Primers for amplification of CP inverted repeats	
23	PRSV-CPIR-F	*GCTCTAGA*GGAGCTAGTGACGGAAA
24	PRSV-CPIR-R	*GCTAGCCCTAGG*CTTGCATTTCGTTATCAT

Non-viral sequences are shown in *italics*, The **ATG** start codon and the complement (**TCA**) to the stop codon TGA are shown in **bold.** Restriction enzymes recognition sites used in cloning are shaded. R, reverse; F, forward; IR, inverted repeat.

CP allowed us to screen for the orientation of insert by restriction digest. For all PCR reactions the Expand High Fidelity PCR system from Roche was used.

The PRSV clones were transformed to *Agrobacterium tumefaciens*, strain ABI. The presence of appropriate plasmids in *Agrobacterium* was confirmed by isolation of plasmid DNA, retransformation into *E. coli* DH5α and restriction enzyme mapping of plasmids isolated from retransformed *E. coli*. This procedure was done to ensure that plasmids were stably maintained in *Agrobacterium*.

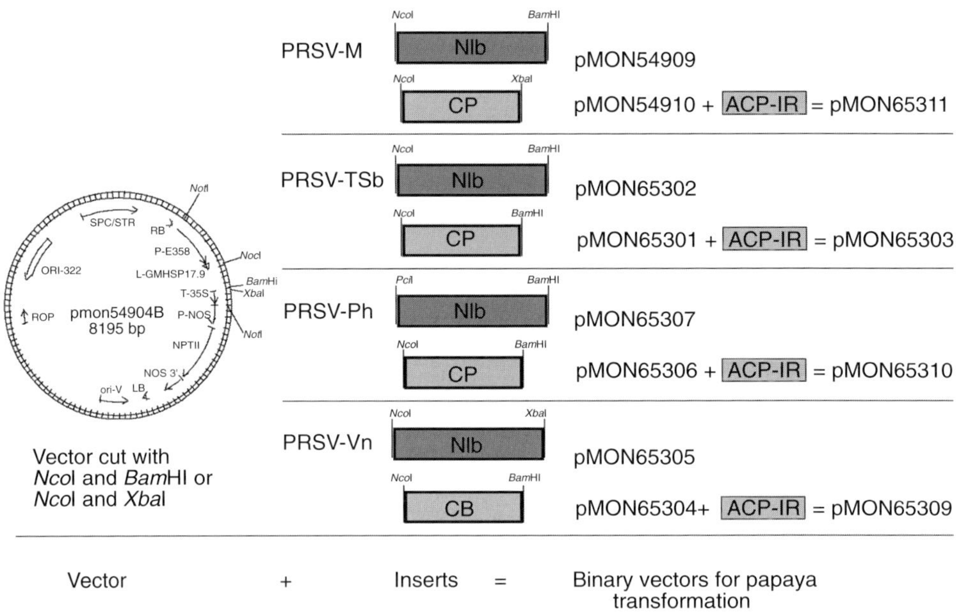

Fig. 14.3. Cloning of the replicase (NIb) and coat protein (CP) genes from four papaya ringspot virus (PRSV) islolates into a binary vector for *Agrobacterium*-mediated papaya transformation. The restriction sites used in cloning are indicated on the figure. The CP inverted repeat fragments (ΔCP-IR) from each isolate was inserted into the *Xba*I site of the corresponding CP vector.

Papaya transformation

For the production of transgenic papaya, a method developed by Ying *et al.* (1999) was used. Somatic embryos obtained from liquid-culture were wounded by vortexing with tungsten M-15 in half-strength MS liquid medium and then co-inoculated with the transformed and induced *Agrobacterium* for 3 days. The next steps include callus initiation on media with 600 mg l^{-1} carbenicillin, and regeneration on media containing 300 and 150 mg l^{-1} kanamycin, and finally rooting. Rooted R_0 plantlets were maintained in the greenhouse and will be used for molecular characterization of transformants. Resistance tests will be done initially on segregating R_1 progeny and confirmed on the homozygous R_2 population. The line selection criteria will be based on virus resistance, the presence of an intact copy of the transgene without any unnecessary sequences, proper R_1 and R_2 segregation, preservation of agronomic characteristics and yield.

Results

PRSV-M full-length sequence comparison

The full-length sequence of the Malaysian isolate was found to be very similar to PRSV-YK from Taiwan (Table 14.2). At the nucleotide level there is more than 95% identity, and at the protein level about 97% identity, between those two isolates. The PRSV-HA was less similar (about 85% at nucleotide level) to both isolates, PRSV-M and PRSV-YK.

The similarity of single proteins was also compared between the PRSV-M and other PRSV isolates as well as tobacco etch virus (TEV), one of the best-characterized potyviruses (Table 14.3). In general, PRSV-M proteins (at the amino acid level) were

Table 14.2. Nucleic acid and protein sequence identity of three fully sequenced papaya ringspot virus (PRSV) isolates.

	Nucleic acid identity (%)		
	PRSV-HA	PRSV-YK	PRSV-M
PRSW-HA	100.0	83.4	85.2
PRSV-YK		100.0	95.0
PRSV-M			100.0
	Polyprotein identity (%)		
PRSW-HA	100.0	90.6	90.8
PRSV-YK		100.0	96.7
PRSV-M			100.0

Table 14.3. Comparison of amino acid composition in Malaysian papaya ringspot virus (PRSV-M) proteins with those of the other PRSV isolates and tobacco etch potyvirus (TEV).

	Amino acid identity %							
Virus	P1	HC-Pro	P3	CI	6K	NIa	NIb	CP
PRSV-YK	92.3	97.8	97.5	98.9	91.2	97.9	97.9	94.8
PRSV-HA	67.1	95.7	93.5	98.1	89.5	93.9	95.9	94.1
TEV	14.3	47.4	32.0	54.3	35.1	46.3	58.5	56.2

very similar to those of PRSV-YK, with helicases (CI) being most conserved (98.9%) and 6K being least conserved (91.2% identity). The CI protein was also very similar between PRSV-M and PRSV-HA (98.1% identity). The most divergent protein between the three isolates of PRSV was the *N*-terminal P1 protein (67.1% identity between isolates M and HA and 92.3% identity between M and YK). The identity of PRSV and TEV was very low, ranging from 58.5% for the replicase protein and 14.3% for the P1 protein.

SE Asian PRSV isolates

The sequences of the NIb and CP genes from four isolates (M, TSb, Ph and Vn) were compared with those of other isolates available from GenBank (isolate from Hawaii, HA, acc. no. S46722; isolate from Taiwan, YK, acc. no. X97251; and isolate from China, SM, acc. no. X96538). The most divergent isolate was PRSV-HA with 85–86% nucleotide identity to other PRSV isolates (Table 14.4, upper part). Nucleotide identity between SE Asian isolates ranged from 89.7 to 91.7% with the exception of PRSV-M and PRSV-YK. Identity between those two isolates was about 95%.

Interestingly, the sequence conservation in the CP gene of PRSV isolates was higher than in the replicase gene (Table 14.4, lower part). When only the CP gene identity was analysed, the difference between Hawaiian and SE Asian isolates was less noticeable.

Figure 14.4 shows the dendrogram of the genetic distance between NIb and CP fragments of the PRSV genome. Again, all SE Asian isolates showed quite similar genetic distance from each other and isolates M and YK clustered together, having the smallest genetic distance between them (Fig. 14.4a).

The PRSV CP sequence comparisons

Many PRSV CP sequences from around the world are available in GenBank. However,

Table 14.4. Identity between replicase (NIb) and coat protein (CP) nucleotide sequence from SE Asian and Hawaiian papaya ringspot virus (PRSV) isolates.

	Nucleic acid identity: NIb+CP (%)						
	PRSV-YK	PRSV-M	PRSV-SM	PRSV-TSb	PRSV-Ph	PRSV-Vn	PRSV-HA
PRSV-YK	100.0	95.0	91.7	90.4	90.0	91.1	85.5
PRSV-M	95.1/94.5	100.0	91.4	90.4	89.7	90.3	85.2
PRSV-SM	91.0/92.4	90.7/92.4	100.0	91.1	90.3	90.4	85.4
PRSV-TSb	89.6/90.7	89.8/90.3	90.8/91.6	100.0	89.9	89.8	85.4
PRSV-Ph	89.9/89.9	89.6/89.8	90.0/91.0	89.4/90.2	100.0	89.7	84.6
PRSV-Vn	90.7/91.4	89.5/91.6	89.4/91.8	88.7/91.2	89.3/89.9	100.0	85.3
PRSV-HA	82.9/89.6	82.9/88.6	83.6/87.8	82.6/88.2	82.4/88.4	82.1/89.7	100.0
	Nucleic acid identity: NIb/CP (%)						

The top part of the table (shaded grey) shows identities of both genes, NIb plus CP. The bottom part shows identities between NIb and CP genes separately.

some of the PRSV CP sequences deposited in GenBank are not full-length, typically they miss a small 5′ portion of the gene. For comparison, we selected 14 isolates including the four sequenced in this study (Table 14.5). Isolates from within a country or location are usually quite similar; examples of this are two isolates from Australia (BD and NT; 98% identity) and Brazil (DFw and DFp; 95% identity). The sequences from Brazil represent two types of PRSV, type P (DFp, papaya-infecting) and type W (DFw, cucurbit-infecting). The lowest sequence identity between CP genes of different isolates was above 88%. Interestingly, the 3′ portion of the CP is

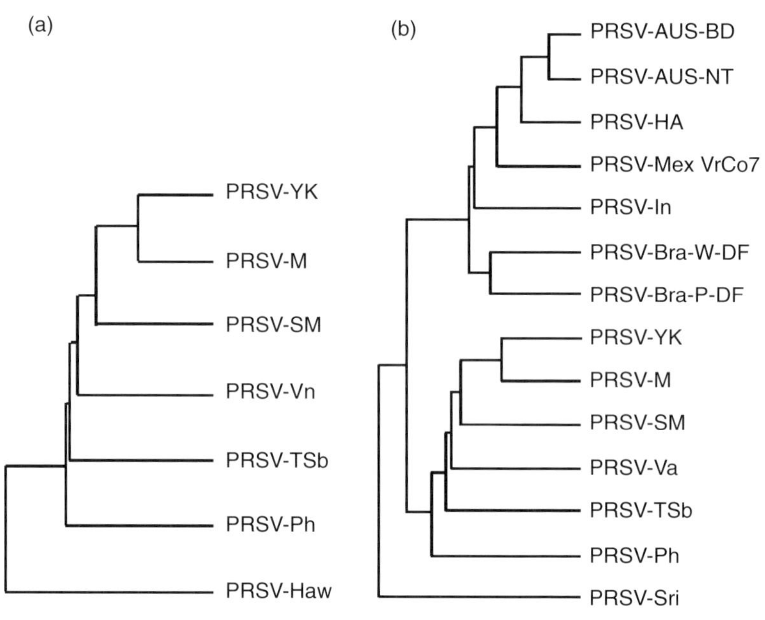

Fig. 14.4. (a) Dendrogram of the genetic distances of replicase (NIb) and coat protein (CP) genes between seven papaya ringspot virus (PRSV) isolates from SE Asia and Hawaii. (b) Dendrogram of genetic distance between CP genes of 14 PRSV isolates from diverse locations. Genetic distances were calculated from the percentage nucleotide differences between isolates.

Table 14.5. Nucleic acid sequence identity of the coat protein (CP) genes of 14 papaya ringspot virus (PRSV) isolates from different parts of the world.

	Nucleic acid identity (%)													
Country	Australia		Hawaii	Brazil		Taiwan	Malaysia	China	Thailand	Philippines	Vietnam	Mexico	India	Sri Lanka
Isolate	BD	NT	HA	DFw	DFp	YK	M	SM	TSb	Ph	Vn	VrC	In	Sri
GB Acc#	U14736	U14744	S46722	AF34469	AF34460	X97251	NA	X96538	NA	NA	NA	AF31957	AF063220	U14741
BD	100.0	98.1	96.8	94.9	94.1	91.0	90.0	89.6	91.1	89.1	92.0	95.4	94.5	88.5
NT		100.0	96.6	94.9	93.9	90.9	90.9	90.0	91.2	89.8	91.8	95.2	94.2	88.7
HA			100.0	94.2	93.7	91.0	90.6	89.8	91.5	89.5	91.9	95.2	94.0	89.0
DFw				100.0	94.8	98.9	90.3	89.4	89.9	88.7	90.4	92.9	93.0	88.0
DFp					100.0	98.7	89.6	89.4	89.5	88.1	90.7	92.9	92.8	87.6
YK						100.0	95.6	93.9	93.3	91.8	93.2	89.2	90.5	89.2
M							100.0	93.0	92.7	91.4	93.0	89.4	90.4	89.7
SM								100.0	92.7	92.1	92.7	89.9	90.6	88.2
TSb									100.0	92.3	92.6	90.1	89.9	89.2
Ph										100.0	92.0	88.9	89.1	87.8
Vn											100.0	91.0	90.9	88.3
VrC												100.0	90.4	88.5
In													100.0	89.7
Sri														100.0

NA, GenBank accession number not available.

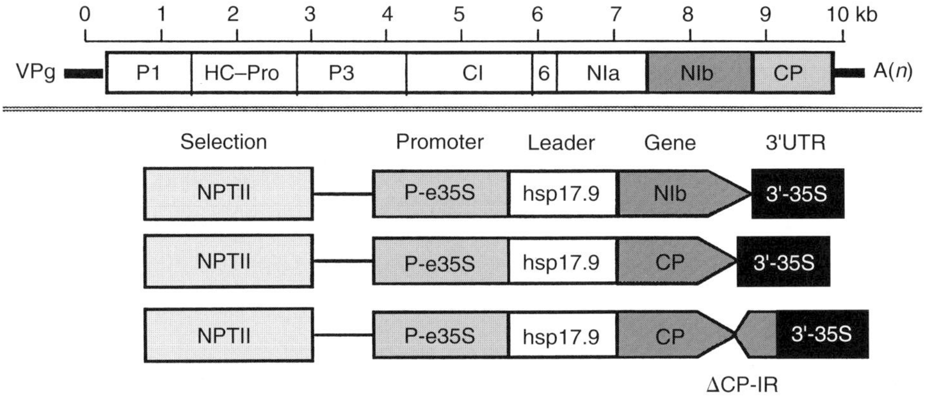

Fig. 14.5. Schematic representation of plant expression binary vectors built for each of the four virus isolates. At the top, genome organization of PRSV is shown and locations of the replicase (NIb) and the coat protein (CP) genes are highlighted. NPTII, neomycin phosphotransferase; P-e35S, promoter from cauliflower mosaic virus (CaMV) with duplicated enhancer; hsp17.9, leader sequence from heat shock protein 17.9 from soybean; 3′-35S, 3′ untranslated sequence from CaMV. The orientation of the gene coding regions is shown by arrowed boxes.

highly conserved with very few nucleotide differences.

The dendrogram shows that all isolates from around the world can be placed into two groups (Fig. 14.4b). The first group, at the top of the dendrogram, consists of sequences with genetic distance from about 2 to 7% and include isolates from different countries and continents (Australia, Hawaii, Brazil, Mexico and India). The second group consists of isolates from SE Asia only. Although those isolates are located in neighbouring countries, genetic distance between them was higher (4.4–12.1%) than in the first group. The isolate from Sri Lanka was equally divergent from the first and second group of PRSV.

Vector for papaya transformation

Three plant expression vectors were constructed for PRSV isolates from Thailand, Vietnam, Philippines and Malaysia (Fig. 14.5). Viral genes in the expression cassettes (driven by enhanced 35S) were sequenced to ensure their correctness. All vectors were stably maintained in *A. tumefaciens* ABI.

Currently these vectors are being used in transformation of papaya somatic embryos as the first step in generation of transgenic papaya plants (Fig. 14.6). Putative transgenic embryos started to grow after being cultured for 3 months on selective medium containing 150 mg l^{-1} of kanamycin. Each kanamycin-resistant line readily grew and

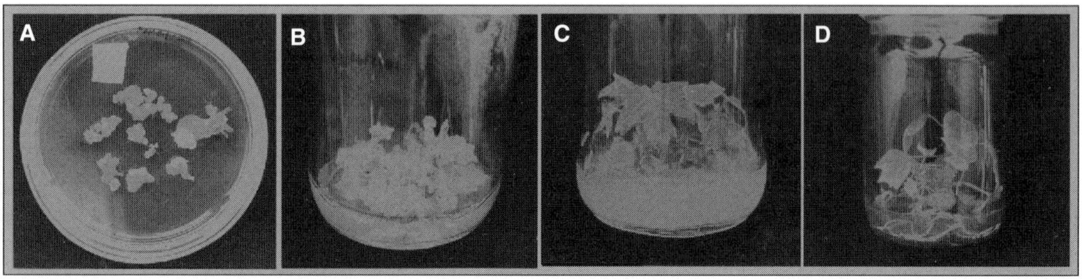

Fig. 14.6. *Agrobacterium*-mediated transformation and regeneration of papaya. (A) Selection of transformed somatic embryos. (B) Germination of resistant lines. (C) Healthy regenerated plantlet. (D) Plantlet on rooting media ready to be transferred to soil.

regenerated on germination and maturation medium. Regenerated shoots were transferred to root developing medium. Vermiculite gave better development of roots in transgenic papaya, compared to other supportive media. The R_0 plants will be analysed for the presence of the NPTII marker gene using ELISA. Evaluation of resistance to PRSV will be done when R_1 plants become available. The results of virus resistance tests will be the most important criteria for line selection.

Discussion

PRSV sequence diversity

The comparison of the sequences of PRSV isolates from different locations indicates that this virus could have been spread naturally during plant evolution and the expansion of papaya cultivation. This scenario may be very applicable to the SE Asia region where genetic distance between isolates is quite high, in some cases more than 10% when the CP gene or both the CP and replicase genes were used in comparisons. The sequence conservation in the CP gene was higher than in the replicase gene. It is very likely that the virus was introduced to the region long ago and evolved independently, with small exceptions. A higher-than-average degree of similarity found between isolate PRSV-YK from Taiwan and PRSV-M from Malaysia was surprising because of the distance between those two countries. This may suggest that PRSV in Malaysia and Taiwan shared evolutionary pathways longer than that for the other isolates. A high degree of similarity between Taiwanese and Malaysian isolates can be seen throughout the entire genome including the N-terminal P1 gene (92.3% identity). In contrast, the P1 gene from the Hawaiian isolate differs from both the YK and M isolates by about 33%. The P1 gene comparison could be very useful for determining similarity and origin of PRSV isolates; unfortunately only sequences of three isolates are now available. The Taiwanese PRSV isolate may have the same origin as PRSV from China, based on similarity of these isolates. Recently Wang and Yeh (1997) reported a detailed comparison of PRSV HA and YK. They concluded that variation in the leader sequence and P1 protein suggests that the two PRSV strains were derived from different evolutionary pathways in different geographic areas. Shukla and Ward (1989) suggested that higher than 90% identity within potyvirus CP sequence may be used to classify isolates as belonging to the same species. Based on complete sequence data of the PRSV-M and PRSV-YK, we conclude that these isolates are strains of the same species. Since the sequence identity between SE Asian isolates in some cases is smaller than 90% for the CP and replicase genes, considering them either PRSV strains or separate virus species according to Shukla and Ward (1989) is debatable. However, biological data (not shown) suggest their classification to the same potyvirus species.

High sequence identity between the CP gene of Australian, North and South American, and Hawaiian PRSV isolates may indicate their common origin despite dramatic distances between these locations. However, PRSV type P and type W from one location are usually more similar than type P isolates from different locations. This suggests that type P may originate from type W as reported by Bateson *et al.* (1994) in Australia. The role of human migration and crop introduction to new regions and a concomitant spread of virus can also be envisioned. For example with the introduction of papaya to Hawaii the virus was also introduced (Gonsalves, 1998).

PRSV sequence diversity in SE Asia is much higher than in other locations. This justifies very well with the use of genes from local virus isolates for the production of resistant transgenic papaya.

Value of the virus resistant transgenic plants

Plant viruses often cause considerable crop damage and significantly reduce yields in the absence of effective protection treatments. In cultivated crop plants, the decrease in yields and crop value caused by virus infection is estimated at US$60 billion per year (Cann, 1997). Virus infected crops are always of inferior quality, which reduces their market value (Waterworth and Hadidi, 1998).

Attempts to maintain crop health create substantial expenses to farmers. Strategies to reduce losses by plant viruses include conventional approaches like seed certification and production, and virus-free planting material, quarantine and pesticide control of virus vectors. Cross protection by infection with an attenuated strain of the virus is still an accepted practice, despite high ecological risk. These procedures are expensive and need to be performed every year. Other than transgenic solutions, only breeding for virus resistance and vector resistance offer long-term solutions to a virus disease problem; however, this process is lengthy and the resistance genes are not always available in the desired species. Where genetic resistance is not available, virus-derived transgenic resistance might be the best option (Kaniewski and Lawson, 1998).

In 1997 Monsanto released transgenic NewLeaf Plus Russet Burbank potatoes resistant to PLRV and Colorado potato beetle (CPB; Lawson *et al.*, 2001). This resistance was achieved by the expression of the full-length unmodified replicase gene (open reading frame 2a/2b) of PLRV and the *cry*3A insect control protein from *Bacillus thuringiensis* var. *tenebrionis* (*Bt*). Extreme resistance of NewLeaf Plus potatoes in field conditions exposed to natural diversity of PLRV strains (Thomas *et al.*, 2000) indicated that this resistance is not strain-specific as reported for other replicase-mediated resistance (Palukaitis and Zaitlin, 1997). Cultivation of the NewLeaf Plus Russet Burbank potatoes results in reduction or elimination of insecticide use (in average of about 60 kg ha^{-1}) that relates to an average saving of $300 per hectare (Riebe and Zalewski, 2002).

In NewLeaf Y potatoes, the resistance to potato virus Y and CPB were engineered by expression of the PVY CP gene and the *Bt* gene *cry*3A. PVY resistance allowed seed growers to pass seed certification more easily and protect their income as well as the income of potato farmers.

Squash with resistance to multiple viruses (CMV, WMV 2 and ZYMV) is another good example of protecting both yield and food quality (Tricoli *et al.*, 1995). Tomato resistant to CMV is ready for commercialization (Kaniewski *et al.*, 1999). Research on transgenic virus resistance is highly advanced in major crops like wheat, maize, cassava, rice, sweet potato, sugarbeet and others.

The introduction of PRSV in Hawaii was an excellent example of a collaborative effort between non-commercial institutions and papaya growers and the goodwill of intellectual property rights owners (Gonsalves, 1998). Transgenic papaya practically saved papaya production in Hawaii (Lius *et al.*, 1997). This initial case of transgenic virus resistance in papaya grown in Hawaii seems to be specific to the Hawaiian strain of PRSV since resistance was not observed for any other PRSV isolates. However, other papaya lines have been produced by the Gonsalves laboratory using a non-translatable CP gene and are highly resistant to a broader spectrum of PRSV isolates (Gonsalves, 1998). In Taiwan, transgenic papaya resistant to a broad spectrum of PRSV isolates was produced by Yeh's laboratory using the CP gene. Recently, a report indicated the development of transgenic papaya in China using the replicase gene (Chen *et al.*, 2001). These data clearly indicate the feasibility of a transgenic approach in controlling PRSV in papaya; however, the technology may be expensive to develop but easy to apply and use in developing countries, which account for 98% of papaya production worldwide (Gonsalves, 1998).

Technology cooperation with developing countries

The purpose of the papaya project is to accelerate the development and safe deployment of PRSV-resistant local papaya cultivars, thereby benefiting small-scale farmers in SE Asia. The project, developed and brokered by ISAAA, with support from Monsanto, will allow SE Asian countries to incorporate PRSV resistance into the most important local cultivars. As network members, participating national research institutes and Monsanto work individually and collectively to address the critical aspects of product development, effective biosafety and food safety regulations, product dissemination, and product acceptance by growers, consumers and the general public.

The approach is to encourage as much sharing of technology as possible between the SE Asian countries through the use of the improved transformation protocols and to produce local papaya varieties suited for each country transformed with vectors containing genes from local virus strains. Advancement in the project includes the sequencing of local PRSV strains, construction of plant expression vectors for *Agrobacterium* transformation, initiation of papaya transformation using a protocol, developed in Mike Davis' laboratory (Ying *et al.*, 1999), and selection of many transgenic papaya plants at the Malaysian Agricultural Research and Development Institute (MARDI) as well as the Kasetsart University, Thailand.

Hands-on training in molecular biology, plant transformation, food safety and regulatory procedures for scientists from Papaya Biotechnology Network was organized by ISAAA. In addition, the ISAAA has organized workshops on specific topics in participating countries since 1998 when the Papaya Biotechnology Network was officially launched (Hautea *et al.*, 1999). In 2001 the biggest challenge will be to scale up the number of transgenic lines produced from local varieties for virus-resistance tests required to move towards commercialization of PRSV-resistant papaya varieties.

Monsanto functions as an active partner in the Papaya Biotechnology Network, and provides technical and general support to the national partners in all aspects of the project. Monsanto is also providing access to proprietary technology and assisting in public education about the benefits of biotechnology for the region. Monsanto has participated in technology cooperation projects and activities for a number of years; and the importance of this effort has been reinforced in the New Monsanto Pledge recently announced by Chief Executive Officer Hendrik Verfaillie (2000). Included in this pledge which will guide the company's ongoing development in biotechnology, is the principle of 'sharing'. A dedicated team has been created within Monsanto to facilitate technology sharing and agricultural development collaborations for the benefit of resource-poor farmers in the developing world. Monsanto's Technology Cooperation programme currently includes several virus resistance projects all over the world (Table 14.6), insect control projects, crop quality improvement projects, as well as sharing the rice genome sequence databases.

It is hoped that projects such as the one described in this manuscript will serve as a useful model for technology cooperation among international organizations, national research programmes and private companies that can work to truly bring the value of biotechnology to developing countries.

Table 14.6. Monsanto participation in virus resistance projects.

Virus resistance project	Crop	Continent/country
PRSV	Papaya	SE Asia (5 countries)
SPFMV	Sweet potato	Kenya, South Africa
PVX and PVY	Potato	Mexico
PVX, PVY and PLRV	Potato	Mexico

Acknowledgements

Short-term fellowships at Monsanto for VMA, CAO, LDN, VP, and KR were sponsored by ISAAA and Monsanto. The authors wish to thank V. Peschke and P. Ouimet for critical reading of the manuscript. Special thanks to Jill Montgomery for her role in coordination of international programmes.

References

Bateson, M.F., Henderson, J., Chaleeprom, W., Gibbs, A.J. and Dale, J.L. (1994) Papaya ringspot potyvirus: isolate variability and the origin of PRSV type P (Australia). *Journal of General Virology* 75, 3547–3553.

Baulcombe, D.C. (1996) Mechanisms of pathogen-derived resistance to viruses in transgenic plants. *Plant Cell* 8, 1833–1844.

Berger, P.H., Barnett, O.W., Brunt, A.A., Colinet, D., Edwardson, J.R., Hammond, J., Hill, J.H., Jordan, R.L., Kashiwazaki, S., Makkouk, K., Morales, F.J., Rybicki, E., Spence, N., Ohki, S.T., Uyeda, I., van Zaayen, A. and Vetten, H.J. (2000) Family Potyviridae. In: van Regenmortel, M.H.V., Fauquet, C.M. and Bishop, D.H.L. (eds) *Virus Taxonomy: Classification and Nomenclature of Viruses: Seventh Report of the International Committee on Taxonomy of Viruses*. Academic Press, San Diego, pp. 703–724.

Cann, A.J. (1997) *Principles of Molecular Virology*. Academic Press, San Diego.

Chen, G., Ye, C.M., Huang, J.C., Yu, M. and Li, B.J. (2001) Cloning of the papaya ringspot virus (PRSV) replicase gene and generation of PRSV-resistant papayas through the introduction of the PRSV replicase gene. *Plant Cell Reports* 20, 272–277.

Cooper, B., Lapidot, M., Heick, J.A., Dodds, J.A. and Beachy, R.N. (1995) A defective movement protein of TMV in transgenic plants confers resistance to multiple viruses whereas the functional analog increases susceptibility. *Virology* 206, 307–313.

De La Rossa, M. and Lastra, R. (1983) Purification and partial characterization of papaya ringspot virus. *Phytopathologische Zeitschrift* 106, 329–336.

Devereux, J., Haeberli, P. and Smithies, O. (1984) A comprehensive set of sequence analysis programs for the VAX. *Nucleic Acids Research* 12, 387–395.

Donson, J., Kearney, C.M., Turpen, T.H., Khan, I.A., Kurath, G., Turpen, A.M., Jones, G.E., Dawson, W.O. and Lewandowski, D.J. (1993) Broad resistance to tobamoviruses is mediated by a modified tobacco mosaic virus replicase transgene. *Molecular Plant–Microbe Interactions* 6, 635–642.

Dougherty, W.G., Lindbo, J.A., Smith, H.A., Parks, T.D., Swaney, S. and Proebsting, W.M. (1994) RNA-mediated virus resistance in transgenic plants: exploitation of a cellular pathway possibly involved in RNA degradation. *Molecular Plant–Microbe Interactions* 7, 544–552.

English, J.J., Mueller, E. and Baulcombe, D.C. (1996) Suppression of virus accumulation in transgenic plants exhibiting silencing of nuclear genes. *Plant Cell* 8, 179–188.

Fuchs, M. and Gonsalves, D. (1995) Resistance of transgenic hybrid squash ZW-20 expressing the coat protein genes of zucchini yellow mosaic virus and watermelon mosaic virus 2 to mixed infections by both potyviruses. *Bio/Technology* 13, 1466–1473.

Gonsalves, D. (1998) Control of papaya ringspot virus in papaya: a case study. *Annual Review of Phytopathology* 36, 415–437.

Gonsalves, D. and Ishii, M. (1980) Purification and serology of papaya ringspot virus. *Phytopathology* 70, 1028–1032.

Goodwin, J., Chapman, K., Swaney, S., Parks, T.D., Wernsman, E.A. and Dougherty, W.G. (1996) Genetic and biochemical dissection of transgenic RNA-mediated virus resistance. *Plant Cell* 8, 95–105.

Guilley, H., Dudley, R.K., Jonard, G., Balazs, E. and Richards, K.E. (1982) Transcription of cauliflower mosaic virus DNA: detection of promoter sequences, and characterization of transcripts. *Cell* 30, 763–773.

Hammond, J. (1998) Resistance to plant viruses: an overview. In: Hadidi, A., Khetarpal, R.K. and Koganezawa, H. (eds) *Plant Virus Disease Control*. APS Press, St Paul, Minnesota, pp. 163–171.

Hautea, R.A., Attathom, S., Kwok, C.Y. and Krattiger, A.F. (1999) *The Papaya Biotechnology Network of Southeast Asia: Biosafety Considerations and Papaya Background Information.* ISAAA Briefs No. 11, ISAAA, Ithaca, New York.

Hellwald, K.H. and Palukaitis, P. (1995) Viral RNA as a potential target for two independent mechanisms of replicase-mediated resistance against cucumber mosaic virus. *Cell* 83, 937–946.

Kaniewski, W. and Lawson, C. (1998) Coat protein and replicase-mediated resistance to plant viruses. In: Hadidi, A., Khetarpal, R.K. and Koganezawa, H. (eds) *Plant Virus Disease Control.* APS Press, St Paul, Minnesota, pp. 65–78.

Kaniewski, W., Ilardi, V., Tomassoli, L., Mitsky, T., Layton, J. and Barba, M. (1999) Extreme resistance to cucumber mosaic virus (CMV) in transgenic tomato expressing one or two viral coat proteins. *Molecular Breeding* 5, 111–119.

Kaniewski, W.K. and Thomas, P.E. (1998) Field testing resistance of transgenic plants. In: Foster, G.D. and Taylor, S.C. (eds) *Plant Virology Protocols: From Virus Isolation to Transgenic Resistance.* Humana Press Inc., Totowa, New York, pp. 497–507.

Kay, R., Chan, A., Daly, M. and McPherson, J. (1987) Duplication of CaMV 35S promoter creates a strong enhancer for plant genes. *Science* 236, 1299–1302.

Lawson, E.C., Weiss, J.D., Thomas, P.E. and Kaniewski, W.K. (2001) NewLeaf Plus® Russet Burbank potatoes: replicase-mediated resistance to potato leafroll virus. *Molecular Breeding* 7, 1–12.

Lindbo, J.A., Silva-Rosales, L., Proebsting, W.M. and Dougherty, W.G. (1993) Induction of a highly specific antiviral state in transgenic plants: implications for regulation of gene expression and virus resistance. *Plant Cell* 5, 1749–1759.

Lius, S., Manshardt, R.M., Fitch, M.M.M., Slightom, J.L., Sandford, J.C. and Gonsalves, D. (1997) Pathogen-derived resistance provides papaya with effective protection against papaya ringspot virus. *Molecular Breeding* 3, 161–168.

Lomonossoff, G.P. (1995) Pathogen-derived resistance to plant viruses. *Annual Review of Phytopathology* 33, 323–343.

Maiti, I.B., Murphy, J.F., Shaw, J.G. and Hunt, A.G. (1993) Plants that express a potyvirus proteinase gene are resistant to virus infection. *Proceedings of the National Academy of Sciences, USA* 90, 6110–6114.

Palukaitis, P. and Zaitlin, M. (1997) Replicase-mediated resistance to plant virus disease. *Advances in Virus Research* 48, 349–377.

Powell-Abel, P., Nelson, R.S., De, B., Hoffmann, N., Rogers, S.G., Fraley, R.T. and Beachy, R.N. (1986) Delay of disease development in transgenic plants that express the tobacco mosaic virus coat protein gene. *Science* 232, 738–743.

Powell, P.A., Sanders, P.R., Tumer, N., Fraley, R.T. and Beachy, R.N. (1990) Protection against tobacco mosaic virus infection in transgenic plants requires accumulation of coat protein rather than coat protein RNA sequences. *Virology* 175, 124–130.

Purcifull, D.E., Edwardson, J.R., Hiebert, E. and Gonsalves, D. (1986) Papaya ringspot potyvirus. In: Brunt, A., Crabtree, K., Dallwitz, M., Gibbs, A. and Watson, L. (eds) *Viruses of Plants: Description and List from VIDE Database.* CAB International, Wallingford, pp. 874–877.

Qaim, M. (1998) *Transgenic Virus Resistant Potatoes in Mexico: Potential Socioeconomic Implications of North-South Biotechnology Transfer.* ISAAA Briefs No. 7, ISAAA, Ithaca, New York.

Riebe, J.F. and Zalewski, J.C. (2002) Pesticide reduction and disease control with genetically modified potato. *Environmental Biosafety Research* (in press).

Sanders, P.R., Winter, J.A., Barnason, A.R., Rogers, S.G. and Fraley, R.T. (1987) Comparison of cauliflower mosaic virus 35S and nopaline synthase promoters in transgenic plants. *Nucleic Acids Research* 15, 1543–1558.

Shukla, D.D. and Ward, C.W. (1989) Identification and classification of potyvirus on the basis of coat protein sequence data and serology. *Archives of Virology* 106, 171–200.

Thomas, P.E. and Kaniewski, W.K. (1998) Agronomic performance of transgenic plants. In: Foster, G.D. and Taylor, S.C. (eds) *Methods in Molecular Biology,* Vol. 81: *Plant Virology Protocols: From Virus Isolation to Transgenic Resistance.* Humana Press Inc., Totowa, New York, pp. 509–518.

Thomas, P.E., Kaniewski, W.K. and Lawson, E.C. (1997) Reduced field spread of potato leafroll virus in potatoes transformed with the potato leafroll virus coat protein gene. *Plant Disease* 81, 1447–1453.

Thomas, P.E., Lawson, E.C., Zalewski, J.C., Reed, G.L. and Kaniewski, W.K. (2000) Extreme resistance to potato leafroll virus in potato cv. Russet Burbank mediated by the viral replicase gene. *Virus Research* 71, 49–62.

Tricoli, D.M., Carney, K.J., Russell, P.F., McMaster, J.R., Groff, D.W., Hadden, K.C., Himmel, P.T., Hubbard, J.R., Boeschore, M.L., Reynolds, J.F. and Quemada, H.D. (1995) Field evaluation of transgenic squash containing single or multiple virus coat protein gene constructs for resistance to cucumber mosaic virus, watermelon mosaic virus 2, and zucchini yellow mosaic virus. *Bio/Technology* 13, 1458–1465.

Verfaillie, H.A. (2000) New Monsanto Pledge. http://www.monsanto.com/monsanto/about_us/monsanto_pledge/default.htm.

Wang, C.H. and Yeh, S.D. (1997) Divergence and conservation of the genomic RNAs of Taiwan and Hawaii strains of papaya ringspot potyvirus. *Archives of Virology* 142, 271–285.

Waterworth, H.E. and Hadidi, A. (1998) Economic losses due to plant viruses. In: Hadidi, A., Khetarpal, R.K. and Koganezawa, H. (eds) *Plant Virus Disease Control.* APS Press, St Paul, Minnesota, pp. 1–13.

White, J.L. and Kaper, J.M. (1989) A simple method for detection of viral satellite RNAs in small plant tissue samples. *Journal of Virological Methods* 23, 83–93.

Wilson, T.M.A. (1993) Strategies to protect crop plants against viruses: pathogen-derived resistance blossoms. *Proceedings of the National Academy of Sciences, USA* 90, 3134–3141.

Yeh, S.D., Jan, F.J., Chiang, C.H., Doong, T.J., Chen, M.C., Chung, P.H. and Bau, H.J. (1992) Complete nucleotide sequence and genetic organization of papaya ringspot virus RNA. *Journal of General Virology* 73, 2531–2541.

Ying, Z., Yu, X. and Davis, M.J. (1999) New method for obtaining transgenic papaya plants by *Agrobacterium*-mediated transformation of somatic embryos. *Proceedings of the Florida State Horticultural Society* 112, 201–205.

Chapter 15

Institutions and Institutional Capacity for Biotechnology – a Case Study of India

Valerie Rhoe, Sivramiah Shantharam and Suresh Babu
International Food Policy Research Institute, 2033 K Street NW, Washington, DC 20006, USA

Introduction

Biotechnology research and commercialization is rapidly growing in several industrialized countries, but similar speed has not been experienced in developing countries. There are several reasons for this lack of progress in biotechnology. Firstly, there is a lack of scientific capacity to perform biotechnology research. Secondly, the adaptation of research into commercialized products is slow and tedious. Thirdly, appropriate policies that promote biotechnology research and commercialization are not effectively implemented, and fourthly, institutions that facilitate the generation and transfer of biotechnology are less efficiently organized.

Agricultural biotechnology and its transfers are important to developing countries for several reasons. First, agricultural biotechnology products can increase agricultural productivity with less agricultural inputs (Flavell, 1999). Higher crop yields may lower the price that farmers receive for their crops, but farmers ultimately have a higher income because of the higher output and less inputs as well as the ability to export excess production (Pinstrup-Andersen, 2001). Second, genetically modified (GM) agricultural products can increase the nutritional quality of the crop, which could prevent and/or treat micronutrient deficiency diseases such as anaemia (Flavell, 1999; Tripp, 2001). Third, GM crops can be physiologically altered to be resistant to diseases and droughts or modified to grow in less favoured lands (i.e. poor soil; Flavell, 1999; Komen *et al.*, 2000). These alterations result in higher output; therefore, the more than 800 million people who are food insecure may not have to go to bed hungry (Pinstrup-Andersen, 1999). Finally, GM crops can conserve natural resources by producing higher yields and reducing the amount of agrochemical used (Singh, 1994).

These potential benefits come along with potential risks. Environmental concerns include the cross-fertilization of GM crops with wild crops, creation of super weeds and modification of insects' immune systems (Hvoslef-Eide and Rognli, 1995; Flavell, 1999). Health concerns involve antibiotic resistance from antibiotic markers and transfer of allergens (Conway, 2000). Some of the negative aspects of technology transfers are inappropriate technologies and cultural conflicts that results in the loss of values (Goulet, 1989). For example, the terminator gene, which makes GM seeds sterile.

Much of the benefit from increasing economic growth and capacity from biotechnology transfers depends on the existing capacity within the recipient country and the degree of technology spillover (Saggi, 2000). Furthermore, Altman (1995) argues that developing countries are at a disadvantage for receiving biotechnology transfers because of the lack of access to capital, labour and support services; limited or no intellectual property rights (IPRs); lack of or excessive restriction on biosafety regulations; and lack of biotechnology information for decision making. For biotechnology transfers to be successful, one must keep in mind the level of human resources available, the capacity to select appropriate technology, and the ability to absorb, scale-up, commercialize and manage technology development (Singh, 1994)

In spite of the potential benefits, the transfers of biotechnology and their adoption in developing countries continue to face several institutional and policy challenges. This paper is an attempt to look at the role of institutions in facilitating the promotion and use of biotechnology for agricultural development. Several questions remain to be answered. What role, if any, do formal and informal institutions play in biotechnology transfers? What policies should be in place to enhance the role of institutions in biotechnology transfers? What can we learn from experiences of developing countries such as India that may be useful for other developing counties entering the field of biotechnology? This paper will first present a conceptual framework that is used to explain the conversion of basic agricultural biotechnology research into commercialized products that reach farmers in developing countries. Following the conceptual framework, what we already know about institutions and biotechnology will be explored. Then a case study of India will illustrate how biotechnology institutions can stall or promote the progress of biotechnology research, development and commercialization. Next, lessons learned from the India case study are mentioned as well as the potential benefits gained from implementing the suggested changes. Finally, policy suggestions are pointed out in the concluding section.

Conceptual Framework

Institutions are a set of formal and informal rules of conduct that facilitate relationships between individuals and organizations. Institutions are needed and used when market failures and imperfect markets exist (Kherallah *et al.*, 2001). Since market failures are common in the research and application of biotechnology, institutions can play an important role in increasing the efficiency of biotechnology development and commercialization. A conceptual framework is developed to analyse the biotechnology institutions for technology development and transfer (Fig. 15.1). If agricultural biotechnology research and product development is expected to occur within a country, then agricultural research institutions and private enterprises must have adequate funding. Funding can be generated through the public sector as well as the private sector; however, the origin of the funding depends on policies that have been established. Once public research institutions have received sufficient funding, basic research (which is usually performed by public institutions) can begin. In order for private institutions and some public institutions to turn basic research into applied research, IPRs need to be established and enforceable. Applied research will then generate applicable agricultural technologies; however, biosafety regulations will need to be implemented before these applicable products can be commercialized. Establishing biosafety regulations will open markets for the export of the new goods. After a product has been commercialized, then farmers can purchase this new biotechnology product from the market or receive the new product through a government programme.

Basically, a similar conceptual framework applies to biotechnology research and product development outside of the developing country. The only difference is the occurrence of technology being transferred

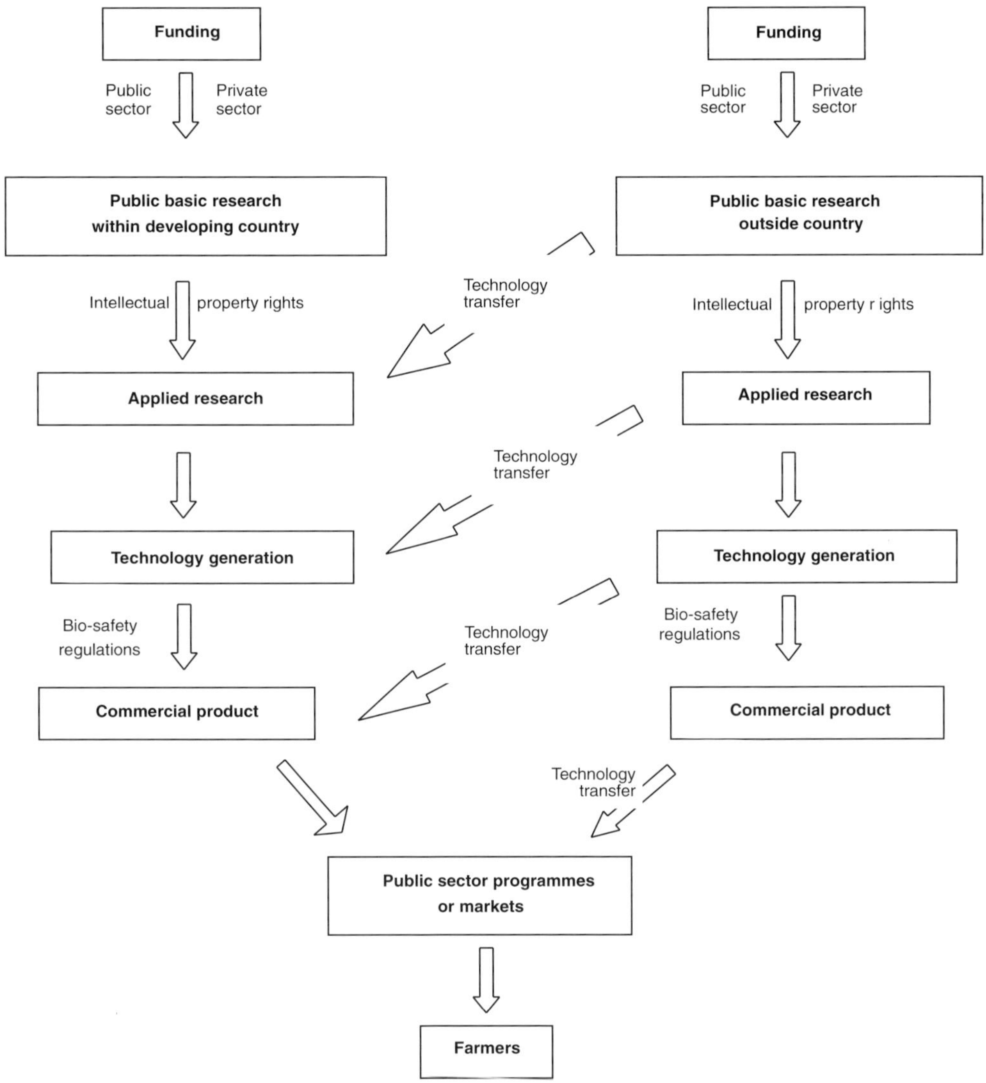

Fig. 15.1. General biotechnology institutional framework: research to farmer's usage.

from one country to another. Basic research, applied research, technology generation and the commercial product can be transferred from one country to another.

What Do we Know about Institutions and Biotechnology?

A vast number of biotechnology studies identified one type of institution or another which has impeded the growth of either biotechnology research or its application. Since government expenditures allocated to agricultural biotechnology research and application have declined over the past several years, governments of developing countries need to allocate more funds to public biotechnology institutions or entice private industry to pursue biotechnology products that will address the social good (poverty, hunger and malnutrition).

How can governments increase the funding of both public and private biotechnology institutes? A practice that has been evolving is the establishment of public/private collaboration. These partnerships will efficiently use the comparative advantage of each sector. The private sector is dependent on the basic research performed by the public sector. Narin *et al.* (1997) found that during 1993–1994, 73% of the scientific papers cited by US industrial patents were from public science sources, while only 37% were authored by industrial scientists. However, the public sector is dependent upon the private sector's managerial expertise for commercialization (further elaborated on later in the chapter).

Byerlee and Fisher (2001) suggest that segmented markets should be used to establish a private/public partnership. Therefore, both sectors can exploit their complementary assets. The public sector will distribute biotechnology products such a hybrid seeds to certain sectors such as resource-poor farmers, while the private sector will supply the same hybrid seeds to commercial farmers. The public sector can also negotiate with the private sector in producing orphan crops. Orphan crops are crops that the private sector would normally not develop, but will develop because it may improve the firm's public image. As one can imagine, there are concerns from the private sector about leakages to areas that are outside of the government's targeted locations, competing firms gaining access to the technology where IPRs are not enforced, and responsibility and liability for risks incurred (Byerlee and Fisher, 2001).

In order for firms to feel that their intellectual property will be protected after a technology is transferred to an institute or government programme, IPRs need to be established and enforceable (James and Krattiger, 1999). Without simple and enforceable IPRs, transaction costs will be high and therefore, licence negotiations will be difficult (Byerlee and Fisher, 2001).

For example, as of 2000, India's parliament has not approved its IPR regulations – the Plant Variety Protection Act (PVPA; Paarlberg, 2001). In 1993, a draft PVPA, which was modelled after UPOV 1978, was submitted to parliament. This revised version was criticized by the private industry as being too weak, while some non-governmental organizations (NGOs) argued that it did not protect the farmers. Again in 1996/1997 a revised PVPA, which addresses farmer rights, was submitted, and it was criticized by NGOs. Then in 1999, a new version was being considered by the parliament for adoption. Even if this version is approved, the IPR policies would be weak, and therefore, limit trade in GM crops (Paarlberg, 2000).

Developing countries do not have adequate mechanisms that promote and facilitate technology transfers (Knorr, 1995). Although joint ventures, licences, and wholly owned subsidiaries are methods of transferring technology from one institution to another, there are limitations to the ability of developing-country private and public institutions to negotiate these transfers. To make technology transfers easier and less expensive for public institutions, governments can establish a centralized technology transfer office, which can aid in harmonizing biotechnology transfers. A centralized technology transfer office will lower transaction costs, facilitate external negotiations and support institutions in skills that are missing (Byerlee and Fisher, 2001).

Trade policies may need to be restructured in order for biotechnology products to be developed for commercialization. For example, Brazil has industrial policies that promote biotechnology, but their trade policy restricts importation of biotechnology inputs (Acharya, 1999). Therefore, industrial policy and trade policy need to be consistent for effective biotechnology transfer and use.

In order for biotechnology products to be traded and accepted by the public, governments need to establish biosafety regulations. Biosafety regulations should be science-based, transparent and harmonized with international protocols. These regulations should specify import and export requirements, and biosafety and food safety institutions need to be credible. For example, India's biosafety guidelines

were initiated in 1989. At the research stage, the guidelines set in 1989 only required screening of GM products that had a scientifically demonstrated risk. However, these guidelines were more stringent on the application of GM products. These guidelines also established the Review Committee of Genetic Manipulation (RCGM) and Genetic Engineering Approval Committee (GEAC). The RCGM is promotional while the GEAC is more cautious about approving projects. Then in 1998, the biosafety guidelines were revised to require GM crops to be tested for toxins and allergens. Although biosafety guidelines have been established, life science companies have had difficulty in gaining
approval roducts because of
trials and lawsuits
ernment agencies.
Os have paralysed
cess (Paarlberg,

999) argue that
lished that cre-
and develop-
governments
es for enter-
d develop-
te venture
C exchange.
Fu stabilize
co orderly
ma private
and nology
prod

An hnol-
ogy re ough
researc 000)
evaluat red
research nd
Klenow, a,
1998; Le
support c
large exp
expected re
and Klenow
SEMATECH,
semiconducto
nating excessiv
fore resulting
Lerner's study c va-

tion Research (SBIR; Lerner, 1998) found that businesses involved in this group grew faster in sales and employment than similar non-supported firms. Although the previous two examples were intended to avoid duplication, the Japanese had a different rendition of the same type of programme (Branstetter and Sakakibara, 1998). The Japanese research consortia's primary goal was to bring together complementary research and development projects. This established a network of professionals working on the same type of project, which has made research and development projects more productive and more profitable. It has also induced learning opportunities, which has enhanced as well as induced research and development projects.

To improve both the research and application of biotechnology, policies need to be established to build capacity at all levels of the conceptual process. In order for appropriate policies to be established, decision makers, policy analysts, and policy advisors need to have appropriate training. Until recently, many developing countries did not have biotechnology courses as a component of their national curricula. Government policy is needed to create this capacity within its country and therefore, government policy should include funding for training in agricultural biotechnology. Capacity can be created by establishing state universities, as in Korea, or scholarships for students to study abroad, as in Taiwan (Acharya, 1999). Furthermore, governments can establish networks that will assist in increasing capacity by permitting information exchange (Knorr, 1995). In addition to building capacity in the policy arena, capacity needs to be built among scientists, which can be achieved through domestic and foreign graduate programmes (Acharya, 1999).

Training is also necessary for the farmers, who will plant the new GM seed. One lesson learned from the Green Revolution is the need to educate the implementers of the technology, i.e. farmers. During the Green Revolution, farmers did not have adequate training to use the new product to its fullest potential. For example, to maximize the output of Green Revolutionary products framers needed to be informed of the additional chemical inputs that were required. If training is not provided, then farmers must learn through trial and error how to manage the crops and the advantages and disadvantages of this new technology, which can take a long time. Furthermore, general knowledge about biotechnology will eliminate the fear of the unknown for the farmer; therefore, making the products more accepted.

Training of consumers may be necessary to encourage consumers to use the GM seed. Creating public awareness will alleviate the fears that may be lying in the back of a consumer's mind. However, outside of satisfying the consumer's fears, training is necessary in order for the benefits of these products to be known. If a new variety is not distinguishable from a non-GM product, then consumers may not realize that these two crops are actually very different, especially if a product is nutritionally enhanced but appears the same (Tripp, 2001).

Moreover, in order for farmers to gain access to new GM seeds, a private seed industry needs to exist. In many developing countries the seed industry has been liberalized; however, adequate incentives have not been established to induce the private sector to step in. Enticing entrepreneurs to invest in a commercial seed industry will provide biotechnology seeds to some farmers; however, many resource-poor farmers will still not have access to this market because they do not have credit. India's private seed sector has been around for 100 years, but it has had a limited role because of restrictive national policies. However, in the 1980s, less restrictive policies took form. These policy changes included private institutes' accessibility to public institutes' breeding lines and improved seeds, the formation of the Department of Biotechnology (DBT), changes in the seed development policy, liberalization of foreign entry restrictions on breeding lines and enhanced seeds, as well as other trade regulations, easier varietal registration procedures and requirements, and implementation of policies that induced foreign collaboration. These policy changes have increased the volume of sales of private seed companies, increased foreign investment and increased collaboration (Selvarajan *et al.*, 1999).

One method for resource-poor farmers to gain access to GM seeds is through government-sponsored small-scale seed projects. Another method is to provide credit; therefore, policies need to be designed so that farmers have access to credit. Past experience has shown us that small farmers did not benefit greatly from Green Revolution technologies because they did not have funds to purchase the inputs or hybrid seeds (Acharya, 1999).

There are also policies that can be established at the production level. The public sectors in developing countries actively need to set up equitable joint ventures with private-sector entities in developed and developing countries. In these ventures, the public sector can provide germplasm, evaluation networks, local knowledge, applied breeding skills and infrastructure, access to a seed delivery system, public sector extension, and a positive perceptions, while the private sector can contribute biotechnology tools and genes, access to international capital markets, flexibility in decision making, and managerial and marketing knowledge to convert the basic research into applied research and eventually products. Joint ventures offer the opportunity for public institutions to license the technology and gain experience with the commercialization of products, which many public institutions lack knowledge in. Establishing a link between the public and private institutions will rectify 'one of the weakest links in the chain of crop production in developing countries' (James and Krattiger, 1999).

In addition to setting up joint ventures, public and private production entities need to contain a strategic management process. For example, in the 1970s, new biotechnology firms (NBFs) began in the USA. These new firms consisted basically of university professors who left the public sphere to try to transfer basic knowledge to the private domain. The goal of these firms was to take a particular technology from beginning to end: basic research, field trials, product development and marketing; however, most of the 1311 firms in 1994 did not achieve this goal. Why? Because the founders of these firms (university professors) did not have training in running a company. Therefore, these companies lacked technical, managerial and marketing skills to make the commercial product as successful as the basic and applied research. Further, the research performed by these firms were focused on a single technology, which restricted expansion into other products. (Acharya, 1999). Establishing a public–private collaboration could convert basic research more quickly into commercial products because private collaborators will have the necessary skills to commercialize the product.

Institutions in Biotechnology, What Role? India Case Study[1]

During the 1990s, 230 million people in India were food insecure as a result of low productivity of agriculture resources. Biotechnology can specifically assist in lowering food insecurity in India through several outlets. Firstly, India produces several dryland crops that are afflicted by pests and diseases. Biotechnology could alter the plants to resist these biotic stresses. Secondly, farmers of dryland crops are clearing new land in India in order to produce more. Genetic engineering could increase the yields of these seeds, therefore reducing the need to expand into fragile lands. Thirdly, insects have plagued, and will continue to plague Indian farmers. Genetically modifying crops to repel or kill specific insects will reduce the amount of insecticide needed. Lowering the number of insecticide applications will not only reduce the damage to the environment, but also to the people spraying the insecticide. Fourthly, the poor in India suffer from vitamin A and iron deficiencies (Paarlberg, 2001). Furthermore, several traditional food crops in India contain substances that are harmful to the human body (Prakas, 1999). If food crops could be genetically modified to increase their levels of micronutrients or decrease their amount of harmful substances, then the health status of the population will improve. Lastly, engineering food crops to have a longer shelf life will reduce the amount of crops lost to spoilage before reaching the market. This modification is pertinent to India because refrigeration, which slows the spoiling process, is not common. Genetic modification of food crops has great potential for improving food security in India (Paarlberg, 2001).

India has demonstrated that it has the capacity to perform basic biotechnology research. This capacity is the result of 29 Indian universities establishing postgraduate training programmes in biotechnology. The DBT has also organized short-term training courses (2–4 weeks) for professionals, who need to learn the latest skills and knowledge about biotechnology techniques. In addition to short-term training courses and postgraduate programmes, the Department of Biotechnology offers 1-year study-abroad fellowships (Acharya, 1999). Each of these methods of training has built and will sustain the biotechnology capacity within India.

Although India has established the capacity to perform basic research and its government has spent US$6 million on plant and molecular biology research between 1989 and 1997, India has had difficulty in turning biotechnology research into commercialized products (Ghosh, 1999). Why has biotechnology not progressed to its full extent in India? What institutions are hindering India's biotechnology growth? The All India Biotechnology Association (AIBA) argues that red tape and lack of transparency of the approval system has hindered the progress of the biotechnology industry and

[1] This section draws heavily from an unpublished consultancy report prepared by Kelman and Shantharam (1995).

has stifled private investment into the research and application of biotechnology (Jayaraman, 2001). AIBA states that India's biotechnology industry failed for six reasons: lack of focus, sub-critical funding, flaws in the approval system, thinly spread resources, a public sector unresponsive to global developments and the needs of the Indian market, and large public sector funding dissuading private investment (Jayaraman, 2001). Although India has established the DBT to develop policy and oversee priority areas of research, difficulties have risen in the allocation of research funds. This case study will focus on one research grant programme offered by the Indian government.

India's biotechnology institutional framework

An institutional framework presented here will reflect Indian institutions in all stages of biotechnology development from basic research to commercialization of a product. The Ministry of Science and Technology created the DBT to support R&D manufacturing, to identify and set up centres of excellence for R&D, to promote large scale use of biotechnology, to integrate programmes for human resource development, to establish facilities to support R&D and production, to serve as a nodal point for the collection and dissemination of information relating to biotechnology, to promote university and industry interaction, to evolve biosafety guidelines, to serve as a nodal point for specific international collaborations, to manufacture and apply cell-based vaccines, and to be responsible for autonomous institutions. This department receives its funding from the Ministry of Finance and the Planning Commission. The Department of Biotechnology established two advisory committees: the Scientific Advisory Committee–Department of Biotechnology (SAC-DBT) and Scientific Advisory Committee-O (SAC-O). SAC-DBT consists of scientists, representatives of industry, NGOs, user ministries and departments, and academic institutions. It advises on priorities, policy, planning and implementation of programmes. The SAC-O consists of native Indians working in foreign laboratories in developed countries. They provide information on current trends, new concepts and areas of biotechnology. The DBT also strengthens capacity in biotechnology by establishing graduate programmes in the state universities. Another component of the DBT that has an important impact on biotechnology research and implementation is the RCGM that was initiated as the result of biosafety guidelines. This committee can either approve or disapprove GM inputs for research and small-scale projects. If a proposed research project is large-scale then it needs approval from the Biotech Research Promotion Committee of the DBT, before the RCGM can review it.

Another branch of India's government that has an impact on biotechnology research and commercialization is the Ministry of Environment and Forest. In 1986, this ministry implemented an Environmental Protection Act that established the Institutional Specific Biosafety Committee (ISBC). The ISBC designed biosafety guidelines that were approved by the parliament. These guidelines established the GEAC, which holds the responsibility of approving field tests for large-scale projects and the importation of GM crops for commercialization. GM agricultural products that are created by life science companies and public institutes must be approved by the GEAC. Furthermore, environmental NGOs influence the decision-making of the GEAC through protests and lawsuits. Once field trials have been approved, they are observed not only by GEAC (Ministry of I&B, 1989; Summary of rules/regulations/guidelines, 1996). Furthermore, environmental NGOs influence the outcome of the field trials because of their attacks on the experimental fields (Paarlberg, 2002). In addition to biosafety guidelines, the Government of India has proposed a Biological Diversity Act that would conserve biodiversity, achieve sustainable use of biological resources, and ensure equitable sharing of the benefits (Kothari, 1999).

The third government sector that has an impact on biotechnology research and commercialization is the Indian Council of Agricultural Research (ICAR). ICAR assists both in strengthening biotechnology capacity and facilitating technology transfers. To facilitate technology transfers, ICAR is attempting to have IPRs passed; however, both private institutions and NGOs representing farmers have stalled this process over the past 8 years.

Once the RCGM approves the research, the GEAC approves the field trials and ICAR gives approval then agricultural biotechnology seeds can be commercialized. Agricultural biotechnology seeds are delivered to farmers either through the market or the Department of Extensions. Then the farmers sell the GM crops to consumers. The institutional framework for biotechnology promotion in India, as derived above, is given in Fig. 15.2.

Review of the Current Agricultural Process Grant Fund (APCF) Programme

ICAR established the APCF programme to support research schemes in agriculture, animal husbandry and fisheries, to solve production problems, and to add new information in high-priority agricultural research. The research schemes are solicited from ICAR institutions, state governments, non-agricultural and agricultural universities, and other qualified institutions. Research schemes to be considered could be concerned with fundamental or applied problems, and they could be single or multi-disciplinary projects.

Constraints in the current programme

Many ICAR and agricultural university scientists have never applied for APCF grants or are reluctant to submit proposals. In

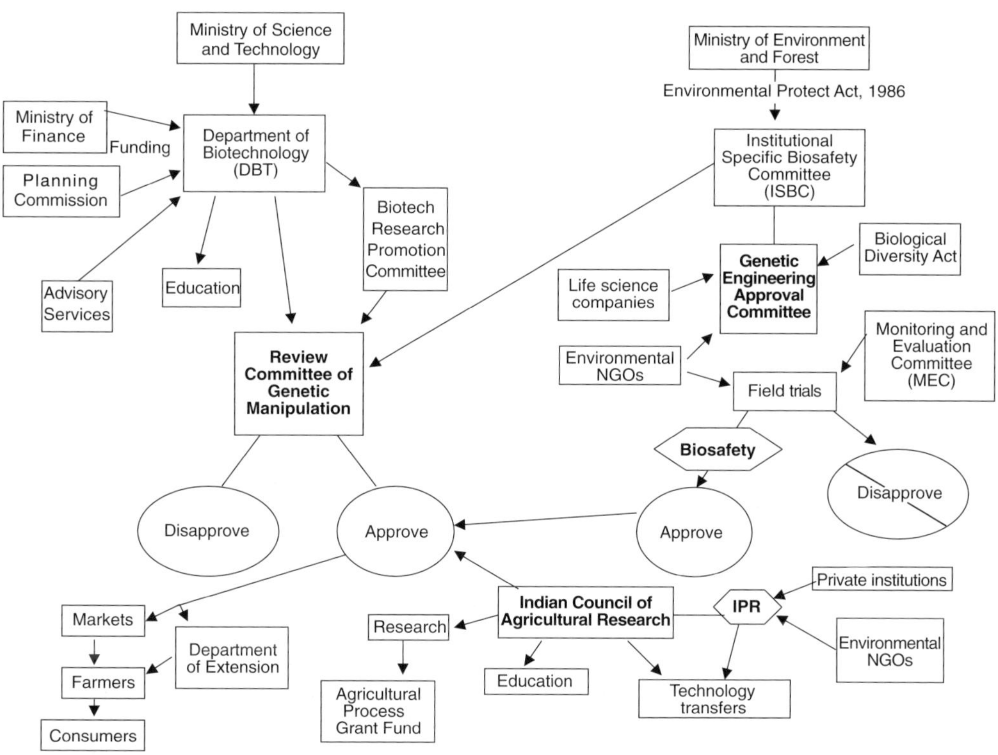

Fig. 15.2. India's biotechnology institutional framework.

part, the reluctance of scientists to apply for APCF grants reflect the perception that there will be a long delay before funds from grants will become available. Although the Project Screening Committees meet at 6-monthly intervals, there is not a scheduled deadline for receipt of proposals in specific subject matter areas. Thus, in some instances 6 months may elapse before a proposal has been considered by a Project Screening Committee. In addition, in many instances the funds provided are not adequate to meet the objectives of the project. Furthermore, there are no standardized guidelines on such important matters as conflict of interest and criteria for evaluation of proposals by ad hoc reviewers and members of the Project Screening Committees. Moreover, some scientists informed the researchers that the comments of the ad hoc reviewers focused too much on minor defects in the proposals and did not provide constructive advice that will assist applicants in improving their research plans.

Furthermore, the ICAR headquarters lacks a computerized database of the grants. Therefore, it is extremely difficult to monitor the operation of the APCF programme. This lack of a computerized database is one of the main reasons why mid-course corrections or policy and administrative changes cannot be made readily to improve efficiency and effectiveness of the system.

Two other major constraints that adversely affect the APCF programme are the lack of logistical support at the ICAR headquarters, which delays the rapid processing of the grant applications and the difficulty in obtaining timely reviews from ad hoc reviewers. Certain institutions of ICAR have their own internal review before the submission of the grant application to an external agency. In some instances, the in-house initial review of proposals contributes to the long delay between completion of a proposal and the submission of the grant to ICAR. Most of all, the financial clearance necessary to disburse the funded projects is a very time-consuming process with many procedural delays.

At many institutions, junior-level scientists are not encouraged to apply for grants; therefore, only senior and principal scientists apply for research grants as principal investigators. Until recently, another constraint was the ineligibility of the ICAR scientists to apply for APCF grants. This factor was cited as one of the reasons for the low number of independent research grants submitted from scientists in ICAR laboratories.

A matter of concern is the lack of financial control of grant funds by the principal investigators, particularly for relatively minor expenditures. Furthermore, principal investigators have no real control over the selection of research personnel employed to work on their grant projects. Another important factor identified as the basis for the lack of enthusiasm or motivation among some scientists to apply for grants is the fact that there is no incentive or reward associated with obtaining competitive research grants.

Steps for efficiently organizing biotechnology grant institutions

The initial step in modifying this programme is the establishment of a separate SAC. It may be advisable to invite several distinguished scientists from other countries to serve on this committee in the initial period. The committee should be charged with making candidate recommendations for the chief scientist with final selection and appointment to be made by the director general of ICAR. The committee, in consultation with the chief scientist, should examine the deficiencies in the current system and the design of the administrative structure of a separate agency for ad hoc grants within ICAR including a finance office for the competitive grant office.

With the appointment of the scientists, the next phase would be the scheduling of meetings to determine which of the priorities established by the NATP and ICAR would be selected for primary focus in the competitive grants programme, to develop the proposal request process and to describe each of the programmes. The third step is to revise the current manuals and procedures with materials and information provided from other granting programmes.

The fourth step can be broken into three phases. The first phase is the development and implementation of a training programme for programme managers. A core group of prospective programme managers as well as the chief scientist could be sent on an internship programme to work in the current Competitive Grants Program of the US Department of Agriculture or the National Science Foundation. The trainees could be assigned to work with staff in the programme for a 2-week period at the time that peer review panels are meeting, and work directly with programme directors and panel managers in the subject areas for which they will have responsibility. The second phase of the training would be the training of selected scientists who would serve as Project Screening Committee chairmen (panel managers) at the time the programme is initiated. The third phase would be to conduct a series of workshops in cooperation with National Academy of Agricultural Research Management (NAARM) at each of the agricultural universities or in selected regions to train scientists who could be selected as potential panel members or *ad hoc* reviewers. Scientists who have had experience in any of the above training programmes could chair these workshops. The fourth phase involves further collaboration with NAARM. NAARM could conduct proposal preparation workshops for scientists at selected regional centres or universities. Young investigators, who had recently completed training programmes at NAARM, indicated that the session concerned with preparation of research proposals was very helpful and well organized. This could be an ongoing process involving staff and initial panel managers or panel members, as well as individuals who have been successful in obtaining grants from ICAR or other agencies.

Other phases in the programme can be implemented as the process develops. It would be essential as these stages progress that a concerted effort be made to provide opportunities for administrators in extension and relevant national agencies concerned with the environment, natural resources and forestry, as well as representatives of growers and commodity groups and agricultural industries to be kept informed of progress and to become involved in the shaping of priorities and objectives.

It would be highly valuable to organize annual grant holder's meetings in selected subject matter divisions to monitor the progress of the funded projects, exchange ideas and develop useful networking among the scientists. Productivity increases are closely tied to integration of extension activities with basic and applied research programmes in agriculture (Huffman and Just, 1994).

A clear vision needs to be presented to all participants in the revision process; the ultimate goal is the enhancement of the overall quality of research in agriculture. This can be accomplished if an effective well-managed competitive peer review system can be established as one component of the ICAR funding system.

Lessons learned from Indian experience

Administration structure

Several lessons could be learnt from the organization and operation of Indian biotechnology institutions. The first lesson is that an efficient administrative structure is needed for grants to be distributed effectively. To create an efficient structure, the size of the review panel should be one panel member for each ten proposals. Furthermore, the wages of the panel members need to compensate for the work performed by these members. Funds should be allocated to provide a salary equivalent to 2–3 months of work for panel managers, which are equivalent to the salaries that they receive from their home institutions plus travel and *per diem* for the time that they consult with programme directors.

Deadline

The second lesson learned from this case study is to set a submission deadline for proposals. This deadline will help the review process to become more efficient. Although there are some disadvantages in scheduling only one period each year for submission and review of proposals, it will facilitate management of the review process and ensure expeditious handling of the grant procedure during a given fiscal year.

Streamlining mechanisms for fund disbursement

Once a grant proposal is approved, a streamlined mechanism for disbursement of grant funds to the principal investigator will encourage researchers to apply. The total process from receipt of proposals to allocation of grant funds should be completed within 6–8 months.

Computerization

It is also essential to computerize the information database of the grants office so that the administration of the review procedures and programme assessment can be facilitated. Computerization will immediately solve the acute problem of grants administration and monitoring. The Competitive Grant Review Program needs to hire an individual with specific skills in use of computers and computer software. This specialist should be responsible for development of the needed software programs. He/she should implement the full utilization of computer technology to reduce paperwork in the programme and assist staff in the development of the needed databases.

Timeliness

Effective measures should be taken so that the procedures and rules governing grant applications and the availability of funds are communicated to the scientific community and research institutions in a timely manner. This awareness will make the process transparent and inform researchers of the available funds early enough for the preparation of good research proposals. An annual report should be prepared listing the abstracts of each of the proposals that have been funded. In addition, information notes or an attractive newsletter should be published at periodic intervals and then distributed to the scientific community and user groups. Both of these activities will better inform the research community.

Freedom to operate funds

A greater degree of freedom and flexibility needs to be granted to the principal investigator for the authority to operate and manage his/her grant funds at the institutional level in order for funds to be used adequately. This flexibility should specifically include the authority and responsibility to purchase necessary equipment and supplies of the research project. In addition, authority should be granted to the principal investigator for travel to meetings, workshops, and to defray publication costs. Current personnel policies for appointment of research associates and assistants on the grants should be revised in order to provide greater freedom to the principal investigators in the selection of the best-qualified candidates.

Instructions

The applicants for grants should be provided with detailed and clear instructions on how to prepare grant proposals in a logical orderly presentation with the basic required information. This will quicken the review process as well as standardize the process.

Guidelines for reviewers

Specific criteria should be provided on review forms to guide the ad hoc reviewers and review panel members in determination of the quality of research proposals. This standardization will allow a fair review of the research proposals. When the review process is initiated, each proposal should be evaluated to ensure that these guidelines are uniformly applied. The guidelines need to be provided to all ad hoc reviewers contacted by mail as well as to panel members.

Pre-proposal

Many scientists have the perception that peer review panels find it difficult to rank proposals involving interdisciplinary teams; however, interdisciplinary teams will improve the review process. The issues involved are admittedly complex, and communication problems do arise among scientists from different disciplines on review panels (Porter and Rossini, 1985). However, direct observation of multidisciplinary panels in other programmes indicate that they can work effectively if care is exercised in the selection of panel members. Because of the effort required in the establishment of multidisciplinary programmes, particularly involving systems approaches, pre-proposals would be required based on prior consultation with the chief scientist in specific instances.

Young investigator programme

A separate category or programme of funding designed specifically to encourage young investigators should be established in order to sustain and enhance the capacity in India. It will be critical to define 'Young Investigators'. The current policy restricting the eligibility of research associates from developing grants should be reevaluated since training in grant application should occur early in the careers of scientists. It was not possible in the period of our visit to gain detailed information on the 'Young Scientists Program' funded by the Department of Science and Technology (DST). However, the DST report on 'Opportunities for Young Scientists' (1993) outlines basic objectives that would be appropriate for a separate programme to provide support for scientists entering research careers in agriculture. Although 227 of the 382 projects were accepted for funding between 1988 and 1992, the total number of projects relevant to agriculture was relatively small. Thus, there is a need for a programme specifically targeted to agriculture. This programme would benefit greatly from the experience of the DST Management Advisory Committee that was established for this programme. Under the APCF, sponsored programme support for travel should be provided to enable young scientists to gain the benefit of participation in international congresses and conferences.

Independent funding for commercialization

An independent catogory of funding based on competitive review should be established to foster creative and innovative ideas for products, processes and commodities that can be commercialized in partnership with the private sector industries and scientists working at the ICAR laboratories and institutions. This funding will quicken the pace of commercialization. The objectives of the programme will be to stimulate technological innovations in the private sector, strengthen the role of industry in development efforts and increase private sector commercialization of innovations derived from ICAR-supported R&D efforts. The basic procedures followed in the evaluation of the quality of proposals should closely parallel the procedure for peer review that will guide decisions of the ICAR competitive grants programme. Guidelines for implementation of these programmes can be developed based on highly successful programmes now in progress in the USA (Small Business Innovation Research Program, 1994). Priorities for funding should also be closely linked to those determined by ICAR and reflected in the current high priority programmes. This effort will provide an important linkage between innovative research and its application.

Financial incentives

Consideration should be given to the prospect of providing financial incentives to recognize individuals highly successful in the quality of their research progress and contributions. These incentives include training in advanced laboratories, travel to training courses and workshops, opportunities to attend national and international research meetings, and sabbatical opportunities. Also, authority should be granted to the principal investigator to subscribe to necessary scientific journals and pay for membership in professional societies.

Sabbatical/exchange programme

A sabbatical or career development grant programme should be established to recognize those research investigators who have clearly demonstrated a capacity for continued professional growth and achievement. This programme will be a great incentive for an active scientist because it helps to break monotony and to rejuvenate his/her interests in research or perhaps, even help seek a change in the direction of research. As in the case of young investigators, travel funds should be provided as part of the sabbatical programme to enable the scientist to attend key workshops and conferences in his subject area during this period. Proposals for sabbatical programmes could be reviewed by the appropriate subject matter Project Screening Committee at the time that they review research proposals.

Guidelines for governing patents

Adequate and fair guidelines governing patents and remuneration for research discoveries will provide incentives and rewards to the investigator(s). This issue will be especially important in view of the prospects of a programme designed to encourage joint projects with companies. Sharing the benefits of a royalty and profits from commercialization, and technical and scientific consulting fees would add to the incentive of the scientists and will act as a great encouraging factor for improved performance. All funded research grants should be 'portable' if a principal investigator moves to a different institution.

Impact assessment

An effective research impact-assessment programme should be established to assess the overall progress of the peer-review system so that mid-course corrections can be made and to be responsive to the needs of the user groups, scientific community and society. This process should be scheduled to operate at regular intervals, and can be greatly facilitated by office automation and computerization.

Priority setting

The programme should be initiated with a small number of priority areas for which research proposals should be solicited. This focus will allow adequate funding. ICAR priorities should involve a multi-component priority-setting mechanisms including particular target zones, productivity factors for commodities as well as economic and research resources data analysis. The major emphasis in the competitive grants programme can be determined once the priority setting process has been completed. At that time, it will also be possible to determine the relative percentage of funds that should be allocated to single versus multidisciplinary research projects.

Women and minority groups

One last lesson is the need for women and minorities to be represented in the committees. Secondly, members of panels need to be distributed with respect to rank and seniority in university, national or industry positions. Thirdly, reasonable efforts should be made to include subject matter experts from different parts of the country. Care should be taken to avoid overrepresentation of panel members from any given institution, agency, university or geographic area. The programme directors (subject matter directors) and panel managers (Project Screening Committee chairmen) in consultation with the chief scientist (executive director) in charge of the proposed Competitive Grants Office should review the composition of their respective panels to ensure that the needed expertise is represented. Typically, two or three positions on each panel should not be filled until late in the panel-member selection process when all the proposals have been received and the specific disciplinary areas can be determined. This representation will provide fair consideration of each research proposal.

Benefits of Efficiently Organized Biotechnology Institutions

There are several major benefits that can result from modifying and strengthening the current peer-review system. The first and the most important benefit of a sound review process is that it provides an opportunity for individual scientists to present new ideas in a cogent manner for review by a select group of competent scientists. After a constructive review, an improved revised proposal (whether approved or rejected and subsequently re-submitted) will lead to an enhancement of the overall quality of the research that will follow. This process in turn will increase the prospects that the research will produce results worthy of publication.

Second, the process examines costly research proposals before the major investment of time and effort is expended. Evaluation of work after completion and when it is too late to correct experimental errors in design or in experimental technique will be non-productive. It stimulates the investigator to organize a research proposal that will have an increased chance of obtaining funding. Following the completion of the review process, the principal investigator is provided by the grant programme technical staff with a rigorous critical review of the proposed research plan with constructive suggestions from the ad hoc reviews and panel. The emphasis should not be solely on possible errors in the plan or deficiencies in the presentation, although these should be identified, but on the strengths of the proposal that can be enhanced by positive suggestions for improvements.

Third, many scientists understandably express concern about the magnitude of time and effort required to prepare a proposal that will meet the requirements of rigorous review. However, if one considers the high cost of research and the importance of making the best use of public tax dollars, this effort is justified.

Fourth, the panel members and ad hoc reviewers are selected because of their qualifications as experts or leaders in the specific area of research to be evaluated. They can provide evaluations that are difficult to obtain at the home institution or university of any individual investigator. It is rare that the number and expertise of scientists located at a given institution would be equivalent to the scientists who would be involved as mail reviewers and panel members of the peer-review system. Moreover, to serve on a Project Review Committee should in itself be considered recognition of a scientist's professional stature.

Fifth, the review process provides intensive training in research evaluation for panel managers and members, informs them of the most recent advances that are being made in their own disciplines, and provides valuable experience in consensus development by all panel members.

Sixth, on a voluntary basis, researchers could provide their administrators with copies of proposal reviews, where they would benefit greatly from reading them. The perspectives obtained could be extremely helpful in evaluating their faculty and staff for promotion and continued financial support with institutional funds in the case of proposals given favourable review, but were not supported because of the lack of funds. This activity would also be helpful in the career improvement of the individual scientists, and it would be an excellent means of evaluating the quality of the review procedure. A failure to receive funding does not necessarily indicate that a proposal lacks merit, but that funds to support all worthy projects are not available. There should be indication in the review as whether an investigator should be encouraged to apply again with an improved proposal.

Seventh, the feedback mechanism should help elevate the quality of agricultural research in general including research supported by core budget funds. The in-depth critical review that could be provided by the proposed modified review system currently is not available to an individual researcher or interdisciplinary team in any department of a college or university (Haribabu, 1991). An additional benefit of participation in the peer review procedure is the training of a large core of scientists each year as panel members and as panel managers in the art of judging the quality of research. The experience is also of great benefit for future administrators.

Eighth, a significant percentage of the funds allocated in grants provides support for undergraduate and graduate students, and postdoctoral associates, who are being trained in the laboratories of senior scientists. This support represents a major investment in human resource development for the future.

Conclusions

This chapter is an attempt to analyse institutional challenges to biotechnology research, development, transfer and commercialization. A conceptual framework on biotechnology transfer in developing countries provided the context for identifying various issues, challenges and constraints to the development and commercialization of biotechnology in developing countries. An in-depth case study of an Indian institution involved in regulation and approval of biotechnology research resulted in a number of lessons for improving the organization of biotechnology institutions. A set of benefits for improving the efficiency of biotechnology institutions and their capacity has also been identified.

In order for biotechnology R&D and commercialization to progress, certain policies need to be implemented. First, IPRs and biosafety regulations need to be implemented. Second, public–private collaborations policies are essential for commercialization. Third, incentives for R&D may be needed. Fourth, policies that enhance capacity are essential. Fifth, policies that provide credit to the resource-poor farmer need to be formed.

At a micro level, institutions need to be well organized, promote collaboration and have strategic management in order for commercialization and research to occur. Institutions need to set priorities and programmes need to be periodically evaluated.

Further research is needed to understand the savings in the transaction costs of better-organized institutions. Unless appropriate institutions are created and organized effectively, the benefits of biotechnology may not be fully realized in developing countries.

References

Acharya, R. (1999) *The Emergence and Growth of Biotechnology: Experiences in Industrialised and Developing Countries.* Edward Elgar Publishing Limited, Northampton.

Altman, D.W. (1995) Issues and problems in the transfer of biotechnology. In: Altman, D.W. and Watanabe, K.N. (eds) *Plant Biotechnology Transfer to Developing Countries.* R.G. Landes Company, Austin, Texas.

Branstetter, L. and Sakakibara, M. (1998) Japanese research consortia: a microeconomic analysis of industrial policy. *Journal of Industrial Economics* 46, 207–233.

Byerlee, D. and Fisher, K. (2001) Accessing modern science: policy and institutional options for agricultural biotechnology in developing countries. *ECAPAPA Electronic Newsletter*, Vol. 4, No. 2–4. ECAPAPA, Entebbe, Uganda.

Conway, G. (2000) Crop biotechnology: benefits, risks and ownership. <http://www.rockfound.org/display.asp?context=1&Collection=4&DocID=141>. Updated 26 February 2001 (accessed: 4 May 2001).

Department of Science and Technology (India) (1993) *Opportunities for Young Scientist: Activities during 1985–1992*. Department of Science and Technology, New Delhi, India.

Flavell, R. (1999) Biotechnology for developing-country agriculture: problems and opportunities: biotechnology and food and nutrition needs. *2020 Focus* Vol. 2 (2). IFPRI, Washington, DC.

Ghosh, P.K. (1999) *Biosafety Guidelines: International Comparisons*. Paper presented on 24 June, Proceedings published by TERI, New Delhi.

Goulet, D. (1989) *The Uncertain Promise: Value Conflict in Technology Transfer*. New Horizon Press, New York.

Haribabu, E. (1991) A large community but few peers: a study of the scientific community in India. *Sociological Bulletin* 40, 77–87.

Huffman, W.E. and Just, R. (1994) Funding, structure, and management of public agricultural research in the United States. *American Journal of Agricultural Economics* 76, 744–759.

Hvoslef-Eide, A.K. and Rognli, O.A. (1995) Environmental issues for plant biotechnology transfer: a Norwegian perspective. In: Altman, D. and Watanabe, K. (eds) *Plant Biotechnology Transfer to Developing Countries*. R.G. Landes Company, Austin, Texas.

Irwin, D. and Klenow, P. (1996) High-tech R&D subsidies – estimating the effects of SEMATECH. *Journal of International Economics* 40, 323–344.

James, C. and Krattiger, A. (1999) Biotechnology for developing-country agriculture: problems and opportunities: the role of the private sector. *2020 Focus* Vol. 2 (4). IFPRI, Washington, DC.

Jayaraman, K.S. (2001) Indian regulatory system stifles industry growth. *Nature Biotechnology* 19, 105–106.

Kherallah, M. and Kirsten, J. (2001) *The New Institutional Economics: Applications for Agricultural Policy Research in Developing Countries*. MSSD Discussion Paper No. 41, International Food Policy Research Institute, Washington, DC.

Klette, J., Møen, J. and Griliches, Z. (2000) Do subsidies to commercial R&D reduce market failures? Microeconomic evaluation studies. *Research Policy* 29, 471–495.

Knorr, D. (1995) Improving food biotechnology resources and strategies in developing countries. *Food Technology* 49, 91–93.

Komen, J., Mignouna, J. and Webber, H. (2000) *Biotechnology in African Agricultural Research: Opportunities for Donor Organizations*. ISNAR Briefing Paper No. 43, International Service for National Agricultural Research, The Hague.

Kothari, A. (1999) Agro-biodiversity: the future of India's agriculture. Mtn-Fourm On-line Library Document. <http://www.mntforum.org/resources/library/kotha99a.htm> (accessed: 11 May 2001).

Lerner, J. (1998) The government as venture capitalist: the long-run impact of the SBIR program. Mimeo (Harvard University). Previously published as NBER WP 5753, 1996.

Narin, F., Hamilton, K. and Olivastro, D. (1997) The increasing linkage between US technology and public science. *Research Policy* 26, 317–330.

Paarlberg, R.L. (2000) *Governing the GM Crop Revolution: Policy Choices for Developing Countries*. 2020 Vision Discussion Paper No. 33, International Food Policy Research Institute, Washington, DC.

Pinstrup-Andersen, P. (1999) The developing world simply can't afford to do without agricultural biotechnology. <http://www.ifpri.cgiar.org/themes/biotech/102899.html>. Updated 28 October 1999 (accessed 3 April 2001).

Pinstrup-Andersen, P. (2001) 'Balancing the benefits of biotechnology.' http://ink.news.com.au/theaustralian/issues/bio1.htm>. Update 23 January 2001 (accessed 3 April 2001).

Saggi, K. (2000) *Trade, Foreign Direct Investment, and International Technology Transfer: a Survey*. Working Paper No. 2349, World Bank, Washington, DC.

Selvarajan, S., Joshi, D.C. and O'Toole, J.C. (1999) *Agro-biotechnology Capacity and Demand: the Indian Private Sector Seed Industry*. Island Publishing House, Inc., Manila, Philippines.

Singh, R.B. (1994) Technology transfer for sustainable agriculture and rural development. In: Kwaschik, R., Singh, R.B. and Paroda, R.S. (eds) *Technology Assessment and Transfer for Sustainable Agriculture and Rural Development in the Asia-Pacific Region*. FAO, Rome.

Small Business Innovation Research Program (1994) Innovation Research Program. Program Solicitation. Cooperative State Research Service, US Department of Agriculture, Washington, DC.

Tripp, R. (2001) Twixt cup and lip – biotechnology and resource-poor farmers. *Nature Biotechnology* 19, 93.

Chapter 16

Social and Economic Impact *Ex Ante* Evaluation of Embrapa's Biotechnology Research Products

Antonio Flavio Dias Avila, Tarcizio Rego Quirino, Elisio Contini and Elíbio Leopoldo Rech Filho

Embrapa, Parquc Estação Biológica, Final Av. W3 Norte 70770-901, Brasilia, DF, Brazil

Abstract

Brazil is one of the developing countries that made significant progress in biotechnology during recent years. The Brazilian Agricultural Research Corporation (Empresa Brasileira de Pesquisa Agropecuária – Embrapa), affiliated with the Ministry of Agriculture, State Research Institutes and Universities are producing the most important and more productive work in agricultural biotechnology research.

The preliminary *ex ante* impact evaluation of the main research projects developed by Embrapa in biotechnology shows that the social and economic benefits are potentially high. Biotechnological research conducted by the institution is in an advanced stage of development and certainly in the next few years should generate new, more productive varieties that use less inputs and have improved resistance to diseases and herbicides. Because of the introduction of new characteristics oriented to the quality of nutrition and to health improvement and because of positive changes in the social relationships when the whole food chain is considered, these new biotech products, to be released by the Embrapa centres, most likely will also generate important social benefits for small producers and for consumers as well.

Potential benefits estimated for each one of the five commodities (soybeans, cotton, potatoes, papaya and beans) included in this study showed that the amount spent annually, directly and indirectly, by the Brazilian society through the Embrapa biotechnology programme (around US$14.4 million, in 2000) will generate high returns. Only the economic impacts expected from the new transgenic varieties of beans will be enough to compensate these investments. In the same way, the social and environmental impacts expected based on preliminary evaluation present a good perspective to be positive and important for consumers and for small producers as well.

It should be emphasized that Embrapa's biotechnology research programme is in its first stages of development and that the perspectives to obtain more expressive results in the future are very high given the projects in progress and the effort to build a new institutional R&D agenda. Investment in human resources, specially in training in this area, continues to be a main priority and investments in infra-structure (laboratories and equipment) made recently or planned for the near future are also important and justify the expectation of significant results. Finally, this chapter has identified the crops new technologies and potential economic, social and environmental parameters within the production chain and its new components. This will be used to generate information for future empirical *ex ante* (and *ex post*) evaluation of impacts as part of the ongoing Embrapa programme of assessment follow-up studies.

The Impact of Agricultural Biotechnology Research

The expected role of biotechnology in a sustainable agriculture is to contribute to the development of new varieties, presenting mainly resistance to diseases, environmental stresses, contribution to recuperation and protection of the environment, decreasing the need for agricultural supplies and producing important metabolites of fundamental importance in the human diet (Carneiro, 2000).

Biotechnology research offers a wide range of possibilities to crop improvement. In the short term its most important contribution is the increase of the quantity and quality of global food, feed and fibre production. Plant biotechnology can already produce transgenic seeds that contribute to increased productivity and to sustainable cropping systems that are an essential ingredient for improving responsible use of natural resources and to safeguard the environment.

A complex list of desirable benefits expected from biotechnological research and its productive use are important parameters that should be carefully analysed as research results come to light and production reaches the market. Impact evaluation is the intellectual tool that should respond to the need of the situation.

Impact Evaluation

Results obtained in Brazil and other countries and presented in the literature by Griliches, Peterson, Evenson, Echeverria and others (see Alston *et al.*, 2001 and Avila, 2001) suggest that investment in agricultural research, such as that made by Embrapa, is highly beneficial, not only to the agricultural community, but also to society as a whole. In the case of agriculture biotechnology research, the first results from the USA and Europe indicate that this also is the case there and that benefits in the form of spillover will spread more and more beyond the boundaries of the farms and will particularly benefit the health and nutrition sectors.

The problem

Presently, in Brazil transgenic products are heavily criticized because it is not clear what benefits they will offer to society. Although one can easily see some advantages for the producers (reduction in production costs, increase in productivity), the groups against these products are posing other questions, especially in terms of impacts at the consumer level or in the environmental area.

Since the transgenic varieties in Brazil are still not released for use by the producers, basically due to the novelty of the biotechnology research programme in progress, the impact will be analysed in potential terms. The absence of a government authorization for the cultivation of transgenic crops in Brazil and consequently the absence of a regular production of genetically modified (GM) products is another factor that contributes to the need of an *ex ante*, rather than an *ex post* evaluation. Therefore, the impact will be analysed in potential terms based on the most advanced results of the research (elite cultivars) and on studies developed in other countries, especially where they are already in use by the producers.

In the *ex ante* case, the evaluation has a strong dose of subjectivity given the complexity for the authors to carry out estimations of potential impacts, in economic, social and environmental terms. The tentative results obtained for each of the commodities selected for study show that the Embrapa biotechnology research in progress at the Brazilian Agricultural Research Corporation (Empresa Brasileira de Pesquisa Agropecuária – Embrapa) will generate important economic and social benefits.

Economic impact evaluation

To evaluate the economic impact of agricultural research, the most popular methods are the economic surplus or input account method, the production function or productivity model and the decomposition method.

The concept of economic surplus is the most used method to analyse the economic impact generated by the agricultural research, especially in the case of Brazil (Avila and Ayres, 1987). In this approach the coefficients of price-elasticities of the demand and supply curves of the product under evaluation, the shift of the supply curve, price changes and the production values of the product area are used. The supply curve would be located to the left-hand side if there were no technological innovation generated by agricultural research. When technological innovation occurs, the consumers benefit with an increase in the supply of products and the producers benefit with a reduction in production costs.

In order to calculate this surplus, it is necessary to determine the rate of supply shift due to the new technology. This rate is calculated by comparing the traditional technology to the new technology (traditional variety vs. improved variety, for example). This rate of shift as a result of agricultural research is computed, in general, using yield increases due to new varieties compared with the traditional varieties and the rate of adoption of new varieties is estimated as a percentage of the cultivated area.

In the case of production or productivity function studies the relationship between research inputs and productivity change is identified by fitting an aggregate production function model (i.e. research, extension and education variables characterize changes in technology). Other explanatory variables can be used according to the production function to be estimated, as for example, price ratios of fertilizers with land, changes of cultivated area, hydric deficiency and soil conditions for crops (Silva, 1984; Evenson and Cruz, 1989).

Finally, there is the decomposition method that is used in two stages. Firstly, the output growth is separated into input-based growth and a multi-factor (total or partial) productivity index is built. In the second stage, this growth index is subjected to a statistical decomposition as in the production function studies (Evenson, 1987). Among the variables introduced in this model to explain changes in the index, agricultural research, rural extension, schooling in rural areas, infrastructure, etc., are included.

In this study, these methodological approaches will not be used, given the nature of the preliminary results obtained by the biotechnology research programme. The potential impact and the related costs of this research will be only superficially quantified and analysed but not measured in terms of internal rate of return (IRR) or net present value (NPV).

For more details regarding these methods and other approaches to evaluate economic impact of agricultural research investments, and also to appreciate the main results of national experiences on economic impact evaluation, see Daniels (1987), Evenson *et al.* (1987), Echeverría (1990) or Alston *et al.* (1995).

Social and environmental impact evaluation

Social and environmental impact evaluation have a less consistent tradition. The main lines of impact assessment focus on *ex ante* or *ex post* evaluations of large interventions on natural settings such as reservoirs, pipelines and highway constructions working on the principle that such impacts focus on a project level that is usually geographically based, such as a planned environmental change. The observation of a set of environmental and social variables that may change (*ex ante* evaluation) or have actually changed as a result of the project (*ex post* evaluation) is the core of the method of comparative diachronic studies. Mitigation procedures, recommendations of alternatives and cumulative effects observations contribute to the satisfaction of the political intent of applicability as a part of the planning process that such studies should usually have (Burdge, 1994).

In the case of biotechnology research products, the social and environmental impacts that are expected from adoption for agricultural production, despite having a clear local basis as far as production is concerned, extends far beyond the farm limits backward and forward.

This paper proposed a methodology that uses two main concepts as guide for identifying possible (*ex ante* evaluation) or actual (*ex post* evaluation) changes derived from the adoption of biotechnological technologies and products. The first is the concept of chain of production directs attention to changes in the procedures that occur during the productive process at the farm level, as well as to those that precede or follow this link of the chain. The second concept is the occupational role that directs attention to the social aspects of production in every link of the chain focusing on the human behavioural aspects of change. The chain can be followed as far as desired in order to identify second generation effects, conflicting interests and beyond.

Knowledge impact evaluation

The image of being a frontier technology that biotechnology research has in the present stage of development demands a special treatment for the link that creates it and is responsible for its impact on our knowledge. This will be achieved by applying the methodology in a special way to the evaluation of the impact on knowledge of the considered research cases.

Embrapa and the Products of Biotechnological Research[1]

An overview

Embrapa, www.embrapa.br, affiliated with the Ministry of Agriculture and state research institutes and universities, mainly those from the state of São Paulo, are producing the most important and more productive work in biotechnology. This effort has made Brazil one of the developing countries making some of the most significant progress in biotechnology. The recently concluded Genome Project (1997–2000) sequenced the genome of *Xylella fastidiosa*, one of the major pathogens of citrus, under the leadership and financial support of FAPESP (Fundação de Amparo à Pesquisa do Estado de São Paulo www.fapesp.br). This brought Brazilian progress in the field to the attention of the international scientific community as it was the first plant pathogen in the world to be completely sequenced. FAPESP invested US$15 million and coordinated an extensive network of laboratories and scientific workers.

Most of the important innovations in agricultural biotechnology originated in the universities. They are utilized to start up new small companies (technology incubators) and are absorbed by the big private transnational companies. Traditional experience in the research arena is that the discoveries and inventions flow from the fundamental science conducted in public institutions to research, development and commercialization, however, almost always it is accomplished by the transnational companies. As most of the private companies

[1] This section heavily relies on a paper, 'Biotechnology in Brazil', prepared by Elíbio Leopoldo Rech Filho for PROCISUR (Rech, 2000).

involved have been restructured into so-called 'life science' companies, their goals were expanded to include basic research, which traditionally was conducted by the public institutions. In a similar manner, public institutions have been expanding their mission to technology transfer activities. As it occurs, the fundamental aim of both public and private companies and institutions is to capture the market value of their discoveries and inventions. Embrapa is one of the public research companies following this path (Rech, 2000).

Embrapa is conducting several projects in partnership with the public and the private sector: among others, Cyanamid, to generate transgenic soybean plants tolerant to the herbicide imazapyr ('Clear Field'); Monsanto, to develop transgenic soybean tolerant to the herbicide glyphosate ('RoundupReady'); and Agrevo, to produce transgenic bean tolerant to the herbicide ammonium glyphosinate ('Liberty Link').

Most of the entities working in genetically modified organisms (GMOs), utilizing recombinant DNA technology, are private-sector companies that produce and market most of the products of biotechnology. However, this does not mean that companies are focused only on short-term objectives. Many companies have longer-term strategic goals in the area of biotechnology. Indeed, one interesting feature of the private-sector role in promoting GMOs is the way that its role is related to major structural changes in the industry.

Because of the large potential of the Brazilian agricultural market, the private sector has shown an interest in participating in this growing market, and in the manipulation of the recombinant DNA technology and all the consequent derived processes and products which will reach the market.

The private companies that work with agricultural biotechnology stress the positive benefits to current farmers as potential adopters. They seek to convince farmers that new herbicide- and insecticide-tolerant crops will allow for more efficient and rational eradication of weeds and pests compared with traditional methods. Farmers can spray any time during the growing season, killing weeds without killing their crops and control insects without the need for expensive pesticides. Biotech companies argue that this is particularly important in view of the fact that up to 40% of the world's food production is currently lost to weeds, pests and diseases (Genetic ID, 1999). In addition, biotech companies argue that genetic engineering will help developing countries to feed their population in ways that spare the environment and improve health.

The research on biotechnology at Embrapa, aiming at the development and adaptation of products and processes of interest for the Brazilian agriculture and agribusiness, has developed considerably in the last few decades. Although many groups have been established in this area, the research is often guided by the interest of the researchers and not by the demand of the market. As a result, the research generated does not always have a useful application for the current problems of the national agriculture. A thorough examination of the potential, and later on the measurement, of their actual impact is important in making decisions between alternative research paths (Rech, 2000).

Embrapa's programme of biotechnology

Embrapa has a national research programme on biotechnology, under the leadership of the Genetic Resources and Biotechnology Center and involving their agricultural research centres and other organizations (universities, state institutes, foundations, etc.). This Program (PBio), establishes an important instrument in redefining a policy for the development of biotechnology at Embrapa and in the National System of Agricultural Research – SNPA. Using the demands of the market as a baseline, PBio represents a dynamic element in national agriculture and aims at improving cooperation between various scientific institutions and the private sector (Rech, 2000).

According to the new system of cost research management, recently introduced through its Finance Department, Embrapa invested US$6.7 million on biotechnology research in 2000, including only the direct expenditures on personnel, operation costs and capital of this programme. If a share of indirect expenditures (the headquarters, national training programme, debt payments of earlier external loans with IDB and World Bank, technology transfer programme, etc.) is added in this cost, the total amount spent on biotechnology research would be around US$14.4 million in 2000. New funds should be added to this amount in 2001 to be spend with new tests on food safety and the environmental impact that will increase these direct expenditures on the biotechnology programme.

The biotechnology programme has collaborating projects, using the knowledge and available infrastructure in different Brazilian laboratories, that should contribute to overcome the present difficulties. These difficulties include the low level of investment in areas of research, training of individuals from the private sector and the incentive for the development of specialized products and new technologies. In addition, they should bestow grounds to facilitate mutual cooperation, which would guarantee a bilateral exchange of information and transfer of technology.

The programme is focusing mainly on basic research and aims to create an incentive for the study of fundamental biological processes, in order to bring about a better understanding of it, and establish and adapt new technologies geared toward the sustainable development of national agriculture.

> The incorporation of modern techniques of molecular and cellular biology in the research activities of private enterprises is of strategic importance, in order for them to remain competitive and to continue producing new products and processes. Nevertheless, this incorporation is made more difficult by the complexity of the matter and the high costs involved. For this reason, interaction between the public sector which possesses the knowledge and infrastructure of research, and the private sector which is capable of adapting the new processes and developing the finished product, becomes extremely desirable for the development of this area in the country. (Rech, 2000)

The objectives of Embrapa biotechnology research include: encouraging the development of new technologies, supporting projects in the area of basic research of fundamental biological processes, adapting already established technologies which exist in other countries and, together with the private sector, carrying on research and development of new products to solve the problems of national agriculture.

The main objectives of the PBio are: (i) to understand the essential biological processes and to develop advanced methods of biotechnology, which are important to make agriculture and forest production competitive, sustainable, and of high quality; (ii) to develop and promote cooperation among national and international institutions, aiming at facilitating the transfer of knowledge and technology in biotechnology; and (iii) to motivate the development and use of modern techniques of biotechnology in Embrapa and SNPA centres, aiming to increase the creation of new products.

According to Rech (2000), the demands and priorities of the biotechnology programme have been concentrated in the following areas: (i) the production of technology for using biological products in agricultural and forest production systems; (ii) development of technology in the area of molecular biology; (iii) development of biotechnology to be used in human nutrition; (iv) development of biotechnological procedures to increase the efficiency of agricultural and forest production systems; and (v) development of biotechnological procedures for native and exotic microorganisms.

Main results of the Embrapa programme

The projects already underway at Embrapa are at different developmental stages related to basic/applied research, such as the development of novel technologies in areas related to seed production. Research has been directed towards the study of apomictic reproduction, control of stature, storage pest control and qualitative seed improvement. The development of technology that allows the control of the reproductive programme of plants would, for example, allow the introduction of an apomictic character in superior crops of grain-producing plants. The production of commercial hybrids would be quite simplified, eliminating the need for progeny tests. Seed would be produced through open pollination, without loss of vigour, conserving the genotype of the apomictic, mother plant (Rech, 2000).

Similarly, the development of molecular markers, for fast selection of individuals carrying desirable traits in a given segregating population, should facilitate the breeding programmes connected to water stress in rice and maize, and aluminium toxicity in maize. The development of new crops adapted to climatic variations should increase the productivity of these crops in the country. Molecular markers will also be developed and utilized as genetic fingerprinting to secure genetic purity of maize seeds.

According to Rech (2000), the biotechnology at Embrapa also offers, through genetic transformation, strategies for the control of pests and diseases, by means of genetic engineering. In national agriculture, virus infestation represents enormous losses in potato and bean productivity. When some commercial varieties of potato are infected, the potato virus X (PVX), together with potato virus Y (PVY), cause synergical losses of up to 70% of the production. The potato leaf roll virus (PLRV) causes a decrease in production of approximately 50% due to tuber necrosis. In the case of the bean golden mosaic virus, the levels of losses are between 40% and 100% in the studies directed at this area, and the quality of the seeds harvested from infected plants is very poor. The use of strategies like antisense RNA, and/or coat protein to control potato viruses, and the endotoxin *Bacillus thuringiensis* for the control of maize insects and pests, could represent, in the short term, an enormous gain in productivity in these crops. The generation of transgenic soybean and bean plants carrying genes which confer tolerance to different herbicides should contribute to a more rational utilization of these. Papaya resistant to the papaya ringspot virus was developed. This particular project should have a significant impact for small-scale producers. Transgenic cotton plants carrying genes conferring herbicide tolerance and insect resistance will add value to the cotton production system.

The population outbreaks of pests are usually related to changes in biotic and abiotic factors. Biotic factors like the reduction of natural enemies, mainly caused by the intensive use of agro-chemical products, and abiotic factors, like irrigation, are related to the outbreaks of grasshoppers/locust. In Brazil, nearly 2,300,000 hectares of crops like rice, sugar cane, corn, and forages are seriously affected by grasshoppers. The development of agents to biologically control grasshoppers and corn borer which is responsible for almost 34% of productivity losses in corn, represent over a short time span, money saved on chemical products and a greater security for the environment. Likewise, the development of bio-herbicides could aid in the control of weeds without damaging the environment, what would also be economically profitable. The worldwide herbicide sales have increased to nearly $5 billion, representing 40% of the total sales of agritoxins all over the world.

Projects have been aimed at the understanding of mechanisms involved in genetic stability, competitiveness, and survival of *Rhizobium* and *Bradyrhizobium* in the soil in light of the development of inoculants for bean and soybean crops, which would allow an increase in productivity and still maintain the current levels of production costs. The use of biological products capable of guaranteeing an adequate supply of nitrogen for crops of agricultural interest is also another viable alternative for the preservation of the environment.

The impact of biotechnology is of great importance in the area of restoration, conservation and characterization of the genetic variability and reforestation. The species of native trees are quite subject to depredation due to unorganized exploitation, planting of pastures, and inadequate use as fuel source.

The use of molecular markers will have an impact on a short term basis in the area of animal health, in determining leucocyte adhesion deficient animals (BLAD disease) in Holstein cattle, in the prevention of bovine leukosis, in the identification of cromossomopathies and sex determination in embryos. In a similar way, growth hormone, kappa-casein, and beta-lactoglobulin gene polymorphism determination in bovines and the use of DNA patterns as a parameter in breeding programs of the Canchim cattle, will provide a positive impact in the intensification of animal protein production systems.

Another area of interest in the animal sector is the development of vaccination technology through the use of DNA. In addition to having a better immune response, this technology avoids the risk of contamination by the use of injected weakened organism or the problem of inducing immune response. The impact of this technology in the control of diseases could be revolutionary, creating efficient vaccines for the control of the most important animal diseases. (Rech, 2000)

The Pilot Study on Impact Evaluation

The present study is a tentative first for identifying and evaluating *ex ante* the five main groups of biotechnology research going on at Embrapa at the present moment. They are not completely concluded, but they are advanced enough so that one can already perceive the pattern of future applications and identify what appears to be their impact on agriculture and society.

Other research results could have been included, such as the most visible of all, namely research with cattle and the transgenic embryo, which have received large coverage from the media. They were excluded from the present study because the presumed impact is still much too conjectural.

Biotechnology research on the selected commodities

Embrapa's research programme in agricultural biotechnology is carried out under the guidance of the Genetic Resources and Biotechnology Research Center. The more advanced research in biotechnology at Embrapa is concentrated in the following areas: (i) identification of genes responsible for resistance to disease and environmental stress in rice; (ii) nitrogen fixation in varieties of soybean and other leguminous plants; (iii) identification of genes and heterosis of importance to maize germplasm, with special reference to disease resistance; (iv) characterization of the molecular structure of germplasm targeting increased cattle productivity and resistance to diseases; (v) production of virus-resistant potatoes, beans and papaya; (vi) production of herbicide-resistant beans and soybeans (glyphosate and imidazolinon); (vii) utilization of soybean and bean plants to produce human growth hormone and insulin; and (viii) production of maize with high nutritional quality (methionine degree).

Among these biotechnology areas and considering the research where the results are more advanced, for this *ex ante* evaluation of the social, economic and environmental impacts the following five commodities were chosen: papaya, soybean, beans, potato and cotton. These commodities are very important for the Brazilian agribusiness with a production value superior to 18% of the total agricultural production with an expressed participation of 27% in terms of exports (Table 16.1). For beans and potatoes, there are also imports. These imports have been extremely important (around US$20 million for potatoes and US$140 million for beans) and are basically used for consumption, except in case of potato, where the imports are also used for seed. This situation should be reversed and the biotechnology products that Embrapa will be releasing in the next 3–5 years, certainly, will be very important for that.

Next, the advances in biotechnology research under way at Embrapa will be

Table 16.1. Brazil: production and export values of the selected commodities in 1999.

Selected commodity	Production (1000 tonnes)	Exports		Production values	
		US$1000	% Total	US$ million	% Total
Cotton	1,444	215,089	1.46	535.07	1.25
Beans	2,896	1,104	0.01	612.22	1.43
Potato	2,843	—	—	1,486.75	3.46
Soybeans	30,765	3,784,357	25.63	4,926.99	11.48
Papaya	1,700	13,578	0.09	126.74	0.30
Sub-total	39,648	4,014,128	27.19	7,687.77	17.92
Other commodities		14,763,119	72.81	35,353.78	82.09
Total		18,777,247	100.00	42,910.89	100.00

Source: (a) Quantity of Production and Exports – CONAB; (b) papaya (production in metric tonnes) – FAO; and (c) production values – Conf. Nacional de Agricultura. *Indicadores Rurais*. 4:27. Brasília, November 2000.

revealed for each of the selected commodities and later, the expected economic, social and environmental impact will be spelled out. An abstract of each of these researches is presented below as they are described in institutional sources, by their own research leaders. Finally, the impacts of the scientific knowledge building will be contemplated.

Papaya

Brazil is the main producer of papaya (*Carica papaya* L.) in the world, producing more than 40% of the world production of this fruit (FAO, 1997). Papaya is the host of several pathogens, where viruses and fungi are the most important constraints for the economic exploitation of this product. Among the viral infections in papaya, the most important is caused by the papaya ringspot virus (PRSV), generating a greater negative impact in this crop.

PRSV was observed for the first time in Brazil during the 1960s and since then it has spread to all the papaya producing areas and regions. Because of this virus, the producing regions of papaya have changed over time and became itinerant. States such as São Paulo and Rio de Janeiro, responsible for a large part of the Brazilian production at the end of the 1970s and the beginning of the 1980s, are now responsible for less than 1%. The presence of PRSV in these states has a direct link to the reduction in the papaya-cultivating area. States such as Bahia and Espirito Santo are now the larger papaya producers in the last 15 years in Brazil. They have maintained a level of production that is economically viable due the displacement of the production within these states and also due to the adoption of a programme of disease control based on the elimination of the infected plants (roguing system). The roguing system is expensive, substantially increases the production costs and depends on the level of virus infestation in the area to be effective.

The absence of natural resistance in the germplasm of this species, the disappointing results obtained from research efforts to transfer a natural gene present in some wild species of papaya *Carica* to the commercial papaya, and the palliative character of other control forms (roguing and cross protection) utilized to combat the disease put genetic engineering as a possible solution. The use of genetic resistance derived from the pathogen (PDR) could have a great potential for controlling the disease.

The development of transgenic papaya expressing the gene pc ('protein cover') in Brazilian isolates of PRSV is presented as a way of controlling this disease in the country, with a large potential for it to be effective and with a chance to remain so for a long period. This research started at the beginning of the 1990s and is now in the final stages where elite cultivars of papaya are being integrated into the breeding programme of the Embrapa Cassava and Fruit National Research Center. Final efforts are to be made in the conventional breeding programme and in the development of biosafety studies needed to release PRSV-resistant varieties into the market in a short time.

Soybean

The soybean research emphasizes the development of a novel system to obtain fertile transgenic soybean plants at a high frequency and the integration of the human growth hormone and insulin genes in soybean genome.

Brazil is the second largest soybean producer in the world, with an annual average production of over 13 million tonnes. There is a worldwide interest in the utilization of the recombinant DNA technology to introduce new traits in soybean, which in turn, should allow cost reduction and productivity to increase among other benefits. Over the past 10 years, several attempts have been made to obtain transgenic soybean plants. The protocols so far published have failed to achieve reproducibility, simplicity in conducting the experiments, high efficiency and variety independence. The combination of: (i) genes which codify herbicide-active polypeptides, capable of translocating systemically and concentrating in the apical meristematic region of the plant and (ii) a short multiple-shooting induction protocol, have allowed the development of a simple and routine system to obtain transgenic soybean plants at a high frequency and variety independence.

The apical meristematic region of mature soybean embryonic axes was excised, and bombarded with the plasmid DNA. Then, the bombarded embryonic axes were transferred to the culture medium containing MS basal salts, sucrose, cytokinin and the selective agent. After 3–5 weeks in culture, putative transgenic shoots were excised and transferred to the greenhouse for further development. The plants were allowed to set seeds and progeny analyses were performed. Four elite commercial soybean varieties (Doko RC; BR16; Celeste and Conquista) were transformed. The frequency of transformation (number of transgenic plants/number of bombarded embryonic axes) varied from 5–20%, depending on the cultivar. Virtually no chimeras were observed. Embrapa is already been utilizing this novel system to generate elite events carrying useful traits and to introduce these plants in the soybean breeding programme. Field release experiments have been carried out over the last 2 years.

The development of technologies for the introduction and expression of foreign genes in plants has allowed studies of gene function, and has resulted in great advances toward plant genetic engineering with enhanced input traits. Recently, several groups of researchers have been actively involved in the evaluation of the potential utilization of plants as novel manufacturing systems, in order to produce different classes of proteins of pharmaceutical value.

Embrapa, with the partnership of the University of Campinas (São Paulo State), has utilized the human growth hormone and insulin genes, under control of the monocot tissue-specific promoter from the sorghum γ-kafirin seed storage protein gene, to generate GM soybean plants by biolistics. The presence of the introduced genes was analysed by polymerase chain reaction (PCR), utilizing specific primers. Studies on the expression of the proteins have been carried out. The results will form the foundation to evaluating the potential commercial utilization of soybean plants to produce human growth hormone and insulin.

Beans

The 'golden mosaic' geminivirus is one of the main diseases affecting the cultivation of beans in Latin America, specially in Brazil. This disease is caused by a geminivirus that is transmitted by the whitefly (*Bemisia tabaci* Gennadius).

The cultivated areas of beans, in several Brazilian states, specially in São Paulo, Paraná, Minas Gerais, Bahia, Goiás and Mato Grosso have been strongly affected by the golden mosaic. This is attributed to the expansion of soybean cultivation in the country, host of this insect. In some regions, due to the high incidence of this disease, the producers have to abandon this crop as the only option. The losses are around 40–85% of the final production and can reach total loss, depending on the variety, the stage of the plants when infected and the type of isolated virus.

A more effective method to control viral diseases has been with the introduction of resistance of the host. Studies with the objective of controlling the golden mosaic geminivirus through the introduction of variety resistance have been carried out in Brazil since the 1970s. Despite the extensive efforts of the bean national research programme, no resistant germplasm has been found. This is also true about research conducted in other countries such as Guatemala, Dominican Republic, Mexico and Argentina, with the guidance of the International Agricultural Research Center for the Tropics (CIAT). The verified low natural resistance to VMDF-BR in species of *Phaseolus* has oriented the Embrapa programme to develop research using genetic engineering to find this resistance, using 'pathogen-derived resistance' to virus PDR.

Products generated by the Embrapa biotechnology programme in the last few years have permitted elite cultivars of beans with resistance to golden mosaic to be obtained. These elite products are actually integrated into the breeding programme at the Embrapa Rice and Beans National Research Center, and biosafety tests are in progress. The aim is to have these new cultivars in a condition to be released in the next 3 or 5 years.

Potatoes

As far as potatoes are concerned, the emphasis is on the study of mechanisms of resistance derived from the pathogen potato leafroll virus (PLRV) and introduction of the gene that gives resistance to the potato virus Y (PVY). This virus belongs to the *Potyvirus* genus (Ward and Shukla, 1991).

PVY infects potato (*Solanum tuberosum*) causing necrosis, mottling or yellowing-vein clearing of leaflets, leaf dropping and premature death (de Bokx, 1990). In fields established with infected tuber seeds, yield reduction is dramatic (De Bokx and Piron, 1990; Brandolini *et al.*, 1992). To avoid it, growers are compelled to renew frequently tuber seed stocks, with lasting effects on production costs, growers' income and final potato prices for consumers. Achat is one important potato cultivar in Brazil (Torres *et al.*, 1999).

Strategies to reduce losses by viral infection are based in the use of virus-free seed potatoes and special cultural practices (Gugerli, 1986). Usually, these procedures do not offer permanent solutions to PVY infection. The development of resistant cultivars in potato can be a more effective strategy but, due to the tetraploid nature of its genome, potato breeding is known to be extremely difficult. However, the ability to transform plants using *Agrobacterium tumefaciens* Ti plasmids has made it possible to produce plants with new traits without severe changes in the genetic background of the cultivar. Coat protein (CP)-mediated protection has been used in several crops to obtain transgenic plants resistant or tolerant to viral infection (Hull and Davies, 1992).

An extremely resistant phenotype against the Brazilian strains PVY-OBR and PVY-NBR was observed in one line. No symptoms or positive enzyme-linked immunosorbent assay (ELISA) results were observed. The stability of the integrated transgenes was examined during several *in vitro* multiplications over a period of 3 years, and no modification in the Southern pattern was observed. The stability of the transgenes, the absence of primary infections and the relatively broad spectrum of resistance suggest that the line obtained in this work can be useful for agricultural purposes.

Cotton

One line of the research on cotton aims to obtain transgenic cotton plants with variety independence and a high frequency, and to develop systems for the introduction and expression of genes of economic interest in elite cotton cultivars.

Due to the economic importance of cotton worldwide, there is considerable interest in utilizing the recombinant DNA technology for the introduction of useful genes. Several laboratories have been involved in the development of a protocol capable of obtaining transgenic cotton.

The combination of genes which codify herbicide-active polypeptides, capable of translocating systemically and concentrating in the apical meristematic region of the plant and a short multiple-shooting induction protocol, have allowed the development of a simple and routine system to obtain transgenic cotton plants at a high frequency and variety independence.

Three elite commercial cotton varieties were transformed. The frequency of transformation (number of transgenic plants/ number of bombarded embryonic axes) varied from 1% to 3%. Virtually no chimeras were observed. Embrapa has been utilizing this novel system to generate elite events carrying useful traits and further introduce these plants into the cotton breeding programme.

Another line of the research on cotton aims to develop cotton seed resistant to herbicides, a demand generated by international private companies and by the Embrapa's system. Herbicides are an important segment in the area of crop protection. As there is this demand in the cotton market, it is necessary to generate elite cultivars of cotton resistant to herbicide to be introduced in the breeding programme for release in the market.

The development of this project is important to the partnership with the Cotton National Research Center, and the gene that will be used is derived from the imidazolinones, which is included in a new class of herbicide developed by Cyanamid. This herbicide acts as an enzymatic system present only in the plants and absent in animals and humans. It has advanced characteristics for environmental safety.

The development of this research should contribute to major advances in basic and applied sciences. The introduction and expression of genes with resistance to herbicides in elite cultivars and the release of transgenic varieties of cotton by Embrapa in the market should have important implications for Brazilian agriculture.

Economic, social and environmental impacts

The impacts of the biotechnology products will be analysed in the context of the productive chain of each one of the commodities involved in this pilot study.

Papaya

ECONOMIC IMPACTS. The release of new varieties with resistance to PRSV will permit a substantial increase in the Brazilian production of papaya due to expansion of the cultivated area to regions abandoned in the past and due to an increase in the productivity. This increase in the production, and the reduction in the production costs, will allow the generation of more revenue for the papaya producers.

Economic impacts will be also important on the government side because of the increase in the volume of tax collected at the producer level and specially in the processing and distribution segments. An economic surplus on the exports sector will also be expected due to the increase in production, and consequently there will be a surplus to be sold on the international market.

Considering, for example, an annual production of 1.7 million tonnes, an increase of 20% in the national production due to the adoption of these new varieties can represent an addition of a minimum of US$25 million to the Brazilian economy.

Also, the abolition of the use of fungicides to control PRSV can represent a reduction in the imports of agricultural inputs or the possibility of exporting more of these inputs no longer used in the papaya production system.

SOCIAL AND ENVIRONMENTAL IMPACTS

Effects on the producers. The introduction of transgenic seeds for the development of papaya trees that are resistant to PRSV will provide important social impacts at the producer level, depending on the farm size (large or small).

For large producers, it will permit sustainable production in the same area by re-establishing the expected triennial rotation of the trees instead of a shorter rotation (1–1.5 years) that is caused by the disease. Disease control will then be carried out by identification and extermination of affected trees. This will eliminate the occupation of 'mosaiqueiro' (hunter of infected trees) that presently produces such a defensive treatment. Manpower will be concentrated on directly productive aspects of the production process.

For large producers it also will diminish the managerial burden of maintaining the quality of the productive process and liberate time and energy for other tasks. It will create a certain dependency on the producer, vis-à-vis the furnisher of GM seeds, and his technical assistance.

The impact of the release of resistant cultivars to PRSV among small producers will be very important, given the concentration of the papaya production on the hands of small producers and family agriculture that have typically had limited resources. The use of biotechnologically modified improved seeds will have a direct influence on the quality of life of these groups as a result of the income increase. It will result in more jobs because of the possibilities for expansion of the productive area. The use of disease-resistant cultivars, which do not require the application of chemicals as control, represents a direct contribution to the wellbeing of human health and so will benefit millions of papaya consumers. It will permit the continuous use of papaya fruits as an important nutritional component for their families, tending to benefit small farmers for longer periods or even in a definitive way in the areas of small production. It will make possible the participation of small papaya producers in the market as it will more frequently generate some consumption surplus. Finally, it will introduce a relationship presently not existing between the distributor of transgenic seeds and the small papaya producers.

Effects on the processing and distribution segments. The effects on these components of the papaya productive chain will be also important given the improvement of quality and the increase in production. The possibility of expansion in the production to new areas also will affect the industries involved with the processing and distribution of papaya in the Brazilian market. These expected social impacts will generate more jobs in these sectors.

Impacts on consumers. As a result of the introduction of these new varieties the production will increase and it will force prices down, quality up and increase further diffusion of the consumption. Among the new consumers, more and more people of the lower social segments shall be included.

The visible split of opinions between consumers who accept GM food products and those who are against them can ensure deep disagreement between them and make the adoption of transgenic papaya seeds less than pacific. One can expect that the political overtones of the discussion about transgenics will be long-lasting and with no unequivocal winners.

Other effects expected. The complexity of seed production technology and quality control will demand highly qualified manpower of medium, superior and doctoral levels as well as efficient communicators to conduct the relationship between genetically modified seed producers and fruit producers. As a result, the use of specialized and general education will intensify and the search for technical information will grow larger.

Brazilian and foreign laws of intellectual property will require tight technical and legal control of the transgenic seed production processes and will induce the participation of law professionals and law firms as a new link in the papaya production chain. This represents a completely new and unusual social impact.

Finally, these new papaya cultivars resistant to PRSV will permit the usage of large old productive areas especially in the states of São Paulo and Rio de Janeiro. These areas were abandoned because of the spread of disease, and their use will intensify competition with the present production areas in the states of Espírito Santo and Bahia.

Soybean

ECONOMIC IMPACTS. According to the estimates made through interviews and personal contacts, the new varieties of soybeans with resistance to herbicides will reduce the production costs by 2–5%. A recent report prepared by the University of Illinois (1999) about the economics of GMOs in agriculture has indicated that an additional benefit for soybean producers was a reduction in the prices of other herbicides. This shows that even the farmers electing not to use GR (glyphosate-resistant) soybeans have benefited from its introduction in the USA.

These new varieties will also allow an increase in the national production due the cleaning of the cultivated area that makes the new varieties more productive. Another benefit will go to the government: more tax will be collected as a consequence of the increase in the revenues of the soybeans producers. An important contribution in the generation of surplus as a result of the increase in soybean exports is also expected.

Given the Brazilian annual production of more than 31 million tonnes, a reduction in the production costs and the increase in the national production (the cleaning of the cultivated area makes the new varieties more productive) can represent an addition of 1 million tonnes in production or more than US$150 million in exports.

In the case of the production of soybean plants capable of producing human growth hormone and insulin, the generation of additional revenues for the producers, industries and for all the Brazilian economy will be very important. If Embrapa succeeds in the generation of these new plants for pharmaceutical use, the value of the production will increase substantially due to the new characteristics of the products and new market niches.

SOCIAL AND ENVIRONMENTAL IMPACTS

Effects on the producers. Herbicide-resistant soybean varieties will increase the gains of the producers, and consequently their quality of life. At the same time, it will make them more dependent on the furnishers of GM seeds and the related technical assistance. This creates a situation of conflicting interests that may characterize the long-term relationship between these two links of the productive chain. The occupational roles will change little, but a few jobs will disappear as the new technology is labour saving. There will be additional income to the producers and technology property owners.

The potential commercial utilization of soybean plants to produce human growth hormone and insulin will create a new and important social organization of the soybean market by adding to it a specialization on pharmaceuticals. The productive chain will probably be organized with the central participation of highly sophisticated small producers that will be able to produce and maintain the quality of the product.

Effects on processing and distribution. No important impact of herbicide-resistant soybean varieties is expected in the present pattern of processing and distribution, except those originating from lower production prices and higher productivity. This can generate a better competitive position for Brazilian soybean production as much as the new technology is not adopted elsewhere.

The new pharmaceutical link of the productive chain will be in constant contact with researchers for the purpose of continuous adoption of new technologies and maintenance of quality of production on the one hand, and with pharmaceuticals, most probably transnational laboratories for the purpose of transforming the soybean new varieties into commercial medicines on the other. This will add a whole new chain segment to the distribution side – the medical/pharmaceutical business.

Until now this business was completely outside the soybean producers' interests. Quality control, the continuity of production and frequent technological actualization should constitute new demands that the productive sector will receive. The laboratories will act as oligopolistic consumers of the pharmaceutical soybean production and as quality and quantity controllers.

Impacts on consumers. The consumers in general will benefit from lower prices and new products. Those that depend on the use of previously expensive health products will be greatly helped by the diffusion and adoption of the research results.

The socially relevant, politically emotional disagreement about transgenics will affect soybean seed adoption as much as it will papaya and any other transgenic product. The discussion about possible environmental impacts will have a damaging influence on adoption of herbicide-resistant soybean varieties as well as on the adoption of transgenic medical products, irrespective of what evidence is already available. This will press scientists, including health and medical specialists, to produce new, more convincing empirical arguments on them.

Other effects expected. The effects on quality and quantity of manpower use, and the increasing value of education for the production of high-tech products will be about the same as in the case of papaya already discussed above. Legal and pharmaceutical professionals will have a previously unsuspected role in the soybean productive chain.

Beans

ECONOMIC IMPACTS. The new transgenic varieties with resistance to golden mosaic virus, in the final process of development at Embrapa and to be released in the next 3 years (if authorized by the government), should represent a very important contribution from this research organization to the Brazilian agricultural sector.

The increase in production and the reduction in the cost of production that will be obtained means an increase in the profitability of this culture, mainly for small producers.

SOCIAL AND ENVIRONMENTAL IMPACTS

Effects on the producers. The small producers are the main growers of beans in Brazil both for self-consumption and for market. The positive results of the research will permit the production of a second yield in the same area at the dry season. It will re-open to the producers the alternative of returning to bean production in areas where the culture was uprooted because of the presence of whitefly. These effects will increase the occupation of the existing manpower.

Effects on processing and distribution. It is expected that more production will result in more participation of the small bean growers on the national market and will increase the local consumption of goods and services. As it was pointed out about other commodities, the bean producers will be more dependent on seed producers because they will have to renew the seed periodically.

Impacts on consumers. The consumers will have access to the goods at lower prices and the supply will become more regular, i.e. there will be more food security and less need for import.

Other effects expected. Controversy about the use of transgenic food will probably continue and some short-term panic, as in the case of the first 'mad cow', may happen if any disaster, true or false, is believed to occur.

Potatoes

ECONOMIC IMPACTS. The new varieties with resistance to PVY will permit a reduction in production cost in Brazil. Recent studies developed in Mexico by Qaim (1998) showed that the production costs can be reduced with the use of transgenic potatoes by 13% for large producers and 32% at the small producers' level.

The reduction expected in the production costs will generate an increase in the revenues of the potato producers and also more tax for the government. For the consumers, this cost reduction should represent a reduction in the price of this commodity.

In 1999, the value of potato production in Brazil was around US$1.5 billion. This means that the release of new varieties with resistance to PVY can represent an important contribution of Embrapa's research to increase the revenues of the potato producers due to the reduction in the production costs. Given the potato production value and supposing a minimum reduction of 10% in the costs, it is expected that there will be a substantial increase in the producers revenues, perhaps more than US$100 million per year.

Finally, the abolition of the use of fungicides for PVY control can also represent an economy in the imports of agricultural inputs or the possibility of exporting more of these inputs no longer used in the potato production system.

SOCIAL AND ENVIRONMENTAL IMPACTS

Effects on the producers. The virus-resistant 'Achat' potato will no longer compel growers to frequently renew tuber seed stocks and will revert the dramatic yield reduction caused by the disease. The harvest forecast will grow in precision and gains in production costs will be offered resulting in growers' higher income levels and a consequent increase in quality of life.

Effects on processing and distribution. The need for frequent renewal of tuber seed stocks will decrease, perhaps almost completely disappearing and making the producer less susceptible to sudden variations. A better quality product will come to the market.

Impacts on consumers. Part of the gains resulting from more favourable conditions of production will go to the consumers, who will have a product of high quality and a more constant supply. The reduction in the use of chemical inputs may result in health quality improvement for consumers and for all those that deal with the product.

Other effects expected. The quality of the new product will increase Embrapa's influence in the potato seed market for benefit of the small growers. The gene that codifies the protein cover is a part of the virus. It is not supposed to be harmful to human beings and animals, as many licensed and largely commercialized GM products have the gene NPT II already. Research abounds and no harm was detected. This can become a good defence argument against any opposition to this transgenic product in particular, and to the class of transgenic products in general.

Cotton

ECONOMIC IMPACTS. The use of the new Embrapa cotton varieties with resistance to herbicides will permit the increase of the Brazilian production of cotton. The main impact will be in the production costs, which are considerably reduced with the introduction of genes resistant to herbicides. This cost reduction is estimated to be 3–5%, which will permit a substantial increase in the revenue per hectare for cotton producers.

This increase in revenues means more profitability for the producers in this activity that permits an expansion in the cultivated area or investments in technology in the same area, making the culture more productive.

Finally, the impact on productivity must be significant due to the cultivation of these cultivars resistant to herbicides. The increase in the productivity will be a consequence of the utilization of a clean cultivated area (without weeds) to cultivate cotton and consequently better conditions to increase the productivity comparatively to a situation where the cotton is cultivated with traditional varieties.

SOCIAL AND ENVIRONMENTAL IMPACTS

Effects on the producers. Herbicide-resistant cotton varieties will positively affect the gains of the producers and also the quality of life. However, as was observed on soybean seeds with this type of resistance, they will become more dependent on the seed furnisher and on technical assistance. Conflicting interests may characterize the long-term relationship between them and affect the behaviour of the productive chain by making internal relations more unstable and conflictive.

A few jobs will disappear since the new technology is labour saving. The profits will increase and their distribution will favour the producers and the technology owners at the expense of labour.

The gene to be utilized for producing herbicide-resistant cotton seeds is a derivation of a new class of herbicide, which behaves as an enzymatic system that is present only in plants, not in animals nor in humans. This corresponds to a very advanced characteristic that favours environmental security.

Effects on processing and distribution. An expansion of the area cultivated with cotton may have an impact on the Brazilian share of the international market and reverse the present position of importer to exporter. This may compensate for the lower job intensity of production by creating new jobs in new productive areas.

Impacts on consumers. The reduction in the production cost will lower cotton prices, and will directly benefit the consumers and consequently their quality of life.

Others effects expected. Relations between the national cotton productive chain and the transnational research laboratories on transgenics will grow closer because of the use of international patent as a basis of the developed technology.

The introduction of new transgenic varieties on cotton production will also expand the interaction of Brazil with the global economy and culture.

Impacts on increasing scientific knowledge

Agricultural research is necessary to provide knowledge about environmental consequences of transgenic technology adoption and also identification and prevention of possible negative impacts.

There are indications that part of Brazilian society tends to reject products identified as transgenically modified. Heavy political overtones have been recently attached to the subject – these are eloquent on the discussion on soybean production and tend to generalize to every new transgenic product as well.

There is an indication that papaya production from GM seeds is becoming contaminated by such a negative general attitude. Researchers have already observed some difficulty in drawing research resources outside the public sector for experimenting with GM papaya seed and technology. This can be a consequence of the hostile attitude and a generalization of the potential risk perception that used to relate to it.

The papaya research, for example, will contribute some very specific results as far as biotechnology is concerned. Among them: (i) an improvement in the knowledge on PMV based on the decodification of its genome; (ii) the production of tools for early diagnosis (PCR and ELISA); (iii) the search for resistant genes (natural and synthetic), aiming the future work on transformation (transgeny); (iv) the development of a technique of gene transfer for elite varieties; and (v) an understanding of the mechanisms of inhibition for replication of the geminivirus by transdominance in beans by studying different alternative solutions.

Results that will contribute to the increase in scientific knowledge in biotechnology research are also expected regarding other commodities.

Concluding Remarks

Conclusions

The preliminary *ex ante* impact evaluation of the main research projects developed by Embrapa in biotechnology show that the social and economic benefits are potentially high. Biotechnological research conducted by the institution is in an advanced stage of development and certainly should generate in the next few years new, more productive varieties, that use less inputs and have improved resistance to diseases and herbicides.

The new biotechnology products to be released in the next 3–5 years by Embrapa will generate important social benefits for the Brazilian economy and society, if authorized by the government. These benefits will also include small producers and consumers, because of the introduction of new characteristics oriented to the quality of nutrition and to health, and because of positive changes in the social relationships between the components of the chains.

Potential benefits estimated for each one of the five commodities (soybeans, cotton, potatoes, papaya and beans) included in this study showed that the amount spent on the Embrapa biotechnology programme (around US$14.4 million in 2000) will have high returns. Only the economic impacts expected from the new transgenic varieties of beans will be enough to compensate for these investments. In the same way, the social and environmental impacts expected also present a good perspective from which to be positive, and will be important for consumers and for small producers as well.

The research investment needs to be complemented by input supply systems, farmer acceptance, distribution networks and retail outlets. Each part of the productive chain stands to gain from the smooth adoption of the new technologies. Thus, structural changes in the private sector are in part a reflection of the new technologies, but structural changes are also about the distribution of those gains and the extent of control over the use of the technologies.

The new biological technologies seem not to have had a large impact on the occupational roles of most of the members of the productive chain, except for introducing a new and powerful link for the producers, or at least greatly reinforcing the importance of previously existing social agents in this position. This link retains the control, guarantees the quality and organizes the distribution of the new technologies at the price of appropriating a part of the value of the production and introducing a locus of decision with an important impact upon the subsequent links of the productive chain.

The productive chain will become more complex, as a presently non-existent link will be introduced just before the 'farm gate', namely the specialist in producing transgenic seed, maintaining and controlling its quality. This new link will revitalize the relevance of research for agricultural production as it is fed by the research institutes and the universities. The techniques of seed production and quality control will be structured as in companies or as in any other organizational arrangement that should be permanent, reliable and available all over the productive regions.

In the case of papaya, some occupational behaviours of those involved in the productive tasks will also change.

There is no guarantee that every element of the private sector will gain from biotech developments. Even without the opposition of consumer and environmental groups, the spread of biotechnology in agriculture and food may have been contentious because of its impact on the structure of the food industry.

On the other hand, significant numbers of consumers identify biotechnological products with products harmful to the environment, human health and the future of the earth. These circumstances cannot be ignored and should not be discarded by any kind of evaluation. The resulting conflictive interests may be the arena where at least a fraction of the future of the biotechnological research will be decided.

Perspectives

Analysis of the economic and social potential impacts of new technology is being made at Embrapa based on a new approach, namely, productive chain analysis, rather than a traditional analysis at the producer or consumer surplus level. Positive and negative impacts are being measured in all the main segments of the chain (industry inputs, rural producers, industry processing, distribution and consumers) and include an analysis of environmental impact.

The approach used in this study has advantages over those used in the past because the evaluation is more complete, offering answers to questions not solved in the traditional approaches (economic surplus, productivity index, production function, etc.) but demanded by society, especially in developing countries. A more formal treatment of the methodology and its use should be the next step in order to develop an efficient tool and to make it able to support comparisons between different instances of impact as well as *ex post* applications.

The research that Embrapa is developing with the University of Campinas on soybean and bean plants to produce human growth hormone and insulin and its preliminary results, indicate that this is another area in which the institution is in a good position to generate important benefits for Brazilian society.

It should be emphasized that Embrapa's biotechnology research programme is in its first stages of development and that the perspectives to obtain more complete results in the future are very high given the projects in progress and the effort to build a new institutional R&D agenda. Investment in human resources, specially in training in this area, continues to be a main priority and investments in infrastructure (laboratories and equipment) made recently or planned for the near future are crucial for that.

Finally, this chapter has identified the crops' new technologies and potential economic, social and environmental parameters within the production chain and its new components. These will be used to generate information for future empirical *ex ante* (and *ex post*) evaluation of impacts as part of the ongoing Embrapa programme of assessment follow-up studies.

Given the critical attitude of various consumer segments towards transgenic products, agricultural research scientists are conscious from the very beginning of the need to establish an empirical base for the social, environmental and economic evaluation of the results, strengthening, as much as possible, the social, economic and ecological components of commodity research programmes.

Acknowledgements

This chapter could not be written without collaboration of many scientists of Embrapa, particularly Francisco Lima Aragão, Manoel Teixeira de Souza, Antonio Carlos Torres and Mauro Carneiro, project leaders dealing with biotechnology research.

References

Alston, M.J., Norton, G.W. and Pardey, P.G. (1995) *Science under Scarcity: Principles and Practice for Agricultural Evaluation and Priority Setting.* ISNAR, Cornell University Press, Ithaca, New York.

Alston, M.J., Chan-Kang, C., Marra, M.C., Pardey, P.G. and Wyatt, T.J. (2001) *A Meta-Analysis of Rates of Return to Agricultural RandD: Ex Pede Herculem Evaluation and Priority Setting.* Research Report 113, IFPRI, Washington. DC.

Avila, A.F.D. and Ayres, C.H.S. (1987) Brazilian experience in ex-post evaluation of agricultural research. In: Evenson, R.E., Cruz, E.R., Avila, A.F.D. and Palma, V. (eds) *Economic Evaluation of Agricultural Research: Methodologies and Brazilian Applications*. Chapter VII. Embrapa. Economic Growth Center, Yale University, New Haven, Connecticut.

Avila, A.F.D. (2001) *Impact Evaluation of the Agricultural Research in Brazil: A Summary Review.* Secretaria de Administração Estratégica, Embrapa, Brasília.

Brandolini, A., Caligari, P.D.S. and Mendoza, H.A. (1992) Combining resistance to potato leafroll virus (PLRV) with immunity to potato viruses X and Y (PVX and PVY). *Euphytica* 61, 37–42.

Brown, L. R. (1994) *State of the World*. Norton, New York.

Burdge, R. J. (1994) *A Conceptual Approach to Social Impact Assessment.* Social Ecology Press, Middleton, Wisconsin.

Carneiro, M. (2000) Introducción y Panorama General. In: Carneiro, M. (Coord.) *Estratégias de biotecnologia agropecuária para el Cono Sur.* PROCISUR, Montevideo, pp. 1–8.

Daniels, D. (1987) *Evaluation in National Agricultural Research: Proceedings of a Workshop*, Singapore, 7–9 July 1986. IDRC, Ottawa, Ontario.

De Bokx, J.A. (1990) Potato virus Y. In: *Compendium of Potato Disease.* APS Press, St Paul, Minnesota, pp. 70–71.

De Bokx, J.A. and Piron, P.G.M. (1990) Relative efficiency of a number of aphid species in the transmission of potato virus Y in the Netherlands. *Netherlands Journal of Plant Pathology* 96, 237–246.

Echeverria, R.G. (ed.) (1990) *Methods for Diagnosing Research System Constraints and Assessing the Impact of Agricultural Research: Assessing the Impact of Agricultural Research*, Vol. II. ISNAR, The Hague.

Evenson, R.E. (1987) Productivity decomposition methods for evaluation of agricultural research systems impacts. In: Evenson, R.E., Cruz, E.R. da, Avila, A.F.D. and Palma, V. (eds) *Economic Evaluation of Agricultural Research: Methodologies and Brazilian Applications.* Embrapa, Economic Growth Center, Yale University, New Haven, Connecticut.

Evenson, R.E. and Cruz, E.R. (1989) T*he Impacts of Technology PROCISUR Program: an International Study.* IICA/BID/PROCISUR, New Haven, Connecticut.

Evenson, R.E., Cruz, E.R. da, Avila, A.F.D. and Palma, V. (eds) (1987) *Economic Evaluation of Agricultural Research: Methodologies and Brazilian Applications.* Embrapa, Economic Growth Center, Yale University, New Haven, Connecticut.

FAO (1997) *The State of Food and Agriculture, 1997.* FAO Agriculture series no. 30, FAO, Rome.

Genetic ID. (1999) *GMF Market Intelligence Newsletter.* http//www.genetic-id.com. Number 32. 1 May 1999.

Gugerli, P. (1986) Potato viruses. In: Bergmeyer, H.U. (ed.) *Methods of Enzymatic Analysis – Antigens and Antibodies, 2*, Vol. 11. VCH Verlagsgesellschaft, Weinheim, pp. 430–446.

James, C. (1999) Preview: global review of commercialized transgenic crops. *ISAAA Briefs* 12, 1–8.

James, C. and Krattinger (1996) Global review of commercialized transgenic crops. *ISAAA Briefs* 12, 1–8.

Hieter, P. and Boguski, M. (1997) Functional genomics: it's all how you read it (Viewpoint). *Science* 278, 601–602.

Hull, R. and Davies, J.W. (1992) Approaches to nonconventional control of plant virus diseases. *Critical Reviews in Plant Sciences* 11, 17–33.

Philips, R.H. (2000) GM crops: an alternative view to Greenpeace. *Feed Compounder* 6, 1–4.

Qaim, M. (1998) Transgenic virus resistant potatoes in Mexico: potential socioeconomic implications of North-South biotechnology transfer. *ISAAA Briefs* No. 7.

RAFI (1998) Seed Industry Consolidation: Who Owns Whom? *Communique*, http:/www.rafi.org.

Rech, E. (2000) La biotecnologia en Brasil. In: Carneiro, M. (Coord.) *Estratégias de biotecnologia agropecuária para el Cono Sur.* PROCISUR, Montevideo, pp. 51–68.

Riechmann, J.L., Zhang, J. and Braun, P. (1999) Plant genomic: the next Green Revolution? *Chemistry and Industry* 12, 468–476.

Silva, G.L.S.P. da (1984) Produtividade agricola, pesquisa e extensão rural. Ensaios econômicos, 40, FIPE/USP, São Paulo.

Souza Junior, M.T. (2001) Desenvolvimento de mamoeiros transgênicos resistentes a fungos. *EMBRAPA Recursos Genéticos e Biotecnologia*, SIGER. (Subprojeto 17.0.99.190.02).

Tanksley, S.D. and McCouch, S.R. (1997) Seed banks and molecular maps: unlocking genetic potential from the wild. *Science* 277, 1063–1066.

Torres, A.C., Ferreira, A.T., Melo, P.E., Romano, E., Campos, M. de A., Peters, J.A., Buso, J.A. and Monte, D. de C. (1999) Plantas transgênicas de batata Achat resistentes ao vírus do mosaico (PVY). *Biotecnologia Ciência e Desenvolvimento* 2, 74–77.

University of Illinois (1999) *The Economics and Politics of Genetically Modified Organisms in Agriculture: Implications for WTO 2000.* University of Illinois Board of Trustees, Bulletin 809.

Ward, C.W. and Shukla, D.D. (1991) Taxonomy of potyviruses: current problems and some solutions. *Intervirology* 32, 269–297.

Chapter 17

Intellectual Property Protection and the International Marketing of Agricultural Biotechnology: Firm and Host Country Impacts

Peter Goldsmith,[1] Gabriel Ramos[1] and Carlos Steiger[2]
[1] *Food and Agribusiness Management Group, Department of Agricultural and Consumer Economics, University of Illinois, 433 Mumford Hall, 1301 West Gregory Drive, Urbana, IL 61801–3605, USA;* [2] *Agribusiness Program, Universidad de Belgrano, Buenos Aires, Argentina*

Introduction

The protection of intellectual property rights (IPRs) has been a contentious issue over the last 20 years. Industrialized nations have moved to knowledge-based economies and simultaneously trade barriers have fallen, making intellectual property vulnerable. Adding to this vulnerability are conflicting international institutional environments, belief systems and economic realities. The debate over IPR protection has become a significant global trade issue pitting the net technology-producing North against the net technology-consuming South. The North has a distinct belief system towards intellectual property (Steidlmeier, 1993; Mittlestaedt and Mittlestaedt, 1997), maintains a comprehensive IPR institutional environment and actively employs enforcement mechanisms. The South, on the other hand, is more conflicted. While in the last 10 years many Southern countries have agreed to multilateral agreements on IPR protection, enforcement and real commitment has been lagging (Thurow, 1997; Levy, 2000).

With this in mind there has been much debate about the impact of alternative IPR regimes (tight or loose) on the welfare of Southern economies. Policy makers in both the South and the North search for arguments to convince recalcitrant Southern countries to follow the Northern model of strict IPR regimes. The South, faced with a dilemma, searches for arguments to justify loose IPR regimes or alternatively to convince its populace that tighter IPR regimes are better for the nation.

While there has been much analytical work, mostly theoretical, conducted on the subject, the final results are inconclusive as to whether a strong IPR regime is better or worse for Southern countries (Vishwasrao, 1994; Sherwood and Braga, 1996; Globerman, 1998). The lack of clarity as to the impact of the IPR regime has been due to both the complexity of problem and the form of analysis. The theoretical models, while being extremely valuable highlighting the drivers of firm and social welfare, are by their nature abstractions.

The empirical models to date suffer from three effects that weaken the impact of their conclusions. The first is that often firms are not able to observe their losses from weak IPR (Fienberg and Rousslang, 1990). Many times the losses are due to investments not made and need to be estimated. (Host country impact analysis too suffers from this problem.) Secondly, firm impacts are generally estimates from surveys of a cross section of firms, representing opinions of impact not factual evidence (Evenson, 1990; Fienberg and Rousslang, 1990; Sherwood, 1990; Braga and Willmore, 1991). Finally, no work to our knowledge has attempted to measure firm and host country impacts directly from weak IPRs. Therefore while there has been some attempt to empiricize welfare impacts, evidence supporting or negating the theory is lagging. The end result for policy makers is there still remains much ambiguity and economic studies have yet to show where the balance should be struck (Dawson, 1987; Alster, 1988).

The objective of this research is to add some empirical clarity of the welfare impacts of weak IPR. To this end we employ a novel methodological design and a unique context. While previous studies have used cross-sectional survey or secondary data, our research employs the critical case study approach (see Yin, 1994). The research design is deductive, in that we use the empirical setting of Pioneer-Argentina, S.A., a seller of bioengineered agricultural seeds, to test the existing theory of weak IPR impacts in a North–South context.

Theoretical background

The dilemma, both domestically and internationally, for IPR protection is the trade-off between short-term costs and long-term benefits. The argument made by Northern countries is that while prices may rise in the short term, new technologies will be available over the long term and will, in turn, raise economic productivity (Stamm, 1993). As the result of protected property rights, the South will gain from new investment (Sherwood and Braga, 1996), the flow of technology (Sherwood and Braga, 1996) and technology spillovers (Zigic, 2000). The preferred mechanism of IPR protection by the net technology-producing countries (North) is through public institutions, not product/process masking by private firms. For the net technology-using countries (South), the significant short-term costs may arise directly from an increase in the cost of the input due to the lack of complete substitutes and indirectly from the administrative and enforcement costs of a Northern-style IPR protection system. Adding to the complexity is the fact that welfare impacts are best understood in a dynamic context, as the short-term losses of strengthening the South's IPR regime are believed to be trumped by the long-term gain from economic growth.

To address the complex question about the welfare impacts of an IPR regime, numerous theoretical models have been developed (Dollar, 1986; Chin and Grossman, 1988; Diwan and Rodrik, 1991; Deardorff, 1992; Helpman, 1993; Taylor, 1993; Maskus and Konan, 1994; Grossman and Helpman, 1995; Zigic, 1998). While it is generally agreed that technology is important for an economy to grow (Dollar, 1986), the theoretical models are not completely successful making the argument that IPR protection in the South improves Southern welfare (Chin and Grossman, 1990; Helpman, 1993; Gould and Gruben, 1996; Zigic, 2000). Additionally, the argument that strong IPRs lead to greater innovation is also questioned (Braga and Willmore, 1991; Gould and Gruben, 1996). The lack of clear benefits from IPR production can be due to a fundamental difference in belief systems about private vs. communal property (Steidlmeier, 1993; Mittlestaedt and Mittlestaedt, 1997; Thurow, 1997; May, 1998), the negative affects of a monopolist on future innovation (Chin and Grossman, 1990; Gould and Gruben, 1996; Zigic, 2000), a 'tit for tat' view in which the South is 'owed' the technology based on a history of Northern resource extraction policies, or the South's assessment of the extreme hardship an enforced patent system would create. Whether it is an intrinsic scepticism about Northern property right regimes or a scepticism based on the realities of the moment, the end result is recalcitrance on the part of Southern countries actively to engage IPR protection.

Chin and Grossman (1990) developed a duopoly competition model of one Northern and one Southern firm. The Northern firm's problem is setting the optimum level of R&D investment. The objective of the model is to analyse how each firm behaves as the IPR regime changes and how the firms' behaviour affects the level of innovation available to society. One important conclusion of the model is that there are certain conditions where the South is better off with weak IPRs, i.e. if the Southern market is small, which it generally is, and the technology jump is moderate. Welfare in the South is improved from the improved product distribution and lower price.

Diwan and Rodrik (1991) follow Chin and Grossman's lead in modelling the welfare balance between the North and the South but diverge in several important ways. They do not assume that holders of protected intellectual property have monopoly power. There is free entry and exit. The result of Chin and Grossman's market power assumption is that the loss to Southern consumers is estimated to be very high due to the high price and reduced product distribution a monopoly connotes, but the direct mapping between a patent and a monopoly price is not axiomatic. There are numerous forces such as a dynamic environment of innovation, product substitutes and market power distribution across the supply chain that drive prices away from the monopoly price towards the competitive price (Goldsmith, 2001). While it is correct to assume that developers of intellectual property gain market power, and the patent system formalizes this position in the marketplace, it is incorrect to assume that holders of patents operate as monopolists. While the monopoly/duopoly model may be appropriate in some cases, it is extreme in numerous situations where markets fall in between monopoly and pure competition. This market structure assumption of course has great impact on welfare impact calculations. In our case of agricultural biotechnology, barriers to entry are relatively low and the availability of substitutes relatively high, thus pure monopoly pricing practices should not be assumed. The setting is more competitive, driving down prices and increasing access. Therefore under such situations the direct welfare gains (losses) to the North (South) from strong IPRs may be much less than anticipated.

Diwan and Rodrik (1991) also do not assume that preferences between the North and South are homogeneous. This is important because it negates the free-riding benefits for the South and at the same time causing a welfare loss from needing to develop its own preferred and necessary intellectual property based products. This is not the case for our situation of marketing biotechnology seed in Argentina where North and South preferences map each other closely.

Deardorff's model (1992) is a monopoly model studying the regionalization of patents as a way of better allocating rents from innovation. He offers the 'Solomonic' strategy of limiting the regional applicability of patents, allowing some monopoly rent extraction while also allowing for greater distribution. The author unrealistically assumes regional patents can be operational, as consumers and producers of innovation are not distinct groups, global information transmission is pervasive and terms of trade will adjust.

Helpman (1993) develops a dynamic general equilibrium model and in doing so attempts to grapple with the real effects of interlinked economies, terms of trade and dynamic effects. His model captures the interplay between the North's rate of innovation and the importance of imitation for providing Northern consumers with a better mix of lower price products and the South with improved terms of trade. The degree of imitation though is critical to the net welfare balance. When the rate of imitation is low, both North and South benefit, as described above. When the rate of imitation is high, as is the case for many pharmaceutical and agricultural innovations such as biotechnology seed, the results do not hold and tighter IPR policy is preferred by the North.

Taylor (1993) developed a partial equilibrium static North–South duopoly model to explore the interaction between masking (North) and imitation (South) and their associated costs. The uniqueness of the model is in the endogeneity of the appropriability regime, where the firm has control, through masking and other private means to affect the rate of imitation. In this way the firm has some control over its ability to appropriate the rents from its intellectual property. Taylor concludes that the current state is pareto-inferior due to the extensive resources devoted to masking and unmasking significant technologies.

Vishwasrao's model (1994) concerns the optimal Northern firm strategy that maximizes profits given that a foreign licensee might pirate the technology and not pay the required royalties. Therefore, while the Northern firm acting as monopolist may prefer to license, it cannot because contracts are not enforceable and licensees pirate the technology. In order to protect the monopoly, the Northern firm internalizes the transactions through a wholly owned subsidiary shifting the net benefits back North.

Zigic (2000) built a duopoly model whereby the South attained its market position through R&D spillovers, the leakage of important technical information, and the North achieved its market position through R&D. The model involves four stages whereby the South chooses its optimal IPR regime in light of the fact that the North, assumed to be a significant trade partner, would respond with high import tariffs, if the South chooses a weak IPR regime. Like Vishwasrao (1994), the model's focus is on how the North addresses the rent appropriability problem given a world of weak IPR.

Theoretical propositions

The theory of welfare and IPR protection hinges on its theoretical propositions about how weak IPRs in the South effects the welfare of firms, consumers and Southern countries as a whole. The following is an overview of nine propositions to be analysed empirically using the Pioneer case study. They are organized into four topics: southern demand, business behaviour, financial impacts and technology flow.

Southern demand

Proposition 1 Welfare in the South improves as preferences between the North and South are more homogeneous. Diwan and Rodrik (1991) and Deardorff (1992) argue that homogeneity allows for free riding and more readily transfers benefits to the South under a weak IPR regime. The South loses from a lack of access to the unique products it prefers. For many products, including our case of marketing agricultural biotechnology in Argentina, demand is somewhat homogeneous. The unique geographic and agronomic features of the Pioneer case provide excellent insight into how willing the Northern firm is to invest in adapting the product to the local market (Argentina) and how willing the Southern consumer is to accept the risk of using a product that is sub-optimally adapted to the local environment. Pioneer can choose to leverage the homogeneity and avoid all adaptation to the Argentinian market or could choose to enhance its products for local use.

Proposition 2 The smaller the demand in the South relative to the North the more advantageous for the South to maintain weak IPRs (Chin and Grossman,

1990; Diwan and Rodrik, 1991; Taylor, 1993). This is due to the South's ability to free ride on the technology combined with the lack of incentive by the North to market to the South even in a world of strong IPR. Pioneer's investment behaviour, which differs across products, not markets will be analysed to assess how they respond to the smaller Argentinian market.

Business behaviour

Proposition 3 Firms will engage in either masking or enforcement to try and protect their intellectual property. Vishwasrao (1994) and Chin and Grossman (1990) raise the issue of masking and its welfare effects. The South is believed to be harmed by masking. While the firm is believed to have a net benefit from masking, there are costs (Globerman, 1988; Braga and Willmore, 1991; Taylor, 1993). The Pioneer case will be used to study the masking, administrative and enforcement (MAE) costs and strategies of the firm, its associated industry group, and the government.

Proposition 4 Weak IPRs reduce investment in the South. Stamm (1993) refers directly to investment and its diversion away from the South to the North. Benko (1987), Globerman (1988), Sherwood (1990), Mansfield (1994) and Vishwasrao (1994) refer to lowered R&D spending in the South, and Benko (1987), Chin and Grossman (1990), Diwan and Rodrik (1991), Helpman (1993) and Mansfield (1994) hypothesize about how innovation is stifled in the South. Zigic (1998) raises the issue of welfare spillovers. The effects on investment are critical components of the indirect impacts from weak IPRs and are necessary for conducting benefit–cost tests. For example, a loose IPR regime might cause a firm to avoid a country. The country not only loses by not having access to the latest technology, the firm would also be withholding investment in the country in support of the innovation. These are the 'pebble in the pond' effects from introducing a new product into a market. They generally do not arrive in a limited fashion but can have broad indirect impacts on the economy. To measure these ripple effects empirically, our case study compares the relative levels of R&D, investment [human (Sherwood, 1990) and physical capital] between Pioneer units whose intellectual property is affected against those whose intellectual property is unaffected by Argentina's weak IPR regime.

Proposition 5 Weak IPRs cause a negative performance bias on the Southern operations of the technology firm. Sherwood (1990) believes that production processes will be outdated; Stamm (1993) states that service levels will be lowered and the distribution system will be less reliable. Agricultural seed production involves three distinct stages beyond R&D; multiplication, processing and distribution. Each one of these will be assessed in terms of the impact of weak IPRs. If theory is correct there should be measurable operational differences between the units due to intellectual property risk.

Financial impacts

Proposition 6 Weak IPRs reduce return on investment (Stamm, 1993). The financial data on Pioneer-Argentina will be analysed estimating the differences in return on investment, as well as other financial metrics, between the two units. Globerman (1988) and Chin and Grossman (1990) expect lower profits. An important factor is the level of technological appropriability (Diwan and Rodrik, 1981; Teece, 1987). This is the degree to which the owner of the intellectual property is able to generate a return and is a function of technological feasibility as well as permissiveness of the IPR environment (Teece, 1987).

If severe financial harm is assumed to occur by these theorists, then there should be evidence in terms of the intensity of the firm's lobbying effort and degree of internalization (Globerman, 1988; Vishwasrao, 1994). The analysis of Pioneer's lobbying behaviour and financial performance will shed light on how bad (good) the second-best outcome is for the firm under weak IPRs. This analysis will provide some insight into Chin and Grossman's (1990) claim that strong IPRs always benefits the firm by answering the question of how 'bad' is a second-best strategy under weak IPRs.

Technology flow

Proposition 7 Weak IPRs allow access to new technologies that improve the welfare of the South (MacLaughlin *et al.,* 1988; Chin and Grossman, 1990; Vishwasrao, 1994; Gould and Gruben, 1996). The case will be used to demonstrate how Pioneer prices the product, how the distribution chain handles the product and how farmers use the product. Results will provide insight into how farmers' welfare is impacted either directly from a change in prices or indirectly from the level of technology.

Proposition 8 Weak IPRs lower the speed on technology's entry into the South (Chin and Grossman, 1990; Stamm, 1993). This has important welfare implications as quality available to the South is reduced (Globerman, 1988). The maize and soybean business units will be compared to look at product offerings. The case is valuable because direct comparisons can be made with Pioneer's product offerings in the North between the two businesses and the lag time before these products reach Argentina.

Proposition 9 Weak IPRs promote diffusion of new technology (Chin and Grossman, 1990; Stamm, 1993). Under weak IPRs prices should fall (Globerman, 1988; Diwan and Rodrik, 1991) allowing for greater distribution of the product. This proposition is juxtaposed against Proposition 8 that supports the idea of less technology rather than more. Thus the lag of the technology roll-out (proposition 8) is a force maintaining the technological disparities between the North and the South while lower prices from weak IPRs promote the rapid diffusion of (dated?) technology; a force for equalization of technology between the North and the South. The case will provide empirical insight into the welfare balance of these two effects.

Empirical studies

As noted above, the theoretical models help to provide a framework to analyse the welfare impacts but are lacking in their conclusiveness. This makes it difficult from theory alone to convince either side of the true welfare impacts. In an attempt to shed more light on to the question of welfare impacts, a few empirical studies have been conducted to try and measure impact. Empirical research in this area too is problematic. Writing in a 1993 survey of the empirical work to date, Helpman concludes that there exists very little evidence on the welfare effects of international infringements of IPRs. Similarly, writing in 1994, Maskus and Konan remark that there is a surprising paucity of empirical evidence concerning the most critical issues at hand. Subramanian (1995) echoes the sentiments of the lack of quantitative estimates of the Southern welfare impacts. Therefore, on the topic of understanding the phenomenon of North–South welfare impacts, the empirical attempts have not been much more successful than the theoretical models.

To date there have been five[1] key studies:

1. Braga and Willmore's use of 1981 survey data of 3000 Brazilian industrial firms;
2. The United States International Trade Commission (USITC) survey conducted in 1986 of 736 US firms;
3. Gadbaw and Richards's 1998 statistical overview of four least-developed-country counties using aggregated secondary data;
4. Mansfield's 1994 survey of 100 US firms; and
5. Pray and Ma's work in China (2001) that departs from the firm-level approach and surveys consumers (farmers) of technology in order to directly assess the benefits to them of exploiting weak property rights.

Braga and Willmore's study comprised a qualitative survey of 3000 Brazilian industrial firms in 1981 linking the lack of a strong property rights to low levels of local technological innovation (Braga and Willmore, 1991). The qualitative survey queried industrial firms about the relationship between the intellectual property environment and their willingness to invest or purchase technology from abroad (Gould and Gruben, 1996). It provided the first empirical evidence of the impact of weak IPRs on Southern domestic firms' willingness to invest.

USITC surveyed 736 US firms in 1986 asking those firms to assess the impact of weak foreign property rights on their profits (USITC, 1988). The methodology used was a mailed structured survey asking 'Fortune 500' firms about the impact on their business of weak foreign IPRs. Losses were estimated at 2.7% of sales, with losses in one industry as high as 21% of sales. Infringement and enforcement costs were estimated at 0.03% of sales.

Feinberg and Rousslang (1990) using the USITC data set expanded its scope of analysis. They attempted to estimate the static welfare consequences of weak foreign property rights on innovating firms, local infringers and consumers. They found that while losses are significant to the legitimate firms, they might be less than the sum total of benefits to consumers and infringers. Consumers benefit from greater price competition and infringers (producing close substitutes) benefit from the inelastic demand for the product and the low marginal costs of infringing. As the authors point out, their study does not account for, among other things, the negative investment impacts arising from weak IPRs in the South.

Gadbaw and Richards (1988) used USITC data as well to estimate the 'right owner revenues' in the absence of piracy, thus the empirical focus is on the loss of sales not investment. The authors admit it is a daunting task as they are using estimates of demand and price elasticities garnered from interviews. Thus like the Braga and Willmore study, the estimate of investment or sales foregone due to piracy is difficult to address because the analysis reflects a partial equilibrium analysis and is subjective and static.

Mansfield (1994), using a similar methodology to Braga and Willmore and the USITC, surveyed 100 major US firms in 1991. The response rate was 94% and

[1] Two other empirical studies were not mentioned. Subramanian (1995) studied prices of specific pharmaceuticals and used previous estimates of price elasticities to estimate welfare effects for the Northern firm and the Southern country. Maskus and Konan (1994) used the estimates of Gadbaw and Richards (1988) to study the welfare effects of licensing in a Southern setting.

respondents were generally patent attorneys, specialists in the firms' international operations or top executives (Mansfield, 1994). He found that the IPR environment had an important effect on some foreign direct investment. More of the effect was felt in R&D and less felt in sales and distribution. Similarly Mansfield found that much of the hesitancy involved the transfer of technology, not necessarily investment in general. This suggests that more of the impact of weak IPRs concerns intellectual property questions, not necessarily investment in general. Our case study will directly address this question to see if investment effects are limited to intellectual property or if there are spillovers to other investments as well.

Most recently, Pray and Ma (2001) studied the adoption of *Bt* cotton in China using a survey of 283 farmers in northern China. Their work documents quite clearly the incentives for local producers to adopt technology when property rights are weak. Firm and host country impacts net of producer impacts were not part of the study.

As Helpman has pointed out, empirical work assessing welfare impacts has been lacking. While Sherwood and Braga (1996) note that the question is essentially empirical, they as well as Maskus and Konan (1994) also admit the difficulty in making welfare estimates. There is not only an inherent complexity to the problem, but the dynamic effects are difficult to capture and modelling in this area requires assumptions that are fundamental to the outcome. Feinberg and Rousslang (1990) find the empirical task difficult because much of the primary data that has been used is self-reported and involves so many estimates.

We offer an alternative empirical approach that recognizes the empirical challenges while at the same time integrating the rich theoretical literature. It is hoped by looking at this question from an alternative methodological perspective a greater understanding of this phenomenon will occur.

Methodology

Numerous propositions about the impact of IPR protection on firm and host country impacts emerge from the theoretical models. A review of the literature of empirical studies attempting to answer some of the fundamental welfare questions about IPR protection provides only weak evidence. Part of the cause for weak evidence is the complexity of the IPR welfare situation. As noted above, there are numerous factors affecting host country and firm welfare and these factors can be contradictory. A second component causing the lack of factual clarity is that empirical measurement is difficult. While the empirical methodologies used to date have been effective in contributing to the debate over IPR protection, they are unable to get at the central issue of measuring and documenting welfare impact. This is because they are at least one degree removed from actors, investments and transactions that comprise the welfare assessment.

Additionally, the empirical effort is complicated because much of the empirical assessment of IPR protection attempts to measure the investment or transaction that was never made. These are the negative effects host country and firms derive from investments purportedly not made. Measuring this is difficult. We think our methodology solves that problem. When researchers query northern firms about the impact of IPR protection they undoubtedly say that it is harmful to their company because revenues are so low from the lack IPR protection making investments in the pirated product untenable (i.e. Illinois Farm Bureau, 1998), but is this true? Can we measure it? Welfare analysis is about weighing benefits and costs, but without measurement how can we perform the analysis? Executives describe transactions not conducted and investments not made, how might they be measured? Once measured then a proper welfare analysis can take place, and the theoretical propositions above can be assessed.

We think we have come up with such a methodology and an empirical setting that will address the problem of measuring the investment not made. The subject of our study is the firm Pioneer-Argentina. Pioneer is a subsidiary of the multinational division Pioneer Hi-Bred International that is part of Dupont de Nemours. Pioneer-Argentina is in the business of producing and selling agricultural seed to farmers. Total sales for 2000 were US$35 million and the firm employs 105 people. The firm sells a variety of cultivars (Table 17.1), but their dominant business is maize and soybean seeds. Herein lies the uniqueness of the empirical setting and its value for addressing the question of the welfare implications of IPR protection.

Maize and soybeans are complements. Agronomic convention holds that neither maize nor soybeans be grown in the same field in a continuous fashion. A producer may be able to get away with 2 years of continuous rotation but beyond that soil fertility suffers and weed and disease impacts increase. Therefore maize and soybeans are grown in rotation with 50% of a farm's acreage in maize and 50% in soybeans. In any given year, though, relative prices and input costs may provide an incentive to move away from a 50–50 split, but, as mentioned above, deviating far from a rotation over a long period of time is not possible. The impact for retail seed suppliers is that offering both maize and soybean seed is a successful strategy as there are very few pure maize farmers or pure soybean farmers. A farmer needs both products. A firm can effectively offer both types of seeds because brands are important and it affords one-stop shopping.

A second feature that makes the case unique and valuable for studying the IPR issue is that Argentinian crop production is very similar to that in the USA. The centre of the maize and soybean area, e.g. the province of Buenos Aires, is the 32nd parallel (south) comparable with the mid-south region of the USA. Thus US varieties and agronomic practices transfer readily to Argentinian producers. This makes the preferences between the North (US) and the South (Argentina) relatively homogeneous. In the year 2000, Argentina cultivated 3,326,000 ha of maize and 10,300,000 ha of soybeans (Secretaria de Agricultura, Ganaderia y Pesca (SAGyP), 2000). Argentina is the world's second largest maize exporter (USDA, 1997) and the world's third leading soybean producer (Elliott, 2000).

To produce seed (for the North or South), a seed company may take one of three general strategies. The first (I), involves extensive R&D to develop seed characteristics that can be introduced into adapted and successful existing germplasm. An example would be the development of transgenic

Table 17.1 Pioneer product mix.

Crop	Category	Number of products
Maize	Elite hybrids	10
	Tropical hybrids	4
	Imidazolinone-resistant hybrids	2
	Insect-resistant (*Bt*) hybrids	5
	Stacked hybrids[a]	1
Total		22
Other crops		
Soybeans	Roundup Ready varieties®	7
Sunflower	Hybrids	4
Sorghum	Hybrids	4
Lucerne	Varieties	5
Total other crops		20

Source: Pioneer-Argentina 2000 Catalogue.
[a] Hybrids that combine insect and herbicide (imidazolinone) resistance.

events, i.e. Roundup Ready® tolerance, and then marketing that technology through the best varieties/germplasm. Therefore Monsanto, as an R&D company, would purchase a company like Asgrow, a producer and marketer of seed, in order to get its technology out into the market (Goldsmith, 2001). This would be the high risk–high return strategy that has dominated the life sciences industry over the last decade (Goldsmith, 2001).

A second strategy (II) and more common in markets of developing countries is for firms to take already developed seed varieties and adapt them to the local environment. This still involves significant investment in seed research trials with extensive breeding programmes and field plots in the local (Southern) environment. As long as the foreign region is relatively homogeneous to that of the central R&D area of the USA, strategy I is unnecessary and strategy II suffices. An example of this is northern Brazil where strategy II is not applicable because of the tropical climate and high aluminium soils (Mcvey *et al.*, 2000). Local investment in R&D by Embrapa, Brazil's agricultural research system, is necessary because direct technology transfers are agronomically inappropriate.

The third strategy (III), involves essentially no investment. A seed company simply exports the seed directly from the North to the South with no adaptation. The more comparable the Northern environment is to the Southern environment the less likely a crop failure would arise due to poor adaptation. It is important to note that a seed firm can never know with certainty that a variety when taken out of its home region will perform exactly the same in a new region. *Ex-ante* the firm may be confident, but only after the seed has been purchased, planted and grown does the adaptiveness reveal itself. Thus older varieties are less risky, but represent older technology.

An anecdotal example of this situation was Pioneer's 'Rio Cuarto' incident in 1994. According to management, and confirmed by competitors, the firm had not been thorough enough in adapting US maize varieties to Argentina. A deadly fungus, named Rio Cuarto (the province where the greatest outbreak occurred) wiped out maize crops planted with Pioneer products, doing tremendous damage to its brand image. Even in the year 2000, the firm was still attempting to rebuild confidence in its products. This highlights the inherent risks in transferring varieties directly from one region to another.

Strategy I (high R&D) is not pertinent to a company like Pioneer-Argentina because the agronomic differences are small and market opportunities specific to Argentina are relatively minor for new technology development. New technologies available to Northern producers in North America and Europe can be successfully introduced in Argentina. Roundup Ready® technology for example is easily introduced into the many local varieties found in the USA, Europe and South America.

Pioneer does have a choice between strategy II (moderate investment) and strategy III (no investment) and this option is the crux of our empirical approach. Pioneer is world leader in maize and soybean seed production and sale. Farmers in Argentina need both products and Pioneer wants to offer both products. The uniqueness of the situation whereby Argentina agronomically is directly comparable to the USA affords a firm like Pioneer the opportunity to choose a type II or type III strategy for either the maize or soybean products. They can both be type II with significant R&D and Argentinean farmers would receive the most advanced technology adapted to their country's environment, or the firm can under-invest and choose a type III strategy and completely free-ride off investments made in the North. Maize and soybeans do not have to be treated the same, i.e. maize can be type II while soybeans kept at type III.

As revealed in our interviews with the company, their objective (not surprisingly) is profitability, which can be translated as return on investment (ROI). They are not wed to one strategy or another or matching a product, i.e. maize, with a certain strategy. They clearly expressed that their objective was profitability and the strategy (either II or III) would be used for the division that best achieved those objectives. Therefore as an endogenous choice problem, the maize or soybean divisions could either involve moderate or no investment. *Ceteris paribus*, according to Pioneer management, high investment is preferred to low investment. Thus if business conditions were ideal, the welfare of the firm is greatest under strategy II, high investment.

The empirical question of this paper is not why Pioneer chooses one strategy or another, but to compare the welfare impacts of a type II strategy vs. a type III strategy. In order to effectively perform the welfare analysis, a comparative case is valuable. In the Pioneer situation, what are the benefits and costs for both the firm and the host country of the high-investment decision and what are the benefits and costs of the low-investment decision?

A final unique and valuable feature of this case is the cross-country differences in institutional environments are controlled. This is because, even though Pioneer's strategic choice is driven by intellectual property concerns and Argentina has a weak IPR regime, maize is agronomically protected from intellectual property piracy and soybeans are not. Therefore our methodology controls for the property rights environment, the market structure and demand (farmers). The reason that the maize division can be operated differently from the soybean division is that maize is a hybrid and soybeans are not. A maize plant is pollinated only by means of another maize plant. If maize seed is saved from one year to next, the maize plant loses its hybrid (cross-pollination) vigour and performs very poorly. Therefore a farmer must return each year to the seller of seed to get a new version of the hybrid that has been properly crossed.

Soybeans on the other hand are self-pollinating and can keep reproducing in perpetuity. A farmer can take seed from the crop just harvested and replant them the following year. In this way a farmer who plants soybeans does not have to return to the seed supplier every year for new seed, dramatically lowering the cost of the seed input. In the USA, 25% (Hayenga, 1998) of the soybean seed is saved seed. Most farmers still have an incentive to purchase new seed every year because new varieties perform better. Saved seed will have a yield drag on average of 2.4% (Purdue University in Illinois *Agrinews*, 2001). Also, purchased seed tends to be more consistent and reliable. In the last few years an added incentive, in the USA and Canada, to purchase soybean seed on a yearly basis has been Monsanto's introduction and enforcement of a grower contract that stipulates that seed cannot be saved as it infringes on Monsanto's patent rights (Goldsmith, 2001). Under the weak property rights conditions of Argentina, this last incentive does not exist. Our interviews with farmers and industry representatives in Argentina feel that the yield drag from saved seed is closer to 1–2% per year for them and well worth absorbing, given that seed costs are so much lower. Because of this unique agronomic feature we are able to study Pioneer's behaviour where the only difference between the business of selling maize and the business of selling soybeans is that maize's intellectual property is naturally protected while soybean's is not.

With the objective of this research being to measure the effects of weak property rights, the Pioneer-Argentina case serves as a valuable empirical setting. Since the two goods are complements and operate side by side, they can be directly compared. Because both divisions operate in the same country any differences in institutional environment is corrected for; because the divisions operate within the same company differences in business behaviour across firms is corrected for. A study of Pioneer-Argentina is a unique opportunity to compare the two divisions and assess how the divisions are operated, where investments are made, what costs are incurred, what revenues are generated, what seed prices are charged and how seed is distributed. Based on theory of IPR protection, the impacts of the two strategies should be different. These differences then will provide evidence of welfare impacts derived from the set of theoretical propositions listed earlier.

To explore this unique empirical situation the case-study method was selected. The lack of empirical evidence generated by previous methodologies in this area led us to believe that a more microeconomic approach was necessary. The case-study method is valuable where depth of analysis is important. The ability to achieve depth is also the case study approach's weakness, in that only 'one' observation is being used. In all of the studies mentioned above numerous observations were used and statistically significant results were estimated. As numerous authors have noted, though, the application of those results to the phenomena has not been significantly illuminating. The case-study method used in this research greatly narrows the focus with the intent of improving the quality of the empirical evidence. The case-study approach's narrow focus and lack of statistical tests are seen as weaknesses as well. Neither broad-based statistical studies nor narrowly focused case studies are the perfect empirical methodology (Yin, 1994; Westgren and Zering, 1998; Gummesson, 2000). Both have their place and we suggest that the case-study method when applied to the situation of Pioneer-Argentina adds important insights into the North–South debate over IPR protection.

Approach

As in quantitative research, there are numerous case-study methodologies. For the purposes of this inquiry, a deductive approach is employed. That is, the case study is used to help provide empirical evidence about a phenomenon that to date has been understood from theoretical, anecdotal and limited empirical perspectives. Yin calls this the critical case-study model and is built upon existing theory and guided by specific propositions. Its goal is to test theory instead of creating theory. Specific questions still remain as to how IPR protection actually affects farmers, firms, supply chains and host countries. Theory, as noted above, abounds about how we think welfare is impacted and the theory yearns for some empirical evidence.

Evidence

The study used the following sources of evidence: key informant interviews, direct observation and quantitative data (financial documents analysis and industry statistics). To conduct the interviews, a semi-structured interview instrument was administered to over 30 key informants representing Pioneer and its various divisions, the Argentinian seed industry, supply chain members, and government (Fig. 17.1). Following Kumar (1989) guidelines for rapid appraisal, these interviews were qualitative and directed to carefully selected subjects.

The instrument comprised over 180 questions drawn and organized thematically from the theory. Depending on the informant's role or organization, some of the questions might not have been asked. In general, questions focused on business operations, investment and intellectual property. The theme was always comparing the maize seed business with the soybean seed business. While an attempt was made to introduce the questions in the same order, it was not uncommon for respondents to shift off topic. The interviewers did keep track of those questions that remained unanswered and worked them back into the interview so that each informant addressed as many of the same questions as possible. This technique allowed for answer triangulation so that any significant answer from one respondent was validated by other informants.

Following Kumar (1989), interviews were conducted with help of a previously designed interview guide taking special care to the way questions were worded in an attempt to maintain as neutral an attitude as possible.

Maintaining an easily retrievable case-study database is critical to assure the validity of a case study (Yin, 1994). In this way, it is possible to re-inspect the data by the author or from other researchers. With this in mind, all interviews were recorded

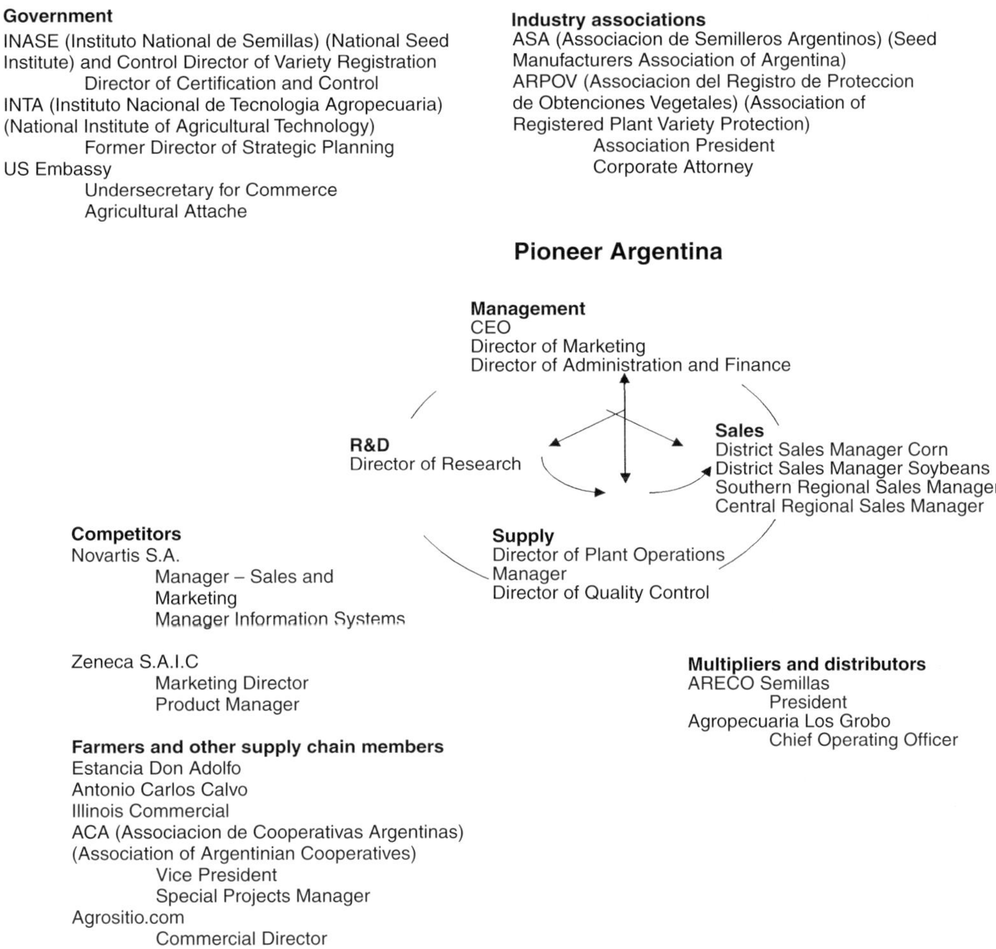

Fig. 17.1. Overview of key informant interviews.

in both audio and digital video formats. Almost all interviews were conducted in Spanish. Spanish-language transcripts were produced and were analysed using a qualitative data analysis software programme called QSR NUD*IST- N5® (QSR International, 2001).

Direct observation

Structured direct observation, according to Kumar (1989), can be extremely useful in the data triangulation process. Armed with theoretical expectations about investment and expenditure differences between the maize and soybean divisions, analysis was made of Pioneer infrastructure, technology, human resources, and advertising and marketing. Our use of the digital video equipment as well as photographs helped to document what investments were made and what equipment was being used for each business unit. To conduct the observations of physical assets visits were made to Pioneer R&D and production facilities and multiplier farms.

Quantitative data

Firm documents

Yin (1994) suggests that the best use for documents is to augment the evidence from other sources. Pioneer-Argentina provided us access to their financial records. Records are maintained separately between the two business units. Therefore an analysis of the balance sheet, income statements and pricing data were made available to the authors. Due to the sensitivity of the material, ratios comparing the maize and soybean units will be used whenever possible. The financial data serves three purposes; first it is useful to corroborate the responses of Pioneer managers as to the state of each of the businesses; second, the data can be used to analyse the propositions pertaining to the difference financial impacts weak IPR protection has on a firm; third, the data helps to quantify the welfare impacts.

Methodological validity

As numerous authors (Yin, 1994; Westgren and Zering, 1998; Gummesson, 2000) have noted, there is no hierarchy of research methodology. Of the many tools available to researchers each has an appropriate place. There is no perfect research methodology that serves all criteria for proper empirical analysis.

Researchers offer several tests of validity with respect to the case-study method. The validity test is: does the empirical evidence in fact correspond to the phenomenon under study (Gummesson, 2000)? Our study employs the single case approach. Therefore, is the case of Pioneer-Argentina and the research design valid for analysing the phenomenon of firm and host country welfare impact from weak IPR protection? In order to make the validity argument there are numerous design options, many of which this research incorporated and some of which it did not.

First, theoretical grounding adds formality and discipline to the research process. Because of our extensive use of the theoretical literature, supported by the more limited empirical evidence, we would argue that Pioneer is a valid context by which to study the phenomenon. Contributing to the methodological validity is our research design and implementation such that our research could be replicated within the same context or transferred to a new context.

The second important aspect of empirical validity with respect to the single case-study approach is its context. Does the study of Pioneer effectively incorporate the phenomenon in question (see Westgren and Zering, 1998)? The explicit discussion above, detailing the uniqueness of the Pioneer case to study the IPR question, constitutes our argument that this case provides an excellent context by which to analyse the phenomenon.

Third is the depth of the research, what Yin calls embeddedness. Cross-sectional data (i.e. USITC, 1988; Braga and Willmore, 1991; Mansfield, 1994) provides a form of breadth of analysis. The single case-study approach, on the other hand, allows depth of analysis. In our case, embeddedness was captured through in-depth interviews, multiple interviews within the firm, quantitative analysis of firm-level and industry-level data.

Fourth is the issue of triangulation, which asks multiple parties the same questions to see if their responses corroborate each other. While all answers need not be 'identical', they should be consistent. If they are not consistent, then a reason needs to be found. Triangulation was achieved by: conducting multiple interviews within the same firm; interviewing competitors, supply-chain members and third parties (i.e. government); analysing quantitative data (firm and industry, and national); and conducting and documenting direct observations. In this way a consistent and reliable picture of the phenomenon is created.

A fifth contributor to empirical validity is the use of multiple cases to analyse the same phenomenon. Additional cases demonstrate reliability and in that way contribute to the robustness of the conclusions. In this way our methodology is lacking. With multiple cases there is always the question of cost and time. More significant in our situation would be replicating the level of intimacy with the company under study. We were very lucky to have had such a high level of access. This depth of access compensates for the lack of additional cases.

Finally, case-study researchers recognize that longitudinal analysis adds power to the results by limiting the possibility of serendipity. Helpman (1993) notes the importance of dynamics when assessing the welfare impacts from weak IPRs. This too is a valid critique of our methodology where more time spent is studying a phenomenon is usually better. Because our analysis is static we are going to have to impute the dynamic implications of the firm's behaviour.

Results and Discussion

At the time of the publication of this draft the results were not completed thus they were not included.

References

Alster, N. (198) New profits from patents. *Fortune,* 25 April.

Benko, R. (1987) *Protecting Intellectual Property Rights: Issues and Controversies.* American Enterprise Institute for Public Policy Research, Washington, DC.

Braga, H. and Willmore, L. (1991) Technological imports and technological effort: an analysis of their determinants in Brazilian firms. *Journal of Industrial Economics* 39, 421–432.

Chin, J and Grossman, G. (1990) Intellectual property rights and North–South trade. In: Jones, R. and Krueger, A. (eds) *The Political Economy of International Trade.* Basil Blackwell, Cambridge, pp. 90–107.

Dawson, L. (1987) Transferring industrial technology to less developed countries. *Industrial Marketing Management* 16, 265–271.

Deardorff, A. (1992) Welfare effects of global patent protection. *Economica* 59, 35–51.

Diwan, I and Rodrik, D. (1991) Patents, appropriate technology and North–South trade. *Journal of International Economics* 30, 27–47.

Dollar, D. (1986) Technological innovation, capital mobility and the product cycle in North–South trade. *American Economic Review* 76, 177–190.

Elliott, R. (2000) Argentine transgenic corn load barred from Brazil. *Reuters,* 5 June.

Evenson, R. (1990) Survey of empirical studies. In: Siebeck, W. (ed.) *Strengthening Protection of Intellectual Property in Developing Countries.* World Bank Discussion Paper No. 112, The World Bank, Washington, DC.

Fienberg, R. and Rousslang, D. (1990) The economic effects of intellectual property infringements.' *Journal of Business* 63, 79–90.

Gadbaw, M. and Richards, T. (eds) (1988) *Intellectual Property Rights: Global Consensus, Global Conflict?* Westview Press, Boulder, Colorado.

Globerman, S. (1998) Addressing international product piracy. *Journal of International Business Studies* 19, 497–504.

Goldsmith, P. (2001) Innovation, supply chain control, and the welfare of farmers: the economics of genetically modified seeds. *American Behavioral Scientist* 44, 1302–1326.

Gould, D. and Gruben, W. (1996) The role of intellectual property rights in economic growth. *Journal of Development Economics* 48, 323–350.

Grossman, G. and Helpman, E. (1991) *Innovation and Growth in the Global Economy.* MIT Press, Cambridge, Massachusetts.

Grossman, G. and Helpman, E. (1995) Technology and trade. In: Grossman, G. and Rogoff, K. (eds) *Handbook of International Economics.* Elsevier Science, Amsterdam, pp. 1279–1335.

Gummesson, E. (2000) *Qualitative Methods in Management Research.* Sage Publications, Thousand Oaks.

Hayenga, M. (1998) Structural change in the biotech seed and chemical industrial complex. *AgBioForum* 1, 43–55.

Helpman, E. (1993) Innovation, imitation and intellectual property rights. *Econometrica* 61, 1247–1280.

Illinois Farm Bureau (1998) Responses to IFB questions, from *Carl Casale/Monsanto.* 22 December.

Kumar, K. (1989) *Methodologies for Assessing the Impact of Agricultural and Rural Development Projects: A Dialogue.* AID Program Design and Evaluation Methodology Report No. 11, Agency for International Development, Washington, DC.

Levy, C. (2000) Implementing TRIPs, a test of political will. Law and policy in international business. *Law and Policy in International Business* 31, 789–795.

MacLaughlin, J., Richards, T. and Kenny, L. (1988) The economic significance of piracy. In: Gadbaw, M. and Richards, T. (eds) *Intellectual Property Rights: Global Consensus, Global Conflict.* Westview Press, Boulder, Colorado, pp. 89–108.

Mansfield, E. (1994) *Intellectual Property Protection, Foreign Direct Investment, and Technology Transfer.* International Finance Corporation Discussion Paper 19, The World Bank, Washington, DC.

Maskus, K. and Konan, D. (1994) Trade-related intellectual property rights: issues and exploratory results. In: Deardorff, A. and Stern, R. (eds) *Analytical and Negotiating Issues in the Global Trading System.* The University of Michigan Press, Ann Arbor, pp. 401–446

May, C. (1998) Thinking, buying, selling: intellectual property rights in political economy. *New Political Economy* 3, 59–78.

Mcvey, M., Baumel, P. and Wisner, R. (2000) Extension of U.S. dams unlikely to reduce competition from Brazil. *Feedstuffs,* September 25, 2001.

Mittlestaedt, J. and Mittlestaedt, R. (1997) The protection of intellectual property: issues of origination and ownership. *Journal of Public Policy and Marketing* 16, 14–25.

Pioneer (2000) Catalogo 2000 (de Semillas). Pioneer-Argentina, Martinez, Buenos Aires.

Pray, C. and Ma, D. (2001) Impact of Bt cotton in China. *World Development* 29.

Purdue University (in Illinois Agrinews) (2001) Bin-run seed offers pitfalls to growers seeking to cut corners. *Agrinews.* Lasalle, Illinois, 16 February, p. D11.

QSR International (2001) *N5-NUD* IST* (Non-numerical Unstructured Data* Indexing Searching and Theorizing). Victoria, Australia.

SAGyPA (Secretaria de Agricultura Ganaderia y Pesca (Secretary of Agriculture, Livestock and Fisheries) (2000) Agricultural Statistics. http://www.sagypa.mecon.gov.ar/agricu/agricultura.htm

Sherwood, R. (1990) *Intellectual Property and Economic Development.* Westview Press, Boulder, Colorado.

Sherwood, R. and Primo Braga, C. (1996) Intellectual property, trade, and economic development: a road map for the FTAA negotiations. *North–South Agenda Papers,* 21.

Stamm, O. (1993) Intellectual property rights and competitive strategy: a multinational pharmaceutical firm. In: Wallerstein, M, Mogee, M. and Schoen, R. (eds) *Global Dimensions of Intellectual Property Rights in Science and Technology.* National Academy Press, Washinghton, DC.

Steidlmeier, P. (1993) The moral legitimacy of intellectual property clams: American business and developing country perspectives. *Journal of Business Ethics* 12, 157–164.

Subramanian, A. (1995) Putting some numbers on the TRIPs pharmaceutical debate. *International Journal of Technology Management* 10, 252–268.

Taylor, S. (1993) TRIPs, trade and technology transfer. *Canadian Journal of Economics* 26, 625–637.

Teece, D.J. (1987) Capturing value from technological innovation: integration, strategic partnering, and licensing decisions. In: Guile, B. and Brooks, H. (eds) *Technology and Global Industry: Companies and Nations in the World Economy.* National Academy Press, Washington, DC, pp. 65–95.

Thurow, L. (1997) Needed: a new system of intellectual property rights. *Harvard Business Review* 75, 94–103.

United States International Trade Commission (1988) *Foreign Protection of Intellectual Property Rights and the Effect on U.S. Industry and Trade.* USITC Publication 2065, Washington DC.

USDA/FAS (1997) Agricultural situation Argentina. http://www.fas.usda.gov/scripts/gain_display_report.exe?Rep_ID=25130148. Accessed May 2000.

Vishwasrao, S. (1994) Intellectual property rights and the mode of technology transfer. *Journal of Development Economics* 44, 381–402.

Westgren, R. and Zering, K. (1998) Case study research methods for firm and market research. *Agribusiness* 14, 415–424.

Yin, R.K. (1994) *Case Study Research: Designs and Methods.* Sage Publications, Newbury Park.

Zigic, K. (1998) Intellectual property rights violations and spillovers in North–South trade. *European Economic Review* 42, 1779–1799.

Zigic, K. (2000) Strategic trade policy, intellectual property rights protection and North–South trade. *Journal of Development Economics* 61, 27–60.

Chapter 18

Efficiency Effects of *Bt* Cotton Adoption by Smallholders in Makhathini Flats, KwaZulu-Natal, South Africa

Yousouf Ismaël,[1] Lindie Beyers,[2,3] Colin Thirtle[2,3] and Jenifer Piesse[3,4]

[1]*University of Reading, Earley Gate, Whiteknights Road, Reading RG6 6AR, UK;* [2]*Imperial College of Science, Technology and Medicine, RSM Building, Prince Consort Road, London, UK;* [3]*University of Pretoria, Pretoria, RSA;* [4]*Birkbeck College, Malet Street, London WC1E 7HX, UK*

Hey, hey, bo-weavil, don't sing them blues no more
Bo-weavil's here, bo-weavil's everywhere you'll go

Bo-Weavil Blues by Ma Rainey, recorded December 1923 (Complete Recorded Works, Vol. 1, 1923–24).

Abstract

The results of a survey of 100 smallholders in the Makhathini Flats region of KwaZulu-Natal give cause for cautious optimism regarding the impact of *Bt* cotton. The farmers who adopted the *Bt* cotton variety in the 1998 and 1999 seasons benefited from the new technology, according to all the measures used. Average yield per hectare and per kilogram of seed was higher for adopters than for non-adopters. The increase in yields and reduction in chemical application costs outweighed the higher seed cost, so that gross margins were also considerably higher for adopters in the second season. This was a bad year, due to unusually heavy rainfall, and the *Bt* adopters suffered far less fall in yields than those who did not adopt.

Both yields and gross margins are partial measures of efficiency, which fail to take account of major inputs such as labour. Thus, they are supplemented by deterministic and stochastic efficiency frontiers, which consider the efficiency with which all inputs are converted into outputs. These methods use only the more reliable input and output quantity data and avoid prices, which are less well recorded or simply non-existent. Both methods confirm the farm accounting results, showing that the *Bt* cotton adopters were considerably more efficient than those who used the non-*Bt* varieties. For 1998, the stochastic frontier results showed that the adopters averaged 88% efficiency, as compared with 66% for the non-adopters. In 1999, the equivalent figures were 74 and 48%. Similarly, the determinist frontier results for both years show that the adopters were over 62% efficient, while the non-adopters averaged only 46%. Finally, there is no evidence that the better-off farmers gained more than the less well off: indeed, income inequality was slightly reduced.

Introduction

Outline

The aim of this study is to quantify the economic impact of genetically modified (GM) cotton production by smallholders in KwaZulu-Natal. The characteristics of the adopters and their reasons for adoption are considered first, in order to determine the probable take up of *Bt* cotton in future years. This identifies some of the constraints to adoption and provides some recommendations on how to improve the access to this technology for small-scale farmers. Then, the economic impacts at the farm level are measured in some detail.

After this introductory section, the chapter proceeds as follows. The next section provides descriptive statistics to give a broad overview of the characteristics of the smallholders in the sample. The following sections analyse adoption and outline the farm accounting results, comparing yields, input levels, costs and gross margins. A later section reports the efficiencies, obtained from fitting stochastic production frontiers, and extends the analysis of returns to scale by using non-parametric frontier techniques. Finally changes in the distribution of income following adoption are examined and this is followed by concluding comments.

Background: GM crops in developing countries

The recent International Fund for Agricultural Development report (IFAD, 2001) makes a strong case that effective use of biotechnology will be essential to the alleviation of rural poverty in developing countries for the foreseeable future. Higher yields, lower levels of labour and pesticide use and higher producer prices for cotton are cited as the main impacts of the adoption of GM crops at the household level (Fernandez-Cornejo and Klotz-Ingram, 1998; Gianessi and Carpenter, 1999; Fernandez-Cornejo *et al.*, 1999; Marra *et al.*, 2000). However these benefits must be set against fears of damage to the environment, the breakdown of resistance, reduction in biodiversity, increased profits for multi-national companies and the impoverishment of small farmers in developing countries.

Herbicide- and insecticide-tolerant traits account for more than 85% of the types of GM crops grown worldwide. Insect resistance has also been a popular target for the GM companies. Here, the focus has primarily been on the transfer of a set of genes controlling production of a natural insecticide in a bacterium called *Bacillus thuringiensis* (*Bt*) to crops. The *Bt*-toxin acts specifically on Lepidoptera (including bollworm in cotton, stem borers in maize), and is harmless to all other insect species.

Currently, the majority of commercial GM crop releases have been in the USA, Canada and some countries in South America. Indeed, the USA, Canada and Argentina account for 99% of the GM crop area in the world. Outside these areas, GM crop release on a commercial scale has been limited. In Africa, for example, commercial scale release of *Bt* cotton and maize is only taking place in South Africa. If the use of *Bt* resistance for control of Lepidoptera pests generates a yield advantage, and *Bt* technology is cheaper than the use of a pesticide with conventional seed, then *Bt* technology should provide farmers with an economic advantage. However, all the studies except Pray *et al.* (2000), which examined *Bt* cotton in China, were conducted in the USA and most of the data comes from the biotechnology industry and is typically based on controlled conditions and extrapolations from small plots. If the small, resource-poor farmers in developing countries reap the same benefits as suggested by the studies carried out thus far, they should have higher incomes, less health hazards and live in a less polluted environment. Thus, the motivation for this study is to provide a sound and impartial account of GM crop adoption based on empirical evidence from a developing country. The focus is on South Africa, but the results should prove relevant to other countries.

GM crops in South Africa

The GMO Act (Genetic Modified Organism Act, Act 15 of 1997), passed in 1997 and implemented in 1999, paved the way for the introduction and commercialization of GM crops in South Africa. The act legislates for the approval to import, use and supply the infrastructure required to utilize and evaluate GM seed in South Africa. Although there have been many crop trials, only *Bt* maize and cotton are grown on a commercial basis.

Approximately 3000 ha of *Bt* maize were planted in 1998 (James, 1999), and up to 50,000 ha of GM maize in 1999 (Thompson, 1999). This is 'yellow' maize, which accounts for 4% of the total crop, and is used for animal feed, cornstarch and maize syrup. *Bt* cotton is grown mostly in the Northern Province with some in KwaZulu-Natal and the Free State. Cotton accounts for 1% (100,000 ha grown by 1530 commercial farmers and 3000 small-scale farmers) of total South African agricultural production. Cotton generates approximately US$50 million per annum (Kock, 2000), mostly grown under dryland conditions.

The Makhathini Flats survey

Since 1998, smallholder farmers in one of the lower potential cotton areas of South Africa have been adopting a GM cottonseed variety (NuCOTN 37-B with Bollgard™). The Makhathini Flats are in KwaZulu-Natal province, where rural households cultivate land allocated to them by their tribal chiefs. Landholders face uncertain tenure arrangements without guaranteed ownership succession within the family. Good-quality cropping land is scarce, unfenced and under threat from livestock that devastate crops due to the communal grazing systems. Labour is also a problem in the rural areas, and because of male migration to the towns, the labour source of farmers is comprised of children, elderly people and female labourers. This influences the type of technology that farmers would readily adopt.

The main focus of the survey was to compare the adoption, productivity and cost of Delta Pineland's GM cotton (Bollgard™) with the non-GM cotton grown in the region. The survey, carried out in November 2000 in Makhathini Flats, covered a stratified sample of 40 non-*Bt* cotton growers and 60 *Bt* cotton growers. Of the 60 *Bt*-cotton farmers, some did not grow *Bt* cotton in the 1998/99 season. The rational for selecting such a sample is to have enough *Bt* and non-*Bt* farmers for comparison. Around 12% of the 4000 farmers in the region have adopted the new GM cotton. Therefore, a random sample would provide inadequate numbers of *Bt* cotton growers for comparison. A questionnaire was used to collect data on household background, farming practices, rational for adopting and not adopting *Bt* cotton, input cost and returns, with the information obtained through personal interview with farmers. Data was obtained for the 1998/99 and the 1999/2000 seasons.

On the smaller plots, there is mixed cropping of an assortment of beans, maize and cotton. GM cotton (NuCOTN 37-B), developed by Delta Pineland, along with varieties such as Akala 90, Opel, Sicala, CA223 and Tetra are grown in this region. Farmers plant the crop just after the rain, i.e. between September and December and even as late as January, and harvest from April to June, some 5–7 months after planting.

Organizational structure

The farmers belong to an association that provides essential support and information through organized meetings where they can discuss mutual concerns and problems. Monsanto owns the *Bt* gene and Delta Pineland developed the *Bt* NuCOTN 37-B cotton variety, that contains the *Bt* gene. Delta Pineland owns the non-GM varieties, Akala 90 and Opel, and Clark Cotton the other non-GM cotton varieties, Sicala and CA223. OTK (Eastern Transvaal Cooperation) developed the Tetra variety, another non-GM variety. VUNISA Cotton is a private company that sells the cottonseed varieties to the farmers in the region

VUNISA also supplies the chemicals and the necessary support for farmers through their extension officers. The Land Bank of South Africa provides credit to the farmers through VUNISA. Credit for land preparation, chemicals and seed is offered to the small-scale farmers based on their credit worthiness. Figure 18.1 summarizes the structure of the industry in the region. VUNISA buys all the cotton from the farmers and grades it according to set criteria. Cotton South Africa fixes the price per kg of cotton.

Descriptive Statistics for the Sample

The sample is comprised of 100 households. The salient characteristics of the smallholders in the sample were determined using cross-tabulations and frequency distributions. These findings are summarized here, beginning with the characteristics of the full sample, followed by the adopters of *Bt* cotton.

Characteristics of the smallholders

General

Respondents, who are the heads of the households, were categorized into four age groups; 76% of the farmers were reported to be 40 years and older. Of the 100 respondents, 48% are female and 52% are male; 27% of households do not own livestock, while the remainder own and use livestock for farming purposes. There was no substantial difference between household ownership of cattle (used for farming purposes) amongst the users and non-users of *Bt* cotton; 25% reported that they earn additional off-farm income from various sources.

Land distribution and use

Most of the farmers owned less than 5 ha of land, with the largest concentration in the category of 2.5–5 ha, as Fig. 18.2 shows.

Only 10% of farmers grew crops for sale, other than cotton, and only 8% used any form of crop rotation. All farmers

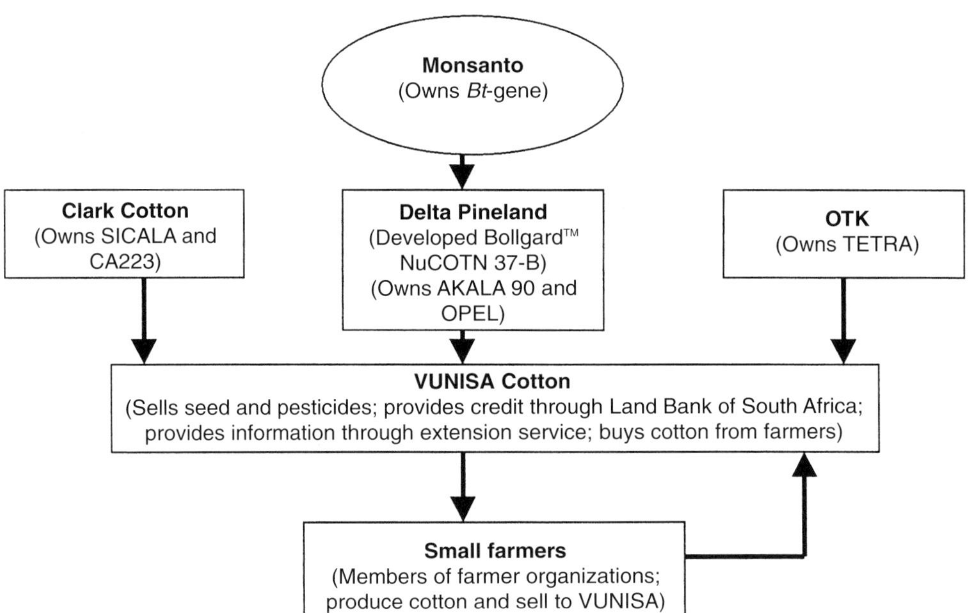

Fig. 18.1. Structure of the farming industry in the Makhathini Flats.

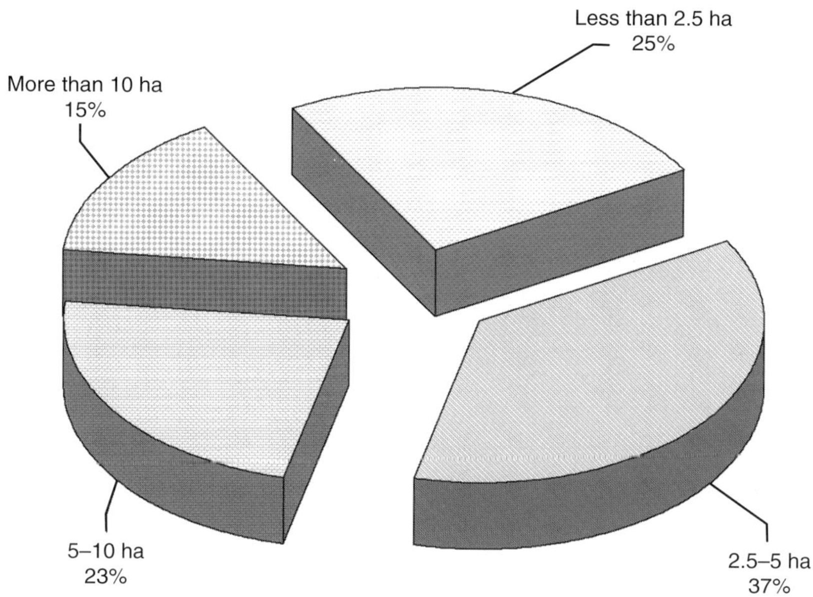

Fig. 18.2. Proportion of producers by farm size (percentage of respondents per category).

claimed to have cultivated at least 1 ha under cotton during the first season (1998/99) and the second season (1999/2000), and the maximum area under cotton was 25 ha in both seasons, as Fig. 18.3 shows.

Labour

Activities on the farm were carried out by family members along with some hired labour. Family labour is mainly female, housewives, elderly people and children, since the younger men take other jobs away from home. This is typical in South Africa, where a considerable percentage of the males are employed away from home, in activities such as mining. Thus, on average, households had more female than male labour, especially on those holdings where *Bt* cotton was planted. In this study, family labour is defined as those who are above 15 years old and not at school. However, children under 15 and those still at school often help with certain activities on the farm such as planting, weeding and harvesting. Labour is hired for specific activities, such as ploughing, spraying and harvesting. All the farms hire a tractor and driver for ploughing, at a cost of Rand 350 ha^{-1}. Most of the small farmers hire labour to spray their cotton, as they cannot afford a knapsack, and for harvest.

Pests

Farmers were asked to cite the main agronomic constraints to cotton cultivation. Pests, excessive rain and drought were ranked as the major or second most important agronomic constraints by 71%, 42% and 12% of the respondents, respectively.

The major pests in the region are the bollworm complex, cotton aphids and jassids, or leafhoppers. Bollworms consist of a complex number of species, which include American bollworm (*Helicoverpa armigera*), red bollworm (*Diparopsis castanea*) and spiny bollworm (*Earias biplaga, Earias insulana*). Their mature larvae feed on developing buds, squares, flowers and cotton bolls. The damage caused to the buds, flowers and bolls results in their shedding from the cotton plant, with consequent yield reduction. Red and spiny bollworms are the two species that cause the most damage.

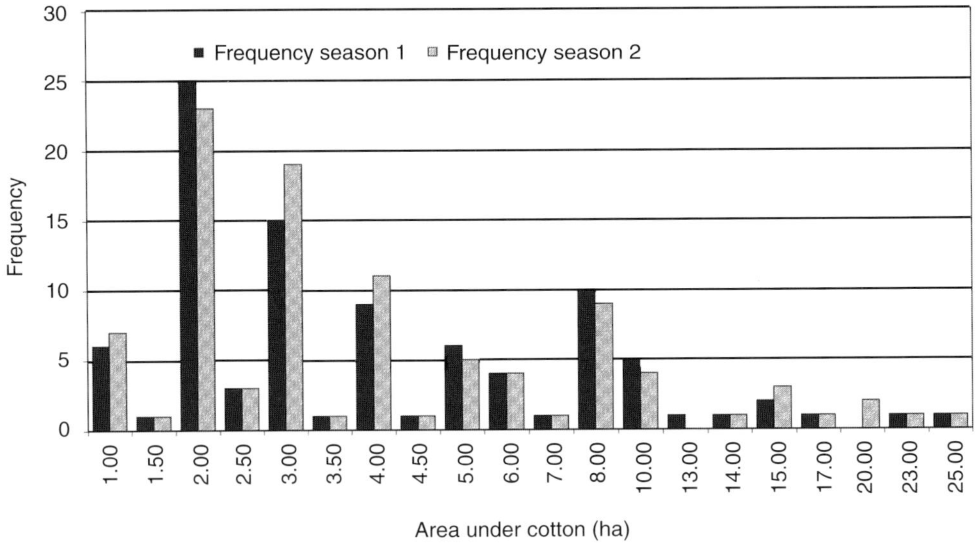

Fig. 18.3. Frequency distribution of the hectares under cotton in seasons 1 and 2.

Cotton aphids (*Aphis gossypii*) are found on young shoots, leaves and growing tissue, where they feed on the plant sap. The cotton aphids release salivary toxins, which cause the leaves to curl consequently reducing respiration, photosynthesis and plant growth. Aphids excrete excess sugar, which they have extracted as honeydew. Sooty mould grows on the honeydew and affects the processing of the lint.

Jassids (*Jacobellia facialis*), also known as leafhoppers, are sap feeders that remove the sap from cotton leaves. In doing so they introduce toxin to the plant sap resulting in curled up and purple leaves. Jassids attack plants that are 6–8 weeks old during particularly wet growing seasons.

Pest control

VUNISA is the sole supplier of four types of pesticides in the region. Monostem (organophosphate), Cypermethrin (pyrethroid), Dices (pyrethroid) and Cruiser (thiamethoxam) are used. Monostem is a systemic insecticide for the control of red, spiny and American bollworm, aphids and red mites, and application takes place as soon as infestation is noted, following regular inspection. Monostem is sold in 5-l containers.

Cypermethrin is an emulsifiable containing a stomach insecticide and is aimed at controlling predominantly bollworms. It is usually used during the period of peak flowering until boll spit. Application depends on the level of infestation. Cypermethrin is sold in litres. Dices, which is the same as Cypermethrin, but with a different trade name, is sold in 2-l containers. Monostem, Cypermethrin and Dices are mixed with water and a knapsack is required for spraying. Cruiser, is a new pesticide, targeting aphids and jassids and is designed to be used solely for *Bt* cotton. Unlike the other post-emergence pesticides, which are mixed with water and then sprayed, Cruiser comes in the form of a powder, which is mixed with the *Bt* cotton seed prior to planting. This is sold in 85-g sachets.

Non-agronomic constraints

All farmers, regardless of their technology decisions, stated that access to capital is the most constraining non-agronomic factor (82%). The majority of respondents have access to credit from VUNISA and use this exclusively, or in combination with household savings.

Farmers mostly received information about planting times and practices and sometimes on recommended spraying techniques (24% and 48% of the respondents, respectively). Most of the farmers received information only once in a growing season (33%), or once a month (27%); 24% were satisfied with the information they received, but 70% did not respond to this question. Producers are generally far away form their seed suppliers: 70% of the respondents farm within 5–20 km from the seed supplier. In terms of the quality grading that the farmers received for their cotton, adopters of *Bt* cotton ranked in the good- to high-quality categories. The majority of farmers (86%) produce cotton of good- to high quality.

Health and environment

Regarding spraying practices, 74% of farmers indicated that they wear protective eyewear and masks when they apply chemicals; and only 3% reported eye problems due to the chemicals, despite wearing protective eyewear; 72% of the farmers were not aware of any health problems incurred due to chemical application (see Fig. 18.4). Fifty-three per cent washed their empty spraying equipment in the field (Fig. 18.5) and 92% disposed of leftover chemicals in the field (Fig. 18.6).

Adoption

Characteristics of *Bt* cotton adopters and reasons for adoption

Farming experience ranged from 1 to 40 years (Fig. 18.7). Those who adopted the technology were amongst the most experienced farmers, particularly those who only adopted in the second season.

Farmers who adopted *Bt* cotton owned more land than those who did not adopt the technology. Eighteen per cent of the farmers used the modified cotton variety (NuCOTN 37-B) in the 1998/99 production year, rising to 60% in the 1999/2000 production year. All the farmers who adopted in the first year continued using *Bt* cotton

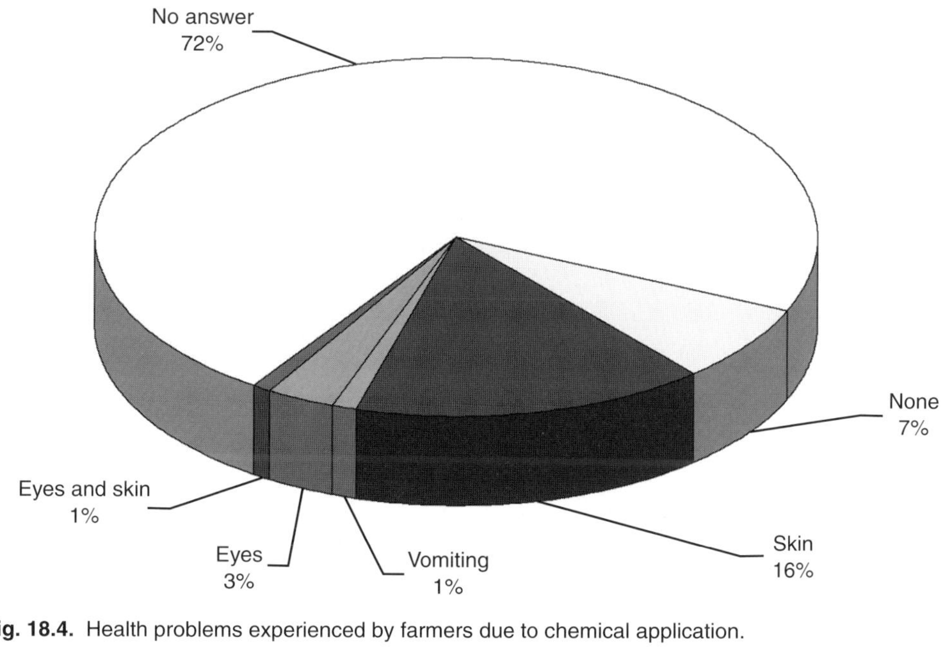

Fig. 18.4. Health problems experienced by farmers due to chemical application.

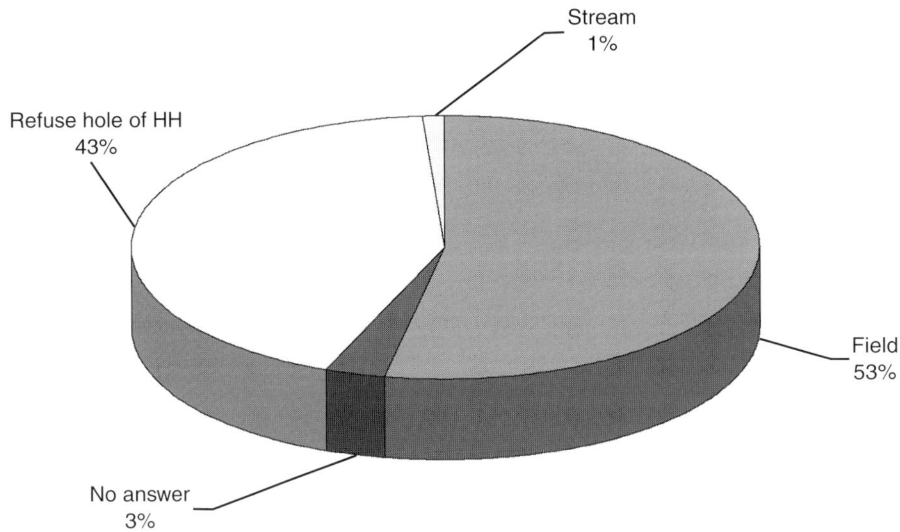

Fig. 18.5. Place where farmers wash spraying equipment after use.

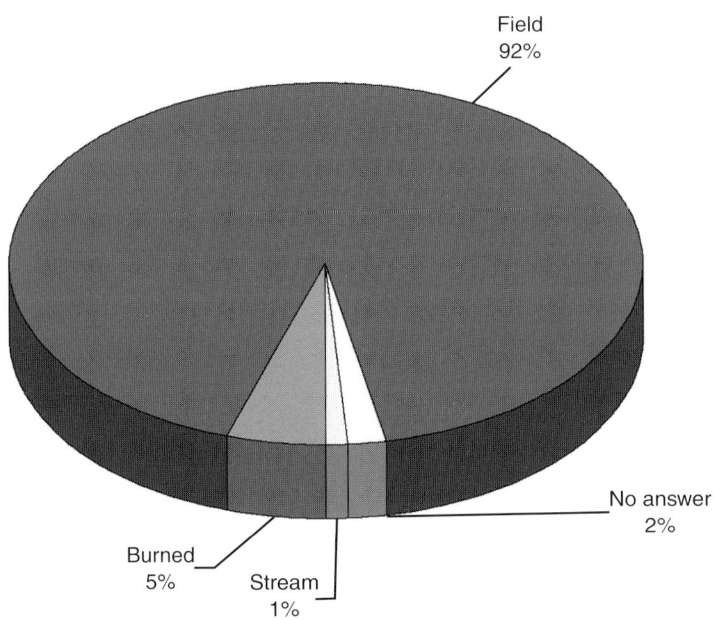

Fig. 18.6. Place where farmers dispose of leftover chemicals and empty chemical containers.

in the second year. This suggests that the farmers were satisfied with the performance of the *Bt* variety. Figure 18.8 shows the change in variety and *Bt* cotton adoption from season 1 to 2.

Farmers were questioned on the reasons they adopted, or what would make them consider adopting in the future; 24% cited expected increases in yields as the main reason, while saving on chemicals and pesticides accounted for 44% of the respondents' main reason; 10% believed that the labour-saving properties of *Bt* cotton were critical in the adoption decision. Adopters

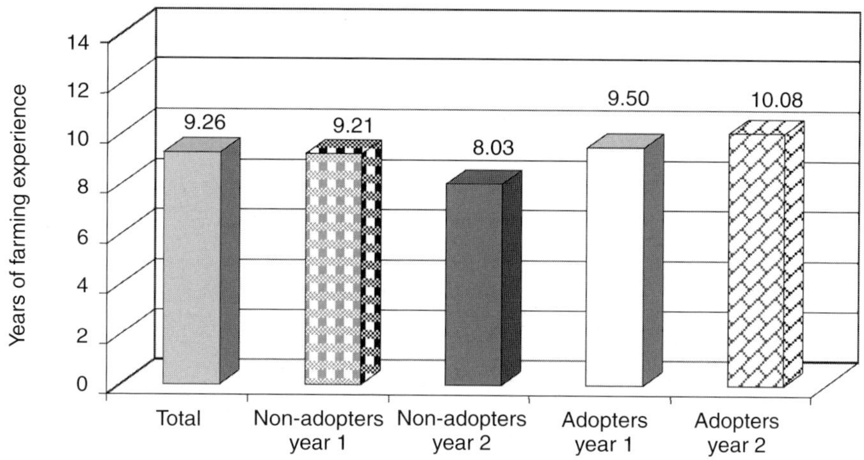

Fig. 18.7. Average years of farming experience for adopters and non-adopters.

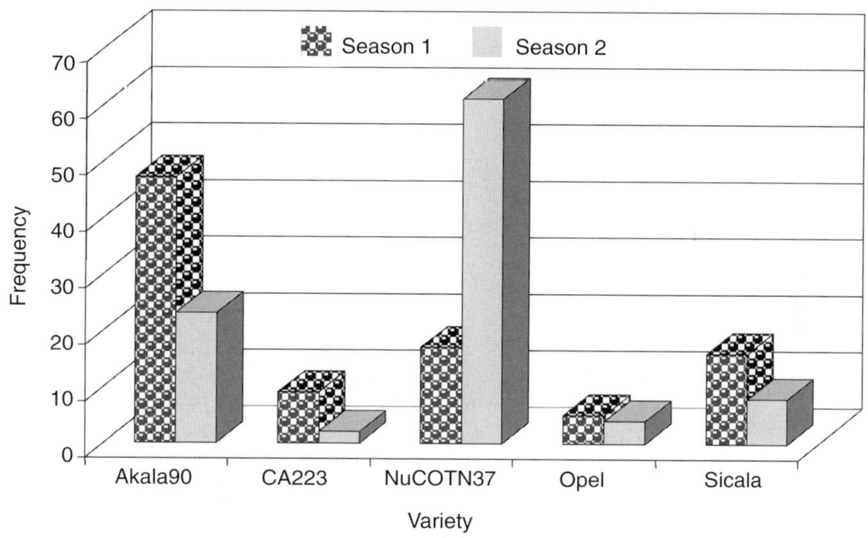

Fig. 18.8. Frequency distribution of cotton varieties planted in seasons 1 and 2.

of *Bt* cotton (35%) felt strongly that pests were their major problem and this prompted them to adopt the bollworm resistant variety. Non-adopters considered excessive rain and weed invasion as their main problems.

Analysis of adoption

This section begins with farmers' perceptions of the problems faced in the Makhathini Flats, since this has a direct bearing on their attitudes towards *Bt* cotton adoption.

Problems reported by the farmers

Figure 18.9 shows that 57% of the farmers rated pests as the dominant agronomic problem in the area, with 24% stating too much rain and 11% stating weeds. With 92% primarily concerned with these three

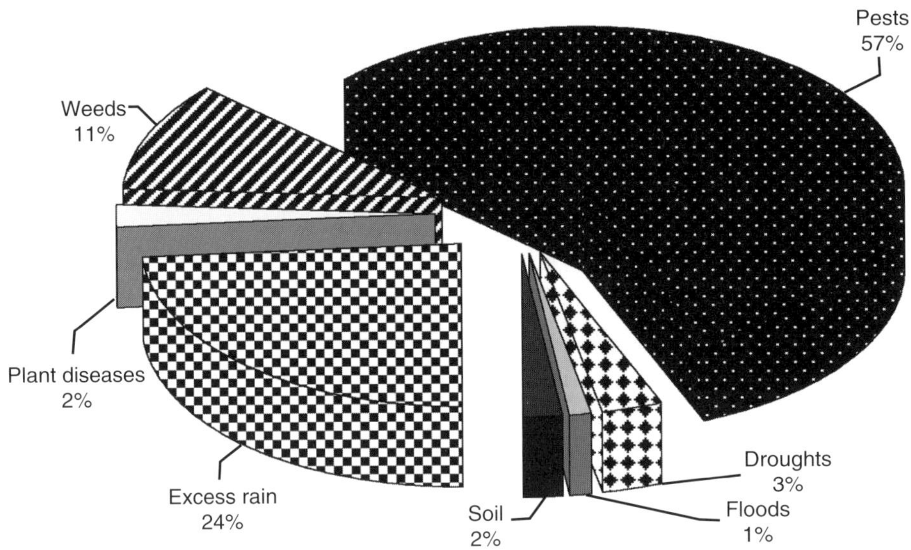

Fig. 18.9. The most important agronomic constraints, as expressed by the farmers.

issues, only 8% were more worried by droughts, floods, soil quality and plant diseases. Figure 18.10 shows that 62% of the farmers regarded the bollworm as the major pest, 21% aphids and 17% jassids. Thus, the area is a prime location for trials of *Bt* cotton, since it has immunity to the most common pest in the region.

Of the farmers surveyed, 69% thought *Bt* cotton was a major benefit, stating that higher yields, savings on pesticides and labour cost were the main benefits they expected, or had realized from *Bt* cotton. Most farmers did not identify any problems with *Bt* cotton, except the high cost of seed. Ninety per cent of non-adopters were willing to adopt the technology, but their main reasons for not adopting were the high technology fee incorporated in the seed cost and the lack of information about the variety.

This is not surprising, for as Fig. 18.11 shows, only 14% of the farmers thought that lack of land was their major problem and as few as 4% were troubled by labour shortages. The vast majority were constrained mostly by lack of capital that restricted their farming activities. Thus, credit availability must be expected to play a major role in adoption.

The adequacy of information on GM crops has been questioned by Tripp (2000), claiming that their qualities are not immediately obvious to farmers. Only 9% of the farmers knew about *Bt* cotton in the 1998/99 season, but by the following year, 50% of farmers were aware of it. VUNISA's extension officers mainly influenced those who adopted *Bt* cotton through the training they organized; 50% of adopters had some form of training, with which they were satisfied, while others were less satisfied. Although the adopters perceived pesticide savings as a major benefit, only 60% of the adopters based their decision on this view.

Limited dependent variable model

To construct a model of adoption, the dependent variable (the decision to adopt) takes the value of either zero or one, and thus a limited dependent variable model is appropriate. The Logit model is from the class of conditional probability analysis (CPA) techniques and yields the probability of the given occurrence rather than dichotomously classifying samples into different groups. It has less demanding assumptions than other discrete choice

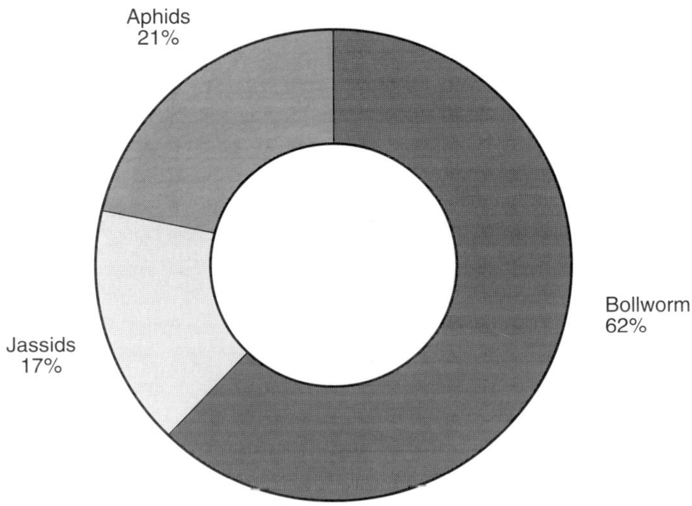

Fig. 18.10. Farmers' opinion of the main pest problem.

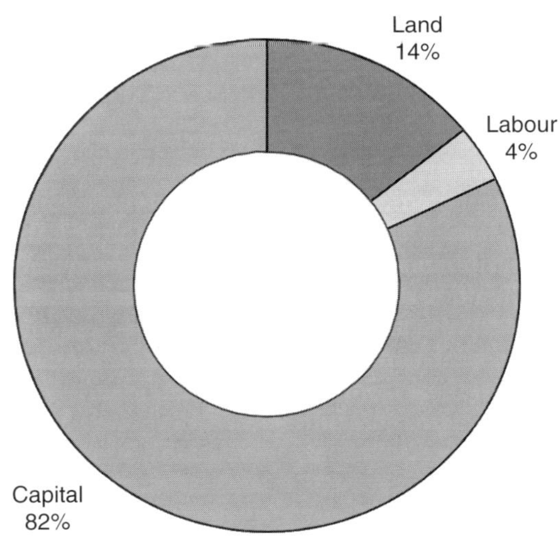

Fig. 18.11. Farmers' perception of the most important non-agronomic constraints to farming.

models, is straightforward to calculate and the output is easy to interpret. In addition, it has a high accuracy rate and a number of comparative studies confirm that Logit models perform at least as well as the alternatives. The basic form of the Logit model is stated:

$$\text{Prob}(Y=1) = \frac{e^{(\beta' X)}}{1+e^{(\beta' X)}} = \frac{1}{1+e^{(-\beta' X)}},$$

where β is a vector of parameters that quantify the behaviour of the exogenous variables in explaining the adoption decision.

The coefficients in a binary choice model are not easy to interpret, apart from their sign. One way to interpret the parameters, and to select amongst different specifications, is to consider the marginal effects, which in any non-linear model are not simply the coefficients. The effect of a change will vary with the values of the independent variables. Thus, the elasticity with respect to each explanatory variable is the marginal effect evaluated most usefully at the mean. The general form of the derivative of the probability $y_i = 1$ with respect to all the x_i is

$$\frac{d \log it(\beta' X)}{d(\beta' X)} = \frac{e^{(\beta' X)}}{1 + e^{(\beta' X)^2}}.$$

Several specifications were tested and only the acceptable results are reported in Table 18.1. No conclusive results were obtained from the first year's data, but for the 1999/2000 season, farm size, non-farm income and female labour jointly determine adoption, but explain only 15% of the variance. The positive signs indicate that owners of larger farms, those with non-farm income and the farms with more female labour were more likely to adopt. The age group of the farmer was also significant at a lower confidence level and cattle ownership could be substituted for this variable, with similar results. The elasticities, evaluated at the means, show that farm size has the greatest impact on the likelihood of adoption. Non-farm income does indicate that an alternative income source can obviously provide the money needed to buy more expensive seed, although the elasticity is low as 75% of the farmers do not have access to additional income. Finally, older farmers, those with larger farms and those with cattle are also both better off and more likely to be granted credit.

Thus, the Logit model confirms the survey responses that lack of capital is the biggest problem, although the credit variable itself was weak, since almost all farmers had some credit, but not enough. Non-farm income may also have a stronger effect because the households with non-farm income are likely to be less risk averse. The only variable that is not easily explained is the presence of female labour. The elasticity at the mean is 0.5, and so clearly this is an important aspect of the adoption decision. Possibly, since spraying is regarded as men's work, a reduction in the need for spraying because of a high proportion of female labour may affect the adoption decision.

These results confirm the qualitative survey responses that suggested that capital is a major constraint to farming in the region. Most farmers are resource poor and do not have the capital to invest in farming; 64% complained of lack of finance as being a major constraint in farming, irrespective of farm size or cotton variety grown. Sixty-three per cent depended on VUNISA for credit to buy their inputs, such as seed and insecticides. Another 27% use their savings and 25% had non-farm income. The remainder relied on group saving schemes and/or family credit. Thus, all the farmers reported some source of credit, but the majority found the levels to be inadequate. Therefore, no seri-

Table 18.1. Logit estimation of the characteristics of smallholders adopting *Bt* cotton ($n = 99$).

Variable	Coefficient	SE	*t*-statistic	Elasticity
Constant	−2.56	0.98	−2.61[b]	–
Farm size	0.11	0.05	2.20[a]	0.836
Non-farm income	1.42	0.54	2.63[b]	0.034
Female labour	0.51	0.19	2.68[b]	0.523
Age group	0.37	0.26	1.42	0.412

Test statistics

Log likelihood	−56.45	Hannan–Quinn criterion	1.29
Restricted log likelihood	−66.37	Avg. log likelihood	−0.57
LR statistic (4 degrees of freedom)	19.85	McFadden R^2	0.15
Probability (LR stat)	0.00	Scale factor	0.14

[a] Significant at the 95% confidence level. [b] Significant at the 99% confidence level.
SE, standard error.

ous barriers to adoption appear to exist, other than finance, which could be resolved. Specifically, there are no agronomic problems that prevent benefits to the rest of the local population from the adoption of the *Bt* variety, if they could afford it.

Production: Yield, Cost and Profitability for Adopters and Non-adopters

The questions on reasons for adoption show that the expectation of better performance was the main reason for adoption, given the higher cost of seed. So, the trade-off between the higher seed costs, vs. the benefits of using the *Bt* varieties, in terms of cost savings and output gains, is the basis of the adoption decision. Farm size, age and technology adoption all explain substantial differences in production efficiency and profitability. These relationships are established using cross-tabulations of yields, seed and chemical costs, for adopters and non-adopters.

The last two rows of Table 18.2 show that in the first season 82% of the farmers surveyed cultivated conventional cotton, compared to 18% who grew the Bollgard™ variety. In the 1999/2000-season, 60% of the farmers grew the Bollgard™ variety. The first row of Table 18.2 shows that in the first season the small sample of adopters had yields that were on average 18 kg higher than the non-adopters' yields. This is partly explained by the seeding rate, which is 22% lower, perhaps because of the cost of the seed. In the second year this situation is reversed and the adopters have a yield advantage of 121 kg ha^{-1}, as can be seen in Fig. 18.12. The seeding rate of the adopters rose a little, but so did that of the non-adopters fall, so the differentials were maintained and still the adopters have a yield advantage of 40%. The lower yields in the second season are attributable to the rainfall in the 1999/2000 season that was 50% above average, causing flooding and delayed planting of cotton. This is the reverse of the previous years in which lower than average rain at the beginning of the season had favoured the cotton crop (KwaZulu-Natal Annual Report, 1998, 1999).

The differences in yields and seeding rates are reflected in the yield per bag of seed, which is 34% higher for the *Bt* users in the first season and 93% higher in the second year. Average seed cost per hectare were 65% higher for adopters in the first season, and 68% higher in the second season. However, as a result of using the *Bt* variety, chemical cost fell by 30% in the first year and by 36% in the second season, as Fig. 18.13 shows.

However, in the first season, the higher

Table 18.2. Per hectare cost and returns based on adoption and farm size.

Average per category	Non-adopters	Adopters
Yield (1998/99) (kg ha^{-1})	457	475
Yield (99/00) (kg ha^{-1})	304	425
Bags (25 kg) of seed ha^{-1} (1998/99)	0.55	0.43
Bags (25 kg) of seed ha^{-1} (1999/2000)	0.57	0.46
Yield (kg) kg^{-1} of seed (1998/99)	36.96	49.56
Yield (kg) kg^{-1} of seed (1999/2000)	22.88	44.20
Seed cost ha^{-1} (1998/99) (Rand)	119.27	197.37
Seed cost ha^{-1} (1999/2000) (Rand)	127.29	213.56
Chemical cost ha^{-1} (1998/99) (Rand)	132.18	93.37
Chemical cost ha^{-1} (1999/2000) (Rand)	129.13	83.15
Gross margin ha^{-1} (1998/99) (Rand)	791.26	781.44
Gross margin ha^{-1} (1999/2000) (Rand)	428.14	677.38
Number of farmers per group (year 1)	82	18
Number of farmers per group (year 2)	40	60

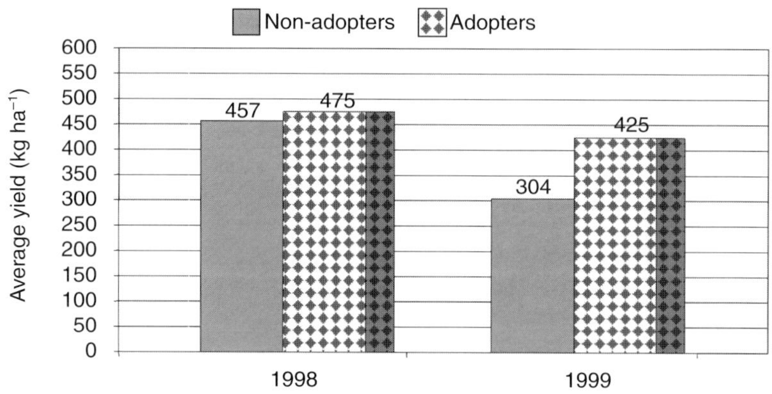

Fig. 18.12. Yields of adopters and non-adopters.

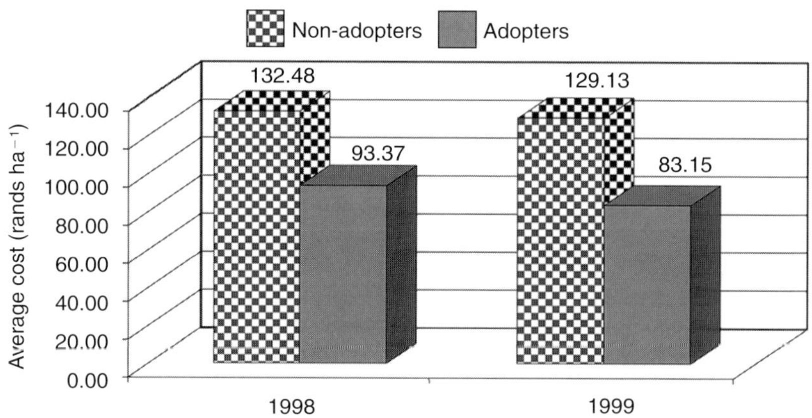

Fig. 18.13. Chemical cost of adopters and non-adopters.

seed cost of the adopters outweighs the yield gain and the savings in chemical cost. Table 18.2 shows that the gross margins of the non-adopters were R10, or 1.3% higher. This can be clearly seen in Fig. 18.14. There is also a saving in labour used for spraying, but the data does not allow this gain to be quantified. The problem is that the smaller farms had higher yields and that larger farms adopted the *Bt* variety, so the unequal adoption across farm sizes disguises the gains to adopters.

The advantage of the *Bt* variety is also apparent in that none of the adopters dropped the *Bt* variety in the second year and a further 42 farmers adopted it in the second season. The gain in output value and the savings in chemical and spraying labour cost were sufficient to compensate for the higher seed cost.

In the second year, the results are far clearer, so that the outcome is entirely unambiguous. The yield gain of 40% and the lower chemical cost easily offset the extra seed cost, so that the gross margins are, on average 58% higher. This is a huge gain by any standards and it does seem that the *Bt* cotton survived the unfavourably wet growing season far better than the non-*Bt* varieties. The yields of those that had adopted in the first year fell from an average of 475 to 425 kg ha^{-1}, which is only 10%, whereas the average yields of the non-adopters fell from 457 to 304 kg,

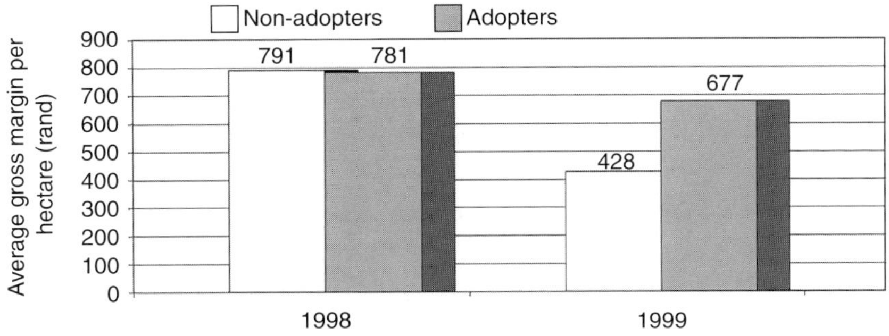

Fig. 18.14. Gross margins of adopters and non-adopters.

which is 33%.[1] Thus, the *Bt* variety performs well in adverse conditions, because spraying for bollworm is a greater problem in the wetter season.

Production Efficiency of Makhathini Flats Cotton Producers

The analysis in the previous section provided clear evidence that in the second season, the *Bt* adopters had higher yields and higher gross margins. These basic farm accounting measures are obviously useful, but say little about the reasons for any observed differences between farms. Yield is a very partial measure of productivity, which is of limited use when the amounts of non-land inputs used, such as labour and fertilizer, differ between farms.

Gross margins take account of intermediate inputs, such as seed and chemicals, but ignore the efficiency with which labour and land are used, and since land and labour are major inputs, this is unsatisfactory. Net margins can be calculated, including land and labour costs, but thus require prices for these inputs, which are not well defined, especially for family labour and land. In addition, neither yields nor margins tell us anything about the existence, or otherwise, of scale economies. Instead, we now resort to production frontiers, which require only input and output quantities and avoid using dubious price data.

Stochastic production frontiers

The measurement of farm level efficiency has become commonplace with the development of frontier production functions. The approach can be deterministic, where all deviations from the frontier are attributed to inefficiency, or stochastic, which discriminates between random errors and differences in inefficiency. This study uses both approaches, since they are complimentary, both in providing different information and in confirming the results from incomplete survey data.

The stochastic frontier model is of the type originally proposed by Aigner *et al.* (1977), extended to include the characteristics of the farm that specifically explain inefficiency levels, following the work of Battese and Coelli (1995). Firstly, the frontier model is constructed to determine the efficiency levels of the sample farms, with respect to those that represent best practice, and then the inefficiencies are explained. The method of maximum likelihood is used to estimate the unknown parameters, with the stochastic frontier and the inefficiency effects estimated simultaneously. Model selection is based on gener-

[1] The yields of those who had adopted in the first year used *Bt* cotton again in the second year are not shown in Table 18.6, but they are close to the results shown for the non-adopters and adopters in total, shown in the table.

alized likelihood ratio (LR) tests that confirm the adequacy of the functional form of the model, whether the model is a frontier or a mean response function and the presence of inefficiencies effects.

Data, estimation and tests

The output and input data is summarized in Table 18.3, which reports the mean, minimum and maximum values and the standard deviations of the variables. The variations in the data are clearly sufficient to allow the estimation of production relationships.

Firstly, a simple production function is estimated, using ordinary least squares (OLS) regressions. Output is bales of cotton, which is a physical measure of output. Inputs are land (in hectares), chemicals (in value terms since it is an aggregate of different types), seed (both in 25 kg bags and costs) and labour. Labour is the number of days of family and hired labour used for spraying, weeding and harvesting, although planting is excluded because all the farms used the available mechanized planting services. All variables are in natural logarithms, so that the coefficients can be interpreted as elasticities, which must have values of between zero and unity to conform to production theory. Since values less than zero make no economic sense, a one-tailed significance test is appropriate here.

The OLS results for the first season are reported in Table 18.4, where the dependent variable is bales of cotton. The adjusted R^2 indicates that the four inputs explain half of the variance in output, which is reasonable for a cross-section of farmers with considerable dispersion of efficiency levels. Indeed, it is rather higher than in similar production studies of smallholder farming in South Africa, such as Piesse *et al.* (1996), which modelled efficiency in the former Homelands of the Northern Transvaal.

The elasticity for land of 0.262 indicates that a 1% increase in land would increase output by 0.26% and the *t*-statistic shows that this estimate is significantly different from zero at the 95% confidence level. Similarly, chemicals have an elasticity of 0.209 and this is significant at the 90% confidence level. Seed has the biggest

Table 18.4. Production function estimation (1998/99) (dependent variable: bales).

Variable	Coefficient	*t*–statistic
Intercept	−3.327	−3.191
Land	0.262	1.774
Chemicals	0.209	1.427
Seed cost	0.426	3.166
Labour	0.392	2.085
Sum of elasticities	1.289	
Adjusted R^2	0.50	

*Critical *t*-value at 95% confidence level = 1.66;
** Critical *t*-value at 90% confidence level = 1.29.

Table 18.3. Summary statistics on output and inputs by season and by type of technology

	Output (kg)		Seed (kg)		Chemicals (Rand)		Labour (days)		Land (ha)	
	Adopt	Non-adopt	Adopt	Non-adopt	v	Non-adopt	Adopt	Non-adopt	Adopt	Non-adopt
Season 1: 1998/99										
Minimum	1244	170	25	25	0	0	4	2	2	1
Average	3059	1726	76	50	526	413	21	19	9	6
Maximum	8011	8878	225	200	2279	2690	48	40	25	20
SD	2024	1743	62	33	529	340	11	8	7	5
Season 2: 1999/2000										
Minimum	340	43	25	25	0	0	6	2	1	1
Average	1983	1183	55	48	317	403	22	19	7	5
Maximum	9333	8500	250	150	1505	2690	51	40	25	15
SD	1598	1495	51	32	267	439	9	9	6	3

impact on output (0.426), followed by labour, which is also highly significant. This is unusually successful, since measuring the labour input at all accurately is a problem in sample surveys of this nature. The sum of the elasticities is 1.289, indicating increasing returns to scale (IRS) in production, since if all the inputs were increased by 1%, output would rise by 1.289%. As these farms appear to be too small it can be inferred that greater efficiency could be achieved if they expanded.

These results are encouraging, but OLS estimation takes the 'average line of best fit' through the observations (hence mean response function) and thus assumes that all the farms are efficient. If there are considerable differences in efficiency levels, this can be misleading. The appropriate approach is to estimate a production frontier, as described above, which will give results that are more accurate and will generate farm-level efficiency levels.

Table 18.5 reports the results of fitting a stochastic frontier model for the first season. All the elasticities are significant, but labour has the biggest impact (0.476), followed by chemicals (0.265), land and seed. The sum of the elasticities again indicates IRS.

The variables that explain inefficiency are the adoption of *Bt* cotton, the planting date and the farmers' experience in growing cotton. Adoption of *Bt* cotton has a *negative sign*, meaning that it *reduces inefficiency*. Thus, using the *Bt* variety increases the efficiency of the farms that adopted. The planting date shows that the later farmers planted, the less output they got, which is a common result in African agriculture. Indeed, a USAID study of Kenya in the mid-1980s showed that timely planting had a greater impact than using fertilizer or improved seed. Lastly, the negative sign on farmer experience means that the more experienced farmers were less inefficient. Taken together, these are predictable results, especially since the survey was conducted during the second season and the first season data relied on the farmers remembering past details.

The γ statistic helps to determine whether this is indeed a frontier model and

Table 18.5. Production frontier and inefficiency model (1998/99).

Variable	Coefficient			*t*-statistic	
Production frontier					
Intercept	−1.874			−2.103	
Land	0.211			1.583	
Chemicals	0.265			2.400	
Seed cost	0.177			1.404	
Labour	0.476			3.538	
Sum of elasticities	1.129				
Inefficiency model					
Adoption	−0.444			−1.383	
Planting date	0.410			1.282	
Experience	−0.118			−1.955	
σ^2	0.89			2.818	
γ	0.81			8.097	
Statistics					
Mean efficiency: total	0.70	Minimum	0.15	Maximum	0.92
Mean efficiency: non-adopters	0.66	Minimum	0.15	Maximum	0.89
Mean efficiency: adopters	0.88	Minimum	0.80	Maximum	0.92

*Critical *t*-value at 95% confidence level = 1.66; ** Critical *t*-value at 90% confidence level = 1.29.

not simply a mean response function. The fact that the γ is close (0.81) to unity and highly significant indicates that the frontier model is appropriate. This implies that one or more of the firms in the sample are fully efficient, that is, they form a frontier of best practice, while the remainder are some measurable distance from this efficiency frontier. Further tests follow the reporting of the results for the second year.

This is not surprising, since the mean level of efficiency of the full sample was 0.70, or 70%. The adopters had a far higher mean efficiency (0.88), as compared with the mean (0.66) for the non-adopters. Even more convincing is that the minimum efficiency level amongst the adopters was 0.80, as compared with 0.15 for the non-adopters. This suggests that despite the slightly lower gross margins for adopters, the *Bt* variety did perform well, when the land and labour inputs are taken into account. This would explain why none of the adopters discontinued the *Bt* variety and many more adopted in the second year.

The same exercise was repeated for the 1999/2000 season, and here the dated fitted better, perhaps because it was more recent and better remembered. This is reflected in the R^2, which shows that 73% (Table 18.6) of the variance in output is now explained, which is unusually high for cross-section data of this nature. The OLS results show that land now has the greatest impact (0.464), followed by labour and seed. Note that seed cost was used in this case, and since *Bt* seed is twice as expensive, this needs to be taken into account. The most interesting result is that with widespread adoption of the bollworm resistant variety, the importance of chemicals falls dramatically. Indeed, apart from the low elasticity of 0.027, the *t*-statistic (0.486) shows that chemicals now seem to have no significant effect. Lastly, the sum of the elasticities falls to 1.099, which suggests that although there are still increasing returns, farms are now more scale efficient.

Table 18.6. Production frontier and inefficiency model (1998/99).

Variable	Coefficient	*t*-statistic
Intercept	−0.104	−0.203
Land	0.464	3.743
Chemicals	0.027	0.486
Seed cost	0.222	1.538
Labour	0.386	2.329
Sum of elasticities	1.099	
Adjusted R^2	0.730	

*Critical *t*-value at 95% confidence level = 1.66;
** Critical *t*-value at 90% confidence level = 1.29.

The stochastic frontier model reported in Table 18.7 shows that labour again has the biggest impact, followed by seeds and land, while the impact of chemicals is far lower than in the first year. However, unlike the OLS model, the frontier shows that all the variables are significant at the 95% level. Summing the elasticities suggests that there is now decreasing returns to scale, meaning that the farms are, if anything, a little too large. However, summing elasticities is a crude means of determining returns to scale, so this issue is pursued further in the next section, which uses a programming approach.

The only variable that was effective in explaining the inefficiencies was adoption of *Bt* cotton, which is now highly significant and has a positive effect on efficiency. There is now a greater dispersion of efficiencies, in this poor season, with a lower mean efficiency level of 0.64 and the minimum efficiency falling to 0.10 for the non-adopters and 0.33 for those who did adopt. The maximum is the same for both groups, but the mean efficiency of the adopters is again far greater.

The γ statistic is even closer (0.94) to unity and is again highly significant (22.171), indicating that the frontier is the preferred model. However, this is now further investigated using more powerful generalized log likelihood ratio tests,[2] reported in Table 18.8. The functional form of the stochastic frontier was determined by testing the adequacy of the log-linear Cobb–Douglas model relative to the less

[2] The likelihood-ratio test statistic, $\lambda = -2\{\log[\text{likelihood } (H_0)] - \log[\text{likelihood } (H_1)]\}$ has approximately χ^2_v distribution with v equal to the number of parameters assumed to be zero in the null hypothesis.

Table 18.7. Production frontier and inefficiency model (1999/2000).

Variable	Coefficient			*t*-statistic	
Production frontier					
Intercept	0.643			1.816	
Land	0.276			2.823	
Chemicals	0.059			1.818	
Seed cost	0.282			2.671	
Labour	0.341			3.002	
Sum of elasticities	0.958				
Inefficiency model					
Adoption	−2.755			−1.640	
σ^2	1.060			1.672	
γ	0.940			22.171	
Statistics					
Mean efficiency: total	0.64	Minimum	0.10	Maximum	0.91
Mean efficiency: non-adopters	0.48	Minimum	0.10	Maximum	0.91
Mean efficiency: adopters	0.74	Minimum	0.33	Maximum	0.91

*Critical *t*-value at 95% confidence level = 1.66; ** Critical *t*-value at 90% confidence level = 1.29.

restrictive translog, which includes cross products and square terms to allow for interactions and non-linearities in the data. For both seasons, the log linear model was accepted as an adequate representation of these data, as the first results in Table 18.8 show. The next test reported is the *t* test on γ which suggests that the frontier model is preferred to a mean response function. The log-likelihood ratio test (LR), which is more powerful than the *t* test on the γ statistic, also confirms that both models are frontiers. In the last test reported, the power of the LR test is increased by testing jointly the null hypothesis that both the frontier parameter and all the inefficiency effects are jointly zero, that is, the inefficiency effects are not present in the model.[3] This proposition is also rejected, which means that the frontier model, with inefficiency terms, is the preferred model for both seasons.

Deterministic frontier programming models

The stochastic frontier model results are entirely acceptable, and supported by deterministic frontier efficiency models. Non-parametric methods have been widely applied to efficiency measurement problems and are particularly appropriate in terms of establishing optimum farm size.

The model used for measuring farm-level efficiency follows the framework introduced by Farrell (1957) and extended by Fare *et al.* (1985), to include the decomposition of overall efficiency into measures of technical and scale efficiency. The method is non-parametric and deterministic, with the best practice frontier constructed by minimizing inputs per unit of output. Then, the efficiency of each farm is measured as a ratio of actual to best practice performance.

There are a number of advantages to this measurement approach. First, it allows the comparison of one farm with a given input–output combination to other farms using inputs in different proportions, neutrally measuring total factor productivity in a multiple input framework. Second, each input and output can be measured separately in its natural physical units, without the need to apply price or share weights in an aggregation procedure. Third, propor-

[3] Since γ takes values between 0 and 1, any LR test involving a null hypothesis which includes the restriction that $\gamma = 0$ has been shown to have a mixed χ^2 distribution, with appropriate critical values (Kodde and Palm, 1986).

Table 18.8. Log-likelihood ratio tests for the frontier and inefficiency model.

	1998/99 season	1999/2000 season
Choice of functional form – H_0: $\beta_{ij} = 0$, $i,j = 1,\ldots,4$. Test statistic: $\chi^2_{v,\,0.95}$, where v= number of additional restrictions = 10		
Test statistic	9.94	16.12
Critical value	18.31	18.31
Test result	Accept H_0: Cobb–Douglas is adequate	Accept H_0: Cobb–Douglas is adequate
Choice of stochastic frontier vs. mean response function – H_0: $\gamma = 0$. Test statistic: one tailed *t*-statistic; 95% confidence level		
Test statistic	8.097	21.17
Critical value	1.96	1.96
Test result	Reject H_0: it is a frontier	Reject H_0: it is a frontier
Presence of inefficiency effects – H_0: all inefficiency coefficients (δ_l) and $\gamma = 0$. Test statistic: mixed-$\chi^2_{v,\,95\%}$ confidence level, where v = number of restrictions (5 in 1998/99 and 3 in 1999/2000)*		
Test statistic	12.25	24.48
Critical value	10.36	7.05
Test result	Reject H_0: it is an inefficiency model	Reject H_0: it is an inefficiency model

[a] Critical values for the mixed χ^2 are from Kodde and Palm (1986).

tional input decreases translate into reduced costs and any input, which is not a constraint on production, will be identified as a slack variable. Therefore, the sources of inefficiency can be identified and policies to procure efficient production can take these into account. Finally, no functional form is imposed and no behavioural assumptions are required.

The radial measure of overall technical efficiency is shown in Fig. 18.15, where the inputs are x_1 and x_2 and the boundary of the set $L(y)$ is the best practice isoquant, defining the minimum combinations of inputs required to produce output level y^*. Observations B and C are efficient and define the isoquant, in this simple example, whereas observation A uses more of both x_1 and x_2 to produce the same output, y^*, and is thus inefficient. The vector OD

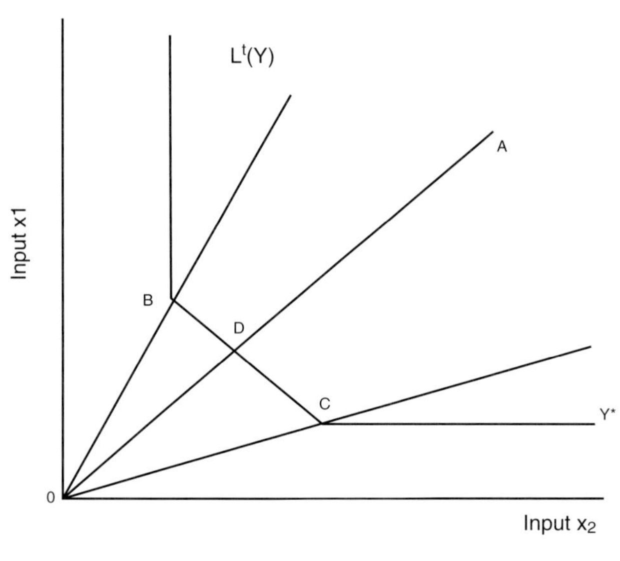

Fig. 18.15. Basic radial measure of overall technical efficiency.

shows the minimum combination of x_1 and x_2 that farm A could use to produce y^* efficiently, using its own factor ratio. OA is the actual combination of inputs used, so the radial measure of the efficiency level of farm A is OD/OA, which will be between zero and unity.

The efficiency results are extended by decomposing the efficiency measure into pure technical efficiency and scale efficiency and then determining whether the farms that are not scale efficient are too small or too large. The results for 1998/99 are summarized in Table 18.9, which shows that these support those from the stochastic frontier. The first column shows that the total technical efficiency is 0.56, as compared with 0.7 for the stochastic frontier, since deterministic model attributes all deviations from the frontier to inefficiency. The adopters are again far more efficient, on average, than the non-adopters and have a much higher minimum efficiency level. The next two rows show that the cause of the superior efficiency of the adopters is that their farms are of optimal size, rather than their level of technical efficiency. This is surprising, since the yield and gross margin analysis showed that the smaller farms had higher yields and gross margins. However, once land and labour are taken into account, it is the larger farmers (a greater percentage of whom adopted *Bt* cotton) who have the higher efficiency levels. This supports the earlier finding in the logit model, that farm size has the highest elasticity.

The efficiency frontier is defined by six farms, only one of which adopted *Bt* cotton. This is a sufficient number to ensure that the frontier is meaningful, rather than simply being determined by observations that might be outliers (which is potentially a serious problem in the deterministic model). The returns to scale results show that apart from the six efficient farms, with an overall efficiency score of unity and scale efficient by definition (shown as constant returns to scale, CRS), all but one of the enterprises is too small. This is why the programme is attributing the greater efficiency of the adopters to farm size: adopters, on average, owned larger farms.

In the second season, reported in Table 18.10, the first column shows that the mean total efficiency was a little lower at 0.50 (as compared with 0.64 for the stochastic model). Change is caused by the greater number of adopters, who have a slightly higher total efficiency than before (compare 0.62 with 0.66), whereas the level for non-adopters is almost unchanged (compare 0.46 with 0.45). Again, the adopters have a higher minimum efficiency level of 0.19 and the minimum for the non-adopters (in this bad season) falls to only 0.05. The next two columns show that the pure technical efficiency level of the adopters is still slightly lower than for the non-adopters, but again much of their advantage is due to scale.

In this season, the frontier is defined by nine farms, seven of which are now adopters. These are scale efficient; however, the level of scale efficiency has decreased to 0.61 for the total sample. This

Table 18.9. Deterministic frontier results for 1998/99.

	Efficiency			Numer of frontier farms	Returns to scale		
	Total	Technical	Scale		IRS	CRS	DRS
Mean: total	0.56	0.78	0.71	6	82	6	1
Mean: non-adopters	0.45	0.84	0.53	5	67	5	1
Mean: adopters	0.62	0.77	0.81	1	15	1	0
Minimum efficiency: total	0.08	0.29	0.08				
Minimum efficiency: non-adopters	0.08	0.29	0.08				
Minimum efficiency: adopters	0.35	0.39	0.51				

IRS, increasing returns to scale; CRS, constant returns to scale; DRS, decreasing returns to scale.

Table 18.10. Deterministic frontier results for 1999/2000.

	Efficiency			Numer of frontier farms	Returns to scale		
	Total	Technical	Scale		IRS	CRS	DRS
Mean: total	0.50	0.83	0.61	9	79	9	3
Mean: non-adopters	0.46	0.84	0.56	2	31	2	0
Mean: adopters	0.66	0.79	0.84	7	48	7	3
Minimum efficiency: total	0.05	0.30	0.10				
Minimum efficiency: non-adopters	0.05	0.30	0.10				
Minimum efficiency: adopters	0.19	0.50	0.19				

Maximum values are all unity
IRS, increasing returns to scale; CRS, constant returns to scale; DRS, decreasing returns to scale.

is reflected in the returns to scale results, which as well as showing more optimal size arms also show that three farms are actually too big. However, the dominant problem is still that 79% of the farms are too small. This shows that the higher yields of the smaller farms do not mean they are more efficient. Land scarcity is not the main problem in KwaZulu-Natal, so to look only at land efficiency makes little sense. Providing credit to allow farmers to adopt *Bt* cotton and increase the area planted is the obvious policy prescription.

Lastly, the 2 years can be compared by pooling the data and re-constructing frontier. Then, there are five farms from the first season on the joint frontier, only one of which is an adopter. However, despite the poorer season in the second year, there are an additional three farms on the frontier, all of which are adopters. This confirms the results of the yield and margin calculations, that all showed that the *Bt* users fared much better than the non-adopters in the wet season.

Income Distribution and Inequality

The introduction of new technologies can have adverse effects on the distribution of income, as the voluminous literature on the green revolution showed. The farmers who have the resources to adopt may become richer, thus increasing inequality, even if the non-adopters do not suffer a reduction in income. If they are disadvantaged and actually lose land to the better off, this situation is exacerbated and their income levels may actually fall. In this case, data is available from the first year of adoption, when very few farmers used the new technology. Thus, 1998/99 can serve as a benchmark for tracking the changes in the distribution of land and incomes that result from the introduction of the *Bt* variety.

The measures of inequality used are the Gini coefficient and the Lorenz curve. The Gini is defined as the ratio of the area between a Lorenz curve and the diagonal and the total area under the diagonal, where the Lorenz curve is the cumulative shares of income/wealth attributable to proportions of the population. The Gini coefficient is bounded by zero and one, where zero identifies absolute equality in the distribution of income/wealth and one represents absolute inequality.

The Lorenz curve for the distribution of household per capita income is shown in Fig. 18.16. The cumulative percentage distribution of per capita income is measured on the vertical axis and that of the population on the horizontal axis. The greater the area between the Lorenz curve and the diagonal, the greater is the level of inequality. The Gini coefficient can be stated: Gini = $A/(A+B)$. For this sample, the Gini has a value of 0.484 for 1998/99 and 0.478 for 1999/2000, suggesting that the per capita distribution of income from cotton in this area is about as unequal as the distribution of per capita incomes in the Western European countries.

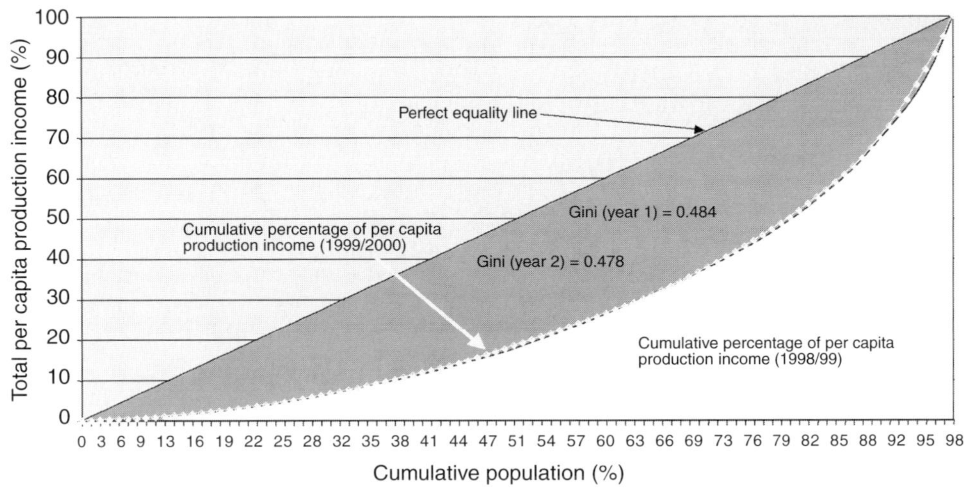

Fig. 18.16. Distribution of per capita production income (years 1 and 2).

The distribution becomes slightly less unequal in the second season, so there is no evidence to suggest that the gains are biased in favour of the better-off households. However, this is still a very early stage in the adoption process, so although there is no cause for concern at present, the results for the 2000/01 season will give a better clue as to the longer-term consequences of the new technology on equality.

Conclusions

The results of this survey of 100 smallholders in the Makhathini Flats region of KwaZulu-Natal give cause for cautious optimism regarding the impacts of *Bt* cotton. The farmers who adopted the *Bt*-cotton variety benefited from the new technology, according to all the measures used. Average yield per hectare and per kilogram of seed was higher for adopters than for the non-adopters. The increase in yields and reduction in chemical application cost outweighed the higher seed cost, so that gross margins were also considerably higher for adopters in the second season. This was a bad year, due to unusually heavy rainfall, and the *Bt* adopters suffered far less of a fall in yields than those who did not adopt.

Both yields and gross margins are useful, but they are partial measures of efficiency, which fail to take account of major inputs such as labour. Thus, they are supplemented by efficiency frontiers, which consider the efficiency with which all inputs are converted into outputs, using only the more reliable input and output quantity data and avoiding prices, which are less well recorded or simply non-existent. Both deterministic and stochastic frontiers were used, the first applying programming techniques and the second relying on econometric estimation. In either case, the results confirm the farm accounting results, showing that the *Bt* cotton adopters were considerably more efficient than those who used the non-*Bt* varieties.

The analysis of the characteristics of adopters and their reasons for adoption was less satisfactory. The tendency was for the older, more experienced farmers and those with larger farms to have higher percentages of adopters. This can be explained by the fact that these were the farmers who were more likely to be granted credit, or be able to finance the higher seed costs from savings or from other income sources. Indeed, almost all in the sample said they would adopt *Bt* cotton if they had the financial resources to do so. There did not appear to be any agronomic or other technical impediments to adoption. All smallholders could benefit, provided that credit is made available.

Typical rural farming constraints hold in this area and farmers require access to input and output markets. Although the local co-operative (VUNISA) plays an important role in facilitating farmers' access to the required markets, the company is the sole supplier of formal credit and production inputs to these smallholders, giving the company a monopolistic advantage.

In conclusion, the comparative analysis and econometric modelling results confirm that *Bt* cotton adopters experience significant benefits from the new technology. However, technological success does not necessarily imply high rates of adoption and overall financial success. Given the short study period (2 years), no definitive conclusions can be drawn about the adoption dynamics in the region. It will be interesting to survey the farmers again over two more production years. Some farmers may have decided to return to non-GM varieties in the long run, if the seed suppliers decide to appropriate a greater share of the benefits by raising their prices. If this were to happen, it would be unfortunate, as this pilot study shows that there are very clear gains at the farm level. However, in any adoption studies of GM crops, the exclusion of some farmers due to the inability to meet the initial higher seed costs and the resulting impact on income inequality must be a crucial part of the analysis.

Acknowledgements

This study could have not been undertaken without the substantial support of several individuals and organizations. We would like to thank the University of Reading for the providing the initial financial assistance, the University of Pretoria for providing all the necessary facilities and human resources to conduct the survey and Imperial College for the analysis of the data and production of this report.

Marnus Gouse warrants our appreciation for helping to contact the relevant authorities in South Africa and helping to conduct the research. Without him, the survey would have been much harder to conduct. Our thanks go to Lwandle and Knassi Mqadi for acting as translators during the survey.

Dr Stephen Morse (Department of Geography) and Dr Richard Bennett (Department of Agricultural and Food Economics) at the University of Reading deserve our gratitude for providing valuable comments, ideas and assistance in designing the questionnaires.

We are indebted to Professor Johann Kirsten for allowing us to conduct this research at the University of Pretoria. Ferdi Meyer, Dr Ravine Poonyth and all the staff at the University of Pretoria are thanked for their support during the survey, and special thanks to Muffy Kock for comments prior to the start of the survey.

We are obliged to all the farmers interviewed and to Patrick Nene and his staff at VUNISA Extension Office in Makhathini, for their interest, patience and the insights they shared during the survey.

Finally we would like to thank Monsanto, VUNISA, Cotton SA and Innovation Biotechnology for providing the necessary links and information and for trusting us, despite their well-founded fears that that strangers asking questions about GM crops are most likely to be Luddites.

References

Aigner, D., Lovell, C.A. and Schmidt, P. (1977) Formulation and estimation of stochastic frontier models. *Journal of Econometrics* 6, 21–37.

Battese, G. and Coelli, T. (1995) A model for technical inefficiency effects in a stochastic frontier production function for panel data. *Empirical Economics* 20, 325–332.

Fare, R., Grosskopf, S. and Lovell, C.A.K. (1985) *The Measurement of Efficiency of Production*, Kluwer-Nijhoff, Boston.

Farrell, M.J. (1957) The measurement of productive efficiency. *Journal of the Royal Statistical Society A* 120, part 3, 253–281.

Fernandez-Cornejo, J. and Klotz-Ingram, C. (1998) Economic, environmental and policy impact of using genetically modified crops for pest management. Selected Papers presented at the 1998 NEREA Meetings. Ithaca, New York, 22–23 June.

Fernandez-Cornejo, J., Klotz-Ingram, C., Jans, S. and McBride, W. (1999) Farm level production effects related to adoption of genetically modified cotton for pest management. http://www.agbioforum.org/vol2no2/klotz.html

Genetic Modified Organism Act, Act 15 of 1997 (1997) Department of Justice, South African Government, South Africa.

Gianessi, L.P and Carpenter, J.E. (1999) *Agricultural Biotechnology: Insect Control Benefits*. National Centre for Food and Agricultural Policy, Washington, DC.

International Fund for Agricultural Development (2001) *Rural Poverty Report 2001*. IFAD, Rome.

James, C. (1999) *Global Review of Commercialised Transgenic Crops:* 1998, No. 9. ISAAA Publication, John Innes Centre, Norwich, UK.

Kock, M. (2000a) Case study: the introduction of GM cotton in South Africa. Paper presented at the ICGEB Workshop: Biosafety 2: Advanced Research and Procedures: Case Studies for Designated Expert.

Kock, M. (2000b) Personal Communication. KwaZulu-Natal Annual Report. (1998). Website: http://agriculture.kzntl.gov.za/publications/annual_report/1998/ner.htm

Kodde, D. and Palm, F. (1986) Wald criteria for jointly testing equality and inequality restrictions. *Econometrica* 54, 1243–1248.

Piesse, J., Thirtle, C., Sartorius von Bach, H. and Van Zyl, J. (1996) Agricultural efficiency in the former South African homelands: measurement and implications. *Development Southern Africa*, 13, 399–414.

Pray, C.E., Ma, D., Huang, J. and Qiao, F. (2000) Impact of *Bt* cotton in China. Paper presented at the Agricultural Economics Society Annual Conference, Manchester, April. (Published in *World Development* 29, May 2001.)

Thompson, A.J. (1999) The genetically modified food debate in South Africa. University of Cape Town Publication. Website: http://www.uct.ac.za/microbiology/gmos.html

Tripp, R. (2000) Can biotechnology reach the poor? The adequacy of information and seed delivery. Paper presented at the 4th International Conference on the Economics of Agricultural Biotechnology, Ravello, Italy, 24–28 August.

Chapter 19
Income and Employment Effects of Transgenic Herbicide-resistant Cassava in Colombia: a Preliminary Simulation

Douglas Pachico,[1] Zully Escobar,[2] Libardo Rivas,[1] Veronica Gottret[1] and Salomon Perez[2]
[1]CIAT (International Center for Tropical Agriculture), Apartado aereo 6713, Cali, Colombia; [2]Universidad del Valle, Cali, Colombia

Introduction

This study makes an economic comparison of the development of a transgenic herbicide-resistant cassava in Colombia with current technology and with two alternative strategies for increasing cassava productivity: improved yield potential through conventional breeding, and the mechanization of cassava planting and harvest. Cassava growers generally constitute some of the poorest of the rural poor in some of the most disadvantaged regions of the low-income tropical countries, including Colombia (Henry and Gottret, 1996). Typically there are few other crop alternatives in the low rainfall regions with poor soils where most of cassava is grown. Cassava farmers critically need cost reducing technology to keep cassava as a competitive food in markets where consumers increasingly have many other alternatives, including other cheap food staples that can include imports as the Colombian economy opens further to economic globalization. Without more productive, lower-cost cassava that could compete in the market, the income and employment prospects of cassava producers will be highly circumscribed.

The potential economic benefits of three strategies for improving cassava productivity are assessed here: transgenic herbicide resistance, improved yield potential through conventional breeding, and mechanization of cassava planting and harvesting. Due to its slow establishment and long growing period, weed control is one of the major costs of cassava production. Moreover, since weeding is done almost exclusively by hand, seasonal labour bottlenecks are a critical constraint on any expansion of cassava production. It has been suggested that a particularly promising avenue of reducing the cost of weed control in cassava, thereby opening up vast new income opportunities for poor cassava growers, would be the introduction of herbicide tolerance into cassava. This would permit low-cost herbicides to be substituted for expensive manual weed control.

Transgenic herbicide-tolerant crops are now being cultivated on a massive scale in temperate agricultural systems. In 1999 this included 21.6 million ha of soybean, 3.6 million ha of maize, 3.5 million ha of canola–rapeseed and 2.4 million ha of cotton (James, 2000). Previous studies have assessed the impacts of transgenic crops (Pray *et al.*, 2000). Whether a similar transgenic herbicide-tolerant cassava for the tropics should be developed depends upon a number of considerations, among them its potential contribution to food security, any environmental risks and any potential hazards to human health. This chapter endeavours to make an initial assessment of the potential income and employment impacts of transgenic herbicide-resistant cassava for the case of the north coast region of Colombia.

Such an approach is necessarily partial. For example, the environmental consequences of increased herbicide use are not specifically analysed. These could include potential negative effects of increased use of toxic herbicide and potential positive effects that could result from reduced soil losses to erosion because the substitution of herbicide for manual weeding would reduce soil disturbance. Likewise, though there are no demonstrated human health risks from transgenic herbicide-resistant crops, neither has it been proven that there are no potential health risks. Other potential environmental risks would have to be considered in a complete analysis, such as gene flow from transgenic cassava to wild South American relatives that could create herbicide-resistant plants, or the potential emergence of herbicide tolerance in weeds due to increased levels of exposure to herbicide over time. Thus, the findings of this chapter are limited to assessing some potential economic consequences of transgenic herbicide resistance in cassava, and as such, this chapter cannot provide a complete basis for assessing whether or not transgenic cassava should be deployed.

There are important alternatives to improving transgenic cassava through other approaches besides transgenics (Plucknett *et al.*, 2000). Improved yield potential through conventional breeding has had an immense impact on the productivity of other tropical crops (Evenson and Gollin, 2001), and this is a strategy for cassava improvement that has been followed for some time (Cock, 1985). The deployment of such technology has not raised the same level of human health and environmental concerns as has been the case with transgenic crops. Nevertheless varieties derived from conventional breeding can also have environmental affects such as higher nutrient demands from the soil or induced changes in natural pest populations.

Improved labour productivity through mechanization has been a major development path for agriculture in the high-income temperate countries, and is likewise spreading in the tropics. Mechanization of cassava sowing and harvesting requires the development of appropriate machinery, essentially adapted from similar machinery for potatoes. It also requires the development through conventional breeding of upright erect cassava with little branching, in order to produce cassava stakes for planting that are straight and even enough for mechanized sowing. Mechanized rather than manual sowing regularizes the distances at which cassava is planted thereby making mechanized harvesting also possible. While the combination of mechanized sowing and harvesting is a potentially feasible way of lowering costs, and increasing productivity, including the returns to labour, there are concerns about labour-displacing technologies exacerbating unemployment. In Colombia, for example, unemployment is already in the order of 20% in the urban economy and even higher in rural areas. Consequently, the economic analysis in this chapter will look not only at economic surplus but also at changes in employment associated with the three technologies.

An economic model based on surplus analysis is used in this paper to estimate the changes in equilibrium output and prices and consumer and producer benefits that would accrue from alternative changes in cassava technology in Colombia. This chapter will proceed by first briefly reviewing the model used here. The data sources are briefly noted. Differences in the costs of cassava production of with the different technologies are considered. Next, consumer and producer surplus are estimated for the alternative technologies, and differences in employment are also estimated. Conclusions will suggest some possible further extensions of this research.

Model

The Dynamic Research Evaluation Model (DREAM) is used in this chapter. The theory underlying this model has been described in detail (Alston *et al.*, 1995) and a user manual is available (Wood and Baixt, 1998). This model is similar to the MODEXC model (Rivas *et al.*, 1999), but offers greater ease in handling multiple regions.

For region i and year t the model specifies linear equations of supply and demand

$$Q_{it} = \alpha_{it} + \beta PP_{it} \quad (1)$$
$$C_{it} = \gamma_{it} + \delta_i PC_{it} \quad (2)$$

where Q_{it} is the quantity produced in region i in period t; PP_{it} is the producer price in region i in period t; C_{it} is the quantity consumed in region i in period t; and PC_{it} is the consumer price in region i in period t.

Other parameters are defined as follows:

$$\beta_{i0} = \varepsilon_{i0}Q_{i0} / PP_{i0} \quad (3)$$
$$\alpha_{i0} = (1-\varepsilon_{i0})Q_{i0} \quad (4)$$
$$\delta_{i0} = \eta_{i0}C_{i0} / PC_{i0} \quad (5)$$
$$\gamma_{i0} = (1-\eta_{i0})C_{i0} \quad (6)$$

where subscript 0 refers to observed values in the initial time period; ε_i is the elasticity of supply; and η_i is the elasticity of demand.

Technical change is modelled through a shift in the supply function

$$K_{it} = c_i A_{it} PP_{i0} \quad (7)$$

where c_i is the reduction in unit costs in region i and A_{it} is the percentage of farmers adopting the cost reducing technology.

Labour employment in the absence of technical change is defined as:

$$E_{itl} = Q_{it}L_{itl} \quad (8)$$

where E_{it} is the level of employment in period t and L_{itl} is the amount of labour per tonne of production with existing technology, so that the amount of labour employed with technical change is E_{itm}.

$$E_{itm} = Q_{it}(1-A_{it})L_{itl} + (Q_{it}A_{it})L_{itm} \quad (9)$$

where L_{itm} is labour use per tonne of production with improved technology.

Data

All the data for this model are from Colombia. Prices and quantities are taken from various government sources (reported in Perez, 2001). Production quantities are reported for six different regions. Costs of production for cassava in the six regions were developed through a combination of secondary sources and key informant interviews (Perez, 2001). Changes in production from the potential new technologies were developed based on interviews with cassava scientists. Elasticities and rates of adoption are from Escobar (2001).

Costs of Cassava Production

Table 19.1 presents costs of production per hectare for cassava under current technology and for three potential alternative technologies for the north coast region of Colombia. Similar costs of production for all four technologies were developed for five other regions (Perez, 2001), but for brevity, only the costs for the north coast are reported here. All costs were originally calculated in Colombian pesos, but are here converted into US dollars at the exchange rate prevailing at the time of the study.

Total costs per hectare are greatest with high yielding varieties developed through conventional breeding, \$815 ha^{-1}, compared with \$592 ha^{-1} with current technology. The largest part of this increased cost

Table 19.1. Cost of cassava production (US\$ ha^{-1})with different technologies in Colombian north coast.

Costs	Current technology	High-yield varieties	Herbicide-resistant varieties	Mechanized planting and harvesting
Land preparation	67	67	14	67
Seed and planting	50	152	50	24
Fertilization	49	49	49	49
Weed control	112	112	11	112
Pest and disease control	—	22	—	22
Harvest	119	158	127	65
Financial cost[b]	143	202	90	122
Land rental	53	53	53	53
Total cost	592	815	429	514
Yield (million t ha^{-1})	15.0	23.0	16.5	17.0
Unit cost (US\$ million t^{-1})	39.5	35.4	26.0	30.2
Cost reduction (%)	—	10.2	34.1	23.3
Number of work days ha^{-1}	62	70	46	39

[a]Prices of 2000. Exchange rate: \$2087.9.
[b]Financial cost equivalent to 36% of total direct cost.
Source: Perez (2001).

is more expensive seed that needs to be purchased in order to obtain the new high-yielding variety. In addition, because yield per hectare increases, there is a need for greater labour at harvest. The higher yields, 23 t ha^{-1} compared to 15 t ha^{-1} with current technology, more than compensate for the increased cost per hectare so that the cost of production per tonne falls from \$39.50 with current technology to \$35.40 with the high-yield variety. This represents a 10.2% decrease in unit costs, and this figure is used as c_i to estimate the shift in the supply function caused by new technology as shown in Equation (7). Due to greater labour at harvest, the number of days of labour needed per hectare rises from 62 with current technology to 70 with higher-yielding varieties. However, in terms of labour per tonne of cassava produced, used in Equations (8) and (9) to estimate the employment effects of the alternative technologies, this falls from 4.13 days t^{-1} with current technology to 3.04 days t^{-1} with higher yielding varieties from conventional breeding.

Herbicide-resistant transgenic varieties would reduce the costs of land preparation and weed control. Weed control costs would fall because the use of relatively inexpensive herbicides would substitute for the intensive use of labour for manual weed control. Land preparation costs would also fall substantially because the herbicide-resistant transgenic varieties would permit the use of minimum tillage. Consequently the costs per hectare of the herbicide resistant varieties would be \$429 compared to \$592 with current technology. The costs of production per tonne would fall 34.1%, from \$35.40 with current technology to \$26.00 with the transgenic herbicide-resistant varieties. Because this cost reduction is achieved principally through a reduction in manual weed control, labour per hectare falls substantially with this technology, 46 days ha^{-1} with transgenic herbicide-resistant cassava compared to 62 days ha^{-1} with current technology.

The third potential new technology appraised here is mechanized sowing and harvest of cassava. The major part of the cost saving from this technology comes from reduced harvest costs, which drop from \$119 ha^{-1} with current technology to \$65 ha^{-1} when mechanized. Planting costs with mechanization fall to \$24 ha^{-1} compared to \$50 ha^{-1} with current technology. Effective yields are also expected to rise slightly with mechanized harvesting which is anticipated to be more thorough than manual harvesting. Total costs per hectare with mechanization are $514 compared with $592 with current technology, while costs per tonne, which provide the estimate of c_i which shifts the supply function, falls 23.3% to $30.20 from $39.50 with current technology.

Similar costs of production have been developed for all four technologies for five other cassava-production regions in Colombia (Perez, 2001). Table 19.2 presents the percentage reduction in unit costs for the three potential new technologies compared to current technology for six cassava-producing regions in Colombia. As noted above, these provide the estimates of c_i, which drive the shift in the supply function in Equation (7). Because the structure of production costs and yields differ among the six regions, the changes in percentage cost reduction also vary among regions for the three technologies. These differences notwithstanding, the transgenic herbicide-resistant cassava leads to the largest cost reduction in all regions. High-yield varieties lead to greater cost reductions than mechanization in the more favoured production zones – the eastern plains, the coffee zone – while in the less favoured regions, mechanization leads to a greater cost reduction than improved yield potential.

Economic Surplus and Employment Effects of New Technologies

Based on the model described above, economic surplus to producers and consumers due to the supply-shifting effect of new technologies was calculated along with the estimated total amount of labour in cassava production. Economic surplus benefits are presented in Table 19.3.

Consistent with the greater unit cost reduction, the transgenic herbicide-resistant cassava yields the greatest total surplus benefits of the three technologies, estimated at a present value of $508 million at a 5% discount rate over the period

Table 19.2. Estimated cost reduction of cassava production in Colombia compared to current technology (%).

	High-yield varieties	Herbicide-resistant varieties	Mechanized planting and harvesting
North coast	10.2	34.1	23.3
Eastern plains	24.7	28.0	21.2
Coffee zone	35.3	42.5	0.0
Cauca – Valle	16.7	25.0	24.8
Huila – Tolima	29.7	41.7	32.8
Santander	6.5	39.6	9.2

Table 19.3. Estimated present value of benefits 2002–2016 from improved cassava production technologies in Colombia (US$ millions; 5% discount rate).

	High-yield varieties	Herbicide-resistant varieties	Mechanized planting and harvesting
North coast	18.0	93.7	48.1
Eastern plains	66.6	59.4	32.9
Coffee zone	19.7	21.0	−1.9
Cauca – Valle	10.6	12.0	10.5
Huila – Tolima	18.8	23.7	13.0
Santander	11.1	118.6	9.9
Others	−17.4	−43.2	−17.1
Consumers	88.1	222.9	86.4
Total	215.4	508.1	181.9

2002–2016. High-yield varieties from conventional breeding would produce a present value of benefits of $215 million while mechanization would produce economic surplus benefits of $182 million. In all cases, consumers would receive approximately 40% so that for consumers, the transgenic herbicide-resistant cassava would result in the largest benefits, some $223 million, while consumer benefits from high-yielding cassava or mechanized planting and harvesting are essentially the same at $88 million and $86 million, respectively.

Among producers, seven groups are shown: six are the regions that would benefit from the potential new technologies while the seventh region is comprised of cassava producers elsewhere in Colombia who are assumed not to be able to utilize or adopt the new technologies. Necessarily, these non-adopting farmers scattered elsewhere in Colombia would be net losers from any of the technical changes because their competitors would be benefiting from cost-reducing technology while they retained their current cost structures. Transgenic herbicide-resistant varieties would lead to the greatest level of benefits from all three technologies except in the favoured region of the eastern plains which would gain most from improved yield potential from conventional breeding.

Estimated employment in cassava production in 2016 for the four technologies in the six adopting regions is presented in Table 19.4. In all cases employment would be greatest with current technology, though it must be stressed that this greater level of employment would come at the cost of forfeiting all the producer and consumer surplus benefits found in Table 19.3. It is hardly surprising that employment is estimated to fall with transgenic herbicide-resistant cassava which would displace labour from land preparation and manual weed control. Likewise it is to be expected that mechanization of planting and harvesting, which would substitute for labour in both these activities, would also reduce employment compared to current technology. Between the two, transgenic herbicide resistance would displace more labour from cassava production than would mechanized planting and harvesting.

Perhaps surprisingly, high-yielding cassava varieties would be similarly displacing of labour and this is estimated to lead to actually a lower level of employment than mechanized planting and harvesting, even though labour use per hectare would rise with the high-yielding cassava due to increased labour at harvest. This occurs because labour use per tonne of cassava production would fall with the high yielding cassava as shown in Table 19.1. Thus, with high-yielding cassava the area needed in cassava production to supply the market would fall and this effect would overwhelm the small increase in per hectare employment with high-yielding technology.

Thus, any of the proposed new production technologies would lead to benefits to producers and consumers, but a lower level of employment. Although lower employment

Table 19.4. Estimated employment in cassava production in Colombia under different technologies in 2016 (thousands of days).

	Current technology	High-yield varieties	Herbicide-resistant varieties	Mechanized planting and harvesting
North coast	4168	3383	3339	3351
Eastern plains	1118	1009	901	974
Coffee zone	269	228	249	269
Cauca – Valle	273	257	225	225
Huila – Tolima	236	220	194	209
Santander	2263	1738	1368	1881
Total	8327	6835	6276	6909

in cassava production might appear to risk jeopardizing the welfare of workers in an economy already characterized by high unemployment, labour productivity, and thus in principle wages, would rise with any of the new technologies. Economic development cannot occur without increases in labour productivity, and in fact increased labour productivity is crucial to improved welfare. Hence, though employment would fall in cassava production, this is an almost essential characteristic of improved labour productivity. The welfare issue should perhaps be seen as the overall macroeconomic performance of the economy with respect to employment creation instead of a matter that can be resolved through a single line of production like cassava.

Conclusions and Need for Further Research

This study has compared the potential benefits from different technologies to improve the productivity of cassava in Colombia. The study shows that transgenic herbicide-resistant cassava would lead to substantially greater producer and consumer benefits than would higher-yielding varieties through conventional breeding or the mechanization of cassava in Colombia. Compared with current technology, all the technical innovations would lead to a reduction in employment in cassava, more or less on the same scale though the herbicide-resistant cassava is the most labour-displacing of the alternatives considered.

These results at the national level mask important differences among regions. For example, labour is a particular bottleneck in the north coast so that the returns to herbicide-resistant cassava and to mechanization are especially high in that region (Table 19.3). In contrast, in the coffee zone, mechanization is not a feasible alternative due to topography and improved yield results in a high share of the benefits. The coffee zone would obtain about 10% of the national benefits from high-yield varieties compared to around 4% of the benefits from herbicide resistance. Because of differences in resource endowments and cost structures between the different regions, no single technology is clearly more attractive in all regions. This suggests that research should aim to produce a variety of technical options that will have differing impacts in different regions.

An important limitation of this study is that it only attempts to estimate the benefits of alternative technologies. A more complete analysis would have to include the different costs of developing the technical innovations, including the amount of time required to develop the technologies and the differing probabilities of success in achieving the different innovations.

Moreover, in addition to considering development costs as well as potential economic benefits, a full analysis of these options would have to also include wider environmental and health issues. For example, the herbicide-resistant cassava would lead to a different pattern in herbicide use which could have negative consequences, but by reducing the frequency of soil tillage it would contribute to reducing the risks of soil erosion. Likewise, high-yielding varieties are certainly going to be more demanding of soil nutrients, and this could risk soil depletion or lead to a greater use of chemical fertilizers which might have undesired secondary effects.

Heightened awareness of the need to include these environmental and potential health consequences in an assessment of transgenic crops has led to the development of new regulatory systems and requirements for transgenic crops. This is embodied both in international convention and national regulatory systems (Convention on Biodiversity, 2000; US Department of State 2000). These regulations require, for example, assessment of possible effects on organisms in the environment; potential to become a weed; potential allergenicity and digestibility. These assessments have not been a requirement for genetic modifications achieved through conventional breeding. Therefore, the high-yielding cassava and the cassava more suitable for mechanization considered in this chapter would not be subject to regulatory review before their use, but considerable research on these environmental and health issues would be required before a transgenic herbicide-resistant cassava could be released into the environment. This in turn would raise the total research costs to the development of transgenic cassava. Clearly, these further issues of costs and environmental and health risks need to be incorporated before drawing conclusions about the relative desirability of the different cassava innovations considered in this chapter.

Acknowledgements

Bernardo Ospina and Hernan Ceballos provided invaluable help in obtaining the data needed for this research. This research was conducted with the financial support of the Consultative Group on International Agricultural Research (CGIAR) and the Latin America Consortium for Cassava Research (CLAYUCA). Neither CIAT, the CGIAR or CLAYUCA are responsible for the interpretations, findings or contents of this chapter which are the sole responsibility of the authors.

References

Alston, J.M., Norton, G.W. and Pardey, P.G. (1995) *Science Under Scarcity.* Cornell University Press, Ithaca, New York.

Convention on Biological Diversity (2000) *Cartegena Protocol on Biosafety.* http://www.biodiv.org/biosafety/protocol

Cock, J. (1985) *Cassava: New Potential for a Neglected Crop.* Westview Press, Boulder, Colorado.

Escobar, Z. (2001) *Analisis sectorial y microeconomico del impacto de la introduccion de cambio tecnologico en la produccion de yuca en Colombia.* Thesis, Universidad del Valle, Cali, Colombia.

Evenson, R.E. and Gollin, D. (eds) (2001) *Crop Improvements and its Effects on Productivity: the Impact of International Agricultural Research.* CAB International, Wallingford, UK.

Henry, G. and Gottret, M.V. (1996) *Global Cassava Trends: Reassessing the Crop's Future.* CIAT, Cali, Colombia.

James, C. (2000) *Global Status of Commercialized Transgenic Crops.* ISAAA, Ithaca, New York.

Perez, S. (2001) *Analisis de la competitividad en la produccion de yuca en Colombia.* Thesis, Universidad del Valle, Cali, Colombia.

Plucknett, D.L., Phillips, T.P. and Kagbo, R.B. (2000) *A Global Development Strategy for Cassava: Transforming a Traditional Tropical Root Crop.* FAO-IFAD, Rome, Italy.

Pray, C.E., Ma, D., Huang, J. and Quia, F. (2000) *Impact of* Bt *cotton in China.* Presented at International Conference on the Economics of Agricultural Biotechnology, Ravello, Italy, August 24–28.

Rivas, L., Garcia, J.A., Seré, C., Jarvis, L.S., Sanint, L.R. and Pachico, D. (1999) *MODEXC Release 4.1-A Friendly Computer Model.* CIAT, Cali, Colombia.

US Department of State (2000) *Food Safety-Regulating Plant Agricultural Biotechnology in the U.S.* http://usinfo.state.gov/topical/global/biotech

Wood, S. and Baitz, W. (1998) *DREAM: Manual Para el Usuario.* IICA-IFPRI-CIAT, San Jose, Costa Rica.

Chapter 20

Estimating the Economic Effects of GMOs: the Importance of Policy Choices and Preferences

Kym Anderson,[1] Chantal Pohl Nielsen[2] and Sherman Robinson[3]
[1]*Centre for International Economic Studies, University of Adelaide, Australia;*
[2]*Danish Research Institute of Food Economics, and University of Copenhagen, Denmark;* [3]*International Food Policy Research Institute, 2033 K Street, NW, Washington, DC 20006, USA*

Introduction

Virtually all new technologies, even when they unambiguously benefit the vast majority of society, are opposed by at least a few people. The new agricultural biotechnologies that are generating transgenic or genetically modified organisms (GMOs), however, are attracting an exceptionally large degree of opposition to their production and trade. Both environmental and food safety concerns have been raised by opponents to the development of transgenic or genetically modified (GM) crops. The vast majority of opponents want at least to have labels on products that may contain GMOs, while the most extreme of them (particularly in Western Europe) want to see GM crops totally excluded from production and consumption in their country. This extreme view contrasts with the more relaxed attitude towards the use of GMOs in pharmaceuticals, and swamps discussions of the positive attributes of the new technology.

Also associated with that view is the idea that we should not try to measure the economic and other effects of GMOs because there is too much uncertainty surrounding the technology. We beg to differ with the latter sentiment, believing that without attempts to quantify the economic effects of GMOs, opinion formation and policy making would be even less well informed because it would have to depend even more on guesswork.

To illustrate the usefulness of quantitative models for informing GMO debates, the present chapter draws on three recent studies by the authors that use existing empirical models of the global economy to examine what the effects of some (non-European) countries adopting the new GMO technology might be without and then with some policy or consumer preference responses.[1] Specifically, the effects of an assumed degree of GMO-induced productivity growth in selected countries are explored for cotton (less controversial because it is not a food), rice (next-least

[1] Nielsen and Anderson (2000b,c) and Nielsen *et al.* (2000).

(eds R.E. Evenson, V. Santaniello and D. Zilberman)

controversial because it is mostly consumed in developing countries), and maize plus soybean (controversial because they are grown extensively in rich countries and are consumed by people there both directly and via animal products). In the maize/soybean case the results are compared with what they would be if: (i) Western Europe chose to ban consumption and hence imports of those products from countries adopting GM technology; or (ii) some Western European consumers responded by boycotting imported GM foods. The final section discusses areas where future empirical work of this sort might focus.

Estimating Economic Effects of GMO Adoption using the GTAP Model

The apparent differences in preferences and views on environmental issues and consumers' right to know about food ingredients are unlikely to disappear in the foreseeable future. The extent to which that could lead to trade disputes depends heavily on the directions and magnitudes of the production, trade and welfare consequences of different responses to the technology by different countries. Theory alone is incapable of determining even the likely direction, let alone the magnitude, of some of the effects of those various responses to GMOs. Hence an empirical modelling approach is called for, to estimate the size of various assumed productivity changes and any associated policy changes and consumer responses. What follows is a summary of some early attempts at doing that for cotton, rice, maize and soybean.

Specifically, this section examines empirically the production, trade and welfare effects of GM crop adoption by selected regions, first without and then with specific policy and consumer responses in Western Europe. This is done mainly using an applied analytical framework involving a global economy-wide model and database known as GTAP (Global Trade Analysis Project).[2] Being a general equilibrium model, GTAP describes both the vertical and horizontal linkages between all product markets both within the model's individual countries and regions as well as between countries and regions via their bilateral trade flows. The database used for all but the last of these applications reflects the global economic structures and trade flows of 1995, and has been aggregated to 16 regions to highlight the main participants in the GMO debate and other key interest groups, and 17 sectors with focus placed on the primary agricultural sectors affected by the GMO debate and their related processing industries.

Currently it is primarily cotton, maize and soybean that are benefiting most from GM technology, although rice is considered to have a similar potential in the near future. Hence the scenarios analysed here assume that GM-driven productivity growth occurs only in the following GTAP sectors and for a subset of countries: plant-based fibres (of which cotton is by far the most important in the countries considered), paddy rice, coarse grain other than wheat and rice (primarily maize) and oilseeds

[2] The GTAP (Global Trade Analysis Project) model is a multi-regional, static, applied general equilibrium model based on neo-classical microeconomic theory. See Hertel (1997) for comprehensive documentation. Markets are characterized by perfect competition, the technology of the profit-maximizing producers exhibit constant returns to scale, and substitution between intermediate inputs is possible. The behaviour of the utility-maximizing consumers is captured in a non-homothetic private demand system. Capital and labour are perfectly mobile between sectors, whilst the total supply of factors of production is fixed within each region. Land is limited to use in the primary agricultural sectors and shifts among these are determined by transformation elasticities. International trade is described by an Armington (1969) specification, which means that products are differentiated by country of origin. This enables a specification of the bilateral trade flows rather than simply net trade. Macro closure of the model is achieved by ensuring equilibrium between global savings and investment. The most recent database available for the model is for 1995. It provides a comprehensive description of the structures of the individual economies, the commodity and services trade between them, and the price and trade policies affecting those flows (Version 4, see McDougall *et al.*, 1998). In its full version the database comprises 50 sectors and 45 countries/regions.

(primarily soybean in the countries considered). Detailed empirical information about the impact of GMO technology in terms of reduced chemical use, higher yields and other agronomic improvements is at this stage quite limited (see e.g. OECD, 1999; Nelson *et al.*, 1999). Available empirical evidence (e.g. USDA, 1999; James, 1997, 1998) does, however, suggest that cultivating GM crops has general cost-reducing effects.[3] The scenarios analysed here are therefore based on a simplifying assumption that the effect of adopting GM crops can be captured by a Hicks–neutral technology shift, i.e. a uniform reduction in all inputs to obtain the same level of production. For present purposes the GM-adopting sectors are assumed to experience a one-off increase in total factor productivity of 5%, thus lowering the supply price of the GM crop to that extent.[4] Assuming sufficiently elastic demand conditions, the cost-reducing technology will lead to increased production and higher returns to the factors of production employed in the GM-adopting sector. Labour, capital and land consequently will be drawn into the affected sector. As suppliers of inputs and buyers of agricultural products, other sectors will also be affected by the use of genetic engineering in GM-potential sectors through vertical linkages. Input suppliers will initially experience lower demand because the production process in the GM sector has become more efficient. To the extent that the production of GM crops increases, however, the demand for inputs by producers of those crops may actually rise despite the input-reducing technology. Demanders of primary agricultural products such as grains and soybean meal for livestock feed or cotton for textiles will benefit from lower input prices, which in turn will affect the market competitiveness of livestock or textile products.

The widespread adoption of GM varieties in certain regions will affect international trade flows, depending on how traded the crop in question is and whether or not this trade is restricted specifically because of the GMOs involved. To the extent that trade is not further restricted and not currently subject to binding quantitative restrictions, world market prices for these products will have a tendency to decline and thus benefit regions that are net importers of these products. For exporters, the lower price may or may not boost their trade volume, depending on price elasticities in foreign markets. Welfare in the exporting countries would go down for non-adopters but could also go down for some adopters if the adverse terms of trade change were to be sufficiently strong. Hence the need for empirical analysis.

The limited experience with the use of GM crops in agricultural production means empirical information about the technology's impact in terms of reduced chemical use, higher yields and other agronomic or nutritional improvements is also rather limited. The empirical studies reported in this paper are therefore necessarily based on no more than assumptions about what the use of GM crops may do in selected sectors and regions. Even so, the analyses are able to highlight the principle economic and trade effects of the technology in different policy environments, and at the same time focus attention on the types of technological parameters needed to do such analysis with more precision.

The GTAP model and database has been aggregated as shown in Table 20.1 for all the scenarios in this section. Eight of the sectors are primary agricultural sectors and four are food-processing sectors.

In order to appreciate the relative importance of these primary agricultural

[3] Nelson et al. (1999), for example, suggest that glyphosate-resistant soybeans may generate a total production cost reduction of 5%, and their scenarios have Bt maize increasing yields by between 1.8% and 8.1%.

[4] Due to the absence of sufficiently detailed empirical data on the agronomic and hence economic impact of cultivating GM crops, the 5% productivity shock applied here represents an average shock (over both commodities and regions). Changing this shock (e.g. doubling it to 10%) generates near-linear changes (i.e. roughly a doubling) in the effects on prices and quantities.

Table 20.1. Regions and sectors used in the model analysis.

Regions		Sectors	
1. Australia and New Zealand	10. Southern Cone	1. Paddy rice	11. Meat and dairy products
2. Japan	11. Rest of Latin America	2. Wheat	12. Vegetable oils and fats
3. East Asian NICs[a]	12. Western Europe	3. Coarse grain	13. Processed rice
4. China	13. Eastern Europe and FSU	4. Vegetables, fruits and nuts	14. Other food products
5. Rest of East Asia	14. South Africa	5. Oilseeds	15. Textiles and clothing
6. India	15. Rest of sub-Saharan Africa	6. Plant-based fibres	16. Manufactures
7. Rest of South Asia	16. Middle East and North Africa	7. Other crops	17. Service
8. North America		8. Livestock	
9. Mexico		9. Forestry and fishing	
		10. Energy and minerals	

[a] Hong Kong, Singapore, South Korea and Taiwan.

sectors and their related processing sectors to the economies of different regions, note from Table 20.2 that plant-based fibre production is of greatest importance to the agricultural sectors of China, India and the poor regions of East and South Asia and sub-Saharan Africa. However, cotton textile and clothing production is very important in the manufacturing sector of many developing countries, so any reduction in the price of the input cotton will impact on such countries. Note also that paddy rice production accounts for between one-quarter and one-sixth of agricultural production in the Asian economies and one-fifth in sub-Saharan Africa, but is negligible in other regions. Coarse grains (particularly maize) and oilseeds (particularly soybean), by contrast, are of equal or greater importance to North American and Western European agriculture as they are to the farm sectors of most developing country regions. Also important to understand are the differing food consumption patterns across regions (Table 20.3) and the various regions' net trading situations in raw and processed forms and export dependence of these products (Table 20.4).

Scenario 1: selected regions adopt GM cotton

To model the effect of cotton farmers adopting GM seeds, Nielsen and Anderson (2000c) assume a 5% primary factor and intermediate input productivity shock applies to plant-based fibres crops in North America, the Southern Cone of Latin America, China, India and the rest of South Asia. As Table 20.5 shows, that shock would induce rather large increases in plant-fibre production in the first two regions, and slightly smaller increases in China and the rest of South Asia. This leads to the USA and developing Asia increasing their exports of plant-based fibres, which lowers the international indicator price of cotton by 4%. That in turn discourages cotton production in Africa where by assumption GMO technology is not adopted, but the cotton price fall also encourages the world's textile industry to expand. The resulting increase in textile and clothing production is largest in China, but it is close to zero in South Asia where exporters of the raw material evidently become relatively more internationally competitive than the textile industry there.

The decrease in price and increase in production of textiles and clothing are both very minor, however: Table 20.5 shows that the major importers of textiles and clothing, North America and Western Europe hardly change the volume of their imports. This is because textile and clothing trade is still severely restricted by 'voluntary' export restraints on key exporters as part of the Multifibre Arrangement. As a result, the estimated welfare gains from this technological boost, shown in the lower part of

Table 20.2. Agricultural and food production structures in selected regions, 1995.

	Japan	China	Rest of East Asia	India	Rest of South Asia	North America	Southern Cone	Western Europe	South Africa	Rest of sub-Saharan Africa
Share of GM-potential crop in agricultural production value (%)										
Paddy rice	25.9	16.8	26.1	18.1	18.1	0.4	3.5	0.2	0.1	19.8
Coarse grain	0.6	5.6	4.8	4.4	4.1	18.5	5.1	5.3	4.7	2.8
Oil seeds	0.1	2.0	2.9	13.9	12.9	7.1	8.0	1.5	0.9	6.4
Plant fibres	0.1	4.2	4.3	6.4	6.7	3.4	2.1	0.5	0.8	4.9
Share of GM-related food production in total value of food production (%)										
Vegetable oils and fats	11.8	11.3	13.4	6.1	6.9	32.8	30.7	36.7	35.3	7.0
Processed rice	7.8	22.0	27.0	3.8	8.3	0.3	3.4	0.4	0.0	7.9
Other foods	79.9	58.5	49.1	67.9	66.4	62.9	59.4	56.7	62.1	73.6

Source: GTAP database, Version 4, McDougall *et al.* (1998).

Table 20.3. Food consumption structures in various regions, 1995.

	Japan	China	Rest of East Asia	India	Rest of South Asia	North America	Southern Cone	Western Europe	South Africa	Rest of sub-Saharan Africa
Structure of selected foods in total food consumption (%)										
Paddy rice	0.0	7.2	1.4	17.9	15.7	0.0	0.5	0.0	0.0	17.6
Processed rice	6.3	7.1	19.5	0.0	1.0	0.2	3.9	0.6	0.1	2.2
Coarse grain	0.0	2.6	1.8	4.9	3.9	0.1	0.5	0.1	0.3	2.5
Oilseeds	0.0	0.1	0.8	6.7	5.4	0.1	0.8	0.1	0.1	3.0
Vegetable oil, fats	0.1	3.3	3.3	5.6	8.2	0.7	3.5	3.5	1.9	3.2
Livestock	1.0	23.6	6.3	19.8	15.6	1.3	3.7	3.2	3.1	13.0
Meat, dairy	11.7	3.5	9.5	1.3	2.6	25.6	26.2	32.6	29.3	3.2
Fish	2.4	6.1	11.8	5.3	5.3	1.1	1.4	2.1	0.0	4.6
Share of food in total value of private household consumption (%)	16.1	49.1	33.9	47.0	46.4	8.0	25.3	13.1	23.9	45.0

[a] For space reasons regions 1, 3, 9, 11, 13, and 16 of Table 20.1 are omitted from this table.
Source: GTAP database, Version 4, McDougall *et al.* (1998).

Table 20.4. Exports of selected products by various regions[b], 1995.

	Japan	China	Rest of East Asia	India	Rest of South Asia	North America	Southern Cone	Western Europe	South Africa	Rest of sub-Saharan Africa
(a) Net exports by various regions as a percentage of world exports of selected products										
Paddy rice (1)[a]	0.4	0.3	7.4	19.2	4.0	37.6	−1.8	−30.7	0.1	−7.3
Coarse grain (16)[a]	−19.3	−6.2	−3.5	0.1	−0.1	56.2	2.6	−2.3	0.5	−0.7
Oilseeds (13)[a]	−18.1	3.0	−3.7	1.1	−0.4	54.8	11.2	−39.5	−0.3	1.3
Plant fibres (10,817)	−7.0	−12.1	−16.1	−1.3	−1.1	33.2	−1.0	−15.4	−0.5	10.7
Vegetable oil, fats (20)[a]	−2.7	−10.2	26.8	−2.0	−6.0	2.8	14.9	−4.5	−1.3	−0.9
Processed rice (6)[a]	0.2	−7.9	18.0	19.0	1.1	8.3	−0.7	−4.5	−2.4	−11.3
Other foods (181)[a]	−12.4	1.4	3.1	0.9	0.1	−0.5	5.0	3.6	0.4	−0.1
Textiles and clothing (314)[a]	−5.4	8.7	3.6	2.6	2.4	−12.1	−0.5	−10.1	−0.1	−0.4
(b) Exports by regions as a percentage of their total production, selected products										
Paddy rice	0.0	0.0	0.4	0.6	0.4	22.1	0.1	17.1	27.2	0.0
Coarse grain	1.9	1.1	2.2	0.2	0.0	18.2	10.5	26.9	43.9	5.6
Oilseeds	0.8	9.2	4.4	0.9	0.7	36.9	13.1	26.8	29.2	4.3
Plant fibres	7.8	0.8	1.0	0.6	9.8	39.1	18.1	54.4	31.8	36.2
Vegetable oil, fats	1.0	2.9	47.2	4.5	1.1	8.2	25.6	11.9	4.8	9.4
Processed rice	0.2	0.2	15.5	97.1	54.0	40.4	3.1	23.5	56.3	0.2
Other foods	0.3	9.2	21.3	10.7	11.0	5.8	9.5	17.9	9.8	16.7
Textiles and clothing	4.6	35.4	44.3	17.1	43.6	7.6	2.6	35.0	12.9	11.8

[a] The value of world exports in US$ billions is shown in parentheses.
[b] For space reasons regions 1, 3, 9, 11, 13 and 16 of Table 20.1 are omitted from this table.
Source: GTAP database, Version 4, McDougall *et al.* (1998).

Table 20.5. Scenario 1: effects of selected regions[a] adopting GM cotton.

(a) Effects on production, domestic prices and trade (percentage changes)

	North America	Japan	China	India	Rest of South Asia	Western Europe	Sub-Saharan Africa
Production							
Plant-based fibres	5.4	−5.7	3.8	1.1	3.4	−6.3	−5.5
Textiles and clothing	0.1	−0.3	2.7	0.0	−0.2	−0.4	−0.1
Market prices							
Plant-based fibres	−4.8	−0.4	−5.4	−5.9	−5.2	−0.7	−0.6
Textiles and clothing	−0.2	−0.0	−0.5	−0.1	−0.1	−0.0	−0.1
Exports[b]							
Plant-based fibres	13.4	−17.3	18.8	31.1	17.7	−10.7	−14.7
Textiles and clothing	0.2	−1.2	4.1	−0.3	−0.4	−0.7	0.2
Imports[b]							
Plant-based fibres	−2.5	0.1	−4.6	−13.1	−5.3	−1.1	4.4
Textiles and clothing	−0.0	1.3	−0.5	0.6	0.3	0.1	0.3

(b) Effects on regional economic welfare

	Equivalent variation (EV)	Decomposition of welfare results, contribution of (US$ million)		
	(US$ million year^{-1})	Allocative efficiency effects	Terms of trade effects	Technical change
North America	350	6	−130	462
Western Europe	140	91	37	0
Japan	107	24	70	0
Other high-income[c]	19	24	21	0
China	389	57	0	375
India	416	33	−4	385
Rest of South Asia	136	14	−3	123
Sub-Saharan Africa	−5	14	−21	0
Other developing and transition economies	177	24	31	150
World	1727	230	0	1497

[a] North America, Southern Cone, China and South Asia. For space reasons, results for numerous regions of Table 20.1 are omitted from this table.
[b] Includes intra-regional trade.
[c] Newly industrialized Asia, Australia and New Zealand.
Source: Nielsen and Anderson's (2000c) GTAP model results.

Table 20.5, are relatively modest: just US$1.7 billion (US) per year,[5] of which $350 million goes to North America alone.[6] If South Asia was unable to adopt GM cotton, Nielsen and Anderson (2000c) note that the global gain would fall to $1.2 billion, of which less than half would accrue to developing countries. Needless to say, if

[5] That global gain would be just $1.2 billion if South Asia did not adopt GMO technology for plant-based fibres.
[6] This general equilibrium estimate of the annual gain of $350 million to North America in 1995 dollars for all plant-based fibres is comparable with that generated using a simpler partial equilibrium model for just cotton for the USA alone for 1996 of $240 million by Falck-Zepeda *et al.* (2000).

the restrictions on textile and clothing trade were removed so that that sector's output was produced in the most efficient locations (meaning much more in developing countries), the gains from GM cotton would be not only greater globally but the share of those gains enjoyed by developing countries also would be greater.[7]

Scenario 2: selected regions adopt GM rice

The importance of rice production for the livelihoods of poor farmers and consumers in a large number of developing countries makes the introduction of even first-generation GMO technology, e.g. insect-resistant rice varieties, of great interest (let alone second-generation GMO technology that could enhance the nutritional attributes of rice, which is not considered in what follows). The rice scenario involves North America, the Southern Cone, China, East Asian non-industrialized countries (NICs), the rest of East Asia, India and the rest of South Asia (but again not Africa, Europe or Japan) adopting productivity-enhancing GM rice seeds. Again the shock is assumed to increase primary factor and intermediate input productivity in the paddy rice sector by 5%.

First, note from Table 20.6 that Chinese and Indian paddy rice production increase by just 0.3% and the rest of South Asia by even less, while the increase is around 2% for North America and the rest of East Asia even though all experience the same assumed productivity increase. These differences are partly explained by the structures of intermediate use of paddy rice: 27% of paddy rice production in China is used to feed livestock and only about half of paddy rice production is further processed, whereas, in the rest of East Asia, paddy rice is not used as livestock feed and almost all is processed. Chinese consumers spend an equal share of total food expenditure on processed and unprocessed rice, whereas consumers in the rest of East Asia almost exclusively purchase processed rice while those in South Asia consume mostly unprocessed rice (Table 20.3). These different degrees of processing mean that the increased productivity in paddy rice production in the rest of East Asia can result in much higher added value than an identical increase in paddy rice productivity in China or South Asia. Moreover, India and the rest of East Asia are major exporters of rice, so a foreign market for their increased production exists to the extent that demand for rice is price elastic – and the average price of processed rice on world markets declines by about 3%. The domestic price of paddy rice within both East and South Asia declines by more: about 6–7%.

Second, as a net exporter of rice, non-adopting sub-Saharan Africa in this scenario has to face a lower export price as a consequence of increased competition from GM-adopting, rice-exporting countries. This leads to a marked decline in sub-Saharan African exports and a rise in the region's imports of rice.

Third, in terms of economic welfare, the world economy is estimated to be better off by $6.2 billion per year in 1995 dollars because of such a technology shock, assuming it has no external effects (lower part of Table 20.6).[8] Western Europe gains mostly because resources move out of producing highly protected rice there, while 'Other high-income' gains partly because Korea and Taiwan are assumed to adopt GM rice and partly because, as in Europe, non-adopting Australia shifts resources out of protected rice production. The only region in that table to lose is North America, and its loss is trivial. The welfare decomposition in Table 20.6 reveals that the reason for its loss is that the deterioration in the terms of trade, because of the fall in international rice prices, outweighs the gain from productivity growth in this (to North America) relatively unimportant crop. All of Asia's

[7] Estimates of those greater gains are to appear in the next revision of Nielsen and Anderson (2000c).
[8] If South Asia did not adopt the new GM technology, the global gains would be reduced by more than one-quarter, to $4.5 billion p.a.

Table 20.6. Scenario 2: effects of selected regions[a] adopting GM rice.

(a) Effects on production, domestic prices and trade (percentage changes)

	North America	China	Rest of East Asia	India	Rest of South Asia	Western Europe	Sub-Saharan Africa
Production							
Paddy rice	1.8	0.3	2.0	0.3	0.1	−11.5	−0.3
Processed rice	−5.2	0.3	2.4	6.8	−3.1	−1.5	−3.7
Domestic prices							
Paddy rice	−5.5	−6.0	−6.7	−7.2	−7.1	−0.9	−0.1
Processed rice	−0.3	−2.7	−4.5	−3.9	−1.8	−0.4	−0.2
Exports[b]							
Paddy rice	10.8	33.6	22.5	28.8	31.7	−23.7	−25.9
Processed rice	−10.2	4.1	10.7	6.8	−5.2	−3.9	−11.6
Imports[b]							
Paddy rice	−0.9	−0.1	−8.9	−25.5	−22.4	−0.3	22.2
Processed rice	9.2	5.5	−5.4	4.4	−1.6	0.0	8.0

(b) Effects on regional economic welfare

	Equivalent variation (EV)	Decomposition of welfare results, contribution of (US$ million)		
	(US$ million year^{-1})	Allocative efficiency effects	Terms of trade effects	Technical change
North America	−30	8	−126	76
Western Europe	295	284	14	0
Other high-income[c]	1427	180	124	1122
China	1715	226	24	1489
Rest of East Asia	804	−232	−87	1120
India	1178	140	−46	1088
Rest of South Asia	389	53	5	328
Sub-Saharan Africa	21	5	15	0
Other developing and transition economies	422	101	77	241
World	6220	765	0	5466

[a] North America, China, Rest of East Asia, India, and Rest of South Asia. For space reasons, results for numerous regions in Table 20.1 are omitted from this table.
[b] Includes intra-regional trade.
[c] Japan, newly industrialized Asia, Australia and New Zealand.
Source: Nielsen and Anderson's (2000c) GTAP model results.

adopting regions gain substantially. All gain from the technological improvement by similar amounts, but the rest of Developing East Asia gains less overall because more resources are attracted to its relatively assisted rice sector.[9]

In this rice scenario, as in the previous cotton scenario, it is assumed that consumers are indifferent as to whether the product is free of GMOs, and hence GM and conventional crops are produced side-by-side and traded in one co-mingled market.

[9] East Asia's NICs also assist their rice industry, but they assist other farm sub-sectors even more so the expansion of rice by the latter does not reduce overall allocative efficiency in their farm sectors, unlike in rest of East Asia.

That is, there are assumed to be no restrictions on trade with GM products, a point taken up below.

Scenario 3: selected regions adopt GM maize and soybean

In modelling the adoption of GMOs in maize and soybean production, Nielsen and Anderson (2000b) assume that GM-driven productivity growth of 5% occurs in two GTAP sectors: coarse grain (excluding wheat and rice) and oilseeds. The productivity shocks are applied to North America, Mexico, the Southern Cone region of Latin America, India, China, the rest of East Asia (excluding Japan and the East Asian NICs), and South Africa. Again the countries of Western Europe, Japan, other sub-Saharan Africa and elsewhere are assumed to refrain from the use of GM crops in their production systems.

The authors consider three maize/soybean scenarios. The first of them (scenario 3) is a base case with no policy or consumer reactions to GMOs, as assumed above also for cotton and rice. The others (scenarios 4 and 5) impose on this base case a policy or consumer response in Western Europe. In scenario 4, Western Europe not only refrains from using GM crops in its own domestic production systems, but the region is also assumed to reject imports of GM oilseeds and coarse grains from GM-adopting regions. Scenario 5 considers the case in which consumers express their preferences through market mechanisms rather than through government regulation.

Table 20.7 reports the results for scenario 3. A 5% reduction in overall production costs in these sectors leads to increases in coarse grain production of between 0.4% and 2.1%, and increases in oilseed production of between 1.1% and 4.6%, in the GM-adopting regions. The production responses are generally larger for oilseeds as compared with coarse grain. This is because a larger share of oilseed production as compared with coarse grain production is destined for export markets in all the reported regions, and hence oilseed production is not limited to the same extent by domestic demand, which is less price-elastic. Increased oilseed production leads to lower market prices and hence cheaper costs of production in the vegetable oils and fats sectors, expanding output there. This expansion is particularly marked in the Southern Cone region of South America where no less than one-fourth of this production is sold on foreign markets (Table 20.4b), thereby allowing for a larger production response to the reduced costs of production in this sector. In North America, maize is also used as livestock feed, and hence the lower feed prices lead to an expansion of the livestock and meat processing sectors there.

Due to the very large world market shares of oilseeds from North and South America and coarse grain from North America (Table 20.4a), the increased supply from these regions causes world prices for coarse grain and oilseeds to decline by 4.0% and 4.5%, respectively. As a consequence of the more intense competition from abroad, production of coarse grain and oilseeds declines in the non-adopting regions. This is particularly so in Western Europe, a major net importer of oilseeds, of which about half comes from North America. Cereal grain imports into Western Europe increase only slightly (0.1%), but the increased competition and lower price are enough to entail a 4.5% decline in Western European production. In the developing countries too, production of coarse grain and oilseeds is reduced slightly. The changes in India, however, are relatively small compared with, e.g., China and the Southern Cone region. This is explained by the domestic market orientation of these sales. That means India's relatively small production increase causes rather substantial declines in domestic prices for these products, which in turn benefits the other agricultural sectors through vertical linkages. For example, 67% of intermediate demand for coarse grain and 37% of intermediate demand for oilseeds in India stems from the livestock sector, according to the GTAP database.

Table 20.7. Scenario 3: effects of selected regions[a] adopting GM maize and soybean.

(a) Effects on production, domestic prices and trade (percentage changes)

	North America	Southern Cone	China	India	Western Europe	Sub-Saharan Africa
Production						
Coarse grain	2.1	1.6	1.0	0.4	−4.5	−2.3
Oilseeds	3.6	4.6	1.8	1.1	−11.2	−1.3
Livestock	0.8	−0.0	0.1	0.4	−0.2	−0.1
Meat and dairy	0.5	0.0	0.1	1.3	−0.1	−0.1
Vegetable oils, fats	1.1	4.5	1.4	0.0	−0.9	−1.2
Other foods	0.2	0.1	0.4	1.5	−0.1	0.0
Market prices						
Coarse grain	−5.5	−5.5	−5.6	−6.7	−0.5	−0.4
Oilseeds	−5.5	−5.3	−5.6	−6.5	−1.2	−0.3
Livestock	−1.8	−0.3	−0.4	−1.4	−0.3	−0.3
Meat and dairy	−1.0	−0.2	−0.3	−1.0	−0.2	−0.2
Vegetable oils, fats	−2.4	−3.1	−2.6	−1.0	−0.5	−0.2
Other foods	−0.3	−0.2	−0.5	−1.0	−0.1	−0.2
Exports[b]						
Coarse grain	8.5	13.3	16.8	37.3	−11.5	−20
Oilseeds	8.5	10.5	8.2	21.5	−20.5	−26.5
Livestock	8.9	−2.0	−3.3	9.4	−1.1	−1.5
Meat and dairy	4.8	−0.9	−0.9	5.8	−0.5	−0.2
Vegetable oils, fats	5.8	14.3	5.6	−3.8	−4.9	−5.3
Other foods	0.2	0.1	1.6	7.6	−0.6	0.1
Imports[b]						
Coarse grain	−1.6	−4.6	−4.2	−20.5	0.1	11.3
Oilseeds	−2.6	−9.2	−1.6	−8.6	2.5	16.5
Livestock	−2.1	1.3	0.9	−5.2	0.2	0.5
Meat and dairy	−1.9	0.2	0.8	−1.7	−0.0	0.1
Vegetable oils, fats	−3.7	−3.6	−1.7	3.1	1.3	3.4
Other foods	0	−0.1	−0.6	−3.1	0.1	−0.1

(b) Effects on regional economic welfare

	Equivalent variation (EV)	Decomposition of welfare results, contribution of (US$ million)		
	(US$ million year^{-1})	Allocative efficiency effects	Terms of trade effects	Technical change
North America	2624	−137	−1008	3746
Southern Cone	826	120	−223	923
China	839	113	66	672
India	1265	182	−9	1094
Western Europe	2010	1755	253	0
Sub-Saharan Africa	−9	−2	−9	0
Other high-income[c]	1186	554	641	0
Other developing and transition economies	1120	171	289	673
World	9859	2756	0	7108

[a] North America, Mexico, Southern Cone, China, Rest of East Asia, India, and South Africa. For space reasons, results for numerous regions in Table 20.1 are omitted from this table.
[b] Includes intra-regional trade.
[c] Japan, newly industrialized Asia, Australia and New Zealand.
Source: Nielsen and Anderson's (2000b) GTAP model results.

Global economic welfare (as traditionally measured in terms of equivalent variations of income, ignoring any externalities) is boosted in this first scenario by $9.9 billion per year, two-thirds of which is enjoyed by the adopting regions (Table 20.7b). It is noteworthy that all regions (both adopting and non-adopting) gain in terms of economic welfare except sub-Saharan Africa. Most of this gain stems directly from the technology boost. The net-exporting GM-adopters experience worsened terms of trade due to increased competition on world markets, but this adverse welfare effect is outweighed by the positive effect of the technological boost. Western Europe gains from the productivity increase in the other regions only in part because of cheaper imports; mostly it gains because increased competition from abroad shifts domestic resources out of relatively highly assisted segments of EU agriculture. The group of other high-income countries, among which are East Asian nations that are relatively large net importers of the GM-potential crops, benefits equally from lower import prices and a more efficient use of resources in domestic farm production.

Scenario 4: selected regions adopt GM maize and soybean plus Western Europe bans imports of those products from GM-adopting regions

In this scenario, Western Europe not only refrains from using GM crops in its own domestic production systems, but the region is also assumed to reject imports of GM oilseeds and coarse grain from GM-adopting regions. This assumes that the labelling requirements of the Biosafety Protocol (UNEP, 2000) enable Western European importers to identify such shipments and that all oilseed and coarse grain exports from GM-adopting regions will be labelled 'may contain GMOs'. Under those conditions, the distinction between GM-inclusive and GM-free products is simplified to one that relates directly to the country of origin,[10] and labelling costs are ignored. This import ban scenario reflects the most extreme application of the precautionary principle within the framework of the Biosafety Protocol.

A Western European ban on the imports of GM coarse grain and oilseeds changes the situation in scenario 3 rather dramatically, especially for the oilseed sector in North America which has been highly dependent on the EU market. The result of the European ban is not only a decline in total North American oilseed exports by almost 30%, but also a production decline of 10%, pulling resources such as land out of this sector (Table 20.8). For coarse grain, by contrast, only 18% of North American production is exported (Table 20.4b) and just 8% of those exports are destined for Western Europe. Therefore the ban does not affect North American production and exports of maize to the same extent as for soybean, although the downward pressure on the international price of maize nonetheless dampens significantly the production-enhancing effect of the technological boost. Similar effects are evident in the other GM-adopting regions, except for India – once again because its production of these particular crops is virtually all sold domestically and so is not greatly unaffected by market developments abroad.

For sub-Saharan Africa, which by assumption is unable to adopt the new GM technology, access to the Western European markets when other competitors are excluded expands. Oilseed exports from this region rise dramatically, by enough to increase domestic production by 4%. Western Europe increases its own production of oilseeds, however, so the aggregate increase in oilseed imports amounts to less

[10] By distinguishing between GMO-inclusive and GMO-free products by country of origin, one concern may be that GM-adopting regions channel their exports to the country or region imposing the import ban (here Western Europe) through third countries that are indifferent as to the content of GMOs and that do not adopt GM technology in their own production systems. The possibility of such trans-shipments is abstracted from in this analysis.

than 1%. Its production of coarse grain also increases, but not by as much because of an initial high degree of self-sufficiency. Europe's shift from imported oilseeds and coarse grain to domestically produced products has implications further downstream. Given an imperfect degree of substitution in production between domestic and imported intermediate inputs, the higher prices of domestically produced maize and soybean mean that livestock feed is slightly more expensive. (Half of intermediate demand for coarse grain in Western Europe stems from the livestock sector.) Inputs to other food processing industries, particularly the vegetable oils and fats sector, are also more expensive. As a consequence, production in these downstream sectors decline and competing imports increase.

Aggregate welfare implications of this scenario are substantially different from those of scenario 3. Western Europe now experiences a decline in aggregate economic welfare of $4.3 billion per year instead of a boost of $2 billion (compare Tables 20.8b and 20.7b). Taking a closer look at the decomposition of the welfare changes reveals that adverse allocative efficiency effects explain the decline. Most significantly, EU resources are forced into producing oilseeds, of which a substantial amount was previously imported. Consumer welfare in Western Europe is reduced in this scenario because, given that those consumers are assumed to be indifferent between GM-inclusive and GM-free products, the import ban restricts them from benefiting from lower international prices. Bear in mind, though, that in this as in the previous scenarios it is assumed citizens are indifferent to GMOs. To the extent that some Western Europeans in fact value a ban on GM products in their domestic markets, that would more or less than offset the above loss in economic welfare.

The key exporters of the GM products – North America, Southern Cone and China – all show a smaller gain in welfare in this as compared with the scenario in which there is no European policy response. Net importers of maize and soybean (e.g. 'Other high-income' which is mostly East Asia), by contrast, are slightly better off in this than in scenario 3. Meanwhile, the countries in sub-Saharan Africa are affected in a slight positive instead of slight negative way, gaining from better terms of trade. In particular, a higher price is obtained for their oilseed exports to Western European markets in this compared with scenario 3.

Two-thirds of the global gain from the new GM technology as measured in scenario 3 would be eroded by an import ban imposed by Western Europe: it falls from $9.9 billion per year to just $3.4 billion, with almost the entire erosion in economic welfare borne in Western Europe (assuming as before that consumers are indifferent between GM-free and GM-inclusive foods). The rest is borne by the net-exporting adopters (mainly North America and the Southern Cone region). Since the non-adopting regions generally purchase most of their imported coarse grain and oilseeds from the North American region, they benefit even more than in scenario 3 from lower import prices: their welfare is estimated to be greater by almost one-fifth in the case of a Western European import ban as compared with no European reaction.

Scenario 5: selected regions adopt GM maize and soybean plus some Western Europeans' preferences shift against GM maize and soybean

As an alternative to a policy response, this scenario analyses the impact of a partial shift in Western European preferences away from imported coarse grain and oilseeds and in favour of domestically produced crops.[11] The scenario is implemented as an exogenous 25% reduction in

[11] See the technical appendix of Nielsen and Anderson (2000a), which describes how the exogenous preference shift is introduced into the GTAP model, a method adopted from Nielsen (1999).

final consumer and intermediate demand for *all* imported oilseeds and coarse grain (that is, not only those which can be identified as coming from GM-adopting regions).[12] This can be interpreted as an illustration of incomplete information being provided about imported products (still assuming that GM crops are not cultivated in Western Europe), if a label only states that the product 'may contain GMOs'. Such a label does not resolve the information problem facing the most critical Western European consumers who want to be able to distinguish between GMO-inclusive and GMO-free products. Thus some European consumers and firms are assumed to choose to completely avoid products that are produced outside Western Europe. That import demand is shifted in favour of domestically produced goods. Western European producers and suppliers are assumed to be able to signal – at no additional cost – that their products are GM-free by, e.g., labelling their products by country of origin. This is possible because it is assumed that no producers in Western Europe adopt GM crops (perhaps due to government regulation), and hence such a label would be perceived as a sufficient guarantee of the absence of GMOs.

As the results in Table 20.9 reveal, having consumers express their preferences through market mechanisms rather than through a government-implemented import ban has a much less damaging effect on production in the GM-adopting countries. In particular, instead of declines in oilseed production as in scenario 4, there are slight increases in this scenario, and production responses in coarse grain are slightly larger. Once again the changes are less marked for India and in part also for China, which are less affected by international market changes for these products. As expected, domestic oilseed production in Western Europe must increase somewhat to accommodate the shift in preferences, but not nearly to the same extent as in the previous scenario. Furthermore, there are in fact minor price reductions for agri-food products in Western Europe in part because (by assumption) the shift in preferences is only partial, and so some consumers and firms do benefit from lower import prices. In other words, in contrast to the previous scenario, a certain link between EU prices and world prices is retained here because we are dealing with only a partial reduction in import demand. The output growth in sub-Saharan Africa in scenario 4, by taking the opportunity of serving European consumers and firms while other suppliers were excluded, is replaced in this scenario by declines: sub-Saharan Africa loses export share to the GM-adopting regions.

The numerical welfare results in this scenario are comparable with those of scenario 3 (the scenario without the import ban or the partial preference shift) for all regions except, of course, Western Europe. Furthermore, the estimated decline in economic welfare that Western Europe would experience if it banned maize and soybean imports is changed to a slight gain in this scenario (although recall that these welfare measures assume consumers are indifferent to whether a food contains GMOs). The dramatic worsening of resource allocative efficiency in the previous scenario is changed to a slight improvement in this one. This is because production in the lightly assisted oilseeds sector increases at the expense of production in all other (more heavily distorted) agri-food sectors in Western Europe.

The welfare gains for North America are more similar in this scenario than in the previous one to those of scenario 3, but even in scenario 4 its gains are large, suggesting considerable flexibility in both domestic and foreign markets to respond to policy and consumer preference changes, plus the dominance of the benefits of the new technology for adopting countries. Given that the preference shift in scenario 5 is based on the assumption that non-adopters outside Western Europe cannot guarantee that their exports to this region

[12] The size of this preference shift is arbitrary, and is simply used to illustrate the possible direction of effects of this type of preference shift as compared with the import ban scenario.

Table 20.8. Scenario 4: effects of selected regions[a] adopting GM maize and soybean *plus* Western Europe bans imports of those products from GM-adopting regions.

(a) Effects on production, domestic prices and trade (percentage changes)

	North America	Southern Cone	China	India	Western Europe	Sub-Saharan Africa
Production						
Cereal grain	0.9	0.0	0.8	0.4	5.3	−2.2
Oilseeds	−10.2	−3.6	−0.8	0.8	66.4	4.4
Livestock	1.2	0.3	0.2	0.4	−0.8	0.0
Meat and dairy	0.8	0.3	0.2	1.4	−0.5	−0.0
Vegetable oils, fats	2.4	8.1	1.6	0.1	−3.4	0.0
Other foods	0.3	0.4	0.5	1.6	−0.5	−0.1
Market prices						
Cereal grain	−6.2	−6.0	−5.6	−6.7	0.8	−0.0
Oilseeds	−7.4	−6.8	−6.0	−6.5	5.8	0.4
Livestock	−2.2	−0.7	−0.4	−1.4	0.5	0.1
Meat and dairy	−1.3	−0.4	−0.3	−1.0	0.3	0.1
Vegetable oils, fats	−3.3	−4.0	−2.7	−1.0	2.0	0.0
Other foods	−0.4	−0.3	−0.5	−1.0	0.1	0.0
Exports[b]						
Cereal grain	0.3	−2.9	5.0	23.4	15.9	−13.1
Oilseeds	−28.8	−69.2	−18.4	−8.7	167.2	105.0
Livestock	13.7	4.0	−1.4	12.6	−3.8	−1.8
Meat and dairy	7.5	2.1	0.1	7.1	−1.4	0.3
Vegetable oils, fats	14.4	26.2	7.0	1.3	−15.0	5.8
Other foods	1.5	1.9	2.0	8.0	−1.4	−0.6
Imports[b]						
Cereal grain	−1.9	−5.3	−2.8	−20	3.3	13.4
Oilseeds	−5.6	−21.9	3.0	−3.7	0.6	22.5
Livestock	−3.2	0.1	0.1	−5.9	0.9	0.5
Meat and dairy	−2.8	−0.5	0.8	−1.8	−0.2	−0.0
Vegetable oils, fats	−7.7	−5.5	−1.7	4.0	5.5	2.4
Other foods	−0.6	−0.6	−0.8	−2.8	0.1	0.2

(b) Effects on regional economic welfare

	Equivalent variation (EV)	Decomposition of welfare results, contribution of (US$ million year^{-1})		
	(US$ million year^{-1})	Allocative efficiency effects	Terms of trade effects	Technical change
North America	2299	27	−1372	3641
Southern Cone	663	71	−303	893
China	804	74	70	669
India	1277	190	−3	1092
Western Europe	−4334	−4601	257	0
Sub-Saharan Africa	42	5	38	0
Other high-income[c]	1371	592	782	0
Other developing and transition economies	1296	101	531	672
World	3419	−3541	0	6966

[a] North America, Mexico, Southern Cone, China, Rest of East Asia, India, and South Africa. For space reasons, results for numerous regions in Table 20.1 are omitted from this table.
[b] Includes intra-regional trade.
[c] Japan, newly industrialized Asia, Australia and New Zealand.
Source: Nielsen and Anderson's (2000b) GTAP model results.

Table 20.9. Scenario 5: effects of selected regions[a] adopting GM maize and soybean *plus* partial shift of Western European preferences away from imports of GM products.

(a) Effects on production, domestic prices and trade (percentage changes)

	North America	Southern Cone	China	India	Western Europe	Sub-Saharan Africa
Production						
Coarse grain	1.8	1.3	1.0	0.4	−2.0	−2.6
Oilseeds	1.0	2.8	1.1	1	8.7	−1.6
Livestock	0.9	0.0	0.2	0.4	−0.4	−0.1
Meat and dairy	0.6	0.1	0.1	1.3	−0.2	−0.0
Vegetable oils, fats	1.2	5.0	1.4	−0.0	−1.1	−1.2
Other foods	0.2	0.2	0.4	1.5	−0.2	0.1
Market prices						
Coarse grain	−5.7	−5.6	−5.6	−6.7	−0.2	−0.4
Oilseeds	−5.9	−5.6	−5.7	−6.5	0.1	−0.3
Livestock	−1.9	−0.4	−0.4	−1.4	−0.1	−0.3
Meat and dairy	−1.1	−0.2	−0.3	−1.0	−0.1	−0.2
Vegetable oils, fats	−2.6	−3.3	−2.6	−1.0	−0.4	−0.2
Other foods	−0.3	−0.2	−0.5	−1.0	−0.1	−0.2
Exports[b]						
Coarse grain	6.6	9.7	13.9	34.1	−29.7	−24.1
Oilseeds	1.4	−4.5	2.1	14.1	−41.5	−32.4
Livestock	9.8	−0.9	−3.0	10.0	−1.8	−1.2
Meat and dairy	5.3	−0.4	−0.8	6.0	−0.7	0.1
Vegetable oils, fats	6.7	15.8	5.5	−4.0	−5.8	−4.9
Other foods	0.4	0.4	1.7	7.6	−0.7	0.1
Imports[b]						
Coarse grain	−1.7	−4.8	−3.9	−20.4	−23.6	11.5
Oilseeds	−2.9	−9.6	−0.7	−7.4	−17.7	17.3
Livestock	−2.3	1.1	0.8	−5.3	0.4	0.2
Meat and dairy	−2.1	0.1	0.8	−1.7	−0.1	−0.0
Vegetable oils, fats	−4.2	−3.8	−1.5	3.4	1.5	3.4
Other foods	−0.1	−0.2	−0.6	−3	0.1	−0.1

(b) Effects on regional economic welfare

	Equivalent variation (EV)	Decomposition of welfare results, contribution of (US$ million)		
	(US$ million year^{-1})	Allocative efficiency effects	Terms of trade effects	Technical change
North America	2554	−100	−1092	3726
Southern Cone	785	109	−246	917
China	834	106	69	672
India	1267	184	−9	1093
Western Europe	715	393	319	0
Sub-Saharan Africa	−5	0	−7	0
Other high-income[c]	1233	567	674	0
Other developing and transition economies	1120	168	293	673
World	8503	1428	0	7081

[a] North America, Mexico, Southern Cone, China, Rest of East Asia, India, and South Africa. For space reasons, results for numerous regions in Table 20.1 are omitted from this table.
[b] Includes intra-regional trade.
[c] Japan, newly industrialized Asia, Australia and New Zealand.
Source: Nielsen and Anderson's (2000b) GTAP model results.

are GMO-free, sub-Saharan Africa cannot benefit from the same kind of 'preferential' access the region obtained in the previous scenario, where coarse grain and oilseeds from just identifiable GMO-adopting regions were banned completely. Hence sub-Saharan Africa slips back to a slight loss in this scenario due to a net worsening of its terms of trade and the absence of productivity gains from genetic engineering techniques. Globally, welfare in this case is only a little below that when there is no preference shift: a gain of $8.5 billion per year compared with $9.9 billion in scenario 3, with Western Europe clearly bearing the bulk of this difference.

Estimating Economic Effects of Adopting GM Maize and Soybean Using an Alternative Model

An alternative modelling framework is used in a recent analysis by Nielsen, Robinson and Thierfelder (2000), hereafter the NRT model, which draws on a model developed by Lewis *et al.* (1999). It involves a more-aggregated multi-region computable general equilibrium (CGE) model consisting of just seven regions and ten sectors but is otherwise similar to the standard GTAP model with one important exception: the coarse grain and oilseed sectors of the NRT model have each been split into two. This split is in order to capture the production and trade effects of segregating maize and soybean markets into GM and non-GM lines of production. In the base data, it is assumed that all regions in the NRT model initially produce some of both the GM and non-GM varieties of oilseeds and coarse grain (in contrast to the assumption in the preceding scenarios that only a subset of countries can or choose to develop GM crops). The assumed shares are as shown in Table 20.10, based on estimates provided in James (1999) and USDA (2000).

The structures of production in terms of the composition of intermediate input and factor use in the GM and non-GM varieties are initially assumed to be identical. The destination structures of exports are also initially assumed to be the same. In the NRT model the authors endogenize the producers' decision to use GM vs. non-GM varieties in production, while at the same time holding the total intermediate demand for oilseeds and coarse grain, respectively, fixed as a proportion of output. In this way, the initial input–output coefficients remain fixed but, for oilseeds and coarse grain, a choice has been introduced between GM and non-GM varieties. Other intermediate input demands remain in fixed proportions to output.

The input–output choice is endogenized for four demanders of coarse grain and oilseeds: livestock, meat and dairy, vegetable oils and fats, and other processed food sectors. The choice between GM and non-GM varieties is determined by a CES function (here shown for intermediate demand for oilseeds, *osd*):

$$IO(osd,j,k) = a(osd,j,k)\cdot\left[\alpha_G(gm\_osd,j,k)\, IO(gm\_osd,j,k)^{-\rho_G(osd,k)} + \alpha_G(ng\_osd,j,k)\cdot IO(ng\_osd,j,k)^{-\rho_G(osd,k)}\right]^{-1/\rho_G(osd,k)} \quad (1)$$

Table 20.10. Assumed initial shares of GM crop varieties used in the NRT model (% of total GM-potential crop production).

	High-income Asia[a]	Low-income Asia	North America	South America	Western Europe	Sub-Saharan Africa	Rest of world
GM coarse grain	10	40	40	40	10	10	10
GM oilseeds	10	60	60	90	10	10	10

[a] Includes Japan, East Asia's NICs, Australia and New Zealand.
Source: Nielsen *et al.* (2000), based on James (1999) and USDA (2000).

where $IO(osd,j,k)$ is sector j in region k's intermediate demand for oilseeds, and $a(osd,j,k)$ is the CES intermediate demand shift parameter. The exponent is defined by the elasticity of substitution between GM and non-GM varieties, $\Phi_G(osd,j,k)$: $\Delta_G(osd,j,k) = (1/\Phi_G(osd,j,k)) - 1$. The CES function share coefficients are $\forall_G(gm\_osd,j,k)$ and $\forall_G(gm\_osd,j,k)$. In the model, the following first-order conditions are included for the four GM-using production sectors mentioned above in all regions – one set of equations for oilseeds (shown here as *osd*) and another for coarse grain:

$$\frac{IO(gm\_osd,j,k)}{IO(ng\_osd,j,k)} = \left[\frac{PC(ng\_osd,k)}{PC(gm\_osd,k)} \cdot \frac{\alpha_G(gm\_osd,j,k)}{\alpha_G(ng\_osd,j,k)}\right]^{1/1+\rho_G(osd,k)} \tag{2}$$

The adding-up constraints are included as follows (again, shown here for *osd*):

$$IO0(gm\_osd,j,k) + IO0(ng\_osd,j,k) = IO(gm\_osd,j,k) + IO(ng\_osd,j,k). \tag{3}$$

The input–output coefficients in all other sectors are assumed fixed, as are the input–output coefficients for the four above-mentioned sectors *vis à vis* other intermediate inputs.

Since the available estimates of agronomic and hence economic benefits to producers from cultivating GM crops are very diverse, NRT simply assume the GM oilseed and GM cereal grain sectors in all regions have a 10% higher level of factor productivity as compared with their non-GM (conventional) counterparts – that is, twice the shock imposed in the above five GTAP scenarios. They introduce a total factor productivity shock in the GM sectors against a variety of base models, which differ in terms of substitution elasticities for GM and non-GM products. To start with, it is assumed that the elasticity of substitution between GM and non-GM varieties is high and equal in all regions. Specifically, σ_G('oilseeds', k) = σ_G('coarse grain', k) = 5.0 for all regions k. Then, in order to reflect the fact that citizens in Western Europe and high-income Asia (particularly in Japan) are skeptical of the new GM varieties, the elasticities of substitution between the GM and non-GM varieties are gradually lowered so that GM and non-GM varieties are seen as increasingly poorer substitutes in production in these particular regions. Citizens in all other regions are basically indifferent, and hence the two crops remain highly substitutable in those production systems. The preferences of citizens in Western Europe and high-income Asia are also reflected in the import demand system. Irrespective of the country of origin of the GM crops, these regions are not very sensitive to price changes for GM crops. The cross-price elasticities between two sources of imports for these crops in the GMO-skeptical regions are therefore set at 0.5 compared with 2.0 in the other regions for all the experiments.

What results should we expect?

Initially, the more-effective GM production process will cause labour, land and capital to leave the GM sectors because lower (cost-driven) GM product prices will result in lower returns to factors of production. To the extent that demand (domestically or abroad) is very responsive to this price reduction, this cost-reducing technology may potentially lead to increased production and hence higher returns to factors. As suppliers of inputs and buyers of agricultural products, other sectors also will be affected by the use of genetic engineering in GM-potential sectors through vertical (or backward) linkages. To the extent that the production of GM crops increases, the demand for inputs by producers of those crops may rise. Demanders of primary agricultural products, e.g. livestock producers using grains and oilseeds for livestock feed, will benefit from lower prices, which in turn will affect the market competitiveness of these sectors.

The other sectors of the economy may also be affected through horizontal (or forward) linkages. Primary crops and livestock are typically complementary in food processing. Cheaper GM crops have the potential of initiating an expansion of food production and there may also be substitution effects. For example, since applying genetic engineering techniques to wheat breeding is apparently more complex compared with maize, the price of wheat will be high relative to other more easily manipulated grains. To the extent that substitutions in production are possible, the food processing industry may shift to the cheaper GM intermediate inputs. Widespread use of GM products can furthermore be expected to affect the price and allocation of mobile factors of production and in this way also affect the other sectors of the economy.

In terms of price effects, there is both a direct and an indirect effect of segregating the markets. Due directly to the output-enhancing productivity effect, countries adopting GM crops should gain from lower cost-driven prices. The more receptive a country is to the productivity-enhancing technology, the greater the gains. There is also an indirect effect, which will depend on the degree of substitutability between GM and non-GM products. When substitutability is high, the price of non-GM crops will decline along with the prices of GM-crops. The lower the degree of substitutability, the weaker will be this effect, and the larger should be the price wedge between GM and non-GM crops. The net effect of these direct and indirect effects on particular countries is theoretically ambiguous, again underscoring the need for empirical analysis.

The widespread adoption of GM varieties in certain regions will affect international trade flows depending on how traded the crop in question is and the preferences for GM vs. non-GM in foreign markets. World market prices for GM products will have a tendency to decline and thus benefit net importers to the extent that they are indifferent between GM and non-GM products. For exporters, the lower price may enable an expansion of the trade volume depending on the price elasticities and preferences in foreign markets. In markets where citizens are critical of GM ingredients in their food production systems, consumers will not fully benefit from the lower prices on GM crops. Furthermore, resources will be retained in the relatively less productive non-GM sectors in these regions. However, as is the case with organic food production, this would simply be a reflection of consumer preferences and hence not welfare-reducing (so long as an appropriate welfare measure is used).

What production and trade results emerge from the NRT empirical analysis?

The expected increase in production of the GM crops is borne out in the empirical results for all regions of the NRT model as a direct consequence of the assumed increase in factor productivity. Due to the relative decline in productivity in the non-GM sectors, production of conventional coarse grain and oilseeds declines. Attention here will focus on the effects on overall trade and bilateral trade patterns for selected regions as citizens in high-income Asia and Western Europe become increasingly critical of GM crops, and hence these crops become correspondingly worse substitutes in production in these two regions.

Being the world's largest exporter of both oilseeds and coarse grain, as well as being particularly dependent on the GM-critical markets for these exports, the North American region is very sensitive to changes in preferences. As is seen in Fig. 20.1, total exports of the GM varieties decline as GM and non-GM substitutability worsens in the GM-critical regions, and this is particularly so for oilseeds because almost 80% of North American oilseeds exports are initially sold in these markets, whereas the share is less than 60% for coarse grain. In response to the changing preferences, exports of the non-GM varieties are boosted. These changes are reflected in North America's production results (Fig. 20.5): once again, with the sensitivity to changing preferences being more vivid for the oilseeds sector. For South America similar effects are at work (Figs 20.2 and 20.6). As in the case of

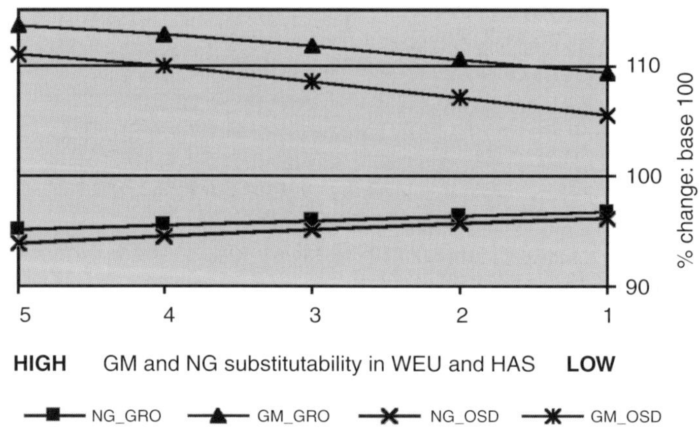

Fig. 20.1. Export changes in North America.

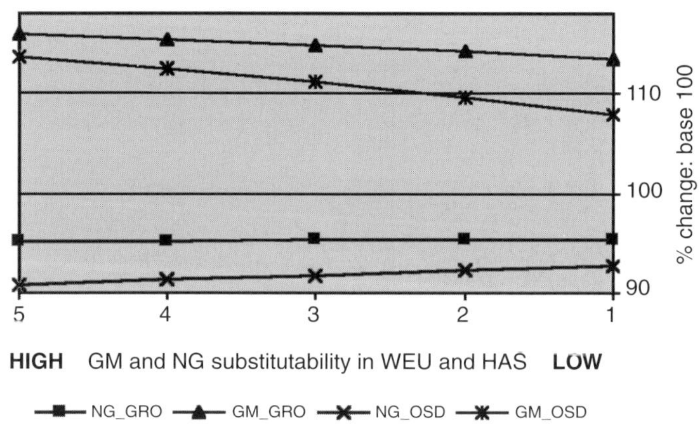

Fig. 20.2. Export changes in South America.

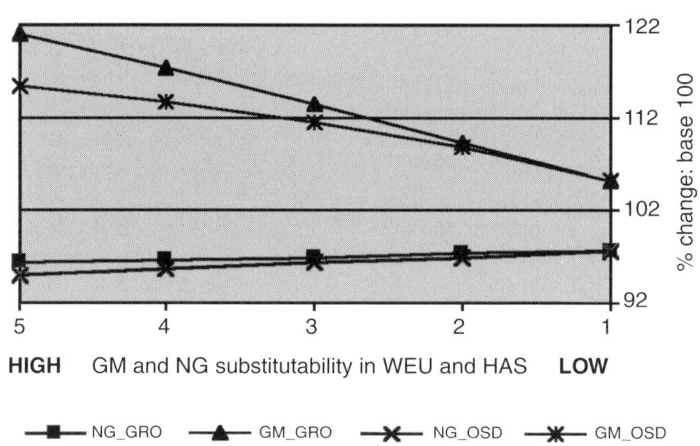

Fig. 20.3. Import changes in Western Europe.

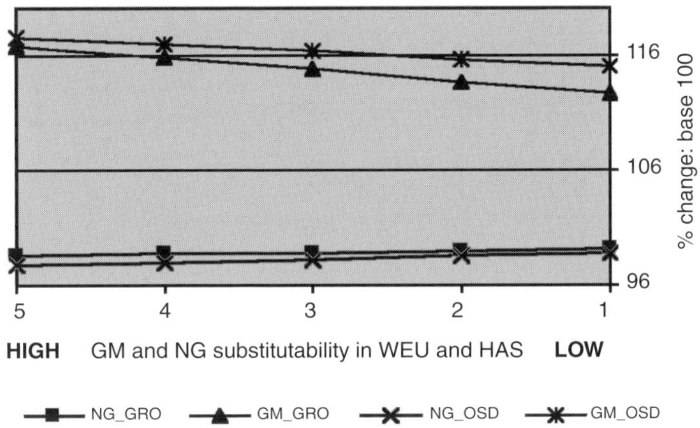

Fig. 20.4. Export changes in sub-Saharan Africa.

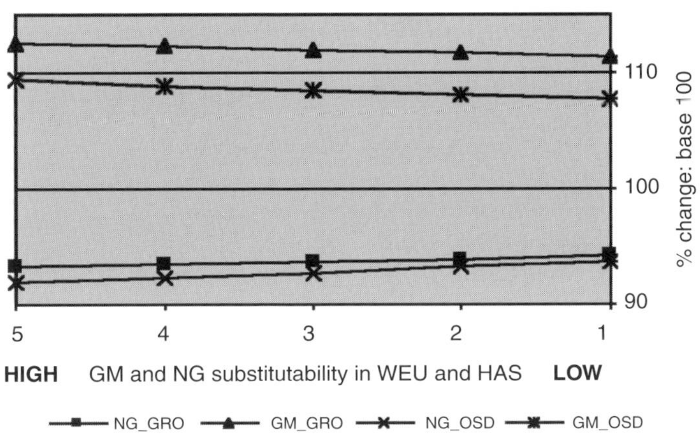

Fig. 20.5. Production changes in North America.

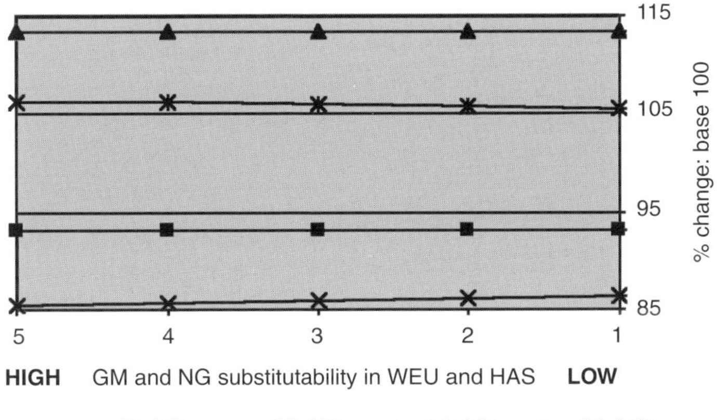

Fig. 20.6. Production changes in South America.

North America, the sensitivity in South American oilseed exports is high because 84% of these exports are initially destined for the GM-critical markets in Europe and high-income Asia.

Given that Western Europe is an important *importer* of oilseeds, Fig. 20.3 depicts the changes in total *imports* of the GM and non-GM varieties into this region. At the extreme, where Western Europeans are unconcerned about the GM or non-GM status of crops used in production, imports increase dramatically as the lower world market prices are exploited. As substitutability is reduced, GM imports and production plunge while non-GM imports and production increase. Note that the reduction of total GM oilseed imports occurs at a slower rate than for total GM cereal grain imports. This is due to the initial high dependence on oilseed imports from regions that are intensive users of GM varieties. Furthermore, as the Western European market becomes more difficult to penetrate, the import prices on GM products faced by Europeans decline. This tendency works against the effect of the preference shift.

Sub-Saharan Africa is not initially a large exporter of either crop, but 8% of its coarse grain absorption is imported. Taking a look at Figs 20.4 and 20.8 reveals that if adopting GM varieties were possible for this region (contrary to the assumption in the earlier GTAP scenarios 1–5), that would boost both production and exports. Note that production of coarse grain does not increase as much as for oilseeds: this region is taking advantage of lower world market prices for GM coarse grain as an importer rather than devoting excessive resources to boosting its own production.

In summary, these results show that production of GM oilseeds in North and South America increases more relative to GM coarse grain since the former is the most (internationally) traded of the two crops. As citizens in high-income Asia and Western Europe become more critical of GM varieties, the split of GM and non-GM varieties in total exports from North and South America changes in favour of the non-GM varieties. This is a clear reflection of the importance of the GM-critical markets to the American producers and hence this is also reflected in their structures of production. sub-Saharan Africa also increases production and exports of the GM varieties and this region's composition of total exports is also sensitive to the preference shift in high-income countries.

What about bilateral trade patterns?

Examining the effects on bilateral trade flows provides an indication as to whether developing countries are able to take advantage of the segregation of oilseed and cereal grain markets in the light of changing preferences in Western Europe and high-income Asia. Given the dominance of North America and Western Europe in world trade in these GM-potential products and their roles in the GMO debate, consider first the changing export and import flows, respectively, in these two regions (Figs 20.9–20.16). Then, Figs 20.17–20.24 show the bilateral trade flows for South America, for whom exports of oilseeds are especially important, and sub-Saharan Africa, for whom the Western European market is important for agricultural exports.

Figures 20.9 and 20.11 clearly illustrate how North American exports of non-GM varieties are redirected towards the GM-critical regions as preferences change there: trade diversion is significant. For the GM-varieties, Figs 20.13 and 20.15 show that these exports are redirected to other regions such as low-income Asia and the 'rest of world' region, as the exports to Europe and high-income Asia drop. Taking a closer look at the sourcing of cereal grain and oilseed imports into Western Europe, Fig. 20.10 suggests that the countries in sub-Saharan Africa do increase their non-GM oilseed exports to Europe as the latter region increasingly reveals its preferences for non-GM crops. However, the steeper curve for North America suggests that this already-dominant exporting region is best able to accommodate to the changing pref-

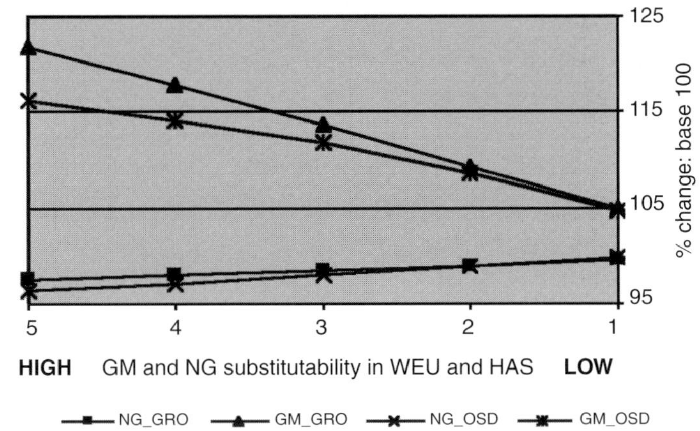

Fig. 20.7. Production changes in Western Europe.

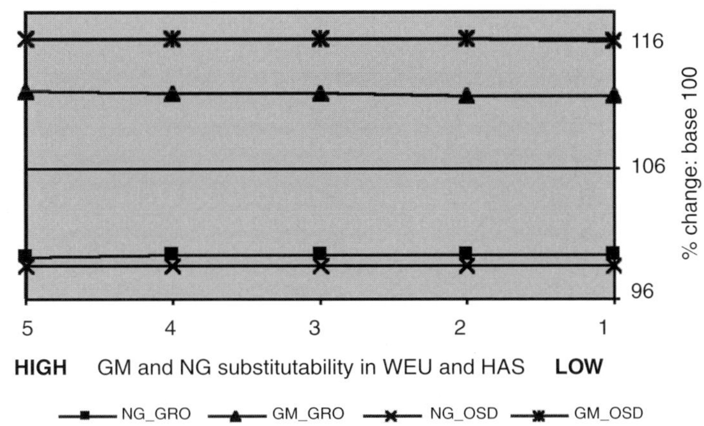

Fig. 20.8. Production changes in sub-Saharan Africa.

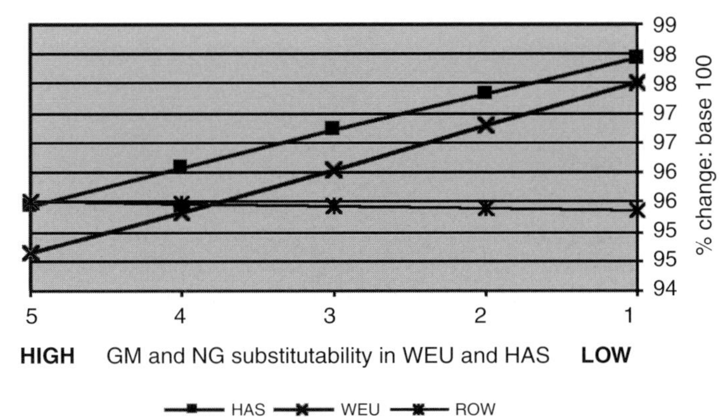

Fig. 20.9. Export changes in North America by destination: NG_OSD.

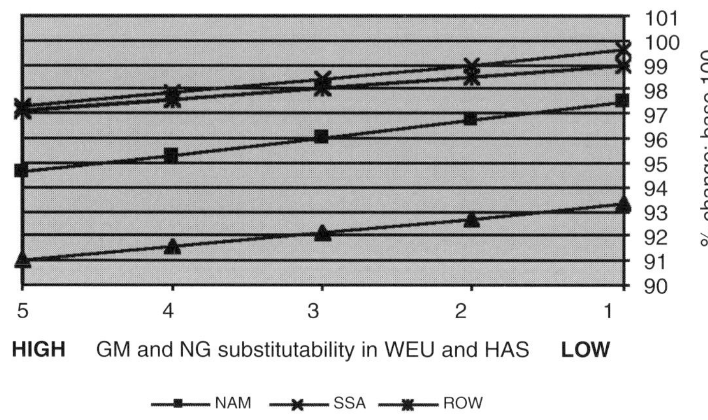

Fig. 20.10. Import changes in Western Europe by origin: NG_OSD.

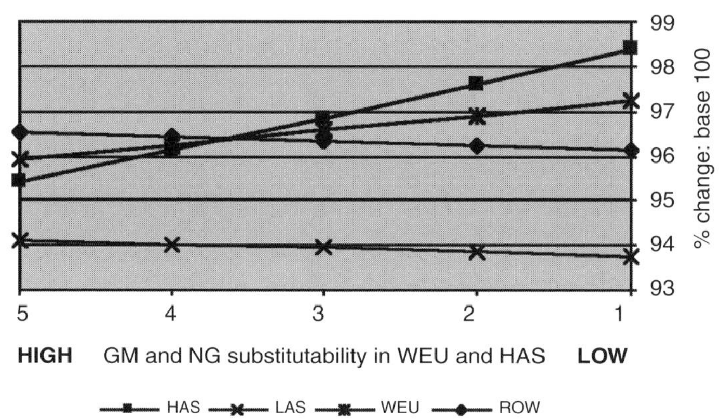

Fig. 20.11. Export changes in North America by destination: NG_GRO.

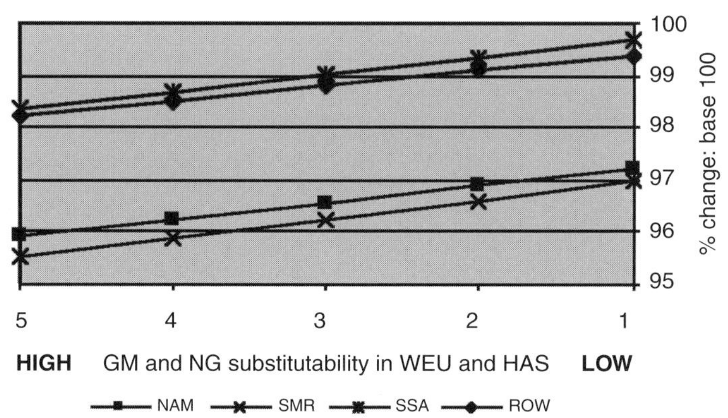

Fig. 20.12. Import changes in Western Europe by origin: NG_GRO.

erences in Europe. For cereal grain exports to Western Europe, Fig. 20.12 shows that although the sub-Saharan African region initially is not hurt as much as North America – relatively speaking and due to very small exports of coarse grain to start with – it seems that the region may again benefit from the segregation of markets, and hence the opportunity of serving a GM-critical market. Figures 20.14 and 20.16 reflect the construction of the experiment, namely that producers in Western Europe find that GM and non-GM crops are increasingly poor substitutes in production, and hence imports from all regions decline in the same percentage changes as preferences strengthen.

The destinations of non-GM oilseed exports from sub-Saharan Africa change in response to the preference changes in Western Europe and high-income Asia, as seen in Fig. 20.17. Non-GM exports are shifted away from the 'rest of world' region and to the GMO-critical regions. The shift is particularly strong for oilseed exports into Western Europe reflecting an initially higher dependence of sub-Saharan African exports to this region as well as a high import general dependence for oilseeds in Western Europe. This is also the case for non-GM coarse grain, although the shift away from other regions is less marked. Furthermore, since grain import dependence is highest in High-income Asia compared with Western Europe, the shift of non-GM grain exports from sub-Saharan Africa is strongest in this direction. For the GM varieties sub-Saharan Africa is also successful in diverting exports into other less critical markets. However, it is also clear that this region will be facing harsh competition from other GM-adopting exporters in these other markets. Similar trade pattern changes are evident for South America, and generally the trade diversion is spread evenly across other trading partners (Figs 20.18, 20.20, 20.22 and 20.24).

In summary, the bilateral trade results show that trade diversion effects are significant. As preferences in high-income Asia and Western Europe turn against GM varieties, trade flows are diverted so that *all* markets – whether GM-indifferent or GM-critical – are served appropriately. That is, markets can be expected to adjust to accommodate the differences in tastes across countries. This favourable outcome is driven by the price differential that results between the two crop varieties. The price wedge for these crops is between 5.5 and 6.7% in Western Europe and high-income Asia, and as low as 3.5–3.8% in North America. In particular, the results generally show that the prices of both the GM and non-GM varieties decline in countries where substitutability between GM and non-GM products is high. The price decline is larger for the GM crops because supplies increase and because exports of these crops are sold on markets where substitutability is poor. In the GMO-critical regions, the prices of non-GM crops do *not* fall, precisely because of the weak links with the GM varieties, and therefore these markets are relatively isolated. However, despite the increase in the price of non-GM crops, and the fact that these regions choose not to partake completely in the productivity gains from GM production, the net effect on aggregate absorption in these countries is close to zero. In all regions where citizens are *not* opposed to GM crops (i.e. substitutability between GM and non-GM crops is high), aggregate absorption and welfare increase.

Conclusions

Lessons

What have we learned? First, the potential economic welfare gains from adopting GMO technology in even just a subset of producing countries for these crops is non-trivial. In the cases considered in the first three scenarios using the GTAP model it amounts to an estimated $1.7 billion per year for cotton, $6.2 billion for rice and $9.9 billion for coarse grain and oilseeds. Moreover, in all three cases, developing countries would receive a sizeable share and possibly the majority of those gains (more so the more of them that are capable

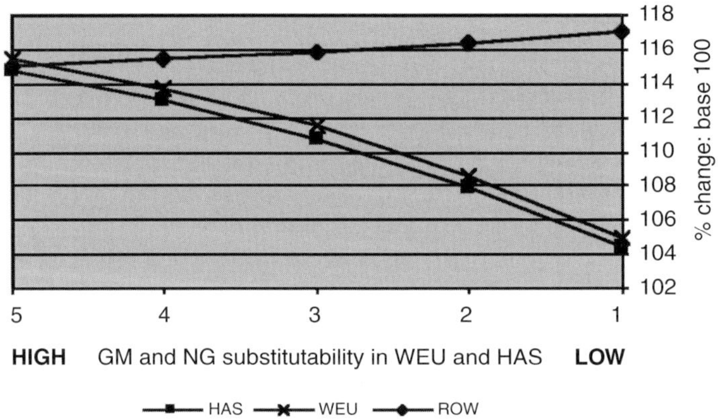

Fig. 20.13. Export changes in North America by destination: GM_OSD.

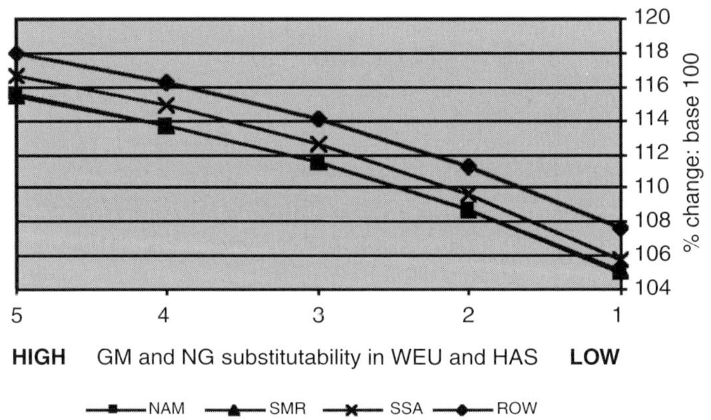

Fig. 20.14. Import changes in Western Europe by origin: GM_OSD.

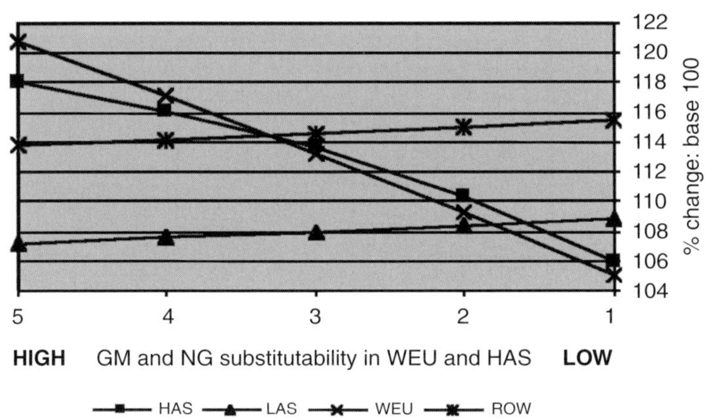

Fig. 20.15. Export changes in North America by destination: GM_GRO.

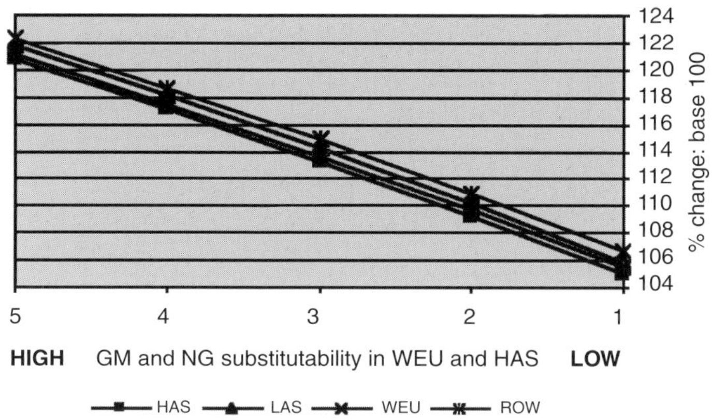

Fig. 20.16. Import changes in Western Europe by origin: GM_GRO.

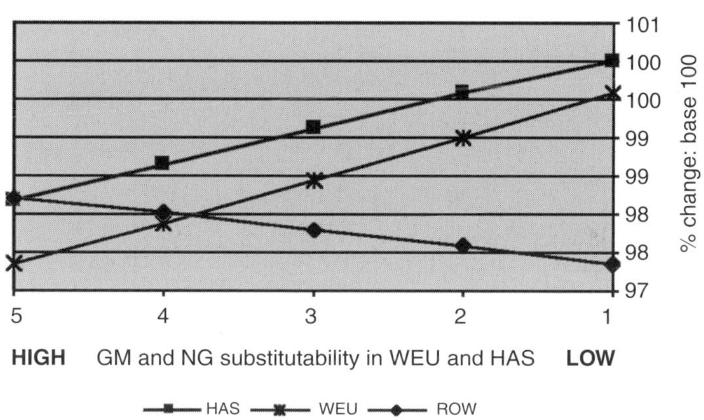

Fig. 20.17. Export changes in sub-Saharan Africa by destination: NG_OSD.

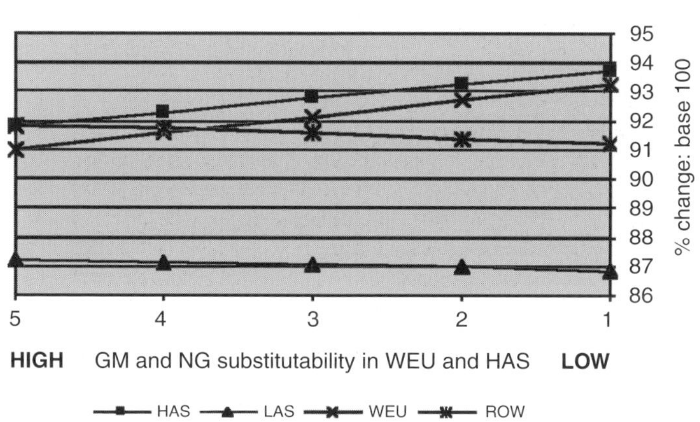

Fig. 20.18. Export changes in South America by destination: NG_OSD.

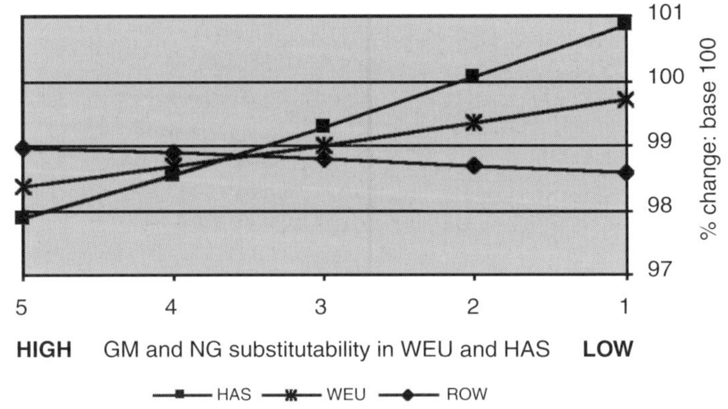

Fig. 20.19. Export changes in sub-Saharan Africa by destination: NG_GRO.

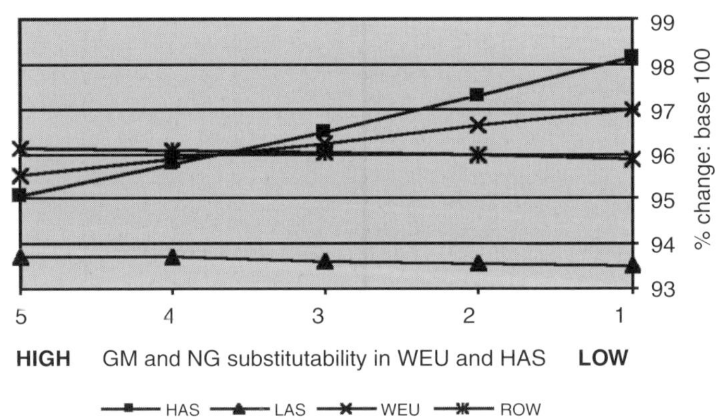

Fig. 20.20. Export changes in South America by destination: NG_GRO.

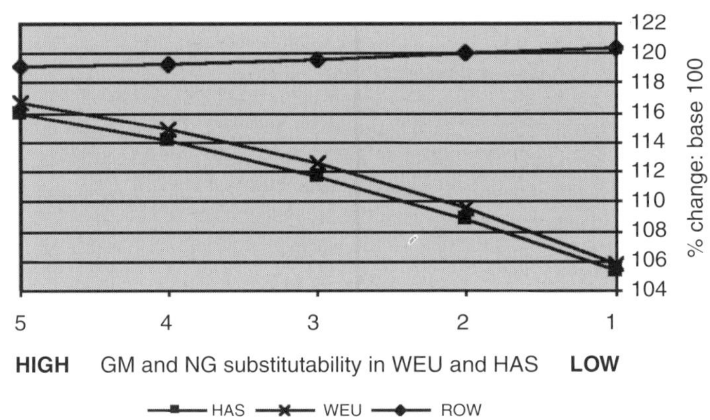

Fig. 20.21. Export changes in sub-Saharan Africa by destination: GM_OSD.

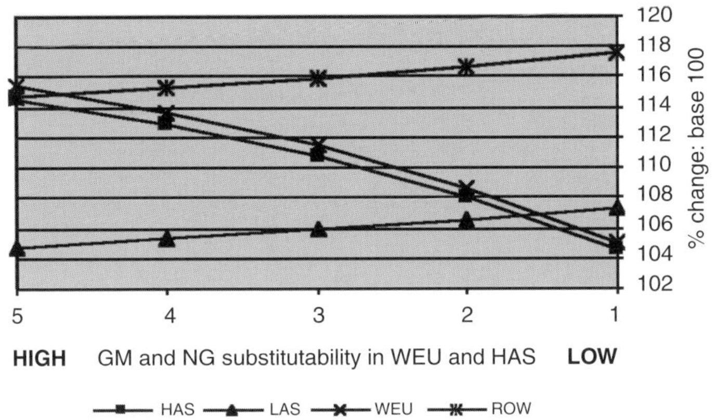

Fig. 20.22. Export changes in South America by destination: GM_OSD.

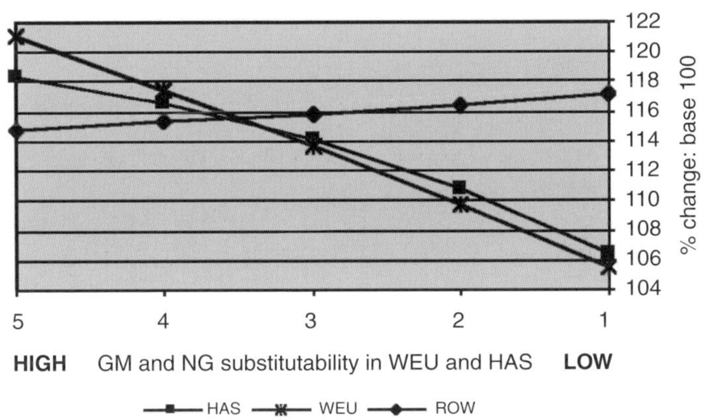

Fig. 20.23. Export changes in sub-Saharan Africa by destination: GM_GRO.

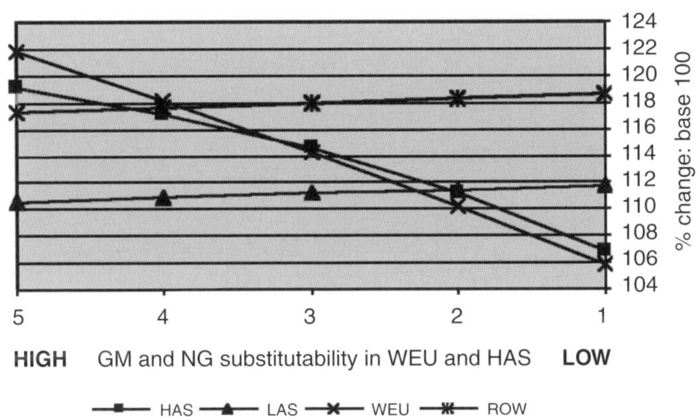

Fig. 20.24. Export changes in South America by destination: GM_GRO.

of introducing the new GM technology). The size of these gains, especially for developing countries, are such that policy makers should not ignore them when considering policy responses to appease opponents of GMO technology.

Second, scenarios 3 to 5 show that the most extreme use of trade provisions by Western Europe, namely an import ban on GM crops, would be very costly in terms of economic welfare for the region itself – a cost which governments in the region need to weigh against the perceived benefits to voters of adopting the precautionary principle in that way. Imposing a ban prevents European consumers and intermediate demanders from gaining from lower import prices, domestic production of maize and soybean would be forced to rise at the expense of other farm production, and hence overall allocative efficiency in the region would be worsened. In the case modelled, the GM-adopting regions still enjoy welfare gains due to the dominating positive effect of the assumed productivity boost embodied in the GM crops, but those gains are reduced by the import ban as compared with the scenario in which GM crops are traded freely. To the extent that some developing and other countries do not adopt GM crops (by choice or otherwise) and they can verify this at the Western European borders, our results suggest it is possible they could gain slightly in gross terms from retaining access to the GMO-free markets when others are excluded. Whether they gain in net terms would depend on the cost of compliance with European regulations.

Third, even if many consumers in Western Europe are concerned about GMOs, the results of the market-based partial preference shift experiment (scenario 5) suggest that letting consumers express that preference through the market reduces the welfare gains from the new technology much less than if (as in scenario 4) a ban on GMOs is imposed in Europe. The results also suggest, however, that developing countries that do not gain access to GM technology may be slightly worse off in terms of economic welfare if they cannot guarantee that their exports entering the Western European markets are GMO-free. For these countries, a complete segregation of GMO-inclusive and GMO-free markets may be a way in which they could reap benefits from selling 'conventional' products to GM-critical consumers in industrialized countries. The NRT model results shows how farmers could be expected to respond to accommodate such a preference swing in favour of GM-free products. However, to do so would require that the relative price premium on these products is sufficient not only to outweigh the productivity growth foregone by not adopting GMOs but also to cover the potentially significant costs of compliance. Recall that the above analyses bravely assume that labelling would enable costless identification of the GM status of any product.

Fourth, large though the estimated welfare gains from the adoption of GM technology are, Nielsen and Anderson (2000b) show they are dwarfed by the welfare gains that could result from liberalizing global markets for farm products and textiles and clothing: around $50 billion per year for rice and textiles alone, again with the majority of that going to developing countries. Should opposition to GMOs lead to the erection of further barriers to trade, as has been sought by numerous groups (Nielsen and Anderson, 2000a), that would simply add to the welfare cost – especially to developing countries – of restrictive trade policies.

Qualifications and areas for further research

The 'realism' of the above quantitative results is of course limited by the lack of empirical data and incomplete knowledge of the effects of GM crops. The analysis has had to be based on no more than assumptions about the productivity impact of introducing GM crops in the agricultural production system. To do better, more data needs to be made available about the nature and size of the productivity gains likely to be made in specific sectors of different countries.

It needs to be kept in mind, too, that the above welfare analysis ignores any of the alleged negative welfare consequences of introducing GMOs, for example due to externalities affecting the natural environment, but it also ignores the positive welfare consequences that would flow to developing countries if, for example, GM rice also was enhanced with vitamins and iron.

The above analyses rely on simplifying assumptions about the productivity impact of adopting GM crops, namely, that it affects each primary factor and intermediate input by the same amount. However, it is a reduction in pesticide and herbicide use that is the main cost-reducing impact of using the first-generation GM crops (see, e.g., Pray *et al.*, 2000). This suggests the above results may be exaggerating the factor re-allocation effects. It also suggests modelling of the productivity impact of GM crops should be amended to take that into account.

In the analysis using the NRT model, substitutability between GM and non-GM varieties has been introduced in the intermediate use of oilseeds and coarse grain by the livestock sector (as feed) and the food processing industries. In most regions, these particular crops are indeed mainly used as intermediate inputs. However, there are some notable exceptions – such as direct consumption of soybeans in high-income Asia – that do not warrant the assumption of fixed *final* consumption shares. Hence, in future work substitutability should be introduced at this level of demand as well. This may also prove to be useful as the GM and non-GM distinction is introduced in other foods that are sold even more for direct consumption, such as fruits and vegetables.

The segregation of production and marketing of GM and non-GM varieties of coarse grain and oilseeds in this analysis is assumed only to bear the cost of the relative productivity difference. This assumption captures the cost of having to preserve the identity of the crop throughout the production and marketing chains, as well as any testing and labelling requirements at national borders. Experience from identity preservation of specialty crops reveals that this can potentially increase the price of such products by between 5 and 15% (Buckwell *et al.*, 1999). It is argued by, e.g., Runge and Jackson (1999) that such a cost-price premium will – in a free market, and in the absence of unsympathetic political reactions – fall on the GM-free products (as now happens with organic food). Whether consumers in Western Europe and elsewhere are in fact willing to pay such a premium is a moot point. Moreover, consumers may demand supplies of *processed* foods that are guaranteed free from GM organisms. To the extent that testing methods and accompanying labelling systems cannot deliver this guarantee, it will be necessary to trace GMO ingredients through feed use and further processing to the final processed food product. This would in turn mean – both in reality and in empirical modelling – that processed foods would also have to be identified as either GMO-inclusive or GMO-free.

Finally, the above type of models could also be used to explore the possible economic effects of second-generation GMO technologies such as nutritional enhancement of grains. Being able to produce rice rich in vitamin A and iron could be a huge boon to the health of literally billions of poor consumers. Whether a country's farmers choose to adopt such crops depends on the extent of consumer resistance or trade barriers to GM products and on the responses of producers in other countries. Empirical economic modelling has the potential to help food-exporting countries assess their prospects under various circumstances.

Acknowledgements

Thanks are due to Australia's Rural Industries Research and Development Corporation and Washington DC's International Food Policy Research Institute for financial assistance.

References

Armington, P.A. (1969) A Theory of demand for products distinguished by place of production. *IMF Staff Papers* 16, 159–178.

Buckwell, A., Brookes, G. and Bradley, D. (1999) *Economics of Identity Preservation for Genetically Modified Crops.* Final report of a study for Food Biotechnology Communications Initiative (FBCI).

Falck-Zepeda, J.B., Traxler, G. and Nelson, R.G. (2000) Surplus distribution from the introduction of a biotechnology innovation. *American Journal of Agricultural Economics* 82, 360–69.

FAO (1998) *Production Yearbook 1997.* Vol. 51. FAO, Rome.

Hertel, T.W. (1997) (ed.) *Global Trade Analysis: Modeling and Applications.* Cambridge University Press, Cambridge and New York.

James, C. (1997) *Global Status of Transgenic Crops in 1997.* ISAAA Briefs No. 5, International Service for the Acquisition of Agri-biotech Applications, Ithaca, New York.

James, C. (1998) *Global Review of Commercialized Transgenic Crops: 1998.* ISAAA Briefs No. 8, International Service for the Acquisition of Agri-biotech Applications, Ithaca, New York.

James, C. (1999) *Global Status of Commercialized Transgenic Crops: 1999.* ISAAA Briefs No. 12, Preview, International Service for the Acquisition of Agri-biotech Applications, Ithaca, New York.

James, C. and Krattiger, A. (1999) The role of the private sector. Brief 4 of 10 in Persley, G.J. (ed.) *Biotechnology for Developing/Country Agriculture: Problems and Opportunities.* Focus 2: A 2020 Vision for Food, Agriculture, and the Environment, International Food Policy Research Institute, Washington, DC.

Lewis, J.D., Robinson, S. and Thierfelder, K. (1999) *After the Negotiations: Assessing the Impact of Free Trade Agreements in Southern Africa.* TMD Discussion Paper No. 46, September, International Food Policy Research Institute, Washington, DC.

McDougall, R.A., Elbehri, A. and Truong, T.P. (1998) (eds) *Global Trade, Assistance, and Protection: The GTAP 4 Data Base.* Center for Global Trade Analysis, Purdue University, West Lafayette.

Nelson, G.C., Josling, T., Bullock, D., Unnevehr, L., Rosegrant, M. and Hill, L. (1999) *The Economics and Politics of Genetically Modified Organisms: Implications for WTO 2000.* With Julie Babinard, Carrie Cunningham, Alessandro De Pinto and Elisavet I. Nitsi, Bulletin 809, College of Agricultural, Consumer and Environmental Sciences, University of Illinois at Urbana-Champaign, November.

Nielsen, C.P. (1999) *Økonomiske virkninger af landbrugets anvendelse af genteknologi: Produktion, forbrug og international handel* (Economic effects of applying genetic engineering in agriculture: Production, consumption and international trade), with an English summary. Report no. 110, Danish Institute of Agricultural and Fisheries Economics (SJFI), Copenhagen.

Nielsen, C.P. and Anderson, K. (2000a) GMOs, Trade Policy, and Welfare in Rich and Poor Countries. Paper presented at the World Bank Workshop on Standards, Regulation and Trade, Washington, DC, 27 April. (Published as Ch. 6 in *Quantifying the Impact of Technical Barriers to Trade*, Maskus, K. and Wilson, J. (eds) (2001), University of Michigan Press, Ann Arbor.)

Nielsen, C.P. and Anderson, K. (2000b) Global market effects of alternative European responses to GMOs. CIES Discussion Paper 0032, Centre for International Economic Studies, University of Adelaide, July. (Published in *Weltwertschaftliches Archiv* 137, 320–346, June 2001.)

Nielsen, C.P. and Anderson, K. (2000c) Global market effects of adopting transgenic rice and cotton. Mimeo, Centre for International Economic Studies, University of Adelaide, July.

Nielsen, C.P., Robinson, S. and Thierfelder, K. (2000) Genetic engineering and trade: panacea or dilemma for developing countries? Paper presented at the Third Annual Conference on Global Economic Analysis, Monash University, Mt Eliza, 28–30 June.

OECD (1999) *Modern Biotechnology and Agricultural Markets: a Discussion of Selected Issues and the Impact on Supply and Markets.* Directorate for Food, Agriculture and Fisheries. Committee for Agriculture, AGR/CA/APM/CFS/MD(2000)2, Paris: OECD.

Pray, C.E., Ma, D., Huang, J. and Qiao, F. (2000) Impact of Bt Cotton in China. Paper presented at a seminar at the International Food Policy Research Institute (IFPRI), 9 May.

Runge, C.F. and Jackson, L.A. (1999) *Labeling, Trade and Gentically Modified Organisms (GMOs): A Proposed Solution.* University of Minnesota, Center for International Food and Agricultural Policy, Working Paper WP99-4, November 1999. (Published in *Journal of World Trade* 34, 111–122, 2000.)

UNEP (2000) *Cartagena Protocol on Biosafety to the Convention on Biological Diversity.* http://www.biodiv.org/biosafe/biosafety-protocol.htm

USDA (1999) Impact of adopting genetically engineered crops in the U.S. – preliminary results. Economic Research Service, USDA. Washington, DC, July.

USDA (2000) *Biotech Maize and Soybeans: Changing Markets and the Government's Role.* 12 April. http://ers.usda.gov/whatsnew/issues/biotechmarkets/

Chapter 21

Smallholders, Transgenic Varieties, and Production Efficiency: the Case of Cotton Farmers in China

Jikun Huang,[1] Ruifa Hu,[1] Scott Rozelle,[2] Fangbin Qiao[2] and Carl E. Pray[3]

[1]*Chinese Center for Agricultural Policy (CCAP), Institute of Geographical Sciences and Natural Resources Research, Chinese Academy of Sciences, China;* [2]*Department of Agricultural Resources and Economics, University of California, Davis, CA 95616, USA;* [3]*Department of Agriculture, Food, and Resource Economics, Rutgers University, 317 Dennison St, Highland Park, NJ 08904, USA*

Abstract

The overall goal of this study is to measure the effect of the impact that genetically modified cotton varieties have had on the production efficiency of smallholders in farming communities in China. We also find that the adoption of *Bt* cotton varieties leads to a significant decrease in the use of pesticides. Hence, we demonstrate that *Bt* cotton appears to be an agricultural technology that improves both production efficiency and the environment. In terms of policies, our findings suggest that the government should investigate whether or not they should make additional investments to spread *Bt* to other cotton regions and to other crops.

Introduction

Farmers in developing countries, including China, have greatly increased production of food and fibre crops during the past several decades in no small part as a result of increases in the use of modern inputs, especially farm chemicals. Particularly after the spread of modern, semi-dwarf, high-yielding varieties in the 1960s and 1970s, China's producers began using increasingly higher levels of pesticides to offset and avoid damage inflicted by insects and diseases. Although the lack of consistent data makes international comparisons difficult, a recent study by the authors argues that since the mid-1990s China has become the largest pesticide user in the world (Huang *et al.*, 2000c).

While the rising level of pesticide use certainly has helped China to raise production, the high, perhaps excessively high, levels of pesticide use may have had a number of adverse consequences. Pesticides may pose a serious danger to the soil and water quality of the agro-ecosystem (Smil, 1993; Rozelle *et al.*, 1997); human health (Rola and Pingali, 1993; Pingali *et*

 Economic and Social Issues in Agricultural Biotechnology (eds R.E. Evenson, V. Santaniello and D. Zilberman)

al., 1994; Huang *et al.*, 2000b); and food safety (Liu *et al.*, 1995). In fact, the negative indirect effects and social costs in some cases may exceed the private cost of purchasing pesticides (Huang *et al.*, 2000b).

Recognizing the negative externalities of excessive pesticide use, China's government has made an effort to regulate pesticide production, marketing and application since the 1970s. The experience with regulation, however, has shown that when officials only promulgate rules, reductions in the use of pesticides, the elimination of banned toxic ones, or the increase in the adoption of safe application productions do not always follow. In many regions of the country and in the case of many crops, farmers still use high levels of sometimes highly hazardous pesticides (ZGNYNJ, 1990–1999; Huang *et al.*, 2000b).

As a result, real reductions in the use of pesticides may have to depend on alternative approaches, such as the adoption of new technologies. For example, the spread of host-plant resistant varieties in the past two decades has effectively reduced pesticide use without affecting yields (Widawsky *et al.*, 1998; Pray *et al.*, 2001; Huang *et al.*, 2000a). China's effort to produce and promote host-plant resistant varieties has successfully extended such varieties to almost 100% of China's rice, wheat and maize area.

Despite such success, challenges remain in China's battle against pests. One study provides evidence that the effectiveness of older rice varieties has fallen over time because of the rising resistance of pests (Widawsky *et al.*, 1998). Interviews with wheat breeders revealed that breeding resistance to certain diseases takes up an increasing part of their breeding effort. In some cases, most notably that of cotton, despite intensive conventional plant breeding efforts, the resistance of pests to the natural defences of resistant varieties has built up to such an extent that crop damage has risen despite increasingly intensive pesticide-spraying campaigns (ERS, 1995).

In response to both the previous successes in traditional plant breeding and the continuing difficulties of mounting resistance, since the late 1980s scientists in China have followed the lead of others in the USA and elsewhere and started developing crops that are genetically engineered to be resistant to important pests (Huang *et al.*, 2001). One of the most successful genes to be inserted into plants is one from a bacterium *Bacillus thuringiensis* (*Bt*). The *Bt* gene has been used as a natural pesticide for decades. Currently, China's breeders are developing and testing about 20 genetically modified (GM) plants (Huang *et al.*, 2001).

Because of a perceived crises in the cotton sector – due to the ineffectiveness of varieties produced by conventional breeding methods and the rising use of pesticides by farmers – in 1997, the Ministry of Agriculture approved the commercial use of cotton varieties that were genetically engineered with a *Bt* gene to produce the toxin that kills bollworms. Monsanto, in a joint venture with the Hebei provincial seed company, introduced an American variety that had been genetically engineered. The Institute of Biotech Research of the Chinese Academy of Agricultural Sciences (CAAS) introduced and extended several local cotton varieties that were engineered to include *Bt* in the same year. The Chinese Cotton Research Institute of CAAS in Henan has also released *Bt* cotton varieties. Various estimates of *Bt* cotton area in 2000 ranged from 400,000 to 700,000 ha. Whatever the estimate, it is clear that cotton producers are among the millions of farmers who are using transgenic varieties.

However, despite the unprecedented release and adoption of GM cotton varieties, little is known about the impact they have had on the farm households using them and on the overall agricultural economy in which they are being extended. Has the adoption of *Bt* varieties of cotton affected the use of pesticides in China? If so, by how much? Once adopted and after accounting for pesticide use, has the adoption of *Bt* cotton affected yields? If so, by how much? Also, more methodologically oriented, how should the impact of *Bt* cotton on yields be best measured?

To meet the above goal and obtain a better understanding of the questions raised above, the rest of the chapter is organized as follows. In the first section, we describe the data set that we collected in 1999 from a farm household survey. A total of 282 cotton farmers were randomly selected from ten villages in five counties from Hebei and Shangdong. Then, an overview of the pest-related crop yield losses and measures to control the pest problems in China is presented, followed by an empirical model that will be used to measure the economy of transgenic crops with resistance to pests. The models then are estimated using our data and the results of econometric estimation are presented. Conclusions and policy implications from this study are provided in the final section.

Data

To examine the impact of biotechnology on pesticide use in the cotton sector, we collected our own data set in 1999. Our own data collection was necessary because China's government does not have a programme to track the cost of production of transgenic crops. In total, we collected data on the production practices of 282 cotton farmers. Since farmers use *Bt* and non-*Bt* varieties, we have information on 382.

The enumeration team put in considerable effort to choose the sample. Since one of our main objectives was to compare the differences in production practices of *Bt* and non-*Bt* varieties (and among *Bt* varieties), we had to carefully select our provinces and counties. In many counties 100% of the farmers were growing *Bt* cotton; in other areas the number of farmers growing *Bt* cotton was less. The coverage of specific varieties tended to be concentrated in certain areas. We chose Hebei Province because it is the only province in which Monsanto varieties had been approved for commercial use in the survey year. Within Hebei province, we selected Xinji County because that is the only area where the newest CAAS genetically engineered variety was being cultivated. We chose the sample counties in Shandong Province because one of CAAS's most successful *Bt* cotton varieties, GK-12, was grown there. Since the *Bt* programme started later in Shandong Province, farmers still had significant area in non-*Bt* cotton varieties. After county selection, we randomly selected the villages and farmers within the villages. The final sample comes from nine villages in five counties in Hebei and Shandong Provinces.

Descriptive statistics illustrate that our sample of farmers are fairly typical of those engaged in cotton production in Hebei and Shandong Provinces (Table 21.1, columns 1 and 2). Farmers cultivate an average of 0.78 ha per household, higher than Hebei and Shandong average (0.43 ha), but nearly the same as the cotton production regions in Hebei and Shandong (0.7 ha). Cotton area accounts for 0.42 ha per household, about 39% of the total sown area in the five counties surveyed in Hebei and Shandong (rows 2 and 3).

Users of *Bt* and non-*Bt* cotton also appear to be fairly similar (Table 21.1, columns 3–6). Although cotton area under *Bt* varieties in the sample region accounts for around 90% of the total cotton area and more than 90% of households in 1999 (bottom row), there are no apparent systematic differences in the type of farmer that is using *Bt* cotton. *T*-tests (between columns 3 and 5) demonstrate that there are no statistically significant differences between *Bt* and non-*Bt* farms in terms of farm size, cotton area, or the age or education of farm household head. Based on these comparisons, it appears as if there is little problem of selection bias in our sample.

Producing *Bt* and non-*Bt* Cotton in China

Yields, prices and the mix of fertilizers used for the *Bt* and non-*Bt* varieties are similar (Table 21.1, rows 6–9). On average, the yield of *Bt* cotton is 5.8% higher than non-*Bt* cotton, but one level is not statistically distinguishable from the other. The prices that farmers get for *Bt* and non-*Bt* varieties

Table 21.1. Summary statistics of *Bt* and non-*Bt* cotton production in sample households in China, 1999.

	All sample		*Bt* cotton		Non-*Bt* cotton	
	Mean	SD	Mean	SD	Mean	SD
Farm size (ha)	0.78	0.35	0.78	0.35	0.77	0.33
Cotton sown area (ha)	0.42	0.21	0.41	0.20	0.51	0.25
Cotton share in total crop sown area (%)	39	17	37	17	47	13
Age (years)	43.1	8.9	42.8	8.9	45.0	9.1
Education (years)	7.5	3.0	7.6	3.0	6.5	2.8
Yield (kg ha^{-1})	3349	627	3371	584	3186	875
Cotton price (yuan kg^{-1})	3.36	0.75	3.37	0.80	3.29	0.14
Ratio of phosphate fertilizer	0.30	0.20	0.31	0.20	0.23	0.14
Ratio of potash fertilizer	0.17	0.17	0.18	0.17	0.13	0.12
Fertilizer use (kg ha^{-1})	399	195	407	200	339	147
Number of pesticide applications (times)	8.1	7.2	6.6	4.2	19.8	12.7
Amount of pesticide use (kg ha^{-1})	17.5	28.9	11.8	13.7	60.7	60.5
Cost of pesticide (yuan ha^{-1})	403	661	261	267	1465	1388
Pesticide price (yuan kg^{-1})	34.5	46.6	35.9	49.4	23.9	8.0
Labor use (days ha^{-1})	530	222	519	223	610	205
Number of observations (*n*)	382		337		45	

SD, standard deviation.
Note: The statistics in the table are from 282 households in five counties of Hebei and Shandong provinces. Some farmers use two or more than two varieties, including both *Bt* and non-*Bt* varieties.

are virtually the same. The mixes of fertilizers (the ratios of phosphates and potash to total fertilizer use) are also nearly the same.

In other ways, the production technologies of *Bt* and non-*Bt* vary sharply (Table 21.1, rows 10–15). For example, *Bt* cotton farmers use more fertilizer. On average *Bt* cotton farmers apply 407 kg ha^{-1} of chemical fertilizer, a level that is nearly 70 kg ha^{-1}, or 20% more, than that used by non-*Bt* cotton farmers.

The largest difference between *Bt* cotton and non-*Bt* cotton production is in the use of pesticides. *Bt* cotton farmers apply pesticide only 6.6 times per season compared to nearly 20 times per season by non-*Bt* cotton farmers. On a per hectare basis, the pesticide use of non-*Bt* cotton production is more than five times higher than *Bt* cotton in terms of both quantity and expenditures. *Bt* cotton farmers spend 261 yuan per season on pesticide for spraying for non-bollworm pests while non-*Bt* cotton users spend 1465 yuan. Because of the reduction of pesticide application in *Bt* cotton, *Bt* cotton farmers reduce their total labour output by 15% when compared to non-*Bt* cotton farmers, including labour saved from pesticide application and pest monitoring in the fields.

Crop production loss and abatement

The frequency of pest outbreaks in the cotton sector has been increasing sharply over time in China, some estimating that the frequency of infestations doubled during the 1990s (ERS, 1995; Huang *et al.*, 2000c). Increases in the intensity of crop production, longer periods of time when the crops are not monitored due to rising wages, and excessive pesticide use have led to higher pest populations and to higher resistance of pests to the pesticides that once effectively controlled them.

Because of the high incidence of pest infestations of China's cotton crop and the high levels of spraying, the amount of loss to the cotton crop and the amount of loss that was abated due to spraying is high and exceeds that of grain (Table 21.2). Nationally, the Ministry of Agriculture's pest prevention teams estimate that cotton yields have been reduced by 5.3–14.0% due to pest infestations in the 1990s (column 2). The levels of loss were higher in some of the important cotton-producing provinces, such as Hebei Province (column 4). In fact, pest infestations from and the losses that such infestations potentially could cause are even more severe (rows 6–10). Had farmers not sprayed, cotton yields in China would have fallen nationally by 19–38.1% (column 2); those in Hebei and Shandong Provinces would have fallen even further (columns 4 and 6). The larger 'gain' (or, more accurately, loss avoidance) of cotton farmers when compared to those of grain farmers come from the fact that pest infestations were more serious and pesticide use was higher than those experienced by grain farmers. For example, pesticide use in cotton production was nearly four times as much as in rice (Huang *et al.*, 2000a).

The data, in fact, are consistent with the observation that increasing pest populations have meant that farmers need to spray increasingly greater amounts of pesticides to control them (Table 21.3). Measured in constant prices, per hectare pesticide use on cotton rose nearly 300% in two decades (row 1). The rise in pesticide use grew faster than the rate of the use of other inputs. The share of pesticide cost in the total cost of production inputs rose from 12–13% in the early 1980s to more than 20% after the mid-1990s (row 2). China's cotton farmers spent more than $500 million annually on pesticides to control pest-related problems in the late 1990s (row 3).

What are the costs of spraying? Without accounting for the effect on human health or the environment, Huang *et al.* (2000b) demonstrate that the gains by farmers from the pesticide use are much higher than the costs farmers paid for the pesticide. Hence, there is a high 'private' incentive for farmers to apply pesticide on crops, particularly on cotton crops.

Table 21.2. Official estimates of pest-related losses and losses abated by pest control efforts in China, 1990–1997.

	National		Hebei		Shandong	
Year	Grain	Cotton	Grain	Cotton	Grain	Cotton
Proportion (%) of losses due to pest infestations						
1990	3.2	5.3	2.9	11.6	5.0	5.1
1992	2.0	14.0	3.3	39.9	3.5	17.0
1994	2.0	11.8	1.9	9.7	3.5	8.9
1996	2.1	6.2	2.2	13.2	3.3	5.9
1997	2.4	6.3	2.2	13.7	3.4	5.1
Proportion (%) of losses to crop production abated by pest control efforts						
1990	7.6	19.0	6.6	32.6	10.1	21.5
1992	6.8	31.1	7.5	77.1	11.1	52.7
1994	7.2	38.1	6.9	43.8	11.4	43.5
1996	7.9	26.6	8.2	51.9	12.1	34.9
1997	9.3	29.1	8.6	73.2	12.5	31.9

Note: Actual crop production loss (a better term is 'official estimate of crop production loss') is due to inability of pest control effort by farmers. Crop production loss abated from the pest is the avoided loss after the existing pest control effort in the farm field.
Source: Computed by authors based on the data from MOA, *Agricultural Yearbook of China*.

Table 21.3. Pesticide use in cotton production in China, 1980–1998.

	1980	1985	1990	1995	1998
Per hectare pesticide use (yuan at 1995 prices)	257	292	381	834	724
Share of pesticide cost in total material costs (%)	13	12	18	22	20
Total value of pesticide applied (million US$)	280	172	356	542	418

Note: Rural retail price index of pesticides is used to deflate the current value.
Source: State Economic Planning Commission and State Statistical Bureau.

The spread of *Bt* cotton

China has pursued a policy that has encouraged the release of *Bt* cotton varieties perhaps because of the high level of pesticide use and the possibility that pests are becoming resistant to popular types of pesticides. By almost all indications, cotton has become the most widespread and aggressive transgenic crop programme for smallholders in the world. In terms of sown areas, *Bt* cotton is the most extensively grown transgenic crop in China today. The official government estimates of *Bt* cotton area in 2000 ranged from 400,000 to 500,000 ha (personal communication with MOA officials). During interviews with a number of industry analysts and executives, estimates had already reached 1 million ha in 1999. Our estimates of *Bt* cotton area, which are based on interviews with provincial agricultural bureaux, extension officials and seed companies, fall in the middle of the official and industrial estimates. Starting from only 2000 ha in 1997, *Bt* cotton sown area grew to around 700,000 ha in 2000 (Huang *et al.*, 2000a). By 2000, we estimate that farmers planted *Bt* varieties on 20% of China's cotton increase. Whatever the source of the estimates, the growth of *Bt* cotton areas has been remarkable in China in the last 3 years.

The expansion of *Bt* cotton across China, however, has not been even. For example, after being the only province to grow *Bt* cotton in 1997, cotton farmers in Hebei account for approximately 30% of the sown area in 2000, 220,000 ha. Shandong Province ranks second in *Bt* cotton sown area at 170,000 ha. In contrast, other provinces, particularly those with lower levels of cotton bollworm infestation, have very little or no area sown to *Bt* varieties.

Models and Estimation

Several economic studies have questioned whether current patterns of pesticide use are economically and socially efficient (e.g. Pimentel and Lehman, 1992; Pingali and Roger, 1995; Yudelman *et al.*, 1998). Some studies show that the costs, both economic and social, related to pesticide use in crop production exceed the gains from the reduction of crop yield losses (Pingali and Roger, 1995). While studies of pesticide productivity are relatively common, few researchers have assessed farmer pesticide adoption behaviour, and no study has been carried out on the productivity of varieties with built-in pesticides such as GM *Bt* varieties.

Damage control production function

In our study, we use a production function approach to estimate the impact of pesticide use and *Bt* cotton variety adoption on crop productivity. It attempts to determine the value and impact on cotton production of two different types of variables: first, abatement inputs such as chemical pesticide use and/or host-plant resistant varieties in particular *Bt* varieties; and second, traditional inputs such as fertilizers and labour. *Ceteris paribus*, the use of chemical pesticides and host-plant resistant varieties does not increase yields *per se*. Instead their primary role is to abate damage or keep output from falling. In contrast, the use of inputs such as fertilizer and labour to yield directly increases.

In our study, we examine two damage abatement inputs: pesticides and *Bt* cotton. Conceptually, *Bt* cotton varieties differ from chemical use only in the way that they control certain pests, since *Bt* cotton is a genetically engineered crop that produces a naturally occuring pesticide: the *Bt* toxin. In this way, *Bt* varieties are acting as an input that can substitute for the use of pesticides. Practically, one of the main production outcome differences between cotton farmers that use *Bt* varieties and those that do not is the difference in the amount of pesticide required to control pests.

When working to model and track empirically the impacts of pesticides and *Bt* varieties on output, special attention needs to be given to the special nature of the inputs. In production function analysis, the effect of damage-abatement inputs must be measured assessing the amount of yield or output that was 'recovered' by the use of damage abatement inputs. Following the studies of Headley (1968) and Lichtenberg and Zilberman (1986), a damage-abatement function can be incorporated into the traditional models of agricultural production. However, unlike all but several previous studies (including our own work on rice – Widawsky *et al.*, 1998), we will include host-plant resistant varieties into our analysis within the damage-abatement approach. We do this primarily by allowing for the interaction between pesticides and *Bt* varieties.

The nature of damage control suggests that the observed crop yield, Y, can be specified as a function of both standard inputs, X, and damage control measures, Z, as:

$$Y = f(X)\,G(Z), \tag{1}$$

where the vector X includes labour, fertilizer, other farm-specific factors that affect yields (such as the human capital characteristics of the farm household that are proxied by the household head's age and education level) and location-specific factors (a set of county dummy variables). The term $G(Z)$ is a damage-abatement function that is a function of the level of control agent, Z (in our case the pesticide used by the farmer to control pests during outbreaks). The abatement function possesses the properties of a cumulative probability distribution. It is defined on the interval of [0, 1]. When $G(.) = 1$, it means that there has been a complete abatement of crop yield losses due to pest-related problems with certain high level of control agent, while when $G(.) = 0$, it means that the crop was completely destroyed by pest-related damage. The $G(.)$ function is non-decreasing in Z and approaches 1 as damage control agent use increases. If we assume a Cobb–Douglas production function, $f(X)$, and if we assume that the damage-abatement function, $G(Z)$, follows a Weibull, Exponential or Logistic specification, then Equation (1) can be written as:

$$Y = a_0\, \Pi_i^n\, X_i^{ai} \left[1-\exp\left(-Z^e\right)\right], \quad \text{(Weibull)} \tag{2}$$

$$Y = b_0\, \Pi_i^n\, X_i^{ki} \left[1-\exp\left(-c\,Z\right)\right], \quad \text{(Exponential)} \tag{3}$$

where a_0, a_i, e in Equation (2), and b_0, k_i, c in Equation (3) are parameters to be estimated. The i indexes inputs, including labour and chemical fertilizer. The variable Z represents pesticide use. The models in equations (2) and (3) could be estimated for *Bt* cotton and non-*Bt* cotton separately.

Because *Bt* cotton differs from non-*Bt* cotton mainly in the pest control efforts that farmers use to control bollworms, it is possible to model the interaction explicitly. To do so, we can pool data on *Bt* and non-*Bt* cotton to estimate a more general damage control production function with the following assumptions on the nature of the *Bt* and pesticide interactions:

$$e = e_0 + e_1\, Bt \tag{4}$$

$$c = c_0 + c_1\, Bt \tag{5}$$

where *Bt* is a dummy variable with a value of 1 for *Bt* variety and 0 otherwise. The models (2) and (3), combined with the working hypotheses (4) and (5), are estimated by non-linear methods. In order to compare the results from the traditional production approach, we estimate a Cobb–Douglas production function using ordinary least squares (OLS), where pesticide use and *Bt* cotton adoption are specified the same as other inputs such as labour and fertilizer.

Marginal impacts of pesticide use on cotton yield for the above models can be estimated as:

$$MP(Z) = a_0 \Pi_i^n X_i^{ai} \left[\exp\left(-Z^e\right) eZ^{e-1}\right], \quad (6)$$

(Weibull)

$$MP(Z) = b_0 \Pi_i^n X_i^{ai} \left[\exp(-cZ)(c)\right]. \quad (7)$$

(Exponential)

The impacts of *Bt* cotton on the marginal products of pesticide use can be examined through Equations (6) and (7) by using the different values of the parameters associated with *Bt* and non-*Bt* varieties from Equations 4 and 5. The optimal pesticide use level can also be estimated for both *Bt* and non-*Bt* cottons based on the assumption that the efficient use of pesticide requires that the value of its MP equals its price.

Finally, the impact of *Bt* cotton on the crop yield can be measured as:

$$DY = a_0 \Pi_i^n X_i^{ai} \left[\exp\left(-Z^e\right)\ln(Z) Z^e e_1\right], \quad (8)$$

(Weibull)

$$DY = b_0 \Pi_i^n X_i^{ai} \left[\exp(-cZ) Z c_1\right]. \quad (9)$$

(Exponential)

Empirical specification and estimation of pesticide use equation

The models specified above do not account for one potential statistical problem: the endogeneity of pesticide use in the production function. Since pesticides are applied in response to pest pressure, which are not controlled for in the analysis high levels of infestations may be correlated with lower yields. Hence, it is possible that the covariance of Z and the residuals of the production function are non-zero, a condition that would bias parameter estimates of the impact of pesticides on output. In other words, pesticides adopted by farmers may be endogenous to production and a systematic relationship among plant pests, pesticide use and cotton yields may exist.[1] Because of the nature of potentially omitted variables and correlations, not accounting for the endogeneity could lead to a downward bias in the coefficient.

To avoid this possible econometric problem, we adopt an instrumental variable (IV) approach. To develop an instrument for pesticide application that is correlated with actual pesticide use but does not affect output except through its impact on pesticides, a pesticide adoption model is estimated first. The predicted values of the pesticide use can then be used in the estimation of models (2) and (3). As long as a set of variables in the pesticide adoption equation exists to explain pesticide use and these variables do not have any independent explanatory power on output, the IV approach should allow us better to examine the impacts of *Bt* and pesticides on cotton output and the interactions of these two pest control technologies.

To implement the IV identification strategy, we hypothesize that in addition to a number of control variables that are also in the yield equation (such as age, education

[1] Theoretically, farmer's adoption of *Bt* cotton should also be treated as the other endogenous variable. However, the adoption of *Bt* cotton in our sampled areas is strongly associated with the commercialization policy of GMO products in China and the public seed distribution system within the region where *Bt* cotton has been approved for commercialization. Estimation of *Bt* cotton adoption was tried, but no robust results were obtained and all damage control models with Bt cotton as the endogenous variable did not converge at a reasonable level of convergence criteria.

and location dummy variables), farmer pesticide use depends on the profitability of pesticide use. Three measures are included to pick up this effect: the price of pesticides (*Price*), the perception by farmers of how severe his or her pest infestation problem is (*Yloss*), and the amount of information the farmer has about infestation from interactions with extension agents (*ExtService*). Although we have only a single cross-section of households, large variations in the price of pesticides exist among the respondents, reflecting the differences in quality, pesticide prices at different times during the cotton growing season and the pesticide composition. *Price* is the unit value price of pesticide, measured as the value of total pesticide use divided by the quantity used, a variable also was created to measure the farmer's expected profitability from pesticide use. During the survey, enumerators asked farmers to provide them with a percentage yield loss that they *typically* would expect to suffer from pest infestation should they not spray. We also asked the farmer about their meetings with extension agents that were charged with pest prevention. *ExtService* is an indicator variable measuring whether or not the farmer was visited by a local pest-prevention agent (*ExtService* = 1) or not (= 0) during the crop year. Logically these variables meet the criteria of IVs and they pass the Hausman–Wu exclusion restriction statistical tests.

In summary, following our above discussion, farmer's pesticide adoption (*Pesticide*) model can be explained by the following equations:

$$\textit{Pesticide} = f(\textit{Yloss, Price, ExtService; Variety Dummy; Age; Education; County Dummies}) \quad (10)$$

where the first three variables on the right hand side of Equation (10) are the instruments, and the others are the control variables. More specifically, in Equation (10), we include *Variety Dummy,* a dummy variable with a value equal to 1 when the farmer uses *Bt* cotton, and 0 otherwise. We also include *Age, Education* and *County Dummies*. In Equation (10), the dependent variable pesticide, is defined in terms of quantity (measured as kg ha^{-1}). An alternative specification in terms of pesticide cost (yuan ha^{-1}) generates similar results. Therefore, only the results from one of these two specifications are presented. The two-equation system model is estimated using a three-stage least-squares estimating approach.

Results

While the focus of the paper is on the impact of pesticides and *Bt* cotton varieties on yields, we begin with a brief discussion of the pesticide equation. In addition to the statistical importance of the estimation of the first stage equation, examining the determinants of pesticide use is interesting in its own right. After discussing the results of the pesticide use equation, we then discuss the cotton yield functions.

Pesticide use

The results of the pesticide equation demonstrate that the first stage of our model generally performed well in explaining pesticide use (Table 21.4, column 1). OLS versions of the same model (not shown) show that the model has a relatively high explanatory power, with adjusted R^2 values that range between 0.50 and 0.60, levels that are reasonable for cross-sectional household data. The results of the alternative functional forms (also not shown) demonstrate that the results are robust, as are most of the results for the different versions of the model using alternative specifications of the dependent variable. Most of the signs of the estimated coefficients of the control variables are as expected.

Most importantly, the regression analysis illustrates the importance of *Bt* cotton in reducing pesticide use (Table 21.4, column 1). The negative and highly significant coefficient on the *Bt* cotton variable means that *Bt* cotton farmers sharply

Table 21.4. Estimated parameters for pesticide use and cotton yield using two-stage least squares and damage-abatement control methods.

Exponential	Amount of pesticide use (kg ha^{-1})	Cobb–Douglas function ln (yield)	Damage control function ln (yield) Weibull	Damage control function ln (yield) Exponential
Intercept	46.98 (9.24)***	7.23 (0.28)***	7.81 (0.28)***	7.20 (0.40)***
Perception of yield loss				
Before flowering	0.04 (0.04)			
After flowering	0.15 (0.03)***			
Average pesticide price	−0.02 (0.03)			
Age	0.06 (0.14)	0.11 (0.05)**	0.12 (0.05)**	0.13 (0.07)*
Education	−0.67 (0.44)	0.01 (0.01)**	0.01 (0.01)**	0.01 (0.01)*
Labour		0.04 (0.03)	0.04 (0.03)	0.06 (0.04)*
Fertilizer		0.01 (0.02)	0.01 (0.02)	0.01 (0.03)
Pest management info. from extension agent (dummy)	−0.09 (0.09)			
Bt cotton variety (dummy)	−35.35 (4.07)***	0.15 (0.04)***		
Predicted pesticide		0.01 (0.01)		
Damage control function parameter estimates				
e_0 (pesticide parameter in Weibull model)			−0.05 (0.02)**	
e_{bt} (*Bt* variety parameter in Weibull model)			0.07 (0.02)***	
c (pesticide parameter in exponential model)				0.22 (0.09)***
C_{bt} (*Bt* variety parameter in exponential model)				5.96 (0.95)***

Notes: Ratios of phosphate and potash fertilizers are specified in linear rather than in log form. The figures in the parentheses are standard errors of estimates. ***, **, * denote significance at 1%, 5% and 10%, respectively. The model includes four county dummy variables to control for county-specific effects, but the estimated coefficients are not included for brevity.

reduce pesticide use when compared to non-*Bt* cotton farmers. *Ceteris paribus*, *Bt* cotton use allows farmers to reduce pesticide use by 35.4 kg ha^{-1}. Given that the mean pesticide use of non-*Bt* cotton producers is 60.7 kg ha^{-1} (Table 21.1), the adoption of *Bt* is associated with a 58% reduction in pesticide use. *Bt* varieties, at least in the sample areas and during the years of their use by farmers that are included in the study, lead to significant pesticide reductions. In other words, with the same set of data, Huang *et al.* (2001) demonstrated that *Bt* cotton adopters spray

67% fewer times and reduce pesticide expenditures by 82%.

Impacts on cotton production

Our analysis of the impact of *Bt* cotton and other pest control methods also shows the effect on cotton production, although the results are more sensitive to the methodological approach. To explore the importance of the choice of methodology, we first present the results that treat pesticide use and *Bt* cotton adoption as traditional inputs using a Cobb–Douglas functional form. We then turn to our non-linear estimate approach, in which we analyse the effect of pest control efforts within a damage-control production function framework. Following the discussion in the methodological section, we use two alternative functional forms of the damage-abatement function.

The production function analysis generates results that are typical of household studies done on China's agricultural sector (Putterman and Ciacu, 1994; Ye and Rozelle, 1994; Li, 1999). In all of the specifications, we find a strong and significant impact of the human capital variables, age and education, on cotton output (Table 21.4, columns 2–5). The coefficients on the labour and fertilizer variables confirm that the output elasticities of both labour and fertilizer are low; our estimated labour elasticities are about 0.04–0.06. Farmers in our sampled areas apply 399 kg ha^{-1} of fertilizer, one of the highest application rates in the world. Labour use also exceeds 500 man-days ha^{-1}. Therefore, such insignificant marginal contributions of fertilizer and labour to cotton production may be expected.

The results of the Cobb–Douglas function approach indicate that although *Bt* varieties raise cotton yields, pesticide use is not effective in raising yields (Table 21.4, column 2). Although the descriptive statistics are statistically indistinguishable (i.e. the unconditional yields of *Bt* cotton users are statistically the same as the unconditional yields of non-*Bt* cotton users), when other inputs and human capital variables are accounted for, *Bt* cotton users get 15% higher yields (see the coefficient for the *Bt* cotton dummy variable in Table 21.4 column 2). The low *t*-ratio on the coefficient of the pesticide, however, can be interpreted to mean that the marginal impact of pesticide use in cotton production is zero when pesticide is treated as a traditional yield-increasing input.

Among the two alternative specifications of the damage control functions, the ones that use the Weibull and exponential damage control functional forms show similar results for the effect of *Bt* cotton (Table 21.4, columns 3–5). If these specifications reflect the true underlying technology, our results suggest that *Bt* cotton is effective in helping pesticides reduce the damage from pest infestations and keeping yields higher than they would have been without *Bt* adoption. In other words, *Bt* cotton increases the technical efficiency of cotton production.

The results of the models that treat pesticides as a damage-abating input produce mixed results. In the model using the exponential function, pesticides are seen to affect yield. In contrast, the coefficient in the equation that uses the Weibull functional form has the wrong sign. In both cases, the marginal impact is small. If our data and econometric approaches are sound, one assessment of the results is that farmers are using so much pesticide, even when they adopt *Bt* cotton, that the marginal effect is near zero.

Using the parameters presented in Table 21.4, the associated output elasticities, average and marginal products of pesticide use, and optimal pesticide applications for both *Bt* and non-*Bt* cottons are computed and presented (Table 21.5).[2] While the point estimates of the marginal products and elasticities vary, the most notable result – for both *Bt* and non-*Bt* varieties – is the gap

[2] The optimal use of pesticides is calculated by solving for the optimal level of pesticide use, given the price of pesticide and the value of its marginal product.

between actual and optimal pesticide use. In all cases, but especially for the case of non-*Bt* varieties, farmers are using pesticides far in excess of their optimal levels. For example, in the case of the estimates that use the exponential functional form, *Bt* cotton users use 10 kg ha^{-1} more than is optimal; non-*Bt* users use nearly 40 kg ha^{-1} more.

Figure 21.1 shows the trend of cotton's marginal product value with respect to pesticide use evaluated at means of all non-pesticide variables. These results show both the over-use of pesticides and the superiority of *Bt* cotton in its ability to lead to lower levels of pesticide use. Increases in the value of an additional kilogram of cotton output approaches zero as pesticide use increases to a level above 20 kg ha^{-1} for *Bt* cotton varieties under a Weibull specification; it approaches zero even more rapidly when using the parameters from the exponential function. For non-*Bt* cotton, the exponential function specification shows that the marginal product value of pesticide use approaches zero after the pesticide use level reaches 30 kg ha^{-1}. These results illustrate that not only are pesticides being over-used by both *Bt* and non-*Bt* users: if users were to use pesticides up to their optimal levels, *Bt* cotton users would use far lower levels of pesticides.

Concluding Remarks

Intensive cultivation and broad adoption of fertilizer responsive varieties have led to widespread pest infestations in China and in every other developing country over the past several decades (Pingali *et al.*, 1997). The extent of pest-related diseases has grown by several times since the early 1980s in China. Rising pest problems and the availability of relatively inexpensive pesticides as China's markets have developed have contributed to the use of pesticides in crop pest management. Although statistics are difficult to compare, China is most probably already the largest pesticide user in the world, and pesticide use is still rising. Among all the major crops in China, cotton producers have traditionally used pesticides in the most intensive ways. Hence, it is important to understand why

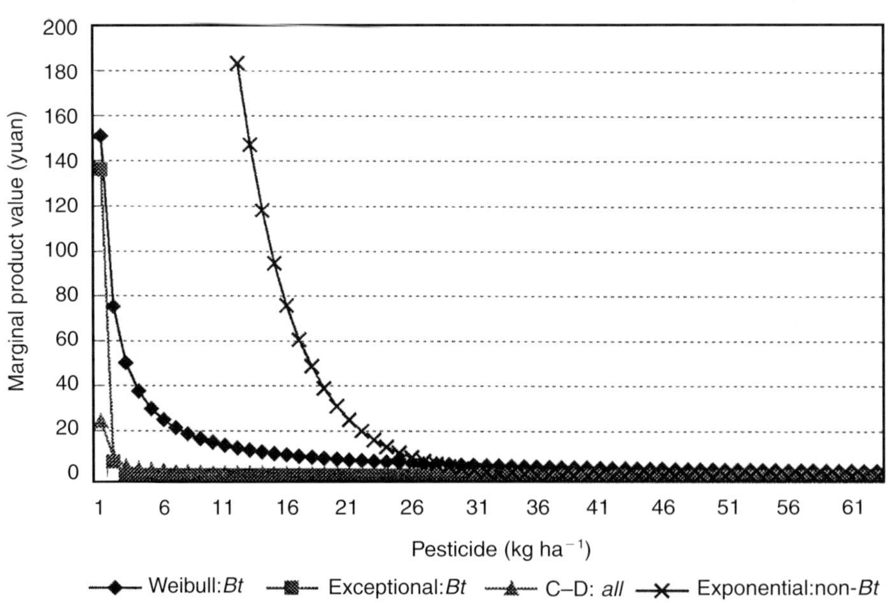

Fig. 21.1. Marginal product values of pesticide use in cotton production. Note: See note to Table 21.5 for description of calculation.

Table 21.5. Estimated productivity measures of pest control management using alternative approaches.

	Cobb–Douglas	Weibull	Exponential
Bt cotton			
Average product	286	286	286
Marginal product	0.315	10.89	11.95
Elasticity	0.001	0.038	0.042
Actual pesticide use (kg ha^{-1})	11.8	11.8	11.8
Optimal pesticide use (kg ha^{-1})	0.34	4.20	1.20
Non-*Bt* cotton			
Average product	52.5	52.5	52.5
Marginal product	0.01	–	7.24
Elasticity	0.000	–	0.138
Actual pesticide use (kg ha^{-1})	60.7	60.7	60.7
Optimal pesticide use (kg ha^{-1})	0.094	–	21.24
Impact of Bt cotton on yield (kg ha^{-1})	514	250	224

Notes: Productivity increases use parameters from Table 21.4. Elasticities, average products, marginal products and optimal pesticide application levels are calculated using means of all variables.

and how cotton producers use pesticides and to explore how alternatives to pesticide use have performed in recent years.

One of the results of our work is that even without alternatives, cotton producers could probably reduce pesticide use without affecting yields or profits. Although a discussion of why farmers over-use pesticides is beyond the scope of this chapter, it is clear that such behaviour is systematic and even exists when farmers use *Bt* cotton varieties. One thought is that farmers may be acting on poor information given to them by the pest control station personnel. In fact, such a hypothesis would be consistent with the findings of work on China's reform-era extension system in general (Huang *et al.*, 2000d).

During the past decade or more, extension agents have had their salaries cut and have been forced to rely on income generated from sales of inputs to farmers, including, in no small way, farm chemicals. Hence, it may be that agents have an incentive to push farmers to apply more than the optimal amount of pesticide as a way to increase their sales and supplement their incomes. Such a hypothesis would also support the observations of foreign seed company managers who report that such agents often resist the spread of *Bt* varieties because of their lower requirement for pesticides. When farmers have adopted *Bt*, such agents also suggest that farmers apply pesticides in the later parts of the season, even though the seed companies' agronomists believe such sprayings are unnecessary.

Our results show the impact of *Bt* cotton varieties on pesticide use, the effectiveness of pesticide's impact on yields and its independent effect on yields. In other studies, we have shown that the recent fall in the provincial use of pesticides in Hebei and Shandong Provinces can almost all be attributed to the spread of *Bt* cotton in these two areas. If the health and environment also improve with the fall in pesticide use, the benefits from extending *Bt* cotton exceed the production efficiency gains found here. In addition, unlike in work that treats pesticides and *Bt* cotton as damage-abatement inputs, we find that *Bt* cotton users also get an independent increase in yields. Although *Bt* cotton is relatively new in China and the long-term effect of *Bt* use in China is not known, it appears to be an agricultural technology that improves both efficiency in production and the environment.

In terms of policies, our findings suggest that the government should invest the money necessary to spread *Bt* to other cotton regions and to other crops. The important caveat is that government investments in the regulation of biotechnology will have to be increased to ensure that widespread use of *Bt* does not lead to the rapid development of resistance.

The second implication of these findings is that the government plant-protection system does not appear to be meeting the goal of reducing pesticide use. This fits with anecdotal evidence that we picked up from seed companies and farmers that the plant protection people often recommend that farmers not use *Bt* cotton and they consistently recommended more pesticide applications than the seed companies that sell *Bt* cotton. One recommendation would be to suggest that the government separates the integrated pest management (IPM) activities and staff of the plant protection system from the pesticide sales activities and staff. Once this is accomplished, the government must give the extension service incentives to promote IPM and the appropriate technology.

References

Colby, W.H. (1995) China's cotton imports surge. *China: Situation and Outlook Report*, Economic Research Service, USDA, Washington, DC.

Headley, J.C. (1968) Estimating the productivity of agricultural pesticides. *American Journal of Agricultural Economics* 50, 13–23.

Huang, J.K., Qiao, F.B., Pray, C. and Rozelle, S. (2000a) Biotechnology as an alternative to chemical pesticides: a case study of *Bt* cotton in China. *Working Paper*, Center for Chinese Agricultural Policy, Chinese Academy of Sciences.

Huang, J.K., Qiao, F.B. and Rozelle, S. (2000b) Farm chemicals, rice production and human health. *Working Paper*, Center for Chinese Agricultural Policy, Chinese Academy of Sciences.

Huang, J.K., Qiao, F.B., Zhang, L.X. and Rozelle, S. (2000c) Farm chemicals, rice production and environment. A Project Report Submitted to EEPSEA, Singapore.

Huang, J.K., Hu, R.F. and Sun, Z.Y. (2000d) *The Reform of Agricultural Extension System. Research on Economic and Technological Development in Agriculture.* China Agricultural Technological Press.

Huang, J.K., Wang, Q.F. and Zhang, Y.D. (2001) Agricultural biotechnology development and research capacity. *Working Paper*, Center for Chinese Agricultural Policy, Chinese Academy of Sciences.

Li, G. (1999) The economics of land tenure and property rights in China's agricultural sector. PhD dissertation, Food Research Institute, Stanford University, Stanford, California.

Lichtenberg, E. and Zilberman, D. (1986) The econometrics of damage control: why specification matters. *American Journal of Agricultural Economics* 68, 261–273.

Liu, H., Cheng, H. and Wang, X. (1995) A general study on Chinese diet: pesticide residue. *Health Research* 24, 6.

Pimentel, D. and Lehman, H. (1992) *The Pesticide Question: Environment, Economics, and Ethics.* Chapman and Hall, New York.

Pingali, P.L. and Roger, P.A. (1995) *Impact of Pesticides on Farmer Health and the Rice Environment.* Kluwer Academic, Massachusetts.

Pingali, P.L., Marquez, C.B. and Palis, F.G. (1994) Pesticides and Philippine rice farmer health: a medical and economic analysis. *American Journal of Agricultural Economics* 76, 587–592.

Pingali, P., Hossein, M. and Gerpacio, R. (1997) *Asian Rice Bowls: The Returning Crisis?* CAB International, Wallingford, UK.

Pray, C.E., Huang, J., Ma, D. and Qiao, F. (2001) Impact of Bt cotton in China. *World Development* 29, 813–825.

Putterman, L. and Chiacu, A. (1994) Elasticities and factor weights for agricultural growth accounting: a look at the data for China. *China Economics Review* 5, 191–204.

Rola, A.C. and Pingali, P.L. (1993) *Pesticides, Rice Productivity, and Farmers' Health: An Economic Assessment.* International Rice Research Institute and World Resource Institute, Los Banos, Philippines, and Washington, DC.

Rozelle, S., Huang, J. and Zhang, L. (1997) Poverty, population and environmental degradation in China. *Food Policy* 22, 229–251.

Smil, V. (1993) *China. Environmental Crisis: an Inquiry into the Limits of National Development.* M.E. Sharpe, New York.

SSB (State Statistical Bureau) *Statistical Yearbook of China*, various issues.

Widawsky, D., Rozelle, S., Jin, S. and Huang, J.K. (1998) Pesticide productivity, host-plant resistance and productivity in China. *Agricultural Economics* 19, 203–217.

Ye, Q. and Rozelle, S. (1994) Fertilizer demand in China's reforming economy. *Canadian Journal of Agricultural Economics* 42, 191–208.

Yudelman, M., Ratta, A. and Nygaard, D. (1998) *Pest Management and Food Production: Looking to the Future.* Food, Agriculture and the Environment Discussion Paper 25, International Food Policy Research Institute, Washington DC.

ZGNYNJ (1980–1998) *Zhongguo Nongye Nianjian* [*China Agricultural Yearbook*]. Ministry of Agriculture, China Statistical Press, Beijing.

Index

Note: Page numbers in **bold** refer to figures in the text; those in *italics* to tables or boxed material. Significant material in the notes is indicated by the letter 'n' after the page number

Browse Read and Buy
www.cabi.org/bookshop
ANIMAL & VETERINARY SCIENCES
BIODIVERSITY CROP PROTECTION
HUMAN HEALTH NATURAL RESOURCES
ENVIRONMENT PLANT SCIENCES
SOCIAL SCIENCES
CABI Publishing
A division of CAB International
Online BOOK SHOP
Subjects
Search
Reading Room
Bargains
New Titles
Forthcoming
Crop Pollination by Bees
Keith S. Delaplane and Daniel F Mayer
A DICTIONARY OF Entomology
G Gordh and D H Headrick
Order & Pay Online!
MasterCard
VISA
AMERICAN EXPRESS
Principles of CATTLE PRODUCTION
C.J.C. Phillips
Seeds
THE ECOLOGY OF REGENERATION IN PLANT COMMUNITIES
2nd EDITION
Edited by Michael Fenner
FULL DESCRIPTION
BUY THIS BOOK
BOOK OF THE MONTH
Tel: +44 (0)1491 832111 Fax: +44 (0)1491 829292